METEORITE MINERALOGY

Meteorites are fascinating cosmic visitors. Using accessible language, this book documents the history of mineralogy and meteorite research, summarizes the mineralogical characteristics of the myriad varieties of meteorites, and explains the mineralogical characteristics of Solar System bodies visited by spacecraft. Some of these bodies contain minerals that do not occur naturally on Earth or in meteorites. The book shows how to recognize different phases under the microscope and in backscattered electron images. It summarizes the major ways in which meteoritic minerals form – from condensation in the expanding atmospheres of dying stars to crystallization in deep-seated magmas, from flash melting in the solar nebula to weathering in the terrestrial environment. Containing spectacular backscattered electron images, color photographs of meteorite minerals, and with an accompanying online list of meteorite minerals, this book provides a useful resource for meteorite researchers, terrestrial mineralogists, cosmochemists, and planetary scientists, as well as graduate students in these fields.

ALAN E. RUBIN is a meteorite researcher who recently retired from the University of California, Los Angeles, where he worked as a research geochemist. He is a fellow of the Meteoritical Society and winner of the Nininger Meteorite Award and seven Griffith Observer science writing awards. He is the namesake of the garnet mineral rubinite and the main-belt asteroid 6227Alanrubin. He is the author of more than 200 research papers and 50 popular science articles. He is also the author of *Disturbing the Solar System* (Princeton, 2004).

CHI MA is a mineralogist at the California Institute of Technology, with research interests in nanomineralogy and the discovery of new minerals in meteorites, especially those representing extreme conditions of formation. He has discovered and led investigations on 47 new minerals, including 14 refractory minerals from the solar nebula and 11 high-pressure minerals. He is a fellow of the Mineralogical Society of America. The oxide mineral machiite was named in his honor. He is the author of more than 100 research papers.

CAMBRIDGE PLANETARY SCIENCE

Series Editors

Fran Bagenal, David Jewitt, Carl Murray, Jim Bell, Ralph Lorenz, Francis Nimmo, Sara Russell

Books in the Series

1. *Jupiter: The Planet, Satellites and Magnetosphere*† Edited by Bagenal, Dowling, and McKinnon 978-0-521-03545-3
2. *Meteorites: A Petrologic, Chemical and Isotopic Synthesis*† Hutchison 978-0-521-03539-2
3. *The Origin of Chondrules and Chondrites*† Sears 978-1-107-40285-0
4. *Planetary Rings*† Esposito 978-1-107-40247-8
5. *The Geology of Mars: Evidence from Earth-Based Analogs*† Edited by Chapman 978-0-521-20659-4
6. *The Surface of Mars*† Carr 978-0-521-87201-0
7. *Volcanism on Io: A Comparison with Earth*† Davies 978-0-521-85003-2
8. *Mars: An Introduction to its Interior, Surface and Atmosphere*† Barlow 978-0-521-85226-5
9. *The Martian Surface: Composition, Mineralogy, and Physical Properties* Edited by Bell 978-0-521-86698-9
10. *Planetary Crusts: Their Composition, Origin and Evolution*† Taylor and McLennan 978-0-521-14201-4
11. *Planetary Tectonics*† Edited by Watters and Schultz 978-0-521-74992-3
12. *Protoplanetary Dust: Astrophysical and Cosmochemical Perspectives*† Edited by Apai and Lauretta 978-0-521-51772-0
13. *Planetary Surface Processes* Melosh 978-0-521-51418-7
14. *Titan: Interior, Surface, Atmosphere, and Space Environment* Edited by Müller-Wodarg, Griffith, Lellouch, and Cravens 978-0-521-19992-6
15. *Planetary Rings: A Post-Equinox View (Second edition)* Esposito 978-1-107-02882-1
16. *Planetesimals: Early Differentiation and Consequences for Planets* Edited by Elkins-Tanton and Weiss 978-1-107-11848-5
17. *Asteroids: Astronomical and Geological Bodies* Burbine 978-1-107-09684-4
18. *The Atmosphere and Climate of Mars* Edited by Haberle, Clancy, Forget, Smith, and Zurek 978-1-107-01618-7
19. *Planetary Ring Systems* Edited by Tiscareno and Murray 978-1-107-11382-4
20. *Saturn in the 21st Century* Edited by Baines, Flasar, Krupp, and Stallard 978-1-107-10677-2
21. *Mercury: The View after MESSENGER* Edited by Solomon, Nittler, and Anderson 978-1-107-15445-2
22. *Chondrules: Records of Protoplanetary Disk Processes* Edited by Russell, Connolly Jr., and Krot 978-1-108-41801-0
23. *Spectroscopy and Photochemistry of Planetary Atmospheres and Ionospheres* Krasnopolsky 978-1-107-14526-9
24. *Remote Compositional Analysis: Techniques for Understanding Spectroscopy, Mineralogy, and Geochemistry of Planetary Surfaces* Edited by Bishop, Bell III, and Moersch 978-1-107-18620-0
25. *Meteoroids: Sources of Meteors on Earth and Beyond* Edited by Ryabova, Asher, and Campbell-Brown 978-1-108-42671-8

† Reissued as a paperback

METEORITE MINERALOGY

ALAN E. RUBIN
University of California, Los Angeles

CHI MA
California Institute of Technology

CAMBRIDGE
UNIVERSITY PRESS

CAMBRIDGE
UNIVERSITY PRESS

Shaftesbury Road, Cambridge CB2 8EA, United Kingdom

One Liberty Plaza, 20th Floor, New York, NY 10006, USA

477 Williamstown Road, Port Melbourne, VIC 3207, Australia

314–321, 3rd Floor, Plot 3, Splendor Forum, Jasola District Centre, New Delhi – 110025, India

103 Penang Road, #05–06/07, Visioncrest Commercial, Singapore 238467

Cambridge University Press is part of Cambridge University Press & Assessment, a department of the University of Cambridge.

We share the University's mission to contribute to society through the pursuit of education, learning and research at the highest international levels of excellence.

www.cambridge.org
Information on this title: www.cambridge.org/9781108484527

DOI: 10.1017/9781108613767

First published 2021
Reprinted 2021 (version 2, September 2023)

Printed in the United Kingdom by TJ Books Limited, Padstow Cornwall, September 2023

A catalogue record for this publication is available from the British Library

ISBN 978-1-108-48452-7 Hardback

For Dorene and Kang

Contents

Preface *page* xiii

1 Minerals and Meteorites: Historical Foundations and Current Status 1

2 Definitions and Explications 44

3 Brief Review of Crystallography and Crystal Chemistry 58
3.1 Crystal Systems and Space Groups 58
3.2 Miller Indices 60
3.3 Chemical Bonding 61
3.4 Coordination and Packing 63
3.5 Pauling's Rules 65

4 Properties of Minerals: Explanations and Applications 66
4.1 Major Physical Properties 66
4.2 Other Physical Properties 75
4.3 Optical Properties 85

5 Identification of Meteoritic Minerals in Reflected Light, by Backscattered Electron Imaging, and by Energy Dispersive X-Ray Spectroscopy, Wavelength-Dispersive X-Ray Spectroscopy, and Electron Backscatter Diffraction Analysis 92
5.1 Reflected Light 92
5.2 Backscattered-Electron Imaging with Energy-Dispersive X-Ray Spectroscopy, Wavelength-Dispersive X-Ray Spectroscopy, and Electron Backscatter Diffraction Analyses 98

6 Meteorite Classification and Taxonomy 101
6.1 History 101
6.2 Taxonomy 103

7 Mineralogy of Major Physical Components of Chondrites 109
7.1 Chondrules 109
7.2 Calcium-Aluminum-Rich Inclusions 116
7.3 Amoeboid Olivine Inclusions 126
7.4 Plagioclase-Olivine Inclusions 127

7.5 Dark Inclusions 127
7.5.1 Dark Inclusions in the Abee EH Chondrite 127
7.5.2 Dark Inclusions in Carbonaceous Chondrites 128
7.6 Carbon-Rich Aggregates 129
7.7 Matrix Material 130
7.7.1 CI Chondrites 130
7.7.2 CM Chondrites 131
7.7.3 CR Chondrites 133
7.7.4 CO Chondrites 133
7.7.5 CV Chondrites 133
7.7.6 CK Chondrites 134
7.7.7 Ungrouped Carbonaceous Chondrites 135
7.7.8 Ordinary Chondrites 136
7.7.9 R Chondrites 137
7.7.10 Enstatite Chondrites 138
7.7.11 Kakangari 138
7.8 Carbide-Magnetite Assemblages 139
7.9 Opaque Phases and Opaque Assemblages 139
7.9.1 Ordinary Chondrites 139
7.9.2 R Chondrites 142
7.9.3 Enstatite Chondrites 142
7.9.4 Kakangari-like Chondrites 144
7.9.5 Carbonaceous Chondrites 144
7.10 Presolar Grains 148
7.11 Calcium-Plagioclase-Bearing Clasts 149
7.12 Eclogitic Clasts 150
7.13 Impact-Melt-Rock Clasts 151
7.14 Impact-Melt Pockets 152

8 Petrologic and Mineralogical Characteristics of Meteorite Groups 153
8.1 Chondrites: Undifferentiated Meteorites 153
8.1.1 Ordinary Chondrites 153
8.1.2 R Chondrites (Prototype Rumuruti) 162
8.1.3 Carbonaceous Chondrites 162
8.1.4 Enstatite Chondrites 170
8.1.5 Kakangari-Like Chondrites (Prototype Kakangari) 171
8.2 Primitive Achondrites 171
8.2.1 Ureilites (Prototype Novo-Urei) 171
8.2.2 Brachinites (Prototype Brachina) 172
8.2.3 Acapulcoites and Lodranites (Prototypes Acapulco and Lodran) 173
8.2.4 Winonaites (Prototype Winona) and IAB Silicates 174
8.3 Asteroidal Achondrites 175
8.3.1 Howardites, Eucrites, Diogenites 175
8.3.2 Angrites (Prototype Angra dos Reis) 176

8.3.3 Aubrites a.k.a. Enstatite Achondrites (Prototype Aubres) 177
8.3.4 Ungrouped Achondrites 178
8.4 Nonasteroidal Achondrites 182
8.4.1 Lunar Meteorites 182
8.4.2 Martian Meteorites 182
8.5 Iron Meteorites 183
8.5.1 Magmatic CC Irons 185
8.5.2 Magmatic NC Irons 186
8.5.3 Non-Magmatic NC Irons 188
8.5.4 Ungrouped Irons 190
8.6 Stony-Iron Meteorites 192
8.6.1 Pallasites 192
8.6.2 Mesosiderites 197

9 Cosmomineralogy 200
9.1 Mercury 200
9.2 Venus 202
9.3 Earth 204
9.4 Moon 206
9.5 Mars 215
9.6 Asteroids 221
9.6.1 Vesta 223
9.6.2 Ceres 224
9.6.3 Eros 225
9.6.4 Itokawa 226
9.6.5 Ryugu 227
9.6.6 Bennu 228
9.6.7 Psyche 229
9.7 Titan 230
9.8 Comets and Kuiper Belt Objects 231
9.8.1 Comet Spectroscopy 231
9.8.2 Stardust 234
9.8.3 Interplanetary Dust Particles 235
9.8.4 Kuiper Belt Objects 235

10 Formation of Meteoritic Minerals in Gas- and Dust-Rich Environments 239
10.1 Condensation in Gaseous Envelopes Around Evolved Stars 240
10.2 Condensation in the Solar Nebula 243
10.3 Crystallization in Ca-Al-Rich Inclusion Melts 246
10.4 Crystallization in Melted Portions of Amoeboid Olivine Inclusions 248
10.5 Crystallization in Chondrule Melts 249
10.6 Exsolution in Opaque Assemblages during the Cooling of Ca-Al-Rich Inclusions 251
10.7 Exsolution during the Cooling of Chondrules 252

10.8 Exsolution during the Cooling of Opaque Assemblages Outside Ca-Al-Rich Inclusions and Chondrules 252
10.9 Annealing of Amorphous Material in the Solar Nebula 253

11 Formation of Meteoritic Minerals on Parent Bodies 254
11.1 Devitrification of Amorphous Material on Parent Bodies 254
11.2 Thermal Metamorphism and Exsolution 254
11.2.1 Heat Sources 254
11.2.2 Ordinary Chondrites 254
11.2.3 Carbonaceous Chondrites 257
11.2.4 R Chondrites 260
11.2.5 Enstatite Chondrites 261
11.2.6 Eucrites 262
11.3 Aqueous Alteration, Hydrothermal Alteration, and Metasomatism 263
11.3.1 CI Chondrites 263
11.3.2 CM Chondrites 264
11.3.3 CR Chondrites 267
11.3.4 CO Chondrites 267
11.3.5 CV Chondrites 268
11.3.6 Ungrouped Carbonaceous Chondrites 270
11.3.7 Ordinary Chondrites 271
11.3.8 R Chondrites 272
11.3.9 Enstatite Chondrites 273
11.3.10 Eucrites 273
11.3.11 Ureilites 274
11.3.12 Mesosiderites 274
11.3.13 Lunar Rocks 274
11.3.14 Martian Meteorites 275
11.4 Precipitation from Asteroidal Brines 275
11.5 Shock Metamorphism 276
11.5.1 Ordinary Chondrites 276
11.5.2 R Chondrites 282
11.5.3 Carbonaceous Chondrites 282
11.5.4 Enstatite Chondrites 285
11.5.5 Howardites, Eucrites, and Diogenites 289
11.5.6 Ureilites 290
11.5.7 Acapulcoites and Lodranites 291
11.5.8 Aubrites 292
11.5.9 Mesosiderites 292
11.5.10 Iron Meteorites 293
11.5.11 Lunar Meteorites 297
11.5.12 Martian Meteorites 297
11.5.13 Earth 300
11.6 Space Weathering 301

11.7 Condensation Within Impact Plumes 302
11.8 Crystallization from Melts in Differentiated or Partially Differentiated Bodies 303
11.8.1 Eucrites 304
11.8.2 Diogenites 304
11.8.3 Howardites 305
11.8.4 Angrites 305
11.8.5 Aubrites 306
11.8.6 Magmatic Iron Meteorites 306
11.8.7 Mesosiderites 307
11.8.8 Pallasites 307
11.8.9 Lunar Meteorites 308
11.8.10 Martian Meteorites 308
11.8.11 Acapulcoites and Lodranites 309
11.8.12 Ureilites 309
11.8.13 Brachinites 310
11.9 Condensation from Late-Stage Vapors in Differentiated Bodies 310
11.10 Exsolution, Inversion, and Subsolidus Redox Effects Within Cooling Igneous Materials 311
11.11 Solar Heating Near Perihelion 315

12 Formation of Meteoritic Minerals in the Terrestrial Environment 317
12.1 Atmospheric Passage 317
12.2 Terrestrial Weathering 318

13 The Strange Case of the Aluminum-Copper Alloys 325

Summary: The Formation of Meteoritic Minerals 328

Epilogue 330

References 331
Index 379
The plate section can be found between pages 210 and 211

Preface

The broad field of mineralogy covers the physical and geochemical characteristics of minerals, mineraloids, and amorphous phases within rocks. The rocks themselves may herald from Earth, the Moon, other planets and satellites, asteroids, and comets. Mineralogical inferences can extend to the realm of exoplanets. Most mineralogical studies are conducted on samples in the lab; in recent years, advanced electron-beam and synchrotron techniques have facilitated the discovery of dozens of new minerals from the microscale to the nanoscale. Important mineralogical information can also be gleaned from telescopic observations, unmanned spacecraft. and thermodynamic calculations. The related field of astromineralogy commonly focuses on astronomical observations of dust in circumstellar disks.

Extraterrestrial field mineralogy was initiated in 1967 by the unmanned lunar lander Surveyor 5. The spacecraft housed an alpha-scattering surface analyzer and determined the local lunar soil to be basaltic; its basic mineralogy could be inferred. In 1972, Apollo 17 geologist Harrison Schmitt visited the Taurus-Littrow Valley at the edge of Mare Serenitatis. He found the observational environment "superb" and had "little difficulty in distinguishing mineralogical and textural differences" among the clean, unaltered rocks at his feet. Since the mid-1960s, robotic vehicles have managed to land on, collide with, crawl along, bounce off, or jab the surfaces of three planets (Venus, Earth, and Mars), two moons (the Earth's Moon and Saturn's Titan), two comets (67P/Churyumov-Gerasimenko and 9P/Tempel), and four asteroids (433 Eros, 25143 Itokawa, 162173 Ryugu, and 101955 Bennu). Samples collected from the coma of Comet 81P/Wild and from the surface of asteroid Itokawa have been brought to Earth. More missions are in the works.

There are many excellent books on terrestrial mineralogy and planetary astronomy but most devote scant attention to meteorites. Books focused on meteorites discuss their mineralogical components but commonly concentrate on whole-rock petrological and geochemical characteristics. Much of the information about meteorite mineralogy can be found only in the primary scientific literature.

This book is focused on the mineralogical properties of meteorites and their petrologic components. It is not simply analogous to stamp collecting or the recitation of many cheerful facts about the square of the hypotenuse. The book explains how to recognize different phases in the microscope and in backscattered electron images. It summarizes the major ways in which meteoritic minerals form: from condensation in the expanding atmospheres of dying stars to crystallization in deep-seated magmas, from flash melting in the solar nebula to weathering in the terrestrial environment. It discusses what is known about the mineralogical properties of celestial bodies visited by spacecraft. Some of these bodies contain minerals that do not occur naturally on Earth or in meteorites.

We acknowledge with deep gratitude the help we received from many colleagues who answered our questions, provided data and samples, and offered potential illustrations. We also received numerous helpful comments on the entire manuscript from Bob Hazen, Mike Zolensky, and Ed Scott. The book is a better product because of their assistance.

1

Minerals and Meteorites

Historical Foundations and Current Status

The use of minerals and rocks by primates for making primitive tools is not confined to our species. Some chimpanzees, long-tailed macaques, and wild bearded capuchin monkeys use stone tools to crack open nuts and fruits, and in the case of coastal-dwelling macaques, shuck oysters. Hominins were using stone tools to scrape flesh from ungulate carcasses 3.4 million years ago. By 1.6 million years ago, hominins had discovered that some rocks (e.g., flint, chert, rhyolite, quartzite) were more suitable than others (e.g., limestone, sandstone, shale) for making hand axes; they presumably developed crude criteria (e.g., color, heft, friability, location) for distinguishing them.

In the Upper Paleolithic and Neolithic eras, modern humans began to produce tools made of flint (opal and chalcedony) and jade (jadeite and nephrite) to manufacture arrowheads and spearpoints. They used gold for ornaments, and native copper for knives, bowls, and cups. Mineral pigments for cave painting and body decoration were made from hematite, red and yellow ocher (hematite mixed with clay), and white chalk. These materials were often used in conjunction with charcoal (from burnt wood) or carbon black (from charred wood, bone, ivory, vines, and stems). The oldest known cave painting (dating more than 40,000 years before the present) is of a four-footed animal, perhaps a banteng (a species of wild cattle), drawn in red ocher on the wall of a cave in Borneo.

Copper mining had begun in Europe by 5400 BCE – there is evidence that miners of the Vinča culture had sunk 20-m shafts at a site at Rudna Glava in Serbia. Within the limestone at that site, miners worked veins of copper ore, mainly malachite ($Cu_2CO_3(OH)_2$) and azurite ($Cu_3(CO_3)_2(OH)_2$), formed by the gradual weathering and decomposition of chalcopyrite ($CuFeS_2$) associated with magnetite (Fe_3O_4). The availability of copper in the Balkans and other regions helped usher in the Bronze Age. On the Iranian Plateau, the Bronze Age began in the fifth millennium BCE when arsenic-laden copper ore was smelted to make arsenical bronze. It took another 2,000 years before bronze was commonly made with tin. [Tin ore, primarily cassiterite (SnO_2), was smelted and added to molten copper to make tin bronze.] The advantage of tin over arsenic is twofold: arsenical bronze had to be work-hardened to become as strong as tin bronze, and it was easier to add specific amounts of tin to molten copper to achieve desired results than to rely on chemically heterogeneous arsenic-bearing copper ore. Both types of bronze are harder than copper and were used to make durable tools, weapons, and armor. New mineral pigments were also adopted: malachite (green), azurite (blue), and cinnabar (red).

There are several examples of metallic iron artifacts from the Bronze Age; specimens that have been analyzed appear to have been manufactured from meteoritic metal. These relics

Figure 1.1 A Chinese early Chou Dynasty bronze weapon with meteoritic iron blade. Photo from Gettens et al. (1971); used with permission from the Smithsonian Institution. (A black-and-white version of this figure will appear in some formats. For the colour version, please refer to the plate section.)

include knives, blades, and axes from China (Figure 1.1), Tutankhamun's dagger from Egypt, an axe from Syria, and needles and bracelets from Europe.

Toward the end of the second millennium BCE, craftsmen began smelting terrestrial iron ore (magnetite, hematite, goethite, limonite,[1] and siderite) and adding small amounts of carbon via local plants to make pig iron. Because iron ores tend to be impure, fluxing agents such as limestone were often used to remove slag. Iron tools and weapons proved superior to those made of bronze and within a few hundred years the technology spread through much of Europe, Asia, and the Middle East.

The Hebrew Scriptures (the earliest parts of which were probably written down in about the sixth century BCE) mention six metals (gold, silver, copper, tin, lead, iron) and one metallic alloy (bronze) as well as about a dozen precious and semiprecious gems (including emerald, topaz, ruby, beryl, turquoise, and several varieties of silica – carnelian, amethyst, agate, onyx, jasper) (New International Version translation).

As humans learned to utilize the resources of their geological environment more effectively, it eventually became apparent to scholars that a systematic approach was necessary to classify minerals and rocks. The more inquisitive yearned to understand the origin of these materials.

The earliest detailed discussions of minerals came from the Greeks. In the fourth century BCE, Aristotle wrote *Meteorologica* (*Meteorology*) and presented his ideas on how metals and minerals were formed: after being heated by the Sun, the Earth produced both moist and dry exhalations. Moist exhalations congealed within dry rocks to form metals such as iron, gold, and copper; dry effluvia may have caused certain rocks to burn and form infusible materials such as realgar, cinnabar, sulfur, and ocher. The idea that Earth emitted gases was well supported by observations of steam and smoke from volcanoes, hot springs, and fumaroles.

Aristotle's student, Theophrastus of Eresus, wrote the first mineralogical treatise, *Peri Lithon* (*On Stones*), in about 314 BCE. He cataloged numerous minerals that were being used and traded in Attica; he also characterized minerals by such physical properties as color, luster, transparency, fracture patterns, hardness, weight (density), and fusibility. He described contemporary techniques for extracting metals and testing their purity.

In the first century CE, the Roman naturalist Pliny the Elder wrote *Naturalis Historia* (*Natural History*), a compendium of the knowledge of the ancient world. In the last five

[1] Although limonite is not an IMA-approved mineral, it is a commonly used term referring to poorly characterized mixtures of hydrous iron oxides such as goethite. The name is often applied to weathering veins and rinds around metallic Fe-Ni grains in chondritic meteorite finds.

volumes of this massive work, he listed numerous minerals and gemstones, reported their crystal shapes, physical properties, and practical uses, and discussed the mining of metals. He cited numerous authorities who had previously written treatises on precious stones, but of these, only Theophrastus's work has survived.

The next known major mineralogical text is *Aljamahir fi Maerifat Aljawahir* (known in English as *Gems*), written 1,000 years later (in the eleventh century CE) by the Persian polymath, Abu Rayhan Muhammad Ibn Ahmad al-Biruni. Al-Biruni discussed the physical properties of minerals and explained how he had constructed a device for measuring specific gravity. He detailed the sources of metals and gemstones, reported their prices, and related anecdotes about specific specimens.

During the 1250s, Albertus Magnus, a German Catholic Dominican friar (later canonized a saint), wrote a monumental work, *Book of Minerals*, covering such topics as the hardness, porosity, density, and fissility of rocks (i.e., their propensity to split along planes of weakness), the properties of gems, the distribution of stones, and the taste, smell, color, and malleability of metals. He discussed whether stones have mystical powers such as curing abscesses, ridding the body of poison, bringing victory to soldiers, and reconciling the hearts of men.

Georgius Agricola (the Latinized name of Georg Bauer) was a sixteenth-century German physician, often called "The Father of Mineralogy." Agricola wrote *De natura fossilium* (*The Natural Minerals*) in 1548, which is essentially the first comprehensive mineralogy textbook. He introduced a systematic classification of minerals, described many new species, and discussed their physical properties and relationships. (The word *fossil* is from the Latin *fossilis* meaning "obtained by digging"; it was often used in this period in reference to minerals.) Agricola's most famous work is *De re metallica* (*On the Nature of Metals*),[2] which was published posthumously in 1556; it covered all aspects of mining including mineral exploration, mine construction, metal extraction, smelting, and refining; he even discussed the legal issues involving mine ownership and labor management. The metals included gold, silver, lead, tin, copper, iron, and mercury. The other mineral categories in Agricola's system were "earths" (mainly powdery argillaceous soils that turned into mud when moistened), "stones" (all manner of hard dry rocks, specifically including limestone, marble, gems, and geodes), and "congealed juices" (consisting of "salts" such as rock salt and alum, and "sulfurs" such as coal and bitumen).

Carl Linnaeus, the Swedish naturalist known as the "Father of Modern Taxonomy," introduced binomial nomenclature for living organisms in the first edition of *Systema Naturae* (*System of Nature*) in 1735. In the ensuing decades, expanded editions were published, eventually leading to the classification of more than 10,000 species. Linnaeus divided the natural world into the animal, plant, and mineral kingdoms. In the 10th edition of his great work (1758), the mineral kingdom was itself divided into three classes: (1) rocks, (2) minerals and ores, and (3) fossils and aggregates. He applied the binomial scheme to minerals, classifying quartz, for example, into white quartz (*Quartzum album*), colored quartz (*Quartzum tinctum*), and clear quartz (*Quartzum aqueum*).

Linnaeus was held in high regard by his contemporaries. The Genevan political philosopher Jean-Jacques Rousseau wrote that he (Rousseau) knew of "no greater man on Earth." Linnaeus

[2] The first English translation of *De re metallica* was made in 1912 by mining engineer and future US President (1929–1933), Herbert Hoover, and his wife, Lou Henry Hoover, a geologist and Latinist. The work was widely acclaimed for its clarity of exposition and informative footnotes.

appears to have shared this view. He often proclaimed "*Deus creavit, Linnaeus disposuit*" ("God created, Linneaus organized") and wrote in his autobiography that "No one has more completely changed a whole science and initiated a new epoch." Linnaeus was ennobled in 1761 and assumed the name Carl von Linné.

One of Linnaeus' most ardent devotees was Abraham Gottlob Werner,[3] a German geologist best known for his theory of neptunism, the subsequently discredited idea that all rocks on the Earth's surface precipitated successively from a deep, hot, viscous, mineral-laden globe-encircling ocean. Rocks of each type were envisioned by Werner as having been deposited all over the Earth at the same time; for example, granites in North America were supposed to be the same age as granites in Europe, Africa and Asia. However, stratigraphic relationships (and, much later, radiometric dating) showed this was not the case. Werner also maintained that basalt had an aqueous origin despite field studies demonstrating it erupted from volcanoes.

Werner's first important work was *Von den äusserlichen Kennzeichen der Fossilien* (*On the External Characteristics of Minerals*), published in 1774. In that treatise he developed a mineral classification scheme, which allowed field geologists to identify minerals accurately by using only qualitative measurements of their external physical properties, e.g., color, hardness, shape, luster, specific gravity, odor, etc. He divided the subject of mineralogy into three major fields of study: (1) identification and classification, (2) distribution, and (3) formation. The book was translated into several languages and used as a field manual by many European and American geologists.

Throughout the eighteenth century, scholars became well acquainted with the physical properties of a wide range of mineral specimens, but the modern science of mineralogy could not flourish until further advances were made in petrography, crystallography, and mineral chemistry. The pioneers in these fields are often honored as founding fathers.

Petrography. The Scottish geologist, William Nicol, invented the polarizing microscope in the early nineteenth entury using Iceland spar (a transparent form of calcite); in 1815 he developed a technique for making thin sections of rocks and minerals. These advances were put to good use by British geologist Henry Clifton Sorby, sometimes called "The Father of Microscopic Petrography" for his detailed studies of terrestrial-rock thin sections in transmitted light with the polarizing microscope (Marvin 2006). Sorby also earned the sobriquet "The Father of Metallography" for his reflected-light microscopic studies of acid-etched iron and steel. He is best known in the meteoritics community for suggesting in 1877 that one possible origin for chondrules is as "droplets of fiery rain" that condensed from interplanetary gas early in the history of the Solar System.

Crystallography. Abbé René-Just Haüy, an eighteenth-century French mineralogist, is often called "The Father of Crystallography." In his seminal 1801 work, *Traité de Minéralogie* (*Treatise on Mineralogy*), he reported examining some crystals that were broken and other crystals that had been deliberately cut into smaller indivisible chunks. He noted their congruent shapes and compared the primitive crystal forms of different classes of minerals. He studied

[3] One of Werner's students was the Prussian naturalist Alexander von Humboldt who eschewed neptunism after studying volcanic rocks and ash in the Andes. Humboldt was among the first to propose that Africa and South America had once been joined, implicitly invoking continental drift. One of Humboldt's close friends was the great German writer Johann Wolfgang von Goethe (of *Faust* fame) who had amassed the largest private collection of minerals in Europe. By the time Goethe died in 1832, he had collected nearly 18,000 rock and mineral samples. The mineral goethite (α-FeO(OH)) is named in his honor.

cleavage planes, measured interfacial angles, and explored pyroelectricity. Haüy explained that "A casual glance at crystals may lead to the idea that they were pure sports of nature, but this is simply an elegant way of declaring one's ignorance. With a thoughtful examination of them, we discover laws of arrangement" (Levin 1990). Ultimately, Haüy described all known minerals by crystal class and chemical composition. It was not until the early twenthieth century that the British father-and-son team of William Henry Bragg and Lawrence Bragg developed X-ray crystallography and explored the structures of crystals in unprecedented detail.

Mineral chemistry. The Swedish chemist, Torbern Bergman, made great advances in the quantitative chemical analysis of mineral species. His 1774 study of a magnesian ankerite ($Ca(Fe,Mg)(CO_3)_2$) may be the first complete chemical analysis of an individual mineral (Hey 1973). Over the next decade, Bergman analyzed other phases and developed a mineral classification scheme based on their chemical and physical properties. In 1784 (the year of Bergman's death), Irish geologist Richard Kirwan published his first edition of *Elements of Mineralogy*, in which he listed the bulk chemical analyses of 74 rocks and minerals. As advances in inorganic chemistry led to an increase in the number of recognized elements (from 23 in 1789 – excluding 10 erroneous entries from a list published by Antoine-Laurent Lavoisier – to 42 in 1800), mineral analyses became more accurate. The first full textbook on mineral chemistry – *Handbuch zur chemischen Analyse der Mineralkörper* (*Handbook on the Chemical Analysis of Mineral Bodies*) – was published in 1801 by the German pharmacist and chemist, Wilhelm August Lampadius.

The Swedish chemist Baron Jöns Jacob Berzelius was the first to designate chemical elements by one- or two-letter symbols (e.g., H, O, Fe, Au), create molecular formulas (e.g., H_2O in modern form), and discover that the constituent elements of pure mineral phases are in constant proportions (e.g., $Ca_1C_1O_3$). He used the formulas to denote chemical reactions, e.g., $H_2SO_4 + Cu \rightarrow CuSO_4 + H_2$. By 1824 he had recognized that the chemical behavior of minerals was influenced more by their anion components (e.g., CO_3, O, S) than their cations (e.g., Ca, Fe, Mg) and divided minerals into groups accordingly, e.g. carbonates, oxides, sulfides, etc. He also identified the elements silicon, selenium, thorium, and cesium. Johan August Arfwedson, a Swedish chemist working in Berzelius's lab, discovered lithium in petalite ore (castorite, $LiAlSi_4O_{10}$) in 1817.

The American mineralogist James Dwight Dana published the first edition of his *System of Mineralogy* in 1837, adopting Linnaean binomial nomenclature (e.g., *Adamas octahedrus* for diamond) and grouping minerals by superficial appearance into higher orders [e.g., diamond, quartz, sapphire, and beryl were lumped into the order *Hyalinea* (hyaline means glassy or transparent)]. However, by the third (1850) and fourth (1854) editions, Dana had revised the nomenclature, coupling the approaches of Berzelius and Haüy. He formulated primary mineral groups: native elements, sulfides, halides, and oxides, and divided oxides into silicates, phosphates, sulfates, and carbonates. This system had been universally adopted by the 1870s, and an expanded version is used today in every mineralogy textbook.

With the development of modern analytical techniques (see McSween and Huss 2010), the number of recognized mineral species jumped from about 200 in 1750 to more than 5670 in early 2021. A periodically updated list of approved minerals is currently available online from the International Mineralogical Association (IAU): www.ima-mineralogy.org; click on "List of Minerals."

Comprehensive mineralogical studies of meteorites had to wait until meteorites were recognized as genuine extraterrestrial objects. There had long been reports of rocks falling from the sky. Joshua 10:11 (New Revised Standard Version) (written in the sixth or seventh century BCE) states: "As [the Amorites] fled before Israel, while they were going down the slope of Beth-horon, the Lord threw down huge stones from heaven on them as far as Azekah, and they died. . ." The passage conflates these huge stones with hailstones: "There were more who died because of the hailstones than the Israelites killed with the sword." Revelations 16:21 (NRSV) (first century CE) states that "And there fell upon men a great hail out of heaven, every stone about the weight of a talent. . ." (King James Version) (i.e., in the range of 33 to 50 kg). The largest authenticated hailstones are only ~1 kg, so stones much more massive than this could not be hail. In the play, *Prometheus Unbound* (attributed to the fifth-century BCE Greek tragedian Aeschylus), an enraged Zeus hurls a shower of stones down to Earth. Falling stones were later discussed by Livy (64 or 59 BCE to 12 or 17 CE), Pliny the Elder (23–79 CE) and Plutarch (46–120 CE). In his book, *Liber Prodigiorum* (*Book of Prodigies*), the pseudonymous fourth-century CE Roman historian, Julius Obsequens, described six events of stones raining on the Italian peninsula between 188 BCE and 94 BCE (Franza and Pratesi 2020).

The idea of rocks falling from the sky was bolstered by numerous observations of meteors and fireballs, but most eighteenth-century CE scientists remained unconvinced. There were reasons for their skepticism. No less an authority than Isaac Newton had declared in 1704 in *Opticks* that interplanetary space was devoid of small solid objects: ". . .to make way for the regular and lasting Motions of the Planets and Comets, it's necessary to empty the Heavens of all Matter, except perhaps some very thin Vapours, Steams, or Effluvia, arising from the Atmospheres of the Earth, Planets, and Comets." Newton's views on the barrenness of space were similar to those of Aristotle and were widely accepted. Also muddling the situation was the fact that actual observations of falling rocks (some with good documentation) were mixed in with fantastic reports of all sorts of objects dropping from the sky: flesh, blood, milk, wool, bricks, paper, money, and gelatin (Burke 1986). It was hard to separate the wheat from the chaff (neither of which was reported to have fallen to Earth).

Some scientists accepted the idea that rocks fell from the sky but averred that they were terrestrial rocks ejected from volcanoes (akin to volcanic bombs), borne aloft by hurricanes, generated by the Northern Lights, or, following Aristotle, precipitated in cold regions of the atmosphere. In 1789, Antoine-Laurent de Lavoisier, often called "The Father of Modern Chemistry," published his seminal textbook, *Traité élémentaire de chimie* (*Elementary Treatise on Chemistry*). He suggested that dust (consisting of stony and metallic particles), entrained in gas, emanated from the Earth and rose high into the atmosphere. There it was ignited by electricity and fused into solid bodies that fell to the ground. [American polymath and Founding Father, Benjamin Franklin (1706–1790), had shown decades earlier that lightning was an electrical phenomenon.]

Five developments in the late eighteenth and early nineteenth centuries finally established the reality that extraterrestrial rocks fall to Earth.

(1) In 1794, the German physicist, Ernst Chladni (already famous as "The Father of Acoustics") published a monograph, *Über den Ursprung der von Pallas gefundenen und anderer ihr ähnlicher Eisenmassen und über einige damit in Verbindung stehende*

Naturerscheinungen (*On the Origin of the Iron Masses Found by Pallas and Others Similar to it, and on Some Associated Natural Phenomena*), postulating that material from space entered the Earth's atmosphere, produced fireballs and dropped meteorites. The book was widely discussed but initially received mixed reviews.

(2) At the suggestion of Chladni, two German physicists, Johann Benzenberg and Heinrich Brandes, simultaneously observed the sky in September and October 1798 from sites 15.6 km apart to determine the height and speed of meteors (Marvin 2006). They made numerous simultaneous observations and concluded that meteors are visible at altitudes ranging from 170 to 26 km and travel at 29–44 km s^{-1}. (In modern usage, the bodies traversing the atmosphere are meteoroids, not meteors.) It was hard to imagine rocks lofted from the Earth reaching those heights or matching those speeds.

(3) In 1802, the British chemist, Edward Charles Howard, published a groundbreaking report in *Philosophical Transactions* titled "Experiments and Observations on Certain Stony and Metalline Substances, Which at Different Times are Said to Have Fallen on the Earth; Also on Various Kinds of Native Iron",[4] showing that several meteoritic stones had similar compositions; they all contained nickel as did all meteoritic irons. This indicated a common origin. [Nickel had been discovered in 1751 in niccolite (NiAs) by the Swedish mineralogist and chemist Baron Axel Fredrik Cronstedt. The mineral cronstedtite ($Fe_2^{2+}Fe^{3+}(SiFe^{3+})O_5(OH)_4$) (found intergrown with tochilinite in many CM chondrites) was named after him.]

(4) There was a spate of well-documented meteorite falls including Siena (Italy, 1794), Wold Cottage (England, 1795), Portugal (from the town of Évora Monte, 1796), Salles (France, 1798), Benares (India, 1798), L'Aigle (France, 1803), and Weston (Connecticut, USA, 1807).

(5) The discovery of asteroids proved that interplanetary space was not empty of small bodies after all: Ceres in 1801, Pallas in 1802, Juno in 1804, and Vesta in 1807. Along with the Moon, asteroids provided a potential source of extraterrestrial material. This idea became more plausible after Wilhelm Olbers suggested in 1803 that Ceres and Pallas were remnants of a planet that had been destroyed after suffering an internal explosion or a catastrophic collision with a comet.

A few scholars acknowledged the existence of extraterrestrial rocks and theorized that they had been blasted out of lunar volcanoes. (At the time, most scientists thought lunar craters were volcanic.) These workers cited the English physicist Robert Hooke who had concluded (after some hesitation) in *Micrographia* in 1665 that lunar craters were volcanic after he examined pits formed at the surface of boiled alabaster. This idea was consistent with sporadic observations of short-lived luminous events on the Moon. The great British astronomer William Herschel (discoverer of Uranus in 1781) reported seeing three luminous red spots beyond the terminator on the night side of the Moon on 19 April 1787; he suggested these were glowing gases disgorged from active lunar volcanoes. Other British, French, and German astronomers made

[4] Associated with this paper is one by Jacques-Louis, Count de Bournon, "Mineralogical Description of the various Stones said to have fallen upon the Earth," that contains the first published descriptions of chondrules ("small bodies, some of which are perfectly globular, others rather elongated or elliptical") as well as fine-grained silicate matrix material (a whitish gray substance with "an earthy consistence"). He also described (although not for the first time) two phases: troilite (characterizing it as nonmagnetic "martial pyrite") and fine-grained metallic iron.

similar observations in this time period. Two centuries after Hooke's monograph, English amateur-astronomer and media personality Patrick Moore termed such luminous events *lunar transient phenomena* (LTPs).

To nineteenth-century scientists, a lunar origin for meteorites was consistent with their chemical similarities (all contained Ni and were presumed to be from a common source) and the fact that the average specific gravity of stony meteorites (~3.34 g cm^{-3}) is the same as that of the bulk Moon.[5]

But there were problems with the suggestion that meteorites were all derived from lunar volcanoes: (1) The velocities of meteors (i.e., meteoroids) were observed to be tens of kilometers per second, seemingly far too great for those objects to have been disgorged from lunar volcanoes. (2) In 1859, American astronomer Benjamin Gould calculated that only 0.00006 percent of rock fragments from lunar volcanoes were likely to reach Earth (and of those few fragments, more than 70 percent would fall in the ocean). The paltry number of expected lunar volcanic ejecta fragments in the hands of scientists seemed grossly inconsistent with the relatively large number of meteorites known at the time (~160). (3) The German physician and astronomer, Franz von Paula Gruithuisen, suggested in 1828 that lunar craters formed by collisions. Grove Karl Gilbert, senior geologist at the United States Geological Survey, wrote an article in 1893 titled "The Moon's Face: A Study of the Origin of its Features," endorsing the impact theory. He measured lunar-crater depth/diameter ratios and explained central peaks as rebounded target material, crater rays as impact ejecta, and terraces inside craters as slumped crater walls.

There *are* volcanic features on the Moon. These include the maria (vast plains of flood basalt), sinuous rilles (collapsed lava tubes), and the Marius Hills (small volcanic domes). But major volcanism on the Moon probably ended more than a billion years ago. No meteorites have been hurled to Earth from lunar volcanoes; the lunar meteorites in our collections (~0.5 percent of samples) were launched by high-velocity collisions of asteroids with the Moon.

[5] The question may arise as to how nineteenth-century scientists knew the density of the Moon. It is a complicated story. The size of the Earth was determined long ago by Eratosthenes. On the summer solstice in c. 230 BCE at local noon, he observed the Sun close to the zenith, i.e., directly over a deep well in Syene, Egypt (now Aswan); in Alexandria at noon on the same day one year later, a vertical rod (a gnomon) cast a shadow. Eratosthenes measured the length of the shadow, and using geometry found that the Sun was 7.2° south of the zenith. The two Egyptian cities were a known distance apart (5,000 stadia, where 1 stade = 184.8 m) and were approximately on a north–south line. He assumed the Earth was a sphere because it cast a curved shadow on the Moon during total lunar eclipses. He divided 7.2° by 360° and found that the distance between these cities was 2 percent of the Earth's circumference. He multiplied the linear distance between the cities by 50 to determine the circumference of the Earth to within about 15 percent of the actual value. Of course, by the nineteenth century, the Earth had been circumnavigated (starting with the Magellan–Elcano expedition, 1519–1522), its size was well known, and accurate maps were available.

In 1686 Isaac Newton published his Law of Universal Gravitation [$F_{grav} = G \cdot ((m_1 \times m_2)/r^2)$], where F_{grav} is the force of gravity, m is mass, r is distance, and G is the gravitational constant; the following year, Newton presented his second law of motion, showing that $F_{grav} = m \times g$, where g is the acceleration of gravity at the Earth's surface. Experiments showed that g is 9.8 m s^{-2}. Using a torsion balance in 1789, Henry Cavendish determined an accurate value for the gravitational constant G. These parameters plus simple algebra allowed measurement of the mass of the Earth.

The distance to the Moon was first estimated by Aristarchus in c. 270 BCE by timing how long it took the Moon to pass through Earth's shadow during total lunar eclipses. He timed it at about 3 hours and calculated that the Earth's shadow was approximately 2.5 times the apparent diameter of the Moon. By using simple geometry, he found that the Moon is about 60 Earth radii away, within 1 percent of the actual value. The Earth–Moon distance can also be computed by parallax if two observers situated a long (and known) distance apart observe the Moon against the background stars at the same time and note the apparent shift in perspective. Trigonometry then provides the distance. Subsequent refinements of the Earth–Moon distance allowed nineteenth-century scientists to determine the Moon's mass from the Law of Universal Gravitation (as the Earth's mass and the value of g were already known).

The Moon subtends an angle of ~0.5° in the sky; because the distance between the Earth and Moon was known, simple geometry yielded the Moon's diameter (and hence its volume). Because density = mass/volume, once the latter values were measured, the Moon's density was easily calculated.

Modern explanations for LTPs include small meteorite impacts (but most are relatively faint and last less than a second as observed in Earth-based telescopes),[6] outgassing through fractures, electrostatic forces, thermoluminescence, terrestrial atmospheric turbulence, bad telescopic optics, and overactive imaginations.

In 1857, the German chemist, Karl Ludwig von Reichenbach, became the first to study meteoritic minerals in the microscope. By 1870, British geologist, Mervyn Herbert Nevil Story-Maskelyne, and his assistant, Austrian chemist Viktor von Lang, had studied the microscopic properties of more than 140 meteorites. Six years later, Austrian mineralogist Gustav Tschermak von Seysenegg initiated a project of photomicroscopy of meteorite thin sections, resulting a decade later in his monograph, *Die mikroskopische Beschaffenheit der Meteoriten* (*The Microscopic Properties of Meteorites*). In that work, Tschermak identified 16 meteoritic minerals as well as maskelynite and igneous glass.

By the middle of the twentieth century, the list of recognized meteoritic minerals had expanded only modestly, reaching 26 in 1960 and 38 in 1962 (Rubin 1997a). In the following decades, the widespread use of reflected light microscopy, the development and continual improvement of analytical techniques down to micro- and nanoscales (e.g., X-ray diffraction, electron microprobe analysis, scanning electron microscopy, and transmission electron microscopy), the recovery of tens of thousands of meteorites from hot and cold deserts, and a sharp increase in the number of meteorite researchers led to the identification of numerous minerals in meteorites. Toward the end of the twentieth century, Rubin (1997a, 1997b) compiled a list of about 300 meteoritic minerals. Two decades later, Rubin and Ma (2017) added another ~135 species to the list.

The number of meteoritic minerals is large because meteorites are derived from many different bodies, each with a distinctive geochemical character. Meteorites are now thought to come from about 100 to 150 asteroids[7] as well as from the Moon and Mars. Micrometeorites are derived mostly from asteroids; a minority (particularly those with ultracarbonaceous compositions) are likely from comets. Because interplanetary dust particles (IDPs) (also known as cosmic dust) are also meteoritic materials, the number of source bodies delivering extraterrestrial material to Earth may be several hundred to several thousand. But these are not the only bodies represented among meteoritic materials: primitive chondrites contain tiny presolar grains

[6] On the night of 20–21 January 2019, a small meteorite or comet, modeled as a mass of ~10 kg, crashed into the Moon during a total lunar eclipse. It produced a brief (0.3-second) yellow-white flash observed by multiple telescopes.

[7] The inference that the vast majority of meteorites are derived from asteroids is based on more than a dozen links pertaining to parent-body size, orbital characteristics, and physical properties: (1) The metallographic cooling rates of many stony and iron meteorites are in the range of 1–100°C/Ma, suggesting they were derived from bodies a few hundred kilometers in size; (2) the presence of solar-wind-implanted noble gases in regolith breccias indicates that these rocks are from bodies too small to retain significant atmospheres; (3) the relatively high concentrations of solar-wind gas are consistent with implantation at about 3 AU away from the Sun; (4) the old formation ages of most meteorites (~4.56 Ga) indicate they came from small bodies that cooled very early in the history of the Solar System; (5) the gravitational influence of Jupiter is expected to perturb some main belt asteroids into resonances that facilitate transfer to the inner Solar System; (6) the cosmic ray exposure (CRE) ages of many stony meteorites are in the range of 30 ka to 70 Ma, consistent with the time inferred for bodies in the asteroid belt to achieve Earth-crossing orbits; (7) the orbits of more than a thousand fireballs (including about three dozen that yielded recovered meteorites) were determined to be very similar to those of Earth-crossing asteroids; (8) the spectral reflectivities of some asteroids match those of meteorites measured in the lab; (9) the brecciated nature of many meteorites is consistent with the extensive cratering evident on many asteroids; (10) material returned from asteroid 25143 Itokawa matches that of LL chondrites; (11) the *Dawn* spacecraft's measurements of the composition and mineralogy of the surface of asteroid 4 Vesta match those of HED meteorites analyzed in the lab; (12) asteroid 2008 TC_3, which crashed into the Sudan ~19 hours after its discovery, yielded the Almahata Sitta polymict ureilite breccia; and (13) asteroid 2018 LA, which crashed into Botswana ~8 hours after discovery, yielded about two dozen howardite specimens.

that formed in the outflows of evolved stars and as supernova ejecta. Most of these particles appear to predate the Solar System by a few hundred million years, but at least 8 percent of the largest grains are more than a billion years older than the Sun (Heck et al. 2020).

Meteorites formed under a variety of conditions: primitive chondrites are interpreted to be products of the processes that occurred in the solar nebula (modified by impact-induced compaction and minor to extensive alteration on their parent asteroids), most iron meteorites formed deep within the cores of differentiated asteroids, regolith breccias formed near the surface of their (atmosphere-less) parent bodies and contain solar-wind-implanted noble gases, most eucrites formed from basaltic lava in the near-surface regions of a single differentiated asteroid (probably 4 Vesta), and martian and lunar meteorites formed as igneous rocks on substantially larger planetary and subplanetary bodies.

Meteorites exhibit diverse oxidation states, ranging from highly oxidized CI carbonaceous chondrites (which contain ~11 wt% H_2O^+ (indigenous water), mainly bound in fine-grained phyllosilicates) to highly reduced enstatite chondrites and aubrites (which contain graphite, Si-bearing metallic Fe-Ni, sulfides bearing Na, Mg, K, Ca, Ti, Cr, Mn, and Fe, and enstatite with very low FeO). The diversity in oxidation state among meteorites is reflected in the set of meteoritic minerals: e.g., elemental C, carbides, and carbonates; alloyed metallic Mo and molybdates; phosphides and phosphates; alloyed metallic Si, silicides, and silicates; elemental S, sulfides, and sulfates; and metallic Fe, wüstite (containing ferrous iron), magnetite (containing both ferrous and ferric iron), and hematite (containing ferric iron).

About 470 minerals have so far been identified in meteorites (Tables 1.1 and 1.2) and more are in the pipeline; this is about 8.3 percent of the total number of well-characterized mineral phases. Meteorite mineral species include native elements, metals and metallic alloys, carbides, nitrides and oxynitrides, phosphides, silicides, sulfides and hydroxysulfides, tellurides, arsenides and sulfarsenides, halides, oxides, hydroxides, carbonates, sulfates, molybdates, tungstates, phosphates and silico phosphates, oxalates, and silicates from all six structural groups (Table 1.1).

Although water ice is not a meteoritic mineral, it may have left traces in the matrices of primitive chondrites in the form of small ultraporous regions. Ice currently occurs on planets (e.g., Earth, Mercury, Mars); dwarf planets (Pluto, Haumea, Eris); moons (e.g., Moon, Europa, Titan); and asteroids. It was detected at the surface of 24 Themis (a 198 km-wide C-asteroid) (Campins et al. 2010; Rivkin and Emery 2010) and found within pyroxene grains from 25143 Itokawa (a subkilometer S-asteroid) (Jin and Bose 2019). Because ice would sublimate quickly at the surface of asteroids in the main belt, there has probably been recent outgassing of water vapor from the interior of Themis and condensation of ice around regolith grains (Rivkin and Emery 2010).

The astronomical menagerie includes such diverse objects as asymptotic giant branch stars, white dwarfs, black holes, neutron stars, Bok globules, Herbig–Haro objects, and planetary nebulae. Our knowledge of the cosmos deepens with the discovery of new members of the menagerie and the discernment of their interrelationships. Our knowledge of the bodies in the Solar System increases with the discovery of new mineral phases and the determination of their formation histories. The study of meteoritic minerals has broadened our understanding of the solar nebula, the geological history of asteroids and comets, the evolution of the Moon and Mars, impact phenomena, alteration and weathering processes, the physics of dying stars, and the nature of the interstellar medium.

Table 1.1 *Minerals in Meteorites*

Mineral	Synonyms and Varieties	Formula	Space Group	Selected References
Native Elements and Metals				
Aluminium		Al	Fm3m	Ma et al. (2017b)
Antitaenite (not approved)		Fe_3Ni	unknown	Rancourt and Scorzelli (1995), Wojnarowska et al. (2008)
Awaruite		Ni_3Fe	Pm3m	Buchwald (1977), Kimura and Ikeda (1995), McSween (1977), Rubin (1990)
Chaoite		C	P6/mmm	Vdovykin (1969), Vdovykin (1972)
Copper		Cu	Fm3m	Ramdohr (1963), Rubin (1994b)
Cupalite		CuAl	unknown	Hollister et al. (2014)
Decagonite		$Al_{71}Ni_{24}Fe_5$	~ $P10_5/mmc$	Bindi et al. (2015)
Diamond		C	Fd3m	Anders and Zinner (1993), Buchwald (1975), Ksanda and Henderson (1939), Russell et al. (1992)
Electrum (not approved)		Au-Ag	Fm3m	McCanta et al. (2008)
Gold		Au	Fm3m	Rubin (2014)
Gold-dominated alloys		(Au,Ag,Fe,Ni,Pt)	Fm3m	Bischoff et al. (1994), Geiger and Bischoff (1995), Schulze et al. (1994)
Graphite-2H	Cliftonite	C	$P6_3/mmc$	Ander and Zinner (1993), Buchwald (1975), Ramdohr (1963)
Graphite-3R		C	R3m	Nakamuta and Aoki (2000)
Hexaferrum		(Fe,Os,Ir,Mo)	$P6_3/mmc$	Ma (2012)
Hexamolybdenum		(Mo,Ru,Fe)	$P6_3/mmc$	Ma et al. (2014a)
Hollisterite	λ-(Al-Cu-Fe)	Al_3Fe	C2/m	Ma et al. (2017b)
Icosahedrite		$Al_{63}Cu_{24}Fe_{13}$	$Fm\overline{35}$	Bindi et al. (2011)
Icosahedrite II	i-phase II	$Al_{62}Cu_{31}Fe_7$	$Fm\overline{35}$	Bindi et al. (2016)

Table 1.1 (*cont.*)

Mineral	Synonyms and Varieties	Formula	Space Group	Selected References
Indium-dominated Alloys		(In,Sn,Pb)	I4/mmm	Wampler et al. (2020)
Iron	Kamacite; Ferrite	α-Fe	Im3m	Afiatalab and Wasson (1980), Ramdohr (1963), Rubin (1990)
Khatyrkite		$CuAl_2$	I4/mcm	Hollister et al. (2014), Ma et al. (2017b)
Kryachkoite	α-(Al-Cu-Fe)	$(Al,Cu)_6(Fe,Cu)$	$Cmc2_1$	Ma et al. (2017b)
Martensite (not approved)		α_2-(Fe,Ni)		Dodd (1981)
Mercury		Hg	R3m	Caillet Komorowski et al. (2012)
Molybdenum (not approved)		Mo		El Goresy et al. (1978)
Nickel		Ni	$P6_3/mmc$	Nyström and Wickman (1991)
Niobium (not approved)		Nb		El Goresy et al. (1978)
Osmium		Os	$P6_3/mmc$	Ma et al. (2014a)
Platinum		Pt	Fm3m	El Goresy et al. (1978)
PGE-dominated alloys		(Pt,Os,Ir,Ru,Re,Rh,Mo, Nb,Ta,Ge,W,V,Pb,Cr,Fe, Ni,Co)	Fm3m	Armstrong et al. (1987), Bischoff and Palme (1987), El Goresy et al. (1978), Wark and Lovering (1978)
Proxidecagonite		$Al_{34}Ni_9Fe_2$	Pnma	Bindi et al. (2018)
Rhenium (not approved)		Re	$P6_3/mmc$	El Goresy et al. (1978)
Rustenburgite		$(Pt,Pd)_3Sn$	Fm3m	Kimura (1996); Schulze et al. (1994)
Ruthenium		(Ru,Os,Ir)	$P6_3/mmc$	El Goresy et al. (1978)
Rutheniridosmine		(Ir,Os,Ru)	$P6_3/mmc$	McSween and Huss (2010)
Selenium		Se	$P3_121$ or $P3_221$	Simpson (1938), Greenland (1965), Akaiwa (1966), www.mindat.org
Steinhardtite		(Al,Ni,Fe)	$Im\bar{3}m$	Bindi et al. (2014)
Stolperite	β-(A-Cu-Fe)	AlCu	$Pm\bar{3}m$	Ma et al. (2017b)
Sulfur		S	Fddd	Buchwald (1977)

	Taenite	Austenite	γ-(Fe,Ni)	Fm3m	Ramdohr (1963)
	Tetrataenite		FeNi	P4/mmm	Clarke and Scott (1980)
	Wairauite		CoFe	Pm3m	Hua et al. 1995
	Zhanghengite		(Cu,Zn)	Im3m	Wang (1986)
Carbides					
	Moissanite		SiC	$P6_3/mc$	Alexander et al. (1991), Anders and Zinner (1993), Bernatowicz et al. (1991), Huss (1990)
	Cohenite		$(Fe,Ni)_3C$	Pbnm	Buchwald (1975), Ramdohr (1963)
	Edscottite		Fe_5C_2	C2/c	Scott and Agrell (1971), Ma and Rubin (2019)
	Haxonite		$(Fe,Ni)_{23}C_6$	Fm3m	Buchwald (1975), Scott and Agrell (1971)
	Molybdenum carbide (not approved)		MoC		Bernatowicz et al. (1991)
	Khamrabaevite		TiC	Fm3m	Ma and Rossman (2009a)
	Ruthenium carbide (not approved)		RuC		Bernatowicz et al. (1996)
	Zirconium carbide (not approved)		ZrC		Bernatowicz et al. (1991), Ott (1996)
Nitrides and Oxynitrides					
	Carlsbergite		CrN	Fm3m	Buchwald (1975), Buchwald and Scott (1971)
	Nierite		$a\text{-}Si_3N_4$	$P3_1C$	Alexander et al. (1989), Alexander et al. (1994), Lee et al. (1995)
	Osbornite		TiN	Fm3m	Bischoff et al. (1993), Ramdohr (1963)
	Roaldite		$(Fe,Ni)_4N$	$P\bar{4}3m$	Buchwald (1975), Nielsen and Buchwald (1981)
	β-Silicon nitride (not approved)		$\beta\text{-}Si_3N_4$		Lee et al. (1995)
	Sinoite		Si_2N_2O	$Cmc2_1$	Andersen et al. (1964)
	Uakitite		VN	Fm3m	Sharygin et al. (2020)

Table 1.1 (*cont.*)

	Mineral	Synonyms and Varieties	Formula	Space Group	Selected References
Phosphides					
	Allabogdanite		$(Fe,Ni)_2P$	Pnma	Britvin et al. (2002)
	Andreyivanovite		FeCrP	Pnma	Zolensky et al. (2008)
	Barringerite		$(Fe,Ni)_2P$	$P\bar{6}2m$	Buchwald (1977), Buseck (1977)
	Florenskyite		(Fe,Ni)TiP	Pnma	Ivanov et al. (2000)
	Melliniite		$(Ni,Fe)_4P$	$P2_13$	Pratesi et al. (2006)
	Monipite		MoNiP	$P\bar{6}2m$	Ma et al. (2014b)
	Nickelphosphide		Ni_3P	$I\bar{4}$	Britvin et al. (1999)
	Ni-Ge phosphide		$Ni_4Ge_{0.33}P_{1.17}$	$P\bar{1}$	Garvie et al. (2021b)
	Schreibersite	Rhabdite	$(Fe,Ni)_3P$	$I\bar{4}$	Ramdohr (1963)
	Transjordanite		Ni_2P	$P\bar{6}2m$	Britvin et al. (2020b)
Silicides					
	Brownleeite		MnSi	$P2_13$	Nakamura-Messenger et al. (2010)
	Carletonmooreite		Ni_3Si	Pm3m	Ma et al. (2018a), Garvie et al. (2021a)
	Gupeiite		Fe_3Si	Fm3m	Yu (1984)
	Hapkeite		Fe_2Si	Pm3m	Anand et al. (2004)
	Linzhiite		$FeSi_2$	P4/mmm	Anand et al. (2004)
	Naquite		FeSi	$P2_13$	Anand et al. (2004); Ma et al. (2017b)
	Perryite		$(Ni,Fe)_8(Si,P)_3$	R3c	Wasson and Wai (1970), Okada et al. (1991)
	Suessite		Fe_3Si	Im3m	Keil et al. (1982)
	Xifengite		Fe_5Si_3	$P6_3/mcm$	Yu (1984), Ma et al. (2017b)
Sulfides and Hydroxysulfides					
	Alabandite		MnS	Fm3m	Mason and Jarosewich (1967)
	Bornite		Cu_5FeS_4	Pbca	El Goresy et al. (1988)
	Brezinaite		Cr_3S_4	I2/m	Satterwhite et al. (1993), Warren and Kallemeyn (1994)
	Browneite		MnS	F43m	Ma et al. (2012b)
	Buseckite		(Fe,Zn,Mn)S	$P6_3mc$	Ma et al. (2012a)

Butianite		Ni_6SnS_2	I4/mmm	Ma and Beckett (2018)
Caswellsilverite		$NaCrS_2$	R3m	Okada and Keil (1982)
Chalcocite		Cu_2S	$P2_1c$	Yudin and Kolomenskiy (1987)
Chalcopyrite		$CuFeS_2$	I42d	Geiger and Bischoff (1995), Ramdohr (1963)
Cinnabar		HgS	$P3_121$, $P3_221$	Ulyanov (1991)
Cooperite		PtS	$P4_1/mmc$	Geiger and Bischoff (1995)
Covellite		CuS	$P6_3/mmc$	El Goresy et al. (1988)
Cronusite		$Ca_{0.2}CrS_2 \cdot 2H_2O$	Rm, R3m or R32	Britvin et al. 2001
Cu-Cr-sulfide (not approved)		$CuCrS_2$		Bevan et al. (2019)
Cubanite		$CuFe_2S_3$	Pcmn	Dodd (1981), Zolensky and McSween (1988)
Daubréelite		$FeCr_2S_4$	Fd3m	Keil (1968), Ramdohr (1963)
Digenite		$Cu_{1.8}S$	R3m	Kimura (1996); Kimura et al. (1992)
Djerfisherite		$K_6(Fe,Cu,Ni)_{25}S_{26}Cl$	Pm3m	Fuchs (1966a)
Erlichmanite		OsS_2	Pa3	Geiger and Bischoff (1989, 1990, 1995)
Ferroan alabandite		(Mn,Fe)S	Fm3m	Keil (1968)
Galena		PbS	Fm3m	Nyström and Wickman (1991)
Gentnerite (not approved)	Cuprian Daubréelite	$Cu_8Fe_3Cr_{11}S_{18}$		El Goresy and Ottemann (1966), Ulyanov (1991)
Greigite		Fe_3S_4	Fd3m	El Goresy et al. (1988)
Heazlewoodite		Ni_3S_2	R32	Buchwald (1977), McSween (1977)
Heideite		$(Fe,Cr)_{1.15}(Ti,Fe)_2S_4$	I2/m	Keil and Brett (1974)
Idaite		Cu_5FeS_6	$P6_3/mmc$	El Goresy et al. (1988)
Isocubanite	Chalcopyrrhotite	$CuFe_2S_3$	Fm3m	Buchwald (1975)
Joegoldsteinite		$MnCr_2S_4$	Fd3m	Isa et al. (2016)
Kalininite		$ZnCr_2S_4$	Fd3m	Sharygin et al. (2020)
Keilite		(Fe,Mg)S	Fm3m	Shimizu et al. (2002), Keil (2007)
Laurite		RuS_2	Pa3	Geiger and Bischoff (1989, 1990, 1995)
Mackinawite		$(Fe,Ni)_{1+x}S$ (x = 0-0.07)	P4/nmm	Buseck (1968)

Table 1.1 (*cont.*)

Mineral	Synonyms and Varieties	Formula	Space Group	Selected References
Marcasite		FeS_2	Pnnm	McSween (1994)
Millerite		NiS	R3m	Geiger and Bischoff (1995)
Molybdenite		MoS_2	$P6_3/mmc$	El Goresy et al. (1978)
Murchisite		Cr_5S_6	P31c	Ma et al. (2011a)
Niningerite		MgS	Fm3m	Keil (1968), Keil and Snetsinger (1967)
Nuwaite		Ni_6GeS_2	14/mmm	Ma and Beckett (2018)
Oldhamite		CaS	Fm3m	Keil (1968)
Pentlandite		$(Fe,Ni)_9S_8$	Fm3m	Ramdohr (1963), Buchwald (1977)
Plagionite		$Pb_5Sb_8S_{17}$	C2/c	Watters and Prinz (1979), www.mindat.org
Pyrite		FeS_2	Pa3	Ramdohr (1963), Nyström and Wickman (1991)
Pyrrhotite		$Fe_{1-x}S$	A2/a	Zolensky and McSween (1988)
Rudashevskyite		(Fe,Zn)S	$F\bar{4}3m$	Britvin et al. (2008)
Schöllhornite		$Na_{0.3}CrS_2{\cdot}H_2O$	R3m?	Okada et al. (1985)
Shenzhuangite		$NiFeS_2$	$I\bar{4}2d$	Bindi and Xie (2018)
Smythite		$(Fe,Ni)_{3+x}S_4$ (x = 0-0.3)	R3m	El Goresy et al. (1988)
Sphalerite		ZnS	F43m	Dodd (1981), El Goresy et al. (1988)
Tochilinite Group				
Tochilinite		$6(Fe_{0.9}S){\cdot}5[(Mg,Fe,Ni)(OH)_2]$	P2, Pm, or P2/m	Zolensky and McSween (1988), Ma et al. (2011a)
Haapalaite		$2[(Fe,Ni)S]{\cdot}1.6[(Mg,Fe)(OH)_2]$	R3m?	Buseck and Hua (1993)
Valleriite		$2[(Fe,Cu)S]{\cdot}1.53[(Mg,Al)(OH)_2]$	R3m	Ackermand and Raase (1973)
Troilite		FeS	$P6_3/mmc$	Ramdohr (1963)
Tungstenite		WS_2	$P6_3/mmc$	El Goresy et al. (1978)

	V,Fe,Cr-rich sulfide		$(V,Fe,Cr)_4S_5$	hexagonal	Ivanova et al. (2019b)
	V-rich brezinaite		$(Cr,V,Fe)_3S_4$	I 2/m	Ivanova et al. (2019b)
	V-rich daubréelite		$Fe(Cr,V)_2S_4$	Fd3m	Ivanova et al. (2019b)
	Violarite		$FeNi_2S_4$	Fd3m	Ulyanov (1991)
	Wassonite		TiS	R3m	Nakamura-Messenger et al. (2012)
	Wurtzite		ZnS	$P6_3mc$	Yudin and Kolomenskiy (1987)
	Zolenskyite		$FeCr_2S_4$	C2/m	Ma (2021)
Tellurides					
	Altaite		PbTe	Fm3m	Karwowski and Muszyński (2008)
	Moncheite	Chengbolite ($PtTe_2$)	$Pt(Te,Bi)_2$	P3m1	Geiger and Bischoff 1995, Connolly et al. (2006), Grady et al. (2015), www.mindat.org
Arsenides and Sulfarsenides					
	Cobaltite		CoAsS	$Pca2_1$	Nyström and Wickman (1991)
	Gersdorffite		NiAsS	$P2_13$	Nyström and Wickman (1991)
	Irarsite		(Ir,Ru,Rh,Pt)AsS	Pa3	Kimura (1996); Schulze et al. (1994)
	Iridarsenite		$(Ir,Ru)As_2$	$P2_1/c$	Geiger and Bischoff 1995
	Löllingite		$FeAs_2$	Pnnm	Geiger and Bischoff 1995
	Maucherite		$Ni_{11}As_8$	$P4_12_12$, $P\,4_32_12$	Nyström and Wickman (1991)
	Nickeline		NiAs	$P6_3/mmc$	Nyström and Wickman (1991)
	Omeiite		$(Os,Ru)As_2$	Pnnm	Geiger and Bischoff 1995
	Orcelite		$Ni_{4.77}As_2$	$P6_3cm$	Nyström and Wickman (1991)
	Rammelsbergite		$NiAs_2$	Pnnm	Nyström and Wickman (1991)
	Safflorite		$CoAs_2$	Pnnm	Nyström and Wickman (1991)
	Sperrylite		$PtAs_2$	Pa3	Geiger and Bischoff 1995
halides					
	Bismuth chloride (not approved)		$BiCl_3$		McCanta et al. (2008)
	Droninoite		$Ni_6Fe^{3+}{}_2Cl_2(OH)_{16}\cdot 4H_2O$	R3m	Chukanov et al. (2009)
	Halite		NaCl	Fm3m	Barber (1981), Berkley et al. (1980)
	Lawrencite		$(Fe^{+2},Ni)Cl_2$	R3m	Keil (1968)
	Sylvite		KCl	Fm3m	Barber (1981), Berkley et al. (1980)

Table 1.1 (*cont.*)

	Mineral	Synonyms and Varieties	Formula	Space Group	Selected References
Oxides					
	Addibischoffite		$Ca_2Al_6Al_6O_{20}$	P1	Ma et al. (2017a)
	Allendeite		$Sc_4Zr_3O_{12}$	R3	Ma et al. (2014a)
	Anatase		TiO_2	$I4_1$/amd	Wopenka and Swan (1985)
	Anosovite (not approved)		$(Ti^{4+},Ti^{3+},Mg,Sc,Al)_3O_5$	Bbmm	Zhang et al. (2015)
	Armalcolite		$(Mg,Fe)Ti_2O_5$	Bbmm	Lin and Kimura (1996)
	Baddeleyite		ZrO_2	$P2_1c$	Davis (1991), Delaney et al. (1984), Krot and Wasson (1994)
	Beckettite		$Ca_2V_6Al_6O_{20}$	$P\bar{1}$	Ma et al. (2016a)
	Bunsenite		NiO	Fm3m	Buchwald (1977)
	Calzirtite		$Ca_2Zr_5Ti_2O_{16}$	$I4_1$/acd	Ma (2020), Xiong et al. (2020)
	Chenmingite	CF-$FeCr_2O_4$	$FeCr_2O_4$	Pnma	Ma et al. (2019c)
	Chlormayenite	Brearleyite	$Ca_{12}Al_{14}O_{32}Cl_2$	I43d	Ma et al. (2011c)
	Chromite		$FeCr_2O_4$	Fd3m	Ramdohr (1963), Zolensky and McSween (1988)
	Corundum		Al_2O_3	R3c	Greshake et al. (1996a), Greshake et al. (1996b), MacPherson et al. (1988)
	Coulsonite		FeV_2O_4	Fd3m	Armstrong et al. (1987), Ulyanov (1991)
	Cuprite		Cu_2O	Pn3m	Ulyanov (1991)
	Dmitryivanovite		$CaAl_2O_4$	$P2_1b$	Mikouchi et al. (2009)
	Eskolaite		Cr_2O_3	R3c	Greshake and Bischoff (1996)
	Feiite		$Fe^{2+}{}_2(Fe^{2+}Ti^{4+})O_5$	Cmcm	Ma and Tschauner (2018a), Ma et al. (2021b)
	Ferropseudobrookite		$FeTi_2O_5$	Cmcm	Kimura (1996); Fujimaki et al. (1981)
	Geikielite		$MgTiO_3$	R3	Lin and Kimura (1996)
	Grossite		$CaAl_4O_7$	c 2/c	Weber and Bischoff (1994a, 1994b)
	Hausmannite		$Mn^{2+}Mn^{3+}{}_2O_4$	$I4_1$/amd	Nakamura et al. (2020)
	Hematite		α-Fe_2O_3	$R\bar{3}c$	Buchwald (1977)

Hercynite	$FeAl_2O_4$	Fd3m	Treiman (1985), Zolensky and McSween (1988)
Hibonite	$CaAl_{12}O_{19}$	$P6_3/mmc$	Dodd (1981), MacPherson et al. (1988)
Hibonite-(Fe)	$(Fe,Mg)Al_{12}O_{19}$	$P6_3/mmc$	Ma (2010)
Ilmenite	$FeTiO_3$	R3	Ramdohr (1963), Snetsinger and Keil (1969)
Kaitianite	$Ti^{3+}{}_2Ti^{4+}O_5$	C2/c	Ma (2019), Ma and Beckett (2020)
Kamiokite	$Fe_2Mo_3O_8$	$P6_3mc$	Ma et al. (2014b)
Kangite	$(Sc,Ti,Al,Zr,Mg,Ca,\square)_2O_3$	Ia3	Ma et al. (2013c)
Krotite	$CaAl_2O_4$	$P2_1/n$	Ma et al. (2011d)
Lakargiite	$CaZrO_3$	Pbnm	Ma (2011)
Lime	CaO	Fm3m	Greshake et al. (1996a), Greshake et al. (1996b), MacPherson et al. (1988)
Liuite	$FeTiO_3$	Pnma	Ma and Tschauner (2018b), Ma et al. (2021b)
Loveringite	$Ca(Ti,Fe,Cr,Mg)_{21}O_{38}$	R3	Ma et al. (2013a), Zhang et al. (2020)
Machiite	$Al_2Ti_3O_9$	C2/c	Krot (2016), Krot et al. (2020)
Maghemite	$Fe_{2.67}O_4$	$P2_13$	Buchwald (1977), Zolensky and McSween (1988)
Magnéli phases	Ti_5O_9 and Ti_8O_{15}	P1	Brearley (1993b, 1995)
Magnesiochromite	$MgCr_2O_4$	Fd3m	Greshake and Bischoff (1996)
Magnesioferrite	$MgFe_2O_4$	Fd3m	Yudin and Kolomenskiy (1987)
Magnesiowüstite	(Fe,Mg)O	Fm3m	Chen et al. (1996)
Magnetite	Fe_3O_4	Fd3m	Buchwald (1977), Kerridge et al. (1979), Ramdohr (1963), Zolensky and McSween (1988)
Majindeite	$Mg_2Mo_3O_8$	$P6_3mc$	Ma and Beckett (2016b)
Nb-oxide	$(Nb,V,Fe)O_2$		Ma et al. (2014b)
Olkhonskite	$Cr_2Ti_3O_9$	unknown	Schmitz et al. (2016)
Panguite	$(Ti,Al,Sc,Mg,Zr,Ca)_{1.8}O_3$	Pbca	Ma et al. (2012c)

Table 1.1 (*cont.*)

Mineral	Synonyms and Varieties	Formula	Space Group	Selected References
Periclase		MgO	Fm3m	Greshake et al. (1996a, 1996b), MacPherson et al. (1988)
Perovskite		$CaTiO_3$	Pnma	Lin and Kimura (1996), MacPherson et al. (1988)
Pseudobrookite		Fe_2TiO_5	Bbmm	Ramdohr (1967)
Pyrolusite		MnO_2	$P4_2/mnm$	Nakamura et al. (2020)
Pyrophanite		$MnTiO_3$	R3	Krot et al. (1993)
Rutile		TiO_2	P4/mnm	Greshake et al. (1996a, 1996b), Lin and Kimura (1996), MacPherson et al. (1988)
Spinel		$MgAl_2O_4$	Fd3m	MacPherson et al. (1988), Zolensky and McSween (1988)
Tazheranite		$(Zr,Ti,Ca,Y)O_{1.75}$	Fm3m	Ma and Rossman (2008)
Thorianite		ThO_2	Fm3m	MacPherson et al. (1988)
Ti^{3+},Al,Zr-oxide (not approved)		$(Ti^{3+},Al,Zr,Si,Mg)_{1.95}O_3$	unknown	Ma and Beckett (2020)
Ti-oxide (not approved)		Ti_3O_5	C2/m	Brearley (1993b)
Ti-rich magnetite		$(Fe,Mg)(Fe,Al,Ti)_2O_4$	Fd3m	Dodd (1981)
Tistarite		Ti_2O_3	R3c	Ma and Rossman (2009a)
Trevorite		$NiFe_2O_4$	Fd3m	Buchwald (1977)
Tschaunerite		$Fe^{2+}(Fe^{2+}Ti^{4+})O_4$	Cmcm	Ma and Prakapenka (2018), Ma et al. (2021a)
Tugarinovite		MoO_2	$P2_1/n$	Ma et al. (2014b)
Ulvöspinel		Fe_2TiO_4	Fd3m	Papike et al. (1991)
V-rich magnetite		$(Fe,Mg)(Fe,Al,V)_2O_4$	Fd3m	Bischoff and Palme (1987), El Goresy (1976), Wark and Lovering (1978)
Vestaite		$(Ti^{4+}Fe^{2+})Ti^{4+}{}_3O_9$	C2/c	Pang et al. (2018)
Wangdaodeite		$FeTiO_3$	R3c	Xie et al. (2016)
Warkite		$Ca_2Sc_6Al_6O_{20}$	P1	Ma et al. (2015a, 2020a)
Wüstite		FeO	Fm3m	Buchwald (1977)

	Xieite	CT-$FeCr_2O_4$	$FeCr_2O_4$	Bbmm	Chen et al. (2008)
	Zirconolite		$CaZrTi_2O_7$	C2/c	MacPherson et al. (1988), Ma and Rossman (2008)
	Zirkelite		$(Ti,Ca,Zr)O_{2-x}$	Fm3m	Kimura (1996); MacPherson et al. (1988)
Hydroxides					
	Akaganéite		β-FeO(OH,Cl)	I2/m	Buchwald (1977), Buchwald and Clarke (1988), Buchwald and Clarke (1989)
	Amakinite		$(Fe,Mg)(OH)_2$	P3m1	Zolensky and McSween (1988)
	Böhmite		AlO(OH)	Cmcm	Bevan et al. (2019)
	Brucite		$Mg(OH)_2$	P3m1	Barber (1981)
	Chlormagaluminite		$Mg_4Al_2(OH)_{12}Cl_2 \cdot 3\ H_2O$	$P6_3/mcm$	Ivanova et al. (2016)
	Feroxyhyte		δ-FeO(OH)	$P\bar{3}m1$	Kimura (1996); Gooding (1981); Buseck and Hua (1993)
	Ferrihydrite		$Fe^{3+}{}_{10}O_{14}(OH)_2$	hexagonal	Tomeoka and Buseck (1988)
	Goethite		α-FeO(OH)	Pbnm	Barber (1981), Buchwald (1977)
	Hibbingite		γ-$Fe_2(OH)_3Cl$	Pnam	Buchwald (1989), Saini-Eidukat et al. (1994)
	Lepidocrocite		γ-FeO(OH)	Amam	Barber (1981), Buchwald (1977)
	Manganite		$Mn^{3+}OOH$	$B2_1/d$	Nakamura et al. (2020)
	Portlandite		$Ca(OH)_2$	P3m1	Okada et al. (1981)
	Pyrochlore		$(Na,Ca)_2Nb_2O_6(OH,F)$	Fd3m	Lovering et al. (1979)
	Zaratite		$Ni_3(CO_3)(OH)_4 \cdot 4H_2O$	isometric	Buddhue (1957)
Carbonates					
	Ankerite		$Ca(Fe^{2+},Mg,Mn)(CO_3)_2$	R3	Zolensky and McSween (1988)
	Aragonite		$CaCO_3$	Pmcn	Endress and Bischoff (1996)
	Barringtonite		$MgCO_3 \cdot 2H_2O$	P1 or P-1	Ulyanov (1991)
	Breunnerite		$(Mg,Fe)CO_3$	R3c	Lee et al. (2014)
	Calcite		$CaCO_3$	R3c	Dodd (1981), Okada et al. (1981), Zolensky and Krot (1996)
	Chukanovite		$Fe_2(CO_3)(OH)_2$	$P2_1/a$	Pekov et al. (2007)
	Dolomite		$CaMg(CO_3)_2$	R3	Zolensky and McSween (1988)
	Hydromagnesite		$Mg_5(CO_3)_4(OH)_2 \cdot 4H_2O$	$P2_1/c$	Velbel (1988), Zolensky and Gooding (1986)

Table 1.1 (*cont.*)

	Mineral	Synonyms and Varieties	Formula	Space Group	Selected References
	Kutnohorite		$CaMn(CO_3)_2$	R3	Zolensky and McSween (1988)
	Magnesite		$(Mg,Fe)CO_3$	R3c	Zolensky and McSween (1988)
	Nesquehonite		$Mg(CO_3)\cdot 3H_2O$	$P2_1/n$	Velbel (1988), Zolensky and Gooding (1986)
	Nyerereite		$Na_2Ca(CO_3)_2$	$Pmc2_1$	Ulyanov (1991)
	Reevesite		$Ni_6Fe^{3+}{}_2(CO_3)(OH)_{16}\cdot 4H_2O$	R3m	Buchwald (1977), White et al. (1967)
	Rhodochrosite		$MnCO_3$	R3c	Ulyanov (1991)
	Siderite		$FeCO_3$	R3c	Buchwald (1977)
	Vaterite		$CaCO_3$	$P6_3/mmc$	Okada et al. (1981)
	Zaratite		$Ni_3(CO_3)(OH)_4\cdot 4H_2O$	unknown (in part amorphous)	Buchwald (1977)
Sulfates					
	Anhydrite		$CaSO_4$	Amma	Brearley (1993a, 1995)
	Baryte	Barite	$BaSO_4$	Pbnm	Nyström and Wickman (1991)
	Bassanite		$CaSO_4\cdot ½H_2O$	B2	Okada et al. (1981), Wentworth and Gooding (1994)
	Blödite		$Na_2Mg(SO_4)_2\cdot 4H_2O$	$P2_1/a$	Zolensky and McSween (1988)
	Celestine		SrSO4	Pnma	Shukolyukov et al. (2002)
	Copiapite		$Fe_5(SO_4)_6(OH)_2\cdot 20H_2O$	P1	Ulyanov (1991)
	Coquimbite		$Fe_2(SO_4)_3\cdot 9H_2O$	P3c	Kimura (1996); Gooding (1981)
	Epsomite		$MgSO_4\cdot 7H_2O$	$P2_12_12_1$	Zolensky and McSween (1988)
	Gypsum		$CaSO_4\cdot 2H_2O$	A2/a	Zolensky and McSween (1988)
	Hexahydrite		$MgSO_4\cdot 6H_2O$	A2/a	Zolensky and McSween (1988)
	Honessite		$(Ni,Fe)_8SO_4(OH)_{16}\cdot nH_2O$	R3m	Buchwald (1977)
	Jarosite		$KFe_3(SO_4)_2(OH)_6$	R3	Buchwald (1977)
	Kieserite		$MgSO_4\cdot H_2O$	C2/c	Kimura (1996); Gooding et al. (1991)

	Melanterite	$FeSO_4\cdot 7H_2O$	$P2_1/c$	Ulyanov (1991)
	Mendozite	$NaAl(SO_4)_2\text{-}11H_2O$	C2/c	www.mindat.org
	Ni-rich blödite	$Na_2(Mg,Ni)(SO_4)_2\cdot 4H_2O$	$P2_1/a$	Brearley (2006)
	Paraotwayite	$Ni(OH)_{2-x}(SO_4,CO_3)_{0.5x}$	Pm	Zubkova et al. (2008), Nickel and Graham (1987)
	Schwertmannite	$Fe^{3+}{}_{16}O_{16}(OH,SO_4)_{13\text{-}14}\cdot 10H_2O$	P4/m	Pederson (1999)
	Slavikite	$NaMg_2Fe^{3+}{}_5(SO_4)_7(OH)_6\cdot 33H_2O$	R3	Kimura (1996); Gooding (1981)
	Starkeyite	$MgSO_4.4H_2O$	$P2_1/n$	Velbel (1988), Zolensky and Gooding (1986)
	Szomolnokite	$FeSO_4\cdot H_2O$	A2/a	Kimura (1996); Gooding (1981)
	Thénardite	Na_2SO_4	Fddd	Brearley (2006)
	Voltaite	$K_2Fe^{2+}{}_5Fe^{3+}{}_3Al(SO_4)_{12}\cdot 18H_2O$	Fd3c	Kimura (1996); Gooding (1981)
Molybdates				
	Powellite	$CaMoO_4$	$I4_1/a$	Ulyanov (1991)
Tungstates				
	Scheelite	$CaWO_4$	$I4_1/a$	MacPherson et al. (1988)
Phosphates				
	Apatite	$Ca_5(PO_4)_3(F,OH,Cl)$	$P6_3/m$	MacPherson et al. (1988), Nyström and Wickman (1991)
	Arupite	$Ni_3(PO_4)_2\cdot 8H_2O$	I2/m	Buchwald (1977)
	Beusite	$(Mn,Fe,Ca,Mg)_3(PO_4)_2$	$P2_1/c$	Ulyanov (1991)
	Brianite	$Na_2CaMg(PO_4)_2$	$P2_1/a$	Buchwald (1977), Bunch et al. (1970), Fuchs et al. (1967)
	Buchwaldite	$NaCaPO_4$	$Pmn2_1$	Buchwald (1977), Olsen et al. (1977)
	Carbonate-fluorapatite	$Ca_5(PO_4,CO_3)_3F$	$P6_3/m$	Nyström and Wickman (1991)
	Cassidyite	$Ca_2(Ni,Mg)(PO_4)_2\cdot 2H_2O$	P1	Buchwald (1977), White et al. (1967)
	Chlorapatite	$Ca_5(PO_4)_3Cl$	$P6_3/m$	Buchwald (1977), Fuchs and Olsen (1965)
	Chladniite	$Na_2CaMg_7(PO_4)_6$	R3	McCoy et al. (1994)

Table 1.1 (*cont.*)

Mineral	Synonyms and Varieties	Formula	Space Group	Selected References
Chopinite		$Mg_3(PO_4)_2$	$P2_1/b$	Grew et al. (2010)
Collinsite		$Ca_2(Mg,Fe,Ni)(PO_4)_2 \cdot 2H_2O$	P1	Buchwald (1977)
Czochralskiite		$Na_4Ca_3Mg(PO_4)_4$	Pnma	Karwowski et al. (2016)
Farringtonite		$Mg_3(PO_4)_2$	$P2_1/a$	Buchwald (1977), Buseck (1977)
Ferromerrillite		$Ca_9NaFe^{2+}(PO_4)_7$	R3c	Britvin et al. (2016)
Fluorapatite		$Ca_5(PO_4)_3F$	$P6_3/m$	Kimura (1996); Kimura et al. (1992)
Galileiite		$NaFe_4(PO_4)_3$	R3	Olsen and Steele (1997)
Graftonite		$(Fe,Mn)_3(PO_4)_2$	$P2_1/c$	Buchwald (1977), Olsen and Fredriksson (1966)
Hydroxylapatite		$Ca_5(PO_4)_3OH$	$P6_3/m$	Fuchs (1969)
Johnsomervilleite		$Na_2Ca(Fe,Mg,Mn)_7(PO_4)_6$	R3	Olsen and Fredriksson (1966)
K-Na-Fe phosphate		$(K,Na)Fe_4(PO_4)_3$		Olsen and Steele (1997)
Keplerite		$Ca_9(Ca_{0.5}\square_{0.5})Mg(PO_4)_7$	R3c	Britvin et al. (2020a)
Lipscombite		$(Fe^{2+},Mn)Fe^{3+}{}_2(PO_4)_2(OH)_2$	$P4_12_12$	Buchwald (1977)
Maricite		$NaFePO_4$	Pmnb	Clarke et al. (1990)
Matyhite		$Ca_9(Ca_{0.5}\square_{0.5})Fe^{2+}(PO_4)_7$	R3c	Hwang et al. (2016a)
Merrillite		$Ca_9NaMg(PO_4)_7$	R3c	Buchwald (1977), Buseck (1977)
Monazite-(Ce)		$(Ce,La,Th)PO_4$	$P2_1/n$	Yagi et al. (1978)
Moraskoite		$Na_2Mg(PO_4)F$	Pbcn	Karwowski et al. (2015)
Na-Ca-Cr phosphate		$Na_4CaCr(PO_4)_3$		Kracher et al. (1977)
Na-Ca-Fe phosphate		$Na_4Ca_3Fe(PO_4)_4$		Kracher et al. (1977)
Na-Mn-Fe phosphate		$Na_4(Mn,Fe)(PO_4)_2$		Kracher et al. (1977)
Na-Fe-Mg phosphate		$Na_2Fe(Mg,Ca)(PO_4)_2$		Litasov and Podgornykh (2017)

Panethite		$(Na,Ca,K)_{1-x}(Mg,Fe,Mn)PO_4$	$P2_1/n$	Buchwald (1977), Bunch et al. (1970), Fuchs et al. (1967)
Sarcopside		$(Fe,Mn)_3(PO_4)_2$	$P2_1/a$	Buchwald (1977), Olsen and Fredriksson (1966)
Stanfieldite		$Ca_4(Mg,Fe)_5(PO_4)_6$	P2/C	Buseck (1977)
Tuite		γ-$Ca_3(PO_4)_2$	$R\bar{3}m$	Xie et al. (2003)
Vivianite		$Fe_3(PO_4)_2{\cdot}8H_2O$	C2/m	Buchwald (1977)
Xenophyllite		$Na_4Fe_7(PO_4)_6$	P1	www.mindat.org
Xenotime-(Y)		YPO_4	$I4_1/amd$	Liu et al. (2016)
Silicates				
Nesosilicates (independent SiO_4 tetrahedra)				
Adrianite		$Ca_{12}(Al_4Mg_3Si_7)O_{32}Cl_6$	$I\bar{4}3d$	Ma and Krot (2018)
Ahrensite		Fe_2SiO_4	$Fd\bar{3}m$	Ma et al. (2016b)
Almandine		$Fe_3Al_2(SiO_4)_3$	Ia3d	Ulyanov (1991)
Andradite		$Ca_3Fe_2(SiO_4)_3$	Ia3d	Kimura and Ikeda (1995)
Bridgmanite	Mg-silicate perovskite	$MgSiO_3$	Pnma	Tschauner et al. (2014)
Britholite-(Ce)	Beckelite	$(Ce,Y,Ca)_5(SiO_4,PO_4)_3(OH,F)$	$P6_3/m$	MacPherson et al. (1988)
Eringaite		$Ca_3Sc_2Si_3O_{12}$	Ia3d	Ma (2012)
Fayalite		Fe_2SiO_4	Pbnm	Krot et al. (1995)
Forsterite		Mg_2SiO_4	Pbnm	Dodd (1981)
Goldmanite		$Ca_3V_2(SiO_4)_3$	Ia3d	Kimura (1996); Simon and Grossman (1992)
Grossular		$Ca_3Al_2(SiO_4)_3$	Ia3d	Kimura and Ikeda (1995)
Hiroseite	Fe-analogue of bridgmanite	$FeSiO_3$	Pnma	Bindi and Xie (2019)
Hutcheonite		$Ca_3Ti_2(SiAl_2)O_{12}$	Ia3d	Ma and Krot (2014)
Kirschsteinite		$CaFe(SiO_4)$	Pbmn	Kimura and Ikeda (1995), Krot et al. (1995)
Laihunite		$(Fe^{3+},Fe^{2+},Mg,\square)_2SiO_4$	$P2_1/b$	Nakamura et al. (2020)

Table 1.1 (*cont.*)

Mineral	Synonyms and Varieties	Formula	Space Group	Selected References
Larnite	Felite; Shannonite	Ca_2SiO_4	$P2_1/n$	Krot (2016) personal communication
Majorite		$Mg_3(MgSi)Si_3O_{12}$	Ia3d	Chen et al. (1996), Dodd (1981)
Monticellite		$CaMgSiO_4$	Pnma	MacPherson et al. (1988), Ma and Krot (2014)
Mullite		$Al_6Si_2O_{13}$	Pbam	Ma and Rossman (2009a)
Olivine	Peridot; Chrysolite	$(Mg,Fe)_2SiO_4$	Pbnm	Buchwald (1977), Dodd (1981), Rubin (1990)
Poirierite		Mg_2SiO_4	Pmma	Tomioka et al. (2021)
Pyrope		$Mg_3Al_2(SiO_4)_3$	Ia3d	Chen et al. (1996)
Reidite		$ZrSiO_4$	$I4_1/a$	Glass et al. (2002)
Ringwoodite		Mg_2SiO_4	Ia3d	Dodd (1981), Price et al. (1979)
Rubinite		$Ca_3Ti^{3+}{}_2Si_3O_{12}$	Ia3d	Ma et al. (2017d)
Sapphirine		$(Mg,Al)_8(Al,Si)_6O_{20}$	$P\bar{1}$	Ulyanov (1991)
Spinelloid silicate	$(Mg,Fe)_3Si_2O_7$	$(Mg,Fe,Si)_2(Si,\square)O_4$	$I4_1/amd$	Ma et al. (2019d)
Spinelloid silicate-II	$(Fe,Mg,Ti,Ca)_3(Si,Cr,Al)_2O_7$	$(Fe,Mg,Cr,Ti,Ca,\square)_2(Si,Al)O_4$	$I4_1/amd$	Ma et al. (2019b)
Sodium-bearing silicate		$(Na,K,Ca,Fe)_{0.973}(Al,Si)_{5.08}O_{10}$		El Goresy et al. (1997)
Tetragonal almandine		$(Fe,Mg,Ca,Na)_3(Al,Si,Mg)_2Si_3O_{12}$	$I4_1/a$	Ma and Tschauner (2016)
Tetragonal majorite		$Mg_3(MgSi)Si_3O_{12}$	$I4_1/a$	Tomioka et al. (2016)
Titanite	Sphene	$CaTiSiO_5$	$P2_1/a$	Delaney et al. (1984)
Tranquillityite		$Fe^{2+}{}_8Ti_3Zr_2Si_3O_{24}$	unknown	Taylor et al. (2001)
Wadalite		$Ca_6Al_5Si_2O_{16}Cl_3$	I43d	Ishii et al. (2010)
Zircon		$ZrSiO_4$	$I4_1/amd$	Buchwald (1977), Ireland and Wlotzka (1992), Marvin and Klein (1964)

Sorosilicates (two isolated SiO_4 tetrahedra sharing one O)				
	Åkermanite	$Ca_2MgSi_2O_7$	$P\bar{4}2_1m$	MacPherson et al. (1988)
	Asimowite	Fe_2SiO_4	I2/m	Bindi et al. (2019)
	Baghdadite	Ca3(Zr,Ti)Si2O9	$P2_1/a$	Ma (2018b)
	Chevkinite-(Ce)	$(Ce,Nd,La,Ca,Th)_4(Ti,Fe,Mg)_5Si_4O_{22}$	C2/m	Liu et al. (2016)
	Gehlenite	$Ca_2Al(SiAl)O_7$	$P\bar{4}2_1m$	MacPherson et al. (1988)
	Melilite	$(Ca,Na)_2(Al,Mg)(Si,Al)_2O_7$	$P\bar{4}2_1m$	MacPherson et al. (1988)
	Paqueite	$Ca_3TiSi_2(Al,Ti,Si)_3O_{14}$	P321	Barber et al. (1994), Paque et al. (1994), Ma and Beckett (2016a)
	Perrierite-(Ce)	$(Ce,Nd,La,Ca,Th)_4(Ti,Fe,Mg)_5Si_4O_{22}$	C2/m	Liu et al. (2016)
	Pumpellyite-(Mg)	$Ca_2(Mg,Fe^{2+})Al_2(Si_2O_7)(SiO_4)(OH)_2{\cdot}H_2O$	A2/m	Ulyanov (1991), Zolensky and McSween (1988)
	Thortveitite	$Sc_2Si_2O_7$	C2/m	Ma et al. (2011b)
	Tilleyite	$Ca_5Si_2O_7(CO_3)_2$	$P2_1/a$	Krot (2020) personal communication
	Wadsleyite	Mg_2SiO_4	I2/m	Ulyanov (1991)
Cyclosilicates (closed rings of SiO_4 tetrahedra)				
	Cordierite	$Mg_2Al_4Si_5O_{18}$	Cccm	MacPherson et al. (1988), Petaev et al. (1993)
	Indialite	$Mg_2Al_3(AlSi_5O_{18})$	P6/mcc	Mikouchi et al. (2016)
	Merrihueite	$(K,Na)_2Fe_5Si_{12}O_{30}$	P6/mcc	Dodd et al. (1965, 1966), Krot and Wasson (1994)
	Osumilite	$KFe_2(Al_5Si_{10})O_{30}$	P6/mcc	Ulyanov (1991)
	Roedderite	$(K,Na)_2Mg_5Si_{12}O_{30}$	P6/mmc	Buchwald (1977), Fuchs (1966b), Krot and Wasson (1994)
	Yagiite	$(K,Na)_2(Mg,Al)_5(Si,Al)_{12}O_{30}$	P6/mmc	Buchwald (1977)

Table 1.1 (*cont.*)

Mineral	Synonyms and Varieties	Formula	Space Group	Selected References
Inosilicates (continuous single or double chains of SiO_4 tetrahedra)				
Aenigmatite		$Na_2Fe^{2+}{}_5TiSi_6O_{20}$	P1	
Al-Ti diopside	Fassaite	$Ca(Mg,Ti,Al)(Si,Al)_2O_6$	C2/c	Dodd (1981), MacPherson et al. (1988)
Akimotoite		$(Mg,Fe)SiO_3$	$R\overline{3}$	Tomioka and Fujino (1999)
Albitic jadeite		$(Na,Ca,\square_{1/4})(Al,Si)Si_2O_6$	C2/c	Ma et al. (2020b)
Anthophyllite		$(Mg,Fe)_7Si_8O_{22}(OH)_2$	Pnma	Brearley (1996)
Augite		$(Ca,Mg,Fe)_2Si_2O_6$	C2/c	Dodd (1981)
Barroisite		$\square NaCa(Mg_3Al_2)(Si_7Al)O_{22}(OH)_2$	C2/m	Dobrica and Brearley (2014)
Burnettite		$CaV^{3+}AlSiO_6$	C2/c	Ma and Beckett (2016a)
Clinoenstatite		$Mg_2Si_2O_6$	P 2_1/c	Lindstrom (1990), Britvin et al. (2008), Pekov (1998)
Davisite	Sc-fassaite	$CaScAlSiO_6$	C2/c	Ma and Rossman (2009b)
Diopside		$CaMgSi_2O_6$	C2/c	Dodd (1981)
Donpeacorite		$(Mn,Mg)Mg(SiO_3)_2$	Pbca	Kimura (1996); Kimura and El Goresy (1989)
Enstatite		$Mg_2Si_2O_6$	Pbca	Dodd (1981), Keil (1968)
Ferrosilite		$Fe_2Si_2O_6$	Pbca	Krot et al. (1997c)
Fluor-richterite		$Na_2Ca(Mg,Fe)_5Si_8O_{22}F_2$	N/A	Bevan et al. (1977), Buchwald (1977), Olsen et al. (1973)
Grossmanite		$CaTi^{3+}AlSiO_6$	C2/c	Ma and Rossman (2009c)
Hedenbergite		$CaFeSi_2O_6$	C2/c	Kimura and Ikeda (1995)
Hemleyite		$FeSiO_3$	$R\overline{3}$	Bindi et al. (2017)
Hornblende		$Ca_2[Mg,Fe,Al]_5[Si,Al]_8O_{22}(OH)_2$	C2/m	Rubin (2014)
Jadeite		$NaAlSi_2O_6$	C2/c	Ulyanov (1991)
Jimthompsonite		$(Mg,Fe)_5Si_6O_{16}(OH)_2$	Pbca	Brearley (1996)

Kaersutite		$(Na,K)Ca_2(Mg,Fe,Ti,Al)_5(Si_6Al_2)O_{22}O_2$	C2/m	Treiman (1985)
Kanoite		$MnMgSi_2O_6$	$P2_1/c$	Kimura (1996); Kimura and El Goresy (1989)
Kosmochlor	Ureyite	$NaCrSi_2O_6$	C2/c	Buchwald (1977), Greshake and Bischoff (1996)
Krinovite		$NaMg_2CrSi_3O_{10}$	P1	Buchwald (1977), Olsen and Fuchs (1968)
Kuratite		$Ca_2(Fe^{2+}{}_5Ti)O_2[Si_4Al_2O_{18}]$	P1	Hwang et al. (2014)
Kushiroite		$CaAlAlSiO_6$	C2/c	Kimura et al. (2009), Ma et al. (2009)
Magnesio-arfvedsonite		$NaNa_2(Mg_4Fe^{3+})Si_8O_{22}(OH)_2$	C2/m	Ivanov et al. (2001)
Magnesiohornblende		$Ca_2(Mg_4Al)(Si_7AlO_{22})(OH)_2$	C2/m	McCanta et al. (2008)
Orthopyroxene		$(Mg,Fe)SiO_3$	Pbca	Buchwald (1977), Dodd (1981)
Pigeonite		$(Mg,Fe,Ca)_2Si_2O_6$	$P2_1/c$	Dodd (1981)
Potassic-chloro-hastingsite		$KCa_2(Fe^{2+}{}_4Fe^{3+})(Si_6Al_2)O_{22}Cl_2$	B2/m	McCubbin et al. (2009)
Pyroxferroite		$FeSiO_3$	P1	Papike et al. (1991)
Rhodonite		$CaMn_4(Si_3O_{15})$	P1	Ulyanov (1991)
Rhönite		$Ca_2(Mg,Al,Ti)_6(Si,Al)_6O_{20}$	P1	Fuchs (1971)
Tissintite		$(Ca,Na,\square)AlSi_2O_6$	C2/c	Ma et al. (2015b)
Wilkinsonite		$Na_2Fe^{2+}{}_4Fe^{3+}{}_2Si_6O_{20}$	P1	Ivanov et al. (2001)
Winchite		$\square NaCa(Mg_4Al)Si_8O_{22}(OH)_2$	C2/m	Dobrica and Brearley (2014)
Wollastonite		$CaSiO_3$	P1	Fuchs (1971)
Phyllosilicates (continuous sheets of SiO_4 tetrahedra)				
Aspidolite		$NaMg_3(Si_3Al)O_{10}(OH)_2$	B2/m	www.mindat.org
Biotite		$K(Mg,Fe)_3(Si_3Al)O_{10}(OH,F)_2$	C2/m	Floran et al. (1978), Johnson et al. (1991)
Chlorite Group				

Table 1.1 (*cont.*)

Mineral	Synonyms and Varieties	Formula	Space Group	Selected References
Chamosite		$(Fe^{2+},Mg,Al,Fe^{3+})_6(Si,Al)_4O_{10}(OH,O)_8$	C2/m	Barber (1981), Zolensky and McSween (1988)
Clinochlore		$(Mg,Fe^{2+})_5Al(Si_3Al)O_{10}(OH)_8$	C2/m	Barber (1981)
Clintonite		$Ca(Mg,Al)_3(Al,Si)_4O_{10}(OH,F)_2$	C2/m	Krot et al. (1995)
Glauconite		$(K,Na)(Mg,Fe^{2+},Fe^{3+})(Fe^{3+},Al)(Si,Al)_4O_{10}(OH)_2$	B2/m	Kyte (1998)
Hisingerite		$Fe_2Si_2O_5(OH)_4{\cdot}2H_2O$		Abreu (2016)
Illite		$K_{\sim0.65}(Al,Mg,Fe)_2(Si,Al)_4O_{10}(OH)_2$	C2/m	Gooding (1992)
Margarite		$CaAl_2(Si_2Al_2)O_{10}(OH)_2$	C2/c	Krot et al. (1995)
Mica		$(K,Na,Ca)(Al,Mg,Fe)_{2-3}(Si,Al,Fe)_4O_{10}(OH,F)_2$	C2/m	Velbel (1988), Zolensky and Gooding (1986)
Muscovite		$KAl_2(AlSi_3O_{10})(OH)_2$	C2/m	Kurat et al. (1981)
Oxyphlogopite		$K(Mg,Ti,Fe)_3[(Si,Al)_4O_{10}](O,F)_2$	C2/m	www.mindat.org
Pecoraite		$Ni_3Si_2O_5(OH)_4$	C2/m	Faust et al. (1973)
Serpentine Group				
Amesite		$Mg_2Al(SiAl)O_5(OH)_4$	C1	Zolensky and McSween (1988)
Antigorite		$Mg_3Si_2O_5(OH)_4$	Bm	Barber (1981)
Berthierine		$(Fe^{2+},Fe^{3+},Mg)_3(Si,Al)_2O_5(OH)_4$	Cm	Barber (1981)
Chrysotile		$Mg_3Si_2O_5(OH)_4$	A2/m	Barber (1981)
Cronstedtite		$(Fe^{2+},Fe^{3+})_3(Si,Fe^{3+})_2O_5(OH)_4$	$P3_1m$	Zolensky and McSween (1988)

Ferroan Antigorite		$(Mg,Fe,Mn)_3(Si,Al)_2O_5(OH)_4$	Bm	Barber (1981)
Greenalite		$(Fe^{2+},Fe^{3+})_{2\text{-}3}Si_2O_5(OH)_4$	unknown	Barber (1981)
Lizardite		$Mg_3Si_2O_5(OH)_4$	P1	Barber (1981)
Smectite Group				
Montmorillonite		$(Na,Ca)_{0.3}(Al,Mg)_2Si_4O_{10}(OH)_2\cdot nH_2O$	C2/m	Krot et al. (1995), Zolensky and McSween (1988)
Nontronite		$Na_{0.3}Fe_2^{3+}(Si,Al)_4O_{10}(OH)_2.nH_2O$	C2/m	Zolensky and McSween (1988)
Saponite		$(Ca,Na)_{0.3}(Mg,Fe)_3(Si,Al)_4O_{10}(OH)_2\cdot 4H_2O$	C2/m	Brearley (1995), Krot et al. (1995)
Sodium-Phlogopite	Aspidolite	$(Na,K)Mg_3(Si_3Al)O_{10}(F,OH)_2$	B2/m	Krot et al. (1995)
Talc		$Mg_3Si_4O_{10}(OH)_2$	C2/c	Barber (1981), Brearley (1996)
Vermiculite		$(Mg,Fe,Al)_3(Si,Al)_4O_{10}(OH)_2\cdot 4H_2O$	C2/m	Ulyanov (1991), Zolensky and McSween (1988)
Tectosilicates (continuous framework of SiO_4 tetrahedra)				
Albite		$NaAlSi_3O_8$	C1	Keil (1968)
Anorthite		$CaAl_2Si_2O_8$	$P\bar{1}$	MacPherson et al. (1988)
Celsian		$BaAl_2Si_2O_8$	I21/c	MacPherson et al. (1988), Dodd (1981)
Chabazite-Na		$(Na_3K)Al_4Si_8O_{24}\cdot 11H_2O$	R3m	Zolensky and Ivanov (2003)
Coesite		SiO_2	C2/c	Weisberg and Kimura (2010), Kimura et al. (2017)
Cristobalite		SiO_2	$P4_12_12$	Dodd (1981), Marvin (1962)
Dmisteinbergite		$CaAl_2Si_2O_8$	P6/mmm	Ma et al. (2013b)
Donwilhelmsite		$CaAl_4Si_2O_{11}$	$P6_3/mmc$	Fritz et al. (2019, 2020)
Feldspar Group		$(K,Na,Ca)(Si,Al)_4O_8$		Buchwald (1977)
Haüyne		$Na_3Ca(Si_3Al_3)O_{12}(SO_4)$	P43n	Flight (1887)
Liebermannite		$KAlSi_3O_8$	I4/m	Ma et al. (2018b)
Lingunite		$NaAlSi_3O_8$	I4/m	Gillet et al. (2000)
Marialite		$Na_4(Si,Al)_{12}O_{24}Cl$	I4/m	Kimura (1996), Alexander et al. (1987)
Maskelynite		$(Na,Ca)(Si,Al)_4O_8$	amorphous	Binns (1967), Rubin (2015b)

Table 1.1 (*cont.*)

	Mineral	Synonyms and Varieties	Formula	Space Group	Selected References
	Nepheline		$(Na,K)AlSiO_4$	$P6_3$	MacPherson et al. (1988)
	Opal		$SiO_2 \cdot nH_2O$		Buchwald (1977)
	Orthoclase		$KAlSi_3O_8$	C2/m	Kerridge and Matthews (1988)
	Plagioclase		$(Na,Ca)(Si,Al)_3O_8$	C1	Dodd (1981)
	Quartz		SiO_2	$P3_1$ 21, $P3_232$	Dodd (1981), Komatsu (2018)
	Sanidine		$KAlSi_3O_8$	C2/m	Floran et al. (1978), Johnson et al. (1991)
	Seifertite		SiO_2	Pbcn or Pb2n	El Goresy et al. (2008)
	Sodalite		$Na_4(Si_3Al_3)O_{12}Cl$	P43n	MacPherson et al. (1988)
	Stilbite-Ca		$NaCa_4(Si_{27}Al_9)O_{72} \cdot 30H_2O$	C2/m	Kimura (1996), Gooding (1981)
	Stishovite		SiO_2	P4/mnm	Chao et al. (1962)
	Stöfflerite		$CaAl_2Si_2O_8$	I4/m	Tschauner and Ma (2017)
	Tridymite		SiO_2	F1	Dodd (1981)
	Zagamiite		$CaAl_2Si_{3.5}O_{11}$	$P6_3/mmc$	Ma and Tschauner (2017), Ma et al. (2017c, 2019a)
	Zeolite Group		$(Na,K)_{0-2}(Ca,Mg)_{1-2}(Al,Si)_{5-10}O_{10-20}.nH_2O$		MacPherson et al. (1988)
Oxalates					
	Whewellite		$CaC_2O_4 \cdot H_2O$	$P2_1/n$	Fuchs et al. (1973)
Phosphate-Silicate					
	Tsangpoite		$Ca_5(PO_4)_2(SiO4)$	$P6_3/m$, $P6_3$, or $P6_322$	Hwang et al. (2016b)

In most cases, the listed citations for the minerals are not exhaustive. Many individual minerals formed by different processes and are found in a variety of meteorites.

Table 1.2 *Alphabetical list of meteoritic minerals*

Addibischoffite	$Ca_2Al_6Al_6O_{20}$
Adrianite	$Ca_{12}(Al_4Mg_3Si_7)O_{32}Cl_6$
Aenigmatite	$Na_2Fe^{2+}{}_5TiSi_6O_{20}$
Ahrensite	Fe_2SiO_4
Akaganéite	β-FeO(OH,Cl)
Åkermanite	$Ca_2MgSi_2O_7$
Akimotoite	$(Mg,Fe)SiO_3$
Alabandite	MnS
Albite	$NaAlSi_3O_8$
Albitic jadeite	$(Na,Ca,\square_{1/4})(Al,Si)Si_2O_6$
Allabogdanite	$(Fe,Ni)_2P$
Allendeite	$Sc_4Zr_3O_{12}$
Almandine	$Fe_3Al_2(SiO_4)_3$
Altaite	PbTe
Al-Ti diopside	$Ca(Mg,Ti,Al)(Si,Al)_2O_6$
Aluminium	Al
Amakinite	$(Fe,Mg)(OH)_2$
Amesite	$Mg_2Al(SiAl)O_5(OH)_4$
Anatase	TiO_2
Andradite	$Ca_3Fe_2(SiO_4)_3$
Andreyivanovite	FeCrP
Anhydrite	$CaSO_4$
Ankerite	$Ca(Fe^{2+},Mg,Mn)(CO_3)_2$
Anorthite	$CaAl_2Si_2O_8$
Anosovite (not approved)	$(Ti^{4+},Ti^{3+},Mg,Sc,Al)_3O_5$
Anthophyllite	$(Mg,Fe)_7Si_8O_{22}(OH)_2$
Antigorite	$Mg_3Si_2O_5(OH)_4$
Antitaenite (not approved)	Fe_3Ni
Apatite	$Ca_5(PO_4)_3(F,OH,Cl)$
Aragonite	$CaCO_3$
Armalcolite	$(Mg,Fe)Ti_2O_5$
Arupite	$Ni_3(PO_4)_2{\cdot}8H_2O$
Asimowite	Fe_2SiO_4
Aspidolite	$NaMg_3(Si_3Al)O_{10}(OH)_2$
Augite	$(Ca,Mg,Fe)_2Si_2O_6$
Awaruite	Ni_3Fe
Baddeleyite	ZrO_2
Baghdadite	$Ca_3(Zr,Ti)Si_2O_9$
Baryte	$BaSO_4$
Barringerite	$(Fe,Ni)_2P$
Barringtonite	$MgCO_3{\cdot}2H_2O$
Barroisite	$\square NaCa(Mg_3Al_2)(Si_7Al)O_{22}(OH)_2$
Bassanite	$CaSO_4{\cdot}½H_2O$

Table 1.2 (*cont.*)

Beckettite	$Ca_2V_6Al_6O_{20}$
Berthierine	$(Fe^{2+},Fe^{3+},Mg)_3(Si,Al)_2O_5(OH)_4$
Beusite	$(Mn,Fe,Ca,Mg)_3(PO_4)_2$
Biotite	$K(Mg,Fe)_3(Si_3Al)O_{10}(OH,F)_2$
Bismuth chloride (not approved)	$BiCl_3$
Blödite	$Na_2Mg(SO_4)_2{\cdot}4H_2O$
Böhmite	$AlO(OH)$
Bornite	Cu_5FeS_4
Breunnerite	$(Mg,Fe)CO_3$
Brezinaite	Cr_3S_4
Brianite	$Na_2CaMg(PO_4)_2$
Bridgmanite	$MgSiO_3$
Britholite-(Ce)	$(Ce,Y,Ca)_5(SiO_4,PO_4)_3(OH,F)$
Browneite	MnS
Brownleeite	MnSi
Brucite	$Mg(OH)_2$
β-Silicon nitride (Not Approved)	β-Si_3N_4
Buchwaldite	$NaCaPO_4$
Bunsenite	NiO
Burnettite	$CaV^{3+}AlSiO_6$
Buseckite	(Fe,Zn,Mn)S
Butianite	Ni_6SnS_2
Calcite	$CaCO_3$
Calzirtite	$Ca_2Zr_5Ti_2O_{16}$
Carbonate-fluorapatite	$Ca_5(PO_4,CO_3)_3F$
Carletonmooreite	Ni_3Si
Carlsbergite	CrN
Cassidyite	$Ca_2(Ni,Mg)(PO_4)_2{\cdot}2H_2O$
Caswellsilverite	$NaCrS_2$
Celestine	SrSO4
Celsian	$BaAl_2Si_2O_8$
Chabazite-Na	$(Na_3K)Al_4Si_8O_{24}{\cdot}11H_2O$
Chalcocite	Cu_2S
Chalcopyrite	$CuFeS_2$
Chamosite	$(Fe^{2+},Mg,Al,Fe^{3+})_6(Si,Al)_4O_{10}(OH,O)_8$
Chaoite	C
Chenmingite	$FeCr_2O_4$
Chevkinite-(Ce)	$(Ce,Nd,La,Ca,Th)_4(Ti,Fe,Mg)_5Si_4O_{22}$
Chladniite	$Na_2CaMg_7(PO_4)_6$
Chlorapatite	$Ca_5(PO_4)_3Cl$
Chlormagaluminite	$Mg_4Al_2(OH)_{12}Cl_2{\cdot}3H_2O$
Chlormayenite	$Ca_{12}Al_{14}O_{32}Cl_2$
Chopinite	$Mg_3(PO_4)_2$

Table 1.2 (*cont.*)

Chromite	$FeCr_2O_4$
Chrysotile	$Mg_3Si_2O_5(OH)_4$
Chukanovite	$Fe_2(CO_3)(OH)_2$
Cinnabar	HgS
Clinochlore	$(Mg,Fe^{2+})_5Al(Si_3Al)O_{10}(OH)_8$
Clinoenstatite	$Mg_2Si_2O_6$
Clintonite	$Ca(Mg,Al)_3(Al,Si)_4O_{10}(OH,F)_2$
Cobaltite	CoAsS
Coesite	SiO_2
Cohenite	$(Fe,Ni)_3C$
Collinsite	$Ca_2(Mg,Fe,Ni)(PO_4)_2{\cdot}2H_2O$
Cooperite	PtS
Copiapite	$Fe_5(SO_4)_6(OH)_2{\cdot}20H_2O$
Copper	Cu
Coquimbite	$Fe_2(SO_4)_3{\cdot}9H_2O$
Cordierite	$Mg_2Al_4Si_5O_{18}$
Corundum	Al_2O_3
Coulsonite	FeV_2O_4
Covellite	CuS
Cristobalite	SiO_2
Cronstedtite	$(Fe^{2+},Fe^{3+})_3(Si,Fe^{3+})_2O_5(OH)_4$
Cronusite	$Ca_{0.2}CrS_2{\cdot}2H_2O$
Cu-Cr-sulfide (not approved)	$CuCrS_2$
Cubanite	$CuFe_2S_3$
Cupalite	CuAl
Cuprite	Cu_2O
Czochralskiite	$Na_4Ca_3Mg(PO_4)_4$
Daubréelite	$FeCr_2S_4$
Davisite	$CaScAlSiO_6$
Decagonite	$Al_{71}Ni_{24}Fe_5$
Diamond	C
Digenite	$Cu_{1.8}S$
Diopside	$CaMgSi_2O_6$
Djerfisherite	$K_6(Fe,Cu,Ni)_{25}S_{26}Cl$
Dmisteinbergite	$CaAl_2Si_2O_8$
Dmitryivanovite	$CaAl_2O_4$
Dolomite	$CaMg(CO_3)_2$
Donpeacorite	$(Mn,Mg)Mg(SiO_3)_2$
Donwilhelmsite	$CaAl_4Si_2O_{11}$
Droninoite	$Ni_6Fe^{3+}{}_2Cl_2(OH)_{16}{\cdot}4H_2O$
Edenite	$NaCa_2Mg_5Si_7AlO_{22}(OH)_2$
Edscottite	Fe_5C_2
Electrum (not approved)	Au-Ag

Table 1.2 (*cont.*)

Enstatite	$Mg_2Si_2O_6$
Epsomite	$MgSO_4·7H_2O$
Eringaite	$Ca_3Sc_2Si_3O_{12}$
Erlichmanite	OsS_2
Eskolaite	Cr_2O_3
Farringtonite	$Mg_3(PO_4)_2$
Fayalite	Fe_2SiO_4
Feiite	$Fe^{2+}{}_2(Fe^{2+}Ti^{4+})O_5$
Feldspar group	$(K,Na,Ca)(Si,Al)_4O_8$
Feroxyhyte	δ-FeO(OH)
Ferrihydrite	$Fe^{3+}{}_{10}O_{14}(OH)_2$
Ferroan alabandite	(Mn,Fe)S
Ferroan antigorite	$(Mg,Fe,Mn)_3(Si,Al)_2O_5(OH)_4$
Ferromerrillite	$Ca_9NaFe^{2+}(PO_4)_7$
Ferropseudobrookite	$FeTi_2O_5$
Ferrosilite	$Fe_2Si_2O_6$
Florenskyite	(Fe,Ni)TiP
Fluorapatite	$Ca_5(PO_4)_3F$
Fluor-richterite	$Na_2Ca(Mg,Fe)_5Si_8O_{22}F_2$
Forsterite	Mg_2SiO_4
Galena	PbS
Galileiite	$NaFe_4(PO_4)_3$
Gehlenite	$Ca_2Al(SiAl)O_7$
Geikielite	$MgTiO_3$
Gentnerite (not approved)	$Cu_8Fe_3Cr_{11}S_{18}$
Gersdorffite	NiAsS
Glauconite	$(K,Na)(Mg,Fe^{2+},Fe^{3+})(Fe^{3+},Al)(Si,Al)_4O_{10}(OH)_2$
Goethite	α-FeO(OH)
Gold	Au
Gold-dominated alloys	(Au,Ag,Fe,Ni,Pt)
Goldmanite	$Ca_3V_2(SiO_4)_3$
Graftonite	$(Fe,Mn)_3(PO_4)_2$
Graphite	C
Greenalite	$(Fe^{2+},Fe^{3+})_{2\text{-}3}Si_2O_5(OH)_4$
Greigite	Fe_3S_4
Grossite	$CaAl_4O_7$
Grossmanite	$CaTi^{3+}AlSiO_6$
Grossular	$Ca_3Al_2(SiO_4)_3$
Gupeiite	Fe_3Si
Gypsum	$CaSO_4·2H_2O$
Haapalaite	$2[(Fe,Ni)S]·1.6[(Mg,Fe)(OH)_2]$
Halite	NaCl
Hapkeite	Fe_2Si
Hausmannite	$Mn^{2+}Mn^{3+}{}_2O_4$

Table 1.2 (*cont.*)

Haüyne	$Na_3Ca(Si_3Al_3)O_{12}(SO_4)$
Haxonite	$(Fe,Ni)_{23}C_6$
Heazlewoodite	Ni_3S_2
Hedenbergite	$CaFeSi_2O_6$
Heideite	$(Fe,Cr)_{1.15}(Ti,Fe)_2S_4$
Hematite	α-Fe_2O_3
Hemleyite	$FeSiO_3$
Hercynite	$FeAl_2O_4$
Hexaferrum	(Fe,Os,Ir,Mo)
Hexahydrite	$MgSO_4{\cdot}6H_2O$
Hexamolybdenum	(Mo,Ru,Fe)
Hibbingite	γ-$Fe_2(OH)_3Cl$
Hibonite	$CaAl_{12}O_{19}$
Hibonite-(Fe)	$(Fe,Mg)Al_{12}O_{19}$
Hiroseite	$FeSiO_3$
Hisingerite	$Fe_2Si_2O_5(OH)_4{\cdot}2H_2O$
Hollisterite	Al_3Fe
Honessite	$(Ni,Fe)_8SO_4(OH)_{16}{\cdot}nH_2O$
Hornblende	$Ca_2[Mg,Fe,Al]_5[Si,Al]_8O_{22}(OH)_2$
Hutcheonite	$Ca_3Ti_2(SiAl_2)O_{12}$
Hydromagnesite	$Mg_5(CO_3)_4(OH)_2{\cdot}4H_2O$
Hydroxylapatite	$Ca_5(PO_4)_3OH$
Icosahedrite	$Al_{63}Cu_{24}Fe_{13}$
Icosahedrite II	$Al_{62}Cu_{31}Fe_7$
Idaite	Cu_5FeS_6
Illite	$K_{\sim 0.65}(Al,Mg,Fe)_2(Si,Al)_4O_{10}(OH)_2$
Ilmenite	$FeTiO_3$
Indialite	$Mg_2Al_3(AlSi_5O_{18})$
Indium-dominated alloys	(In,Sn,Pb)
Irarsite	(Ir,Ru,Rh,Pt)AsS
Iridarsenite	$(Ir,Ru)As_2$
Iron	α-Fe
Isocubanite	$CuFe_2S_3$
Jadeite	$NaAlSi_2O_6$
Jarosite	$KFe_3(SO_4)_2(OH)_6$
Jimthompsonite	$(Mg,Fe)_5Si_6O_{16}(OH)_2$
Joegoldsteinite	$MnCr_2S_4$
Johnsomervilleite	$Na_2Ca(Fe,Mg,Mn)_7(PO_4)_6$
Kaersutite	$(Na,K)Ca_2(Mg,Fe,Ti,Al)_5(Si_6Al_2)O_{22}O_2$
Kaitianite	$Ti^{3+}{}_2Ti^{4+}O_5$
Kalininite	$ZnCr_2S_4$
Kamiokite	$Fe_2Mo_3O_8$
Kangite	$(Sc,Ti,Al,Zr,Mg,Ca,\square)_2O_3$
Kanoite	$MnMgSi_2O_6$

Table 1.2 (*cont.*)

Keilite	(Fe,Mg)S
Keplerite	$Ca_9(Ca_{0.5}\square_{0.5})Mg(PO_4)_7$
Khamrabaevite	TiC
Khatyrkite	$CuAl_2$
Kieserite	$MgSO_4{\cdot}H_2O$
Kirschsteinite	$CaFe(SiO_4)$
K-Na-Fe phosphate	$(K,Na)Fe_4(PO_4)_3$
Kosmochlor	$NaCrSi_2O_6$
Krinovite	$NaMg_2CrSi_3O_{10}$
Krotite	$CaAl_2O_4$
Kryachkoite	$(Al,Cu)_6(Fe,Cu)$
Kuratite	$Ca_2(Fe^{2+}{}_5Ti)O_2[Si_4Al_2O_{18}]$
Kushiroite	$CaAlAlSiO_6$
Kutnohorite	$CaMn(CO_3)_2$
Lakargiite	$CaZrO_3$
Laihunite	$(Fe^{3+},Fe^{2+},Mg,\square)_2SiO_4$
Larnite	Ca_2SiO_4
Laurite	RuS_2
Lawrencite	$(Fe^{2+},Ni)Cl_2$
Lepidocrocite	γ-FeO(OH)
Liebermannite	$KAlSi_3O_8$
Lime	CaO
Lingunite	$NaAlSi_3O_8$
Linzhiite	$FeSi_2$
Lipscombite	$(Fe^{2+},Mn)Fe^{3+}{}_2(PO_4)_2(OH)_2$
Liuite	$FeTiO_3$
Lizardite	$Mg_3Si_2O_5(OH)_4$
Löllingite	$FeAs_2$
Loveringite	$Ca(Ti,Fe,Cr,Mg)_{21}O_{38}$
Machiite	$Al_2Ti_3O_9$
Mackinawite	$(Fe,Ni)_{1+x}S$ (x = 0-0.07)
Manganite	$Mn^{3+}OOH$
Maghemite	$Fe_{2.67}O_4$
Magnéli phases	Ti_5O_9 and Ti_8O_{15}
Magnesio-arfvedsonite	$NaNa_2(Mg_4Fe^{3+})Si_8O_{22}(OH)_2$
Magnesiochromite	$MgCr_2O_4$
Magnesioferrite	$MgFe_2O_4$
Magnesiohornblende	$Ca_2(Mg_4Al)(Si_7AlO_{22})(OH)_2$
Magnesiowüstite	(Fe,Mg)O
Magnesite	$(Mg,Fe)CO_3$
Magnetite	Fe_3O_4
Majindeite	$Mg_2Mo_3O_8$
Majorite	$Mg_3(MgSi)Si_3O_{12}$
Marcasite	FeS_2

Table 1.2 (*cont.*)

Margarite	$CaAl_2(Si_2Al_2)O_{10}(OH)_2$
Marialite	$Na_4(Si,Al)_{12}O_{24}Cl$
Maricite	$NaFePO_4$
Martensite (not approved)	α_2-(Fe,Ni)
Maskelynite	$(Na,Ca)(Si,Al)_4O_8$
Matyhite	$Ca_9(Ca_{0.5}\square_{0.5})Fe^{2+}(PO_4)_7$
Maucherite	$Ni_{11}As_8$
Melanterite	$FeSO_4{\cdot}7H_2O$
Melilite	$(Ca,Na)_2(Al,Mg)(Si,Al)_2O_7$
Melliniite	$(Ni,Fe)_4P$
Mendozite	$NaAl(SO_4)_2{\cdot}11H_2O$
Mercury	Hg
Merrihueite	$(K,Na)_2(Fe,Mg)_5Si_{12}O_{30}$
Merrillite	$Ca_9NaMg(PO_4)_7$
Mica	$(K,Na,Ca)(Al,Mg,Fe)_{2\text{-}3}(Si,Al,Fe)_4O_{10}(OH,F)_2$
Millerite	NiS
Moissanite	SiC
Molybdenite	MoS_2
Molybdenum (not approved)	Mo
Molybdenum carbide (not approved)	MoC
Monazite-(Ce)	$(Ce,La,Th)PO_4$
Moncheite	$Pt(Te,Bi)_2$
Monipite	MoNiP
Monticellite	$CaMgSiO_4$
Montmorillonite	$(Na,Ca)_{0.3}(Al,Mg)_2Si_4O_{10}(OH)_2{\cdot}nH_2O$
Moraskoite	$Na_2Mg(PO_4)F$
Mullite	$Al_6Si_2O_{13}$
Murchisite	Cr_5S_6
Muscovite	$KAl_2(AlSi_3O_{10})(OH)_2$
Na-Ca-Cr phosphate	$Na_4CaCr(PO_4)_3$
Na-Ca-Fe phosphate	$Na_4Ca_3Fe(PO_4)_4$
Na-Fe-Mg phosphate	$Na_2Fe(Mg,Ca)(PO_4)_2$
Na-Mn-Fe phosphate	$Na_4(Mn,Fe)(PO_4)_2$
Naquite	FeSi
Nb-oxide	$(Nb,V,Fe)O_2$
Nepheline	$(Na,K)AlSiO_4$
Nesquehonite	$Mg(CO_3){\cdot}3H_2O$
Nickel	Ni
Ni-Ge phosphide	$Ni_4Ge_{0.33}P_{1.17}$
Ni-rich blödite	$Na_2(Mg,Ni)(SO_4)_2{\cdot}4H_2O$
Nickeline	NiAs
Nickelphosphide	Ni_3P
Nierite	α-Si_3N_4
Niningerite	MgS
Niobium (not approved)	Nb

Table 1.2 (*cont.*)

Nontronite	$Na_{0.3}Fe_2^{3+}(Si,Al)_4O_{10}(OH)_2 \cdot nH_2O$
Nuwaite	Ni_6GeS_2
Nyerereite	$Na_2Ca(CO_3)_2$
Oldhamite	CaS
Olivine	$(Mg,Fe)_2SiO_4$
Olkhonskite	$Cr_2Ti_3O_9$
Omeiite	$(Os,Ru)As_2$
Omphacite	$(Ca,Na)(Mg,Fe,Al)Si_2O_6$
Opal	$SiO_2 \cdot nH_2O$
Orcelite	$Ni_{4.77}As_2$
Orthoclase	$KAlSi_3O_8$
Orthopyroxene	$(Mg,Fe)SiO_3$
Osbornite	TiN
Osmium	Os
Osumilite	$KFe_2(Al_5Si_{10})O_{30}$
Oxyphlogopite	$K(Mg,Ti,Fe)_3[(Si,Al)_4O_{10}](O,F)_2$
Panethite	$(Na,Ca,K)_{1-x}(Mg,Fe,Mn)PO_4$
Panguite	$(Ti,Al,Sc,Mg,Zr,Ca)_{1.8}O_3$
Paqueite	$Ca_3TiSi_2(Al,Ti,Si)_3O_{14}$
Paraotwayite	$Ni(OH)_{2-x}(SO_4,CO_3)_{0.5x}$
Pecoraite	$Ni_3Si_2O_5(OH)_4$
Pentlandite	$(Fe,Ni)_9S_8$
Periclase	MgO
Perovskite	$CaTiO_3$
Perrierite-(Ce)	$(Ce,Nd,La,Ca,Th)_4(Ti,Fe,Mg)_5Si_4O_{22}$
Perryite	$(Ni,Fe)_8(Si,P)_3$
PGE-dominated alloys	(Pt,Os,Ir,Ru,Re,Rh,Mo,Nb,Ta,Ge,W,V,Pb,Cr,Fe,Ni, Co)
Phlogopite	$KMg_3(Si_3Al)O_{10}(OH,F)_2$
Pigeonite	$(Mg,Fe,Ca)_2Si_2O_6$
Plagioclase	$(Na,Ca)(Si,Al)_3O_8$
Plagionite	$Pb_5Sb_8S_{17}$
Platinum	Pt
Poirierite	Mg_2SiO_4
Portlandite	$Ca(OH)_2$
Potassic-chloro-hastingsite	$KCa_2(Fe^{2+}{}_4Fe^{3+})(Si_6Al_2)O_{22}Cl_2$
Powellite	$CaMoO_4$
Proxidecagonite	$Al_{34}Ni_9Fe_2$
Pseudobrookite	Fe_2TiO_5
Pumpellyite-(Mg)	$Ca_2(Mg,Fe^{2+})Al_2(Si_2O_7)(SiO_4)(OH)_2 \cdot H_2O$
Pyrite	FeS_2
Pyrochlore	$(Na,Ca)_2Nb_2O_6(OH,F)$
Pyrolusite	MnO_2
Pyrope	$Mg_3Al_2(SiO_4)_3$

Table 1.2 (*cont.*)

Pyrophanite	$MnTiO_3$
Pyroxferroite	$FeSiO_3$
Pyrrhotite	$Fe_{1-x}S$
Quartz	SiO_2
Rammelsbergite	$NiAs_2$
Reevesite	$Ni_6Fe^{3+}{}_2(CO_3)(OH)_{16}\cdot4H_2O$
Reidite	$ZrSiO_4$
Rhenium (not approved)	Re
Rhodochrosite	$MnCO_3$
Rhodonite	$CaMn_4(Si_3O_{15})$
Rhönite	$Ca_2(Mg,Al,Ti)_6(Si,Al)_6O_{20}$
Ringwoodite	Mg_2SiO_4
Roaldite	$(Fe,Ni)_4N$
Roedderite	$(K,Na)_2Mg_5Si_{12}O_{30}$
Rubinite	$Ca_3Ti^{3+}{}_2Si_3O_{12}$
Rudashevskyite	(Fe,Zn)S
Rustenburgite	$(Pt,Pd)_3Sn$
Ruthenium	(Ru,Os,Ir)
Ruthenium carbide (not approved)	RuC
Rutheniridosmine	(Ir,Os,Ru)
Rutile	TiO_2
Safflorite	$CoAs_2$
Sanidine	$KAlSi_3O_8$
Saponite	$(Ca,Na)_{0.3}(Mg,Fe)_3(Si,Al)_4O_{10}(OH)_2\cdot4H_2O$
Sapphirine	$(Mg,Al)_8(Al,Si)_6O_{20}$
Sarcopside	$(Fe,Mn)_3(PO_4)_2$
Scheelite	$CaWO_4$
Schöllhornite	$Na_{0.3}CrS_2\cdot H_2O$
Schreibersite	$(Fe,Ni)_3P$
Schwertmannite	$Fe^{3+}{}_{16}O_{16}(OH,SO_4)_{13\text{-}14}\cdot10H_2O$
Seifertite	SiO_2
Selenium	Se
Shenzhuangite	$NiFeS_2$
Siderite	$FeCO_3$
Silica with ZrO_2-like structure (not approved)	SiO_2
Sinoite	Si_2N_2O
Slavikite	$NaMg_2Fe^{3+}{}_5(SO_4)_7(OH)_6\cdot33H_2O$
Smythite	$(Fe,Ni)_{3+x}S_4$ (x = 0-0.3)
Sodalite	$Na_4(Si_3Al_3)O_{12}Cl$
Sodium-bearing silicate	$(Na,K,Ca,Fe)_{0.973}(Al,Si)_{5.08}O_{10}$
Sodium-phlogopite	$(Na,K)Mg_3(Si_3Al)O_{10}(F,OH)_2$
Sperrylite	$PtAs_2$

Table 1.2 (*cont.*)

Sphalerite	ZnS
Spinel	$MgAl_2O_4$
Spinelloid silicate	$(Mg,Fe,Si)_2(Si,\square)O_4$
Spinelloid silicate-II	$(Fe,Mg,Cr,Ti,Ca,\square)_2(Si,Al)O_4$
Stanfieldite	$Ca_4(Mg,Fe)_5(PO_4)_6$
Starkeyite	$MgSO_4.4H_2O$
Steinhardtite	(Al,Ni,Fe)
Stilbite-Ca	$NaCa_4(Si_{27}Al_9)O_{72}{\cdot}30H_2O$
Stishovite	SiO_2
Stöfflerite	$CaAl_2Si_2O_8$
Stolperite	AlCu
Suessite	Fe_3Si
Sulfur	S
Sylvite	KCl
Szomolnokite	$FeSO_4{\cdot}H_2O$
Taenite	γ-(Fe,Ni)
Talc	$Mg_3Si_4O_{10}(OH)_2$
Tazheranite	$(Zr,Ti,Ca,Y)O_{1.75}$
Tetragonal almandine	$(Fe,Mg,Ca,Na)_3(Al,Si,Mg)_2Si_3O_{12}$
Tetragonal majorite	$Mg_3(MgSi)Si_3O_{12}$
Tetrataenite	FeNi
Thénardite	Na_2SO_4
Thorianite	ThO_2
Thortveitite	$Sc_2Si_2O_7$
Ti^{3+},Al,Zr-oxide	$(Ti^{3+},Al,Zr,Si,Mg)_{1.95}O_3$
Ti-oxide	Ti_3O_5
Ti-rich magnetite	$(Fe,Mg)(Fe,Al,Ti)_2O_4$
Tilleyite	$Ca_5Si_2O_7(CO_3)_2$
Tissintite	$(Ca,Na,\square)AlSi_2O_6$
Tistarite	Ti_2O_3
Titanite	$CaTiSiO_5$
Tochilinite	$6(Fe_{0.9}S){\cdot}5[(Mg,Fe,Ni)(OH)_2]$
Tranquillityite	$Fe^{2+}{}_8Ti_3Zr_2Si_3O_{24}$
Transjordanite	Ni_2P
Trevorite	$NiFe_2O_4$
Tridymite	SiO_2
Troilite	FeS
Tsangpoite	$Ca_5(PO_4)_2(SiO_4)$
Tschaunerite	$Fe^{2+}(Fe^{2+}Ti^{4+})O_4$
Tugarinovite	MoO_2
Tuite	γ-$Ca_3(PO_4)_2$
Tungstenite	WS_2
Uakitite	VN

Table 1.2 (*cont.*)

Ulvöspinel	Fe_2TiO_4
V,Fe,Cr-rich sulfide	$(V,Fe,Cr)_4S_5$
V-rich brezinaite	$(Cr,V,Fe)_3S_4$
V-rich daubréelite	$Fe(Cr,V)_2S_4$
V-rich magnetite	$(Fe,Mg)(Fe,Al,V)_2O_4$
Valleriite	$2[(Fe,Cu)S]\cdot 1.53[(Mg,Al)(OH)_2]$
Vaterite	$CaCO_3$
Vermiculite	$(Mg,Fe,Al)_3(Si,Al)_4O_{10}(OH)_2\cdot 4H_2O$
Vestaite	$(Ti^{4+}Fe^{2+})Ti^{4+}{}_3O_9$
Violarite	$FeNi_2S_4$
Vivianite	$Fe_3(PO_4)_2\cdot 8H_2O$
Voltaite	$K_2Fe^{2+}{}_5Fe^{3+}{}_3Al(SO_4)_{12}\cdot 18H_2O$
Wadalite	$Ca_6Al_5Si_2O_{16}Cl_3$
Wadsleyite	Mg_2SiO_4
Wairauite	CoFe
Wangdaodeite	$FeTiO_3$
Warkite	$Ca_2Sc_6Al_6O_{20}$
Wassonite	TiS
Whewellite	$CaC_2O_4\cdot H_2O$
Wilkinsonite	$Na_2Fe^{2+}{}_4Fe^{3+}{}_2Si_6O_{20}$
Winchite	$\square NaCa(Mg_4Al)Si_8O_{22}(OH)_2$
Wollastonite	$CaSiO_3$
Wurtzite	ZnS
Wüstite	FeO
Xenophyllite	$Na_4Fe_7(PO_4)_6$
Xenotime-(Y)	YPO_4
Xieite	$FeCr_2O_4$
Xifengite	Fe_5Si_3
Yagiite	$(Na,K)_{1.5}Mg_2(Al,Mg)_3(Si,Al)_{12}O_{30}$
Zagamiite	$CaAl_2Si_{3.5}O_{11}$
Zaratite	$Ni_3(CO_3)(OH)_4\cdot 4H_2O$
Zeolite group	$(Na,K)_{0\text{-}2}(Ca,Mg)_{1\text{-}2}(Al,Si)_{5\text{-}10}O_{10\text{-}20}\cdot nH_2O$
Zhanghengite	(Cu,Zn)
Zircon	$ZrSiO_4$
Zirconium carbide (not approved)	ZrC
Zirconolite	$CaZrTi_2O_7$
Zirkelite	$(Ti,Ca,Zr)O_{2-x}$
Zolenskyite	$FeCr_2S_4$

2
Definitions and Explications

Every academic field has its share of confusing technical terms. Their redeeming feature is that they lend precision to scholarly discussion, but in some cases, a word may have different meanings in different contexts. For example, the word *abduction* has specific definitions in the legal realm (a kidnapping); formal logic (a syllogism with a major premise that is certain and a minor premise that is merely probable); and anatomy (the action of moving a part of the body away from the midline). Technical terms also bedevil meteoritics; some terms don't have exactly the same meaning as in common parlance. In some cases, the following definitions differ from those approved by the International Astronomical Union (IAU), but the set presented here is precise and self-consistent. We recommend its adoption by astronomers, geologists, cosmochemists, and planetary scientists. Terms are arranged topically, not alphabetically.

Meteorite. In the nineteenth century, meteorites were commonly defined as solid objects that fell to Earth, but since Sputnik 1 reentered the atmosphere in January 1958, man-made objects (or at least scorched debris derived from such objects) joined natural bodies falling from the sky. The identification of a couple of chondritic meteorites among Apollo samples and of some iron meteorites (and probably a few stony meteorites) photographed by rovers on the martian surface demonstrate that meteorites can fall onto bodies other than Earth. They can also hit spacecraft. In addition, lunar and asteroidal materials very similar to lunar meteorites and known groups of chondrites have been retrieved by manned and unmanned spacecraft, i.e., transported to Earth by artificial means. Such complications mandate a more comprehensive definition of "meteorite" (modified from Rubin and Grossman 2010):

> *A meteorite is a natural solid object, larger than 1 μm in size, derived from a celestial body (e.g., asteroid, moon, comet, planet), that was transported by natural means (e.g., impact excavation) from the body on which it formed to a region outside the dominant gravitational influence of that body (i.e., to interplanetary space) and that later collided with a natural or artificial body larger than itself (e.g., Earth, Moon, Mars, an asteroid, a comet, or the International Space Station).*

The most studied meteorite in history is named Allende (Figure 2.1); it fell in Mexico in 1969. Our current understanding of the history of the solar nebula was greatly enhanced by extensive studies of the components of this meteorite.

Micrometeorite. Because micrometeorites smaller than 4 μm have been observed (Figure 2.2), the lower limit for a micrometeorite has been set to 1 μm. *A micrometeorite is a meteorite 1 μm to 2 mm in size.*

Figure 2.1 Cut fusion-crusted individual of the Allende CV3 chondrite. The stone contains numerous CAIs of various sizes and shapes (white and light gray) and less-discernable millimeter-sized chondrules (medium-dark gray) surrounded by fine-grained silicate-rich matrix material (dark gray). Image courtesy of Darryl Pitt/Macovich Collection. (A black-and-white version of this figure will appear in some formats. For the colour version, please refer to the plate section.)

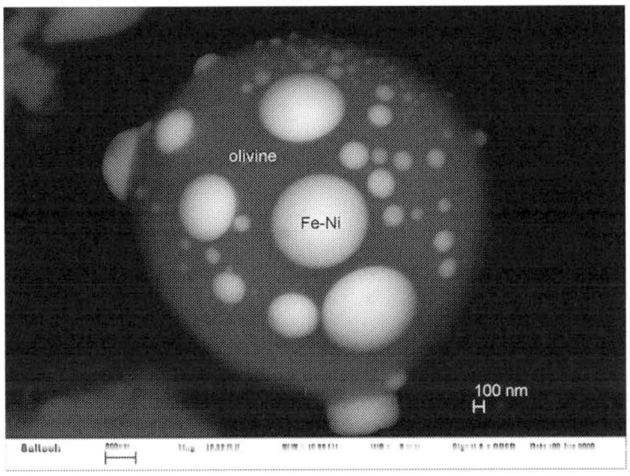

Figure 2.2 One micrometeorite grain from the Allende matrix, showing metallic Fe-Ni on an olivine sphere. BSE image

Meteorwrong (a.k.a. pseudometeorite). *A substance mistakenly thought to be a possible meteorite.* This category includes all types of terrestrial rocks (e.g., basalt, quartzite, scoria, limestone, granite), iron ore (magnetite and hematite), chunks of sulfide, manganese nodules, metallic slag, ball bearings, machine parts, rusted spikes, bottle glass, petrified wood, even occasional twigs, pottery shards, and gelatin. Most meteorwrongs are found on the ground; a few are alleged to be observed falls. Some are family heirlooms. The Moon and every planet in the Solar System have been implicated as the parent body of one meteorwrong or another. The term "meteorong" (without the "w") appears to have been coined by Fredrick C. Leonard of UCLA sometime before 1950. Discussions and some photos of meteorwrongs can be found in

Olsen (1979), chapter 9 of Norton and Chitwood (2008), and at the website https://sites.wustl.edu/meteoritesite/items/some-meteorite-realities/.

Dark flight. *The final stage of descent of a meteoroid through an atmosphere after the object has slowed sufficiently so that it no longer produces light generated by frictional heating with the surrounding gas.*

Meteoroid. *A meteoroid is a natural solid object, 1 μm to 1 m in size, moving in interplanetary space.* Meteoroids can be of any composition; for example, they may consist mainly of silicate, metallic Fe-Ni, sulfide or ice. A meteoroid or a meteoroid remnant that survives collision with a larger body becomes a meteorite.

Micrometeoroid. *A micrometeoroid is a meteoroid 1 μm to 2 mm in size.* A micrometeoroid that survives atmospheric passage becomes a micrometeorite or dust.

Meteoric smoke. *Tiny particles, typically tens of nanometers in size, that condensed from meteoroid material that was vaporized during atmospheric passage.*

Dust trail. *An atmospheric train of small meteoroid particles that ablated but did not vaporize during atmospheric passage.* The particles move along the meteoroid's trajectory before dispersing and settling.

Dust. *A dust grain is a particle larger than a molecule and less than 1 μm in size.* A 1-μm- to 1-m–sized porous aggregate of dust particles is a micrometeoroid or meteoroid. Many small interplanetary dust particles do not melt and do not produce meteors when they pass through an atmosphere; they eventually settle to the body's surface.

Asteroid. *An asteroid is a natural subplanetary object, between 1 m and 1,000 km in diameter, in orbit around its central star; it does not show the characteristics of an active comet. It does not currently experience appreciable outgassing (sufficient to produce a coma and tail and cause orbital deviations); it would not undergo appreciable outgassing upon close approach to its central star; and it did not previously undergo appreciable outgassing.* Older definitions of asteroids had placed the minimum size at 10 m, but the recent detection of smaller asteroids necessitated the change. These small interplanetary bodies include the meteorite-droppers 2008 TC_3 (4.1 ± 0.3 m) and 2018 LA (2.6–3.8 m), asteroid 2014 AA (2–4 m) that plunged into the Atlantic, as well as two objects that did not collide with Earth: 2008 TS_{26} (0.61–1.36 m) and 2011 CQ_1 (0.8–2.6 m). Some asteroids can survive atmospheric passage and land as intact bodies with mean diameters exceeding 1 m; these large objects are considered to be meteorites as well as asteroids (e.g., the $2.7 \times 2.7 \times 0.9$-m-sized IVB-iron-meteorite Hoba; the ~$3 \times 2 \times 1.3$-m-sized ungrouped iron Willamette). The upper limit on the diameter of an asteroid is rounded up from the mean diameter of Ceres (946 km). Bodies in interplanetary space with diameters exceeding 400 km are considered planets or dwarf planets; a body between 400 and 1,000 km in diameter (such as Ceres) can be both an asteroid *and* a dwarf planet. Asteroid Bennu (d ~492 m) is the target of a NASA sample-return mission (Figure 2.3).

Comet. *A comet is a natural volatile-rich solid subplanetary object, at least 1 m in size, in orbit around its central star that exhibits (or would exhibit) extensive outgassing and develop a coma and tail upon close approach to the star; a comet is also a body that can be inferred to have been volatile rich and undergone extensive outgassing in the past.* Icy planetesimals that remain far from the Sun are comets even though they do not exhibit outgassing. These include (a) Kuiper Belt Objects (KBOs) and bodies in the outlying scattered disk (all of which are sometimes called trans-Neptunian objects, TNOs); these regions are the sources of short-period

Figure 2.3 Asteroid 101955 Bennu. (Credits: NASA/Goddard/University of Arizona)

comets with periods $\leq$ 200 years, and (b) Oort Cloud objects (the source of long-period comets, some of which have periods exceeding 1 million years). Most TNOs and Oort Cloud objects never enter the inner Solar System. In contrast, extinct comets that exhausted their supply of volatiles during numerous past approaches to the Sun are still considered comets, even though they have come to resemble asteroids.

Planet. With the discovery of the TNO Eris (mean diameter 2,326 km) in 2005, it became clear that the then-accepted-planet Pluto (mean diameter 2,377 km) was not the only substantial body orbiting the Sun in this region of the Solar System. The situation was analogous to one that had occurred two centuries earlier. The discoveries of the asteroids 2 Pallas (mean diameter 512 km) in 1802, 3 Juno (mean diameter 233 km) in 1804, and 4 Vesta (mean diameter 525 km) in 1807 indicated that the then-accepted-planet Ceres (mean diameter 946 km) was not the only substantial body orbiting the Sun between Mars and Jupiter. (Modern values for the sizes of these bodies are given here.) Just as Ceres lost its planetary status in the 1850s (along with Pallas, Juno, and Vesta) after dozens of bodies had been discovered in the asteroid belt, Pluto followed suit in the early twenty-first century after a handful of large TNOs were discovered. In 2006, the IAU redefined a planet in our Solar System as a body that orbits the Sun, is sufficiently massive for gravity to have molded it into a round shape, and has managed to clear its neighborhood of comparably sized objects. An analogous definition could be applied to exoplanets (although it is not known if all bodies so labeled have cleared their neighborhoods of siblings). The eight planets of our Solar System are illustrated in Figure 2.4.

In principle, a planet could be orbiting a star, a black hole, a pulsar or a white dwarf.[1] Planet-sized objects orbiting substellar objects should be considered dwarf planets, moons or substellar objects in a binary or multiple-object system. Planets do not lose their planethood if they are ejected from their system by gravitational forces.

[1] The 600-km-diameter remnant metallic Fe core of a planet orbiting the white dwarf SDSS J1228+1040 (Manser et al. 2019) should be considered a planet in its own right.

Figure 2.4 The eight planets of the Solar System. (Credit: NASA) (A black-and-white version of this figure will appear in some formats. For the colour version, please refer to the plate section.)

However, planet-sized objects may have formed directly by gas collapse (perhaps with the aid of magnetic fields) in interstellar space, never having been part of a planetary system; these bodies should not be considered planets. It is often difficult to determine the origin of a low-mass isolated body. Such objects have been variously named "escaped planets," "rogue planets," "free-floating planets," "Steppenwolf planets," "sub-brown dwarfs," and "planetary-mass brown dwarfs." Brown dwarfs have been defined by the IAU as having masses ranging between 13 times that of Jupiter (13 M_J) and 0.08 times that of the Sun (0.08 $M_\odot$, equivalent to ~80 M_J). They may be undifferentiated, fully convective bodies. Those with >13 M_J may fuse deuterium ($^2D + p^+ \rightarrow {}^3He$); those with >65 M_J may fuse lithium ($^7Li + p^+ \rightarrow 2\ {}^4He$). One astronomer suggested that sub-brown dwarfs should be defined as free-floating bodies with masses between 1 and 13 times that of Jupiter. The smallest known star, a red dwarf labeled OGLE-TR-122b, has a mass of about 0.1 solar masses (equivalent to ~100 M_J). The estimate for the smallest mass of a star capable of the proton–proton chain reaction is ~75 M_J.

An important question we need address is the size range of planets. If Ceres had been alone in its region, it would have retained its planetary status; it seems likely that if Juno had been the sole substantial object in its neighborhood, it also would have been judged a planet. This suggests that the minimum diameter for a planet could be set at 200 km. Bodies with diameters less than 200 km should be considered asteroids, comets, or meteoroids even if there are no comparably sized bodies in their neighborhoods. [The smallest exoplanet found to date is PSR B1257+12A (a.k.a. Draugr), orbiting a pulsar in Virgo. It has a mass of ~0.02 $M_\oplus$ (where $M_\oplus$ = 1 Earth mass); if its bulk density is 3 g cm^{-3}, it would be ~4,000 km in diameter.]

The maximum size of a planet is more difficult to define. Jupiter has a mean diameter of 139,822 km; the red dwarf star OGLE-TR-122b has a diameter of ~334,000 km. We suggest that a body can be categorized as a planet if it has a diameter up to about twice that of Jupiter, rounded up to 300,000 km; larger bodies should be considered stars or substellar objects. This definition can hold for bodies orbiting central stars, but there will still be ambiguity for bodies that are free floating in interstellar space; this is because of the overlap in sizes between escaped planets and objects formed by gas collapse.

We can now formulate a definition: *A planet is a natural body that (1) is orbiting (or was previously orbiting) a star or former star, (2) has sufficient mass for gravity to have molded it into a round shape, (3) has cleared its neighborhood of comparably sized objects (if it is still in orbit around a star or substellar object), and (4) has a mean diameter between 200 km and 300,000 km.*

Exoplanets can be classified into six groups by mass range (Abron 2019): Mercurian, $\leq$0.1 $M_\oplus$; subterran, 0.1–0.5 $M_\oplus$; terran, 0.5–2 $M_\oplus$; super-Earths or mini-Neptunes, 2–10 $M_\oplus$; Neptunian, 10–30 $M_\oplus$; and Jovian, >30 $M_\oplus$. [The nearest star outside our Solar System is Proxima Centarui (4.2 light years away); it appears to shine on two exoplanets – one terran body and one super-Earth. The next closest star is Barnard's Star, a red dwarf; it is orbited by a super-Earth (Barnard b) of about 3 $M_\oplus$.]

Our meteorite and rock collections currently contain specimens from only two planets: Earth and Mars. There is a small chance there are misidentified meteorites from Mercury in our collections (Love and Keil 1995) but almost no chance we have meteorites from Venus (Gladman et al. 1996) and no chance of samples from the outer planets.

Dwarf planet. We follow here a modification of the IAU definition, taking into account the above definition of planet. *A dwarf planet is a natural body that (1) is in orbit (or was once in orbit) around a star, former star, or substellar object, (2) has sufficient mass for gravity to have molded it into a round shape, (3) has a mean diameter exceeding 400 km, and (4) has not cleared its neighborhood of comparably sized objects (if it is still in orbit around a star or substellar object.)* Smaller objects are asteroids and meteoroids. The minimum diameter of 400 km would allow asteroid 10 Hygiea (d $\approx$ 430 km) to be included in the definition; new images from the Very Large Telescope at the European Southern Observatory in Chile show Hygeia to be quasispherical. An object $\geq$400 km in diameter can be both a dwarf planet and an asteroid (e.g., 1 Ceres). There is no formal limit on the maximum size of a dwarf planet; however, if bodies are sufficiently massive (Earth-sized, for example), almost certainly they would have cleared their zone of comparably sized objects. In principle, a dwarf planet that has not cleared its region of comparably sized bodies could be larger than a planet in another, otherwise-unoccupied, region of the same planetary system.

Moon (a.k.a. natural satellite). *A moon is a natural body, at least 1 m in diameter, which is in orbit around a larger planetary body, dwarf planet, asteroid, comet, or substellar object.* Bodies smaller than 1 m are meteoroids even if they are in orbit around a planet. A moon can be larger than a planet (but not the one it is orbiting); in our Solar System, the moons Ganymede (diameter 5,268 km) and Titan (diameter 5,151 km) are both larger than the planet Mercury (diameter 4,879 km).

Our rock and meteorite collections currently contain material from only one moon – ours. Specimens include several hundred lunar meteorites, 382 kg of Apollo samples from six manned missions, 0.32 kg of material returned by three unmanned Soviet Luna spacecraft, and ~2 kg from the Chang'e-5 mission. Calculations indicate that it is possible that meteorites could reach Earth from the martian moons Phobos and Deimos (Wiegert and Galiazzo 2017), although none has yet been identified.

Meteor. *A meteor is the light phenomenon produced by a solid object transiting an atmosphere at high velocity.* Collisions with atmospheric molecules cause the outer layers of the incoming particle to sputter away, typically producing a vapor of metal atoms. Light is emitted when excited electrons in the orbits of these atoms fall back to lower-energy states. The color is dependent on the element: e.g., Na – orange-yellow; Fe – yellow; Mg – blue-green; Ca^+ ions – violet. Ionized oxygen and nitrogen atoms in the atmosphere can produce red light. The

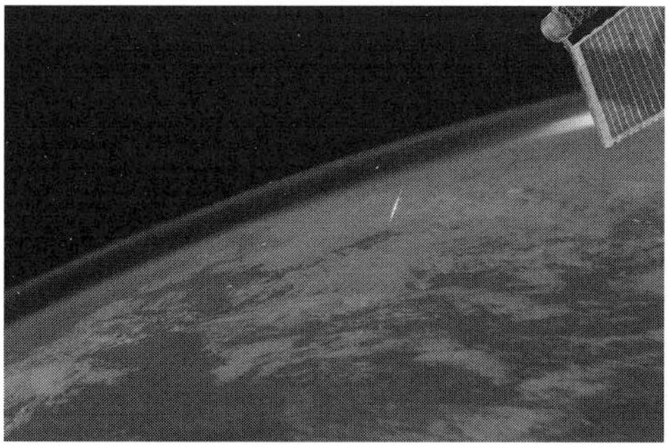

Figure 2.5 A meteor. Photograph ISS028-E-24847 taken on August 13, 2011, from the International Space Station. (Credits: NASA/ISS) (A black-and-white version of this figure will appear in some formats. For the colour version, please refer to the plate section.)

solid object transiting the atmosphere could be natural or artificial; the atmosphere could be one surrounding any solid body (typically a planet, but also including a gas-shrouded moon like Titan). Meteors are commonly referred to as "shooting stars" or "falling stars" (Figure 2.5).

Meteor train. *The trail of light or ionization left in the wake of a meteoroid during atmospheric passage.*

Meteor shower. Dust grains and larger particles lost by comets (primarily) or asteroids (in one or two cases)[2] typically escape at low relative velocities and spread out along their parent-body's orbit. If the Earth passes through that orbital plane, our planet will encounter the ejected debris. As the dust particles, micrometeoroids, and meteoroids are frictionally heated by the Earth's atmosphere, they produce meteors; if many particles enter the atmosphere within a short interval of time (hours to weeks), they create a meteor shower. Peak rates are typically 5–80 meteors per hour, but every 33 years, the Leonids can produce several thousand meteors per hour. Because nearly all the particles in a meteor shower are transiting the atmosphere at the same angle, they appear from the Earth's surface to radiate from the same point in the sky.

Meteor showers are named after the constellation from which they appear to radiate; e.g., the Perseids (peaking around August 13) appear to emanate from Perseus (Figure 2.6), the Orionids (peaking around October 21) from Orion, and the Leonids (peaking around November 17) from Leo. There are also daytime meteor showers, detected mainly by radar observations and radio scatter (e.g., Arietids, peaking around June 8 and emanating from Aries).

The Perseids are derived from the debris of comet 109P/Swift-Tuttle, the Leonids from comet 55P/Tempel-Tuttle, the Orionids from the famous comet 1P/Halley, and the Arietids possibly from asteroid 1566 Icarus. The 33-year period of high Leonid-meteor counts corresponds to the period of comet 55P/Tempel-Tuttle and is due to large concentrations of dust in the comet's proximity. *A meteor shower is the occurrence within a short period of time (typically a few days or weeks) of numerous meteors appearing to radiate from a particular point in the sky.*

[2] The Geminid meteor shower (peaking around December 14) appears to be derived from the weird blue asteroid 3200 Phaethon (Whipple 1983). This object is in a very eccentric orbit, and, although it has no coma, exhibits a small dust tail near perihelion (0.14 AU). Its dust trail was imaged in 2019 by the WISPR camera aboard the *Parker Solar Probe*; the trail is estimated to be ~23 million km long with a total mass of about a billion tons. The Taurid–Perseid meteor shower may originate from asteroid 1566 Icarus (Ohtsuka et al. 2007).

Figure 2.6 Time-lapse image of the Perseid meteor shower of August 2009. (Credits: NASA/JPL) (A black-and-white version of this figure will appear in some formats. For the colour version,please refer to the plate section.)

They can occur above any celestial body with a sufficiently thick and sufficiently transparent atmosphere; meteor showers should also be visible on Venus and Mars.

Meteoroid stream. *A population of meteoroids in interplanetary space that have similar orbits and were derived from the same parent object.* When a meteoroid stream enters an atmosphere, it can produce a meteor shower.

Fall. A meteorite can crash through your roof while you're watching TV [Wethersfield (1982)] or taking a nap (Sylacauga); it can dent your mailbox (Claxton), break into your garage (Benld, Canon City), damage your car (Benld, Peekskill, Neagari, Orlando), land in your rice field (Kamiomi), plop into your cooking pot (Juancheng), hit your dog house (Aguas Zarcas), or plunk you on the head (Mbale). A meteoroid can explode in the atmosphere, creating a shock wave that can shatter window glass and masonry and send you to the hospital (Chelyabinsk). *A fall is a recovered meteorite that was (a) responsible for a fireball observed by people (e.g., L'Aigle, Revelstoke), fireball-network cameras (e.g., Lost City, Bunburra Rockhole), Doppler radar (e.g., Mifflin), seismometers (e.g., Vilna), and/or satellites (e.g., Berduc); (b) responsible for thunderous sonic booms (e.g., Dong Ujimqin Qi, Nogata), whistling or hissing sounds (probably produced by very low frequency radio waves, i.e., electrophonic sounds) (e.g., Guangmingshan) or noise produced when the object strikes a structure or the ground (e.g., Novato); (c) inferred to have fallen recently from the damage it caused to structures (e.g., Tahara, Dunbogan), the ground (e.g., Woolgorong), or living creatures (Valera); or (d) found on a surface where recent prior observation had not indicated its presence (e.g., Sayama).* Most falls belong to more than one observational category.

People who have picked up freshly fallen stony meteorites have often described the samples as cold. Four factors contribute to the low temperature: (1) The temperatures of chondritic meteoroids in near-Earth orbits are ~250 K: (2) the heated outer surface of the incoming meteoroid ablates away during atmospheric passage; (3) the period of surficial heating of the meteoroid as it transits the atmosphere is brief, typically a few seconds (too short a time to allow much heat to be conducted to the interior); (4) silicate material is a poor conductor of heat in any case.

Terms often associated with observed falls include **fireball** (*a very bright meteor*) and **bolide** (*a fireball-producing meteoroid that breaks apart in the atmosphere*).[3] The IAU defined bolides

[3] Hartmann et al. (2018) suggested that the Biblical event witnessed by Saul of Tarsus on the road to Damascus described in the Acts of the Apostles may have been an actual natural phenomenon caused by a bright fireball and associated electrophonic sounds. However, the initial description of the event (Acts 9:1–19) differs in important details from two retellings of the event (Acts 22:3–21 and Acts 26:1–18), casting doubt on its historicity.

and fireballs to be brighter than absolute visual magnitude -4 at a distance of 100 km. A **superbolide** is a bolide brighter than absolute magnitude -17 at a distance of 100 km.

Find. *A find is a meteorite that is discovered at the surface of a target body without having been observed to fall.* It may have been discovered by accident or during an intentional search. There are 50 times as many meteorite finds as falls on Earth. A find could be the sole meteorite recovered from a site or could be one of several dozen to several thousand from a dense collection area such as Antarctica, the Sahara, or desert dry lakes in the southwestern United States. Meteorites are routinely named after the place where they were recovered: a geomorphological feature (e.g., a mountain, lake, river, or valley) or a geographic locale (e.g., a town, city, or county). Meteorites from dense collection areas share a name followed by a unique number (e.g., Allan Hills 84001; Northwest Africa 5000; Cayote Dry Lake 033). Dense collection areas have also been identified on Mars: Aeolis Mons and Aeolis Palus. Many finds on Earth have been altered texturally, mineralogically, and chemically by interaction with water, a process known as *terrestrial weathering*.

Meteorite Shower. Many meteoroids break apart in the atmosphere during deceleration. The fragments continue along the original object's trajectory, but small fragments (with their greater surface/volume ratios) encounter more effective air resistance, decelerate more rapidly, and hit the ground sooner. The fallen fragments are distributed within an ellipse, known as a **strewn field**, with the larger fragments concentrated at one end. The orientation of the ellipse is governed by the trajectory of the original meteoroid. In the case of multiple fragmentation events, a series of overlapping strewn fields can be formed. *A meteorite shower is the result of the collision with a large gas-shrouded body of a swarm of fragments produced in the atmosphere by the breakup of an incoming meteoroid or small asteroid.* The 1912 Holbrook L/LL6 meteorite shower (Navajo County, Arizona) produced about 14,000 individuals with a total recovered mass of ~220 kg. By one estimate, the 1868 meteorite shower associated with the Pultusk H-chondrite regolith breccia (Warsaw, Poland) produced 180,000 individuals, although the recovered mass is only ~200 kg.

Pairs. Meteorites are collected as individual objects. In those cases where two or more objects can be shown to have originated from the same parent meteoroid that fragmented in the atmosphere, those objects are considered paired. Pairing among finds can be inferred from similarities in texture, patina, mineralogy, bulk chemical composition, bulk isotopic composition, cosmic ray–exposure (CRE) age, and proximity of discovery sites; in some cases, individual fragments fit together. Some meteorites, given separate names when first found, were later shown to be paired. Freshly fallen meteorites recovered from a strewn field can also be paired. *Pairs are separate meteorites that are inferred to have been derived from the same parent meteoroid that fragmented in the atmosphere or from a larger individual meteorite that broke apart on the ground.*

"Launch pairs" are meteorites (particularly lunar or martian meteorites) inferred to have been impact launched from the same crater at the same time even though they may have reached Earth at different times and landed in different locations. These inferences are based on similarities to one another in texture, mineralogy, bulk chemical composition, crystallization age and CRE age (and dissimilarities to other samples from the same meteorite group in these characteristics).

Fusion Crust. As a meteoroid traverses the atmosphere at high velocity, friction with the surrounding air heats the meteoroid's surface to a temperature exceeding its melting point. The

Figure 2.7 An individual specimen with fusion crust of the Chelyabinsk meteorite fall. (A black-and-white version of this figure will appear in some formats. For the colour version, please refer to the plate section.)

wind generated by the downward plunge of the meteoroid pushes the melt away from the projectile, exposing a fresh surface, which is then subject to additional frictional melting. As this process is repeated, a meteoroid can lose more than 95 percent of its initial mass during descent. At some point during deceleration, frictional heating has been greatly reduced and the temperature of the meteoroid surface falls below its melting temperature. *The fusion crust of a meteorite is the final melt produced on the surface of its parent meteoroid during atmospheric passage.* An individual meteorite can have different generations of fusion crust produced after fragmentation at different altitudes. Many amateurs confuse desert varnish with fusion crust, but these features are readily distinguishable. Desert varnish appears as a dark stain on the rock surface; it is generally only ~1 µm thick and is composed of clay particles along with iron and manganese oxides. Fusion crust is a distinct melt layer, typically ~1 mm thick, atop the surface of the unmelted meteorite (Figure 2.7).

Mineral. *A mineral is a naturally occurring homogeneous solid, formed in most cases by inorganic processes, that has a definite (although typically not fixed) chemical composition and an ordered atomic arrangement.* Although substances identical in form, appearance, and composition to minerals can be created in the laboratory, these synthetic products are not *naturally occurring* and are not considered minerals. Synthetic crystals include precious gemstones (diamond, emerald, ruby, sapphire), semiprecious gemstones (e.g., garnet, amethyst) as well as more mundane substances (e.g., ice, salt). Ice in your kitchen freezer is not a mineral, but winter frost on your car windshield is. Minerals are *homogeneous solids* in that they cannot be physically divided into separate crystalline substances. Although most minerals are formed by *inorganic processes*, i.e., without the interaction of living organisms, some mineralogists accept a few organically produced substances as genuine minerals. The most prominent example is calcium carbonate from oyster shells and pearls that is chemically, physically, and optically identical to aragonite. Because rocks are formed from minerals and some legitimate rocks are composed of organic materials (e.g., coal, coquina), consistency demands that legitimate minerals include those formed by organisms.

Minerals have a *definite chemical composition* that can be written with a specific chemical formula (e.g., SiO_2; $CaCO_3$; NaCl; Cu_9S_5; SiC; TiN; TiS; FeCrP; Fe_3Si). However, most minerals do not have a *fixed chemical composition* – they may contain impurities that can vary from sample to sample; these impurities are additional elements not expressed in the ideal formula. For example, orthopyroxene in the enstatite–ferrosilite solid solution series ($MgSiO_3$–$FeSiO_3$) can contain up to ~10 mol% Al, Ca, Mn, Fe^{3+}, Ti, Cr, and Ni; melilite (the intermediate portion of the gehlenite–åkermanite solid solution series: $Ca_2Al(Si,Al)_2O_7$–$Ca_2MgSi_2O_7$) can contain up to ~7 wt% total of Ti, Fe^{3+}, Fe^{2+}, Mn, Na, and K. These minor-element impurities substitute for the major elements in specific crystallographic sites. Minerals have an *ordered atomic arrangement*, indicating they are crystalline solids; their constituent atoms[4] or ions are arranged in a geometric structural framework. However, mercury is an exception; it is a liquid (and thus lacks atomic order) at room temperature.

The terms *mineral species* and *mineral phase* refer to the entire set of samples of a particular mineral (e.g., diopside; troilite; diamond; calcite; rutile; halite), defined by a unique set of compositional and structural properties. Although there are more than 5,670 approved minerals as of this writing, only about 470 have been identified in meteorites, and only about 100 minerals are relatively common on Earth. A number of reduced meteoritic minerals (e.g., some that are present in enstatite chondrites and aubrites) and some refractory minerals (e.g., panguite and kangite in carbonaceous chondrites) are not known to occur in terrestrial rocks.

Mineraloid. *A mineraloid is a naturally occurring amorphous solid, lacking long-range atomic order, but possessing short-range order.* Examples include maskelynite (a shock-produced glass that can have the same chemical composition and external form as its crystalline plagioclase precursor) and opal (a hydrated amorphous variety of silica, typically containing 6–10 wt% H_2O). Terrestrial mineraloids also include obsidian (dark-colored, extrusive volcanic glass), tektites and related impact glasses (derived from fine-grained materials melted by meteorite impacts), jet (a form of coal derived from decaying wood under high pressure), amber (fossilized tree resin), and tar and related materials (pitch, asphalt, and bitumen) (highly viscous liquid or semisolid petroleum products formed as residues after volatilization of more-volatile hydrocarbons).

Crystal. *A crystal is a largely homogeneous solid (either naturally occurring or synthetic) that possesses long-range three-dimensional internal atomic order.* Its three-dimensional, periodically repeating atomic pattern is called a **lattice**. The internal symmetry of the crystal conforms to one of seven crystal systems, depending on the relative lengths of its three axes and the angles between the axes. Many individual coarse crystals exhibit planar faces arranged in geometric forms; they grace the covers of mineral magazines and are coveted by collectors. Because the internal atomic arrangements of these crystals do not differ from those of their malformed congeners (crystals that may have encountered external obstacles or unusual ambient conditions while growing), it is clear that the internal structure of a crystal is a more fundamental feature than the external shape (habit). **Euhedral crystals** are bounded by well-formed faces; **anhedral crystals** are bounded by no crystallographically formed faces; and **subhedral crystals** are bounded by some well-formed and some poorly formed faces or bounded completely by only partially developed faces.

[4] The concept of atoms as tiny indivisible particles is commonly attributed to the pre-Socratic Greek philosophers Democritus (c. 460–c. 370 BCE) and his mentor Leucippus as well the ancient Indian philosopher Kanada (who lived sometime between the sixth and second centuries BCE). Kanada proposed that minute invisible, indivisible, and eternal particles (parmanu) combined in different arrangements to create complex common materials.

A crystal form is the set of symmetrically equivalent crystal faces. Crystal forms may be closed (e.g., cubes and dipyramids, wherein the faces entirely enclose the crystal volume) or open (e.g., pinacoids and pyramids, wherein the faces do not enclose the entire volume). The number of faces characteristic of a particular crystal form is dependent on the crystal class to which the mineral belongs and on the orientation of particular crystal faces to the symmetry axes and planes.

In polyatomic crystals, the atoms of different elements generally occupy distinct locations within the lattice. For example, nesosilicates contain isolated SiO_4 tetrahedra with four O^{2-} anions located at the apices of a regular tetrahedron and one Si^{4+} cation at the center. The nesosilicate monticellite ($CaMgSiO_4$) is composed of stacked layers, individually consisting of octahedra cross-linked by SiO_4 tetrahedra; there are two distinct octahedral sites: *M1* (a moderately distorted site) contains Mg, and *M2* (a generally undistorted site) contains Ca.

Natural crystals are not perfectly homogeneous; they contain various defects. Point defects include vacant lattice sites, interstitial defects (wherein atoms occupy sites that are typically vacant), antisite defects (wherein atoms of different elements exchange sites in the lattice), and topological defects (wherein atoms are arranged in different patterns than that of the bulk of the crystal). An important parameter is the Burgers vector (*b*), which represents the magnitude and direction of the lattice distortion. Line defects include edge dislocations (wherein a plane of atoms abruptly terminates within the body of a crystal and the Burgers vector is perpendicular to the dislocation line) and screw dislocations (wherein the planes of atoms are arranged in a helix around a dislocation line and the Burgers vector is parallel to the dislocation line). Common planar defects include grain boundaries (typically responsible for changes in crystal orientation), stacking faults (changes in the normal sequence of adjacent layers), and twin boundaries (which represent domains related by a mirror plane, a rotation axis, or an inversion axis). Bulk defects include fractures, clusters of vacancies, inclusions, and pores.

Although crystals can contain different numbers of cations and anions, the total electrical charge of each unit cell is zero. In halite (NaCl), there is an equal number of Na^+ cations and Cl^- anions; in quartz (SiO_2), there are twice as many O^{2-} anions as Si^{4+} cations; in spinel ($MgAl_2O_4$), charge balance is maintained by four O^{2-} anions, two Al^{3+} cations, and one Mg^{2+} cation. Real crystals also contain vacancies (where atoms are absent from the ideal crystal lattice); the number of vacancies tends to increase with temperature. Common types of vacancies in crystals are called Schottky defects; they comprise unoccupied cation and anion sites that are in a stoichiometric ratio, thereby maintaining charge neutrality.

The sizes of natural crystals vary enormously and are studied by different techniques: 10 nm to 100 nm–sized grains are routinely imaged with the transmission electron microscope (TEM); 0.05–500-μm-sized grains are observed with backscattered electrons (BSEs); ~0.2-μm- to 1 cm-sized grains with the optical microscope; and larger grains are observed visually or with hand lenses. Among the largest known crystals are those of selenite ($CaSO_4 \cdot 2H_2O$) from the Giant Crystal Cave in Naica, Mexico; these enormous crystals range up to 12 m in length and 4 m in width. They crystallized slowly over a period of ~0.5 Ma from ground water heated by a magma chamber located a few kilometers below the cave. The largest known crystal may be a beryl ($Be_3Al_2(SiO_3)_6$) from the Malakialina pegmatite, Madagascar; the crystal is 18 m long and 3.5 m wide (Rickwood 1981).

Among meteorites, crystals range from submicrometer-sized (e.g., every known grain of wassonite (TiS)) to a ~50-cm-sized low-Ni metallic iron (kamacite) grain in the Coahuila IIAB hexahedrite (Buchwald 1975). (Hexahedrites are a structural type of iron meteorite discussed in Section 8.5). Some hexahedrites may contain single kamacite grains approaching 1 m in maximum dimension (Fesenkov 1958). The Agpalilik specimen of the Cape York IIIAB iron meteorite crystallized as a single crystal of taenite at least 2 m across, but the crystal developed a Widmanstätten pattern (consisting of numerous separate kamacite and taenite grains) during slow subsolidus cooling.

Quasicrystal. *A quasicrystal is a solid substance (either naturally occurring or synthetic) that has an ordered atomic arrangement but is not periodic.* The internal crystalline patterns of the material do not occur at regular intervals, i.e., there is no translational symmetry. They can display diffraction patterns with otherwise forbidden rotational symmetry, e.g., fivefold, eight-fold, and more axes of symmetry (Socolar et al. 1985). The internal structure of a quasicrystal is intermediate between that of an amorphous substance and a crystalline solid. Quasicrystals were first conceptualized and synthesized in 1984 (Shechtman et al. 1984; Levine and Steinhardt 1984). Although three quasicrystals have been reported in the CV3 Khatyrka meteorite including two icosahedrites ($Al_{63}Cu_{24}Fe_{13}$ and $Al_{62}Cu_{31}Fe_7$) plus decagonite ($Al_{71}Ni_{24}Fe_5$) (Bindi et al., 2011, 2015, 2016) and approved by the IMA-CNMNC, their validity as natural species remains controversial.

Rock. *A rock is a naturally occurring solid aggregate of one or more minerals and/or mineraloids.* Concrete is a hard composite material but is not naturally occurring and is therefore not a rock. Although naturally occurring, some beachrocks contain artifacts such as coins and bottle glass; the presence of these man-made components does not disqualify beachrocks from being legitimate rocks.

The three principal rock types are **igneous rocks** (formed by the cooling and solidification of molten or partly molten material), **sedimentary rocks** (formed by the mechanical deposition and consolidation of material or by chemical precipitation from a fluid), and **metamorphic rocks** (formed by mineralogical, textural, and/or chemical alteration of preexisting rocks in response to significant ambient changes in temperature, pressure, chemical environment, and/or stress). Terrestrial sedimentary rocks include some composed of inorganic material (e.g., sandstone, claystone, evaporites), some with high concentrations of organic material (e.g., coal, oil shale), and some made mainly of fossilized organic material (e.g., coquina – bioclastic limestone; chert – derived ultimately from the skeletal remains of diatoms, silicoflagellates, and radiolarians).

All three major rock types are present among meteorites: igneous meteorites include basalts, dunites, orthopyroxenites, magmatic irons, and impact-melt rocks; sedimentary meteorites include primitive chondrites (e.g., LL3.00 Semarkona, which formed principally by the agglomeration of individual constituents in the solar nebula); metamorphic meteorites include type 4–6 H, L, LL, EH, EL, R and CK chondrites, and many eucrites.

Common terrestrial rocks that are mainly monomineralic include sedimentary rocks (e.g., salt, ice, limestone, and some sandstones such as the St. Peter Sandstone), metamorphic rocks (e.g., quartzite and marble), and igneous rocks (e.g., dunite and anorthosite), although essentially all of them contain minor amounts of additional minerals. Common terrestrial rocks that are polymineralic include shale (sedimentary rock composed mainly of clay minerals such as kaolinite, montmorillonite, and illite, with minor to accessory quartz, calcite, hematite, goethite, and/or mica), schist (metamorphic rock commonly composed of mica, quartz, and feldspar, and in some cases containing minor garnet, hornblende, talc, chlorite, graphite, or glaucophane), granite (igneous rock composed of major quartz, K-feldspar, and plagioclase with minor mica and amphibole), and basalt (igneous rock composed of major plagioclase and pyroxene with accessory to minor olivine, magnetite, ilmenite, and/or ulvöspinel).

Igneous rocks made of crystals large enough to be seen by the unaided human eye are labeled *phaneritic*. Rock textures can be described as very-coarse-grained (>3-cm-sized grains), coarse-grained (5 mm to 3 cm), medium-grained (1–5 mm), and fine-grained (<1 mm). By these criteria, nearly all meteorites are mostly fine grained. If a rock contains numerous grains, too small to be seen visually, it is called *aphanitic*.

Breccia. *A breccia is a rock composed of rock, mineral, and/or mineraloid fragments held together by a fine-grained or glassy matrix.* Breccias can be monomict (consisting of a single lithology), dimict (two lithologies), polymict (more than two lithologies), or genomict (consisting of clasts of a single rock type but with different metamorphic grades) (e.g., Bischoff et al. 2006). Most breccias are fragmental breccias: cemented assemblages of angular clasts. Terrestrial fragmental breccias include those formed in sedimentary, tectonic, volcanic, and hydrothermal environments as well as those associated with impact craters. Impact-melt breccias contain shocked but unmelted rock fragments surrounded by a major glassy or fine-grained component that solidified from a cooling impact-derived melt. Regolith breccias, known from the Moon, Mars, and asteroidal meteorites, formed in the near-surface regions of these bodies; they are commonly composed of rock fragments admixed with glass and/or fine-grained material and typically contain relatively high concentrations of solar-wind-implanted noble gases. They include some mineral grains whose crystal lattices were damaged by energetic cosmic ray particles. Many regolith breccias also contain projectile fragments (unrelated to the principal rock components) that survived collision with the parent-body's surface. Examples include (1) CM carbonaceous-chondrite clasts constituting ~5 vol% of the Kapoeta howardite and ~1–3 vol% of the Abbott H-chondrite regolith breccia, (2) a shocked LL5 clast in the Dimmitt H-chondrite regolith breccia, (3) a 7 mm eucritic clast in the NWA 869 L-chondrite regolith breccia, and (4) two dark H-chondrite clasts in the St. Mesmin LL-chondrite regolith breccia.

3

Brief Review of Crystallography and Crystal Chemistry

It is not the purpose of this chapter to present a comprehensive overview of crystallography and crystal chemistry. Excellent discussions of those topics can be found in the classic *Crystallography and Crystal Chemistry: An Introduction* by F. D. Bloss (1971) as well as in several more recent texts including those by Klein and Dutrow (2007) and Nesse (2012). Here we cover only the most salient points about crystallography to facilitate understanding of topics and terms presented elsewhere in this book without explanation.

3.1 Crystal Systems and Space Groups

Let's start by imagining a single point in an infinite plane. Move the point repeatedly a fixed distance in one direction to generate a series of dots. Take each of those dots and move it repeatedly a fixed distance in a different direction. The pattern is akin to that of an infinite sheet of polka dot wallpaper. Now, connect each of the four adjacent points in these two directions with line segments to make an infinite series of identical quadrilaterals. Each line segment in the first direction will be called a; each line segment in the second direction, b, and the angle between a and b will be called γ. There are five different quadrilaterals (or plane lattices) that can be produced by these procedures for different values of a, b and γ :

(1) $a = b$ and $\gamma = 90°$; the figure is a square.
(2) $a \neq b$ and $\gamma = 90°$; the figure is a rectangle.
(3) $a \neq b$ and $\cos \gamma = a/(2b)$; the figure is a diamond.
(4) $a = b$ and $\gamma = 120°$; the figure is a rhombus.
(5) $a \neq b$ and $\gamma \neq 90°$; the figure is a parallelogram.

If we draw the smallest possible rectangle atop the diamond pattern in case 3, there will be one point at the center of the rectangle, producing a face-centered quadrilateral. If we connect diagonal points in each rhombus in case 4, we will create a repeating hexagonal pattern.

We can take these two-dimensional figures and extend them infinitely in the third dimension, to generate a series of 14 distinct space lattices, known as *Bravais lattices* (after the French physicist and crystallographer, Auguste Bravais). We can imagine replacing the points in the space lattices with atoms to produce a set of parallelepipeds (three-dimensional figures formed by six parallelograms). The smallest group of atoms that conform to the Bravais-lattice parallelepipeds are called **unit cells**; these are typically 3–20 Å in size.

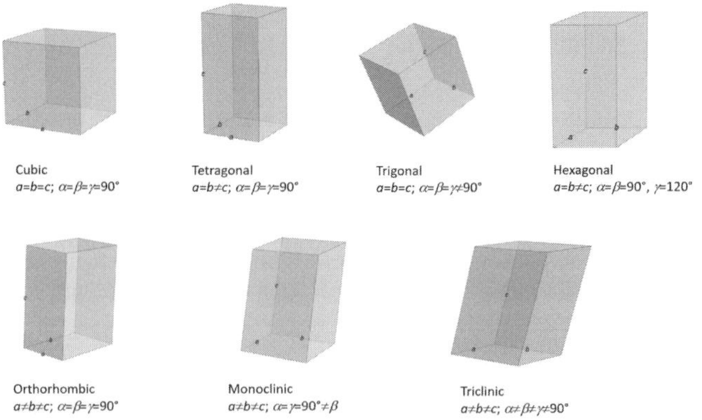

Figure 3.1 The seven crystal systems. (Credit: WOLFRAM Demonstrations Project) (A black-and-white version of this figure will appear in some formats. For the colour version, please refer to the plate section.)

There are seven possible ways in which these parallelepipeds can be stacked in three dimensions without producing gaps; these correspond to the seven crystal systems defined in the following paragraphs (Figure 3.1). By convention, the three axes of the unit cells in these systems are labeled *a*, *b*, and *c*; if two or three axes are equal, they are sometimes designated a_1, a_2, and a_3. The angles between the axes are labeled α (located between axes *b* and *c*), β (between *c* and *a*), and γ (between *a* and *b*); if two or three angles are equal, they are sometimes designated, for example, β_1, β_2, and β_3. Every crystalline material belongs to one of these crystal systems defined by the relationships among these axes and interaxial angles.

Triclinic ($a \neq b \neq c$; $\alpha \neq \beta \neq \gamma \neq 90°$). There is no external symmetry, i.e., it has a 1-fold axis. This means that the unit cell must be rotated 360° (i.e., back to its original position) to make the unit cell appear as it did initially.

Monoclinic ($a \neq b \neq c$; $\alpha = \gamma = 90°$; $\beta \neq 90°$). This has one 2-fold axis, i.e., the unit cell appears as it did originally twice in a 360° rotation. (It need be rotated only 180° to be reproduced.)

Orthorhombic ($a \neq b \neq c$; $\alpha = \beta = \gamma = 90°$). This has three 2-fold axes.

Tetragonal ($a_1 = a_2 \neq c$; $\alpha = \beta = \gamma = 90°$). This has one 4-fold axis; i.e., the unit cell appears as it did originally four times in a 360° rotation. (It need be rotated only 90° to be reproduced.)

Hexagonal ($a_1 = a_2 = a_3 \neq c$; $\beta_1 = \beta_2 = \beta_3 = 90°$; $\gamma = 120°$). This has one 6-fold axis (the unit cell appears as it did originally six times in a 360° rotation, i.e., every 60°).

Trigonal or Rhombohedral ($a = b = c$; $\alpha = \beta = \gamma \neq 90°$). This has one 3-fold axis (the unit cell appears as it did originally three times in a 360° rotation, i.e., every 120°). This system is often considered a subclass of the hexagonal crystal system and many texts list only six crystal systems.

Isometric ($a = b = c$; $\alpha = \beta = \gamma = 90°$). This has four 3-fold axes. In face-centered cubic lattices, atoms are located at all eight corners of a cube as well as the center of each face. In body-centered cubic lattices, atoms are located at all eight corners of a cube as well as the internal center of the lattice.

New lattices can be constructed from the Bravais lattices by *rotation* through a certain angle around an imaginary axis or by a combination of rotation and translation. These symmetry operations produce congruent patterns that can be superimposed on each other. Rotation axes involve an imaginary line through which a crystal could be rotated one, two, three, four, or six times, respectively, repeating itself every 360°, 180°, 120°, 90°, or 60°.

Other symmetry operations produce incongruent patterns (known as enantiomorphs) that cannot be superimposed on each other. Such operations include *rotoinversion* axes (an imaginary line though which a crystal could be rotated and then inverted through the center of the crystal), *center of symmetry* or *center of inversion* (an imaginary line that passes from any point on the surface of the crystal through the center of the crystal so that the original point is reproduced on the other side of the crystal at an equal distance from the center), and *mirror planes* (an imaginary plane through which a crystal is divided in half so that each side is the mirror image of the other).

The symmetry elements of rotation, inversion, and mirror reflection can be combined to make 32 *point groups*, each group representing ways in which the initial appearance of the unit cell of the crystal is reproduced after movement. These point groups are analogous to the 32 *crystal classes*, i.e., the 32 possible external symmetries of crystals. Each crystal system is compatible with a restricted number of point groups. For example, in a triclinic system there are two point groups; the unit cell can be reproduced only by a 1-fold (i.e., 360°) rotation or by inversion. Tetragonal systems have seven point groups, the most of any crystal system; these point groups represent different combinations of mirror reflections, rotations, and inversions.

The 32 point groups can be subject to additional symmetry operations involving translation, movement along a glide plane, and movement around a screw axis to produce a total of 230 unique three-dimensional structures known as *space groups*. In principle, every mineral crystal can be assigned to a space group. In some cases, however, all known occurrences of a particular mineral phase are very small and there are insufficient crystallographic data to allow assignment of the mineral to a space group.

3.2 Miller Indices

Macroscopic crystals can display external faces that are not parallel to the planes that constitute the unit cell. The locations of these phases relative to the unit cell can be described by a coordinate system known as a *Miller index* (after the British mineralogist William Miller). The index is described by three integers enclosed in parentheses (h,k,l), wherein h, k, and l are derived from the unit cell axes a, b, and c.

To explain the Miller indices, we first define a point at one corner of the unit cell as the origin; in most illustrations, this is the back bottom left corner. The cell extends along each axis from the origin to a length equivalent to 1 unit, even though the axes may be of different absolute lengths. A plane transecting the unit cell at an arbitrary angle will intersect the three axes at different points. The Miller indices are equal to the reciprocal of the points of intersection. For example, if the plane intersects the a axis at 1, the b axis at 1, and the c axis at ¼, the Miller indices will be: $h = 1/1$, $k = 1/1$, $l = 1/¼$, expressed as (114). If a different plane intersects the unit cell at $a = 1$ and $c = 1$ and is parallel to the b axis, then $h = 1/1$, $k = 1/\infty$, $l = 1/1$, expressed as (101) because $1/\infty = 0$. Planes can extend to negative values behind the original unit cell; Miller indices express negative numbers with a bar over them (e.g., $\bar{1}$, $\bar{2}$), which are read as "negative one" or, more commonly, as "bar one," and "negative two" or "bar two," etc.

For example, if a plane intersects the crystal relative to the original unit cell at $a = ½$, $b = -¼$, $c = ½$; the Miller indices would be $h = 1/½$, $k = 1/-¼$, $l = 1/½$, expressed as $(2\bar{4}2)$ and read as "two, bar four, two."

As stated earlier, a crystal form is the set of symmetrically equivalent crystal faces. In crystallography, forms are denoted by braces around the Miller index of one of the faces of the unit cell. Any face of a crystal form can serve as a template for symmetry operations that duplicate that face and produce all the faces of the crystal form. For example, given the (001) face of a cube (which has six identical sides), symmetry operations will reproduce the original face; the crystal form is given in braces by {001}. Given the (100) face of a tetragonal crystal, symmetry operations can duplicate that face and produce all the faces of the form, which is given the symbol {100}.

Crystal symmetry can be lowered by ordering the constituent atoms. For example, tetrataenite (ordered FeNi) is a tetragonal phase that forms when Fe and Ni atoms order on alternate (002) planes of cubic taenite (e.g., figure 1 of Clarke and Scott 1980). The same tetragonal structure occurs in synthetic FeNi.

3.3 Chemical Bonding

Electron shells that are full or half-full are stable, symmetric, low-energy configurations. Noble gases have filled shells and tend to be unreactive. Helium, for example, forms very few compounds with other elements; exceptions include (a) HeH^+ (probably the first molecule to form in the early universe, found recently in a planetary nebula), (b) compounds formed under extremely high pressure (e.g., Na_2He; FeHe), and (c) compounds formed at very low temperatures (e.g., HeN_2^+). Neutral elements that are not noble gases tend to bind with other elements to achieve filled or half-filled electron shells and thereby assume lower-energy configurations. It requires relatively little energy to strip the valence electrons from neutral metals, turning them into cations. Neutral nonmetals have electron subshells that typically need only a few electrons to become filled; these elements tend to acquire electrons and become anions.

Many common chemical bonds involve the gain, loss, or sharing of valence electrons, resulting in a compound with a filled or half-filled outer electron shell. There are three such common bonding types: ionic, covalent, and metallic.

In **ionic bonding**, there is an electrostatic attraction between cations and anions caused by the cation donating one or more electrons from its outer shell and the anion acquiring those electrons. Each ion thereby assumes the configuration of a filled electron shell. For example, in halite (NaCl), the Na atom donates its sole 3*s* subshell electron (thereby assuming the electron configuration of Ne) and the Cl atom acquires that electron, filling its 3*p* subshell and acquiring the electron configuration of Ar. The Na ion acquires a net charge of +1, the Cl ion acquires a net charge of −1, and the two ions experience an electrostatic attraction, forming the ionic bond.

The Na and Cl ions bind together in a compact face-centered cubic lattice in which every Na ion is surrounded by six Cl ions and every Cl ion is surrounded by six Na ions. The distance between the centers of the ions is ~2.8 Å. When stress is applied to ionically bonded minerals, they tend to exhibit brittle behavior (i.e., they rupture); this is because any slippage of atomic layers in the crystal would result in bringing ions of like charge together, a condition that would be resisted by electrical repulsion. Ionically bonded minerals have moderately high melting points and are poor conductors of electricity.

In **covalent bonding**, the electron orbitals between adjacent atoms overlap. Because the electrons are attracted to the positively charged nuclei of both atoms, an electrostatic bond is formed, as the atoms share the orbiting electrons. The larger the overlap of electron orbitals, the stronger the covalent bond. Diamond consists of C atoms with greatly overlapping hybrid orbitals. Each C atom has four vacancies in its $2p$ shell and thus shares electrons with four other C atoms. This allows the electronic configuration of C to assume the filled electron configuration of Ne. The extreme hardness of diamond is a result of its strong covalent bond.

In **metallic bonding**, the metal cations are compactly arranged in the crystal lattice; their valence electrons are no longer confined to shells around a single atom and become *delocalized*, moving freely throughout the structure. The criteria required to form metallic bonds include (1) weakly held valence electrons, (2) electron shells that require a large number of additional electrons to achieve a noble gas configuration, and (3) vacant energy levels into which electrons can be placed. The electrostatic attraction between the freely floating electrons and the metal cations produces the metallic bonding. However, the bonding forces are weak, causing metals to be soft and malleable. The freely floating electrons are responsible for the high electrical conductivity of metals with metallic bonds. Metals are generally also good conductors of heat because the delocalized electrons can transfer energy more quickly than the more tightly bound electrons in many nonmetallic compounds.

Two other types of bonding do not involve the exchange or sharing of valence electrons: van der Waals and hydrogen bonding. These are weak bonds produced by weak electrostatic interactions.

In **van der Waals bonding** (named after Dutch physicist Johannes Diderik van der Waals), electrostatic attraction results from the asymmetric distribution of electric charge. For example, graphite (a polymorph of diamond) is covalently bonded to three adjacent C atoms within individual C layers. Each C atom contributes an additional valence electron that becomes delocalized, freely floating through the C layer. At any one time, there may be more electrons wandering on one side of a C layer than the other, temporarily producing a net charge. The negatively charged side of this layer induces the near side of the adjacent layer to become positively charged; the opposite situation occurs on the other side of the first layer. The electrostatic attraction between these layers is van der Waals bonding. These are weak bonds, allowing adjacent C layers to slip by one another easily and separate when relatively low stress is applied. These bonds cause graphite to be weak and soft, facilitating its use as pencil lead and as a lubricant for locks and bearings. The delocalized electrons account for graphite's electrical conductivity.

Hydrogen bonding occurs in water ice but few other minerals. The two H atoms in a water molecule are covalently bonded with the O atom, sharing their electrons; however, because O has a high electronegativity (its nucleus has a +8 charge), it has a much stronger affinity for the electrons than the H atoms (each of which has a +1 charge). In normal ice (Ice 1h), this leads to an asymmetric charge distribution at the corners of a tetrahedron – positive charge in the vicinity of the two H atoms and negative charge near the O atom. There are currently about 18 identified crystalline forms of water ice, more than for any other known substance. Most of these phases are stable or metastable only at high pressure; all of them are hydrogen bonded. Hydrogen bonding is moderately stronger than van der Waals bonding.

3.4 Coordination and Packing

Crystal lattices can be modeled as polyhedral structures consisting of a group of atoms packed around a central atom. Ionically bonded crystals can be modeled as polyhedra with large anions surrounding a central void that, in many cases, contains a smaller cation. Larger voids are surrounded by greater numbers of anions. The *coordination number* (CN) is the number of atoms in a polyhedron surrounding a central void that may house a cation.

The shape of the polyhedron in an ionically bonded crystal is determined by the number of cations.

(1) A cation surrounded by three anions that are situated at the corners of an equilateral triangle (more accurately, a trigonal planar figure or hyperbolic triangle) has a coordination number of 3. Example: CO_3 in carbonates.
(2) A cation surrounded by four anions at the corners of a tetrahedron has a coordination number of 4. Example: SiO_4 in silicates. (Charge balance is maintained because the O anions are shared between adjacent SiO_4 tetrahedra.)
(3) A cation surrounded by six anions at the corners of an octahedron has a coordination number of 6. Example: Fe^{2+} and Mg^{2+} in olivine ($(Mg,Fe)_2SiO_4$). (In olivine, there are two, differently oriented, octahedral sites known as *M1* and *M2*. *M1* is somewhat distorted and 2.16 Å in size; *M2* is less distorted and a little larger, 2.19 Å. The *M1* sites form distorted columns parallel to the *c* axis of the crystal. The Mg^{2+} and Fe^{2+} cations randomly occupy both octahedral sites.) Another example: Na^+ in halite (NaCl). (The central Na^+ cation is surrounded by six Cl^- anions.)
(4) A cation surrounded by eight anions at the corners of a cube has a coordination number of 8. Example: Cs^+ in CsCl. (Cesium chloride is an inorganic compound present in low concentrations in sylvite and a few K-rich evaporite minerals) Another example: Ca^{2+} in fluorite (CaF_2). (The central Ca^{2+} cation is surrounded by eight F^- anions.)
(5) A cation within a face-centered cubic structure, surrounded by 12 anions located at the corners of a cube and at the center of each side has a coordination number of 12. Example: taenite (γ-(Fe,Ni)).

The coordination number of a crystal is a function of the relative sizes of its anions and cations, expressed as the *radius ratio*: R_{cation}/R_{anion}. If a crystal has more than one cation (e.g., $MgSiO_3$), each cation will have its own radius ratio. Ionic radii are usually measured in Å or pm; e.g., the effective ionic radii of Na^+ and Cl^- are 1.02 and 1.81 Å, respectively. In general, the higher the radius ratio of an ionically bonded crystal, the larger the void and the higher the coordination number. For example, the radius ratio of halite (CN = 6) is 0.56 while that of fluorite (CN = 8) is 0.75.

In the case where all the ions in a lattice are the same size, the radius ratio would be 1 and the central ion would be surrounded by six other ions. This atomic arrangement can be modeled as analogous to stacked layers of marbles, ping-pong balls, or spheres. A single layer **A** of closely packed spheres leaves trigonal voids or holes between any three adjacent spheres. [Although this type of void is often described as triangular shaped, its projection in two dimensions actually has the shape of a hyperbolic triangle (with angles that sum to $<180°$).] Half of the trigonal voids point in one direction ("up"); these are designated the B voids. The other half of

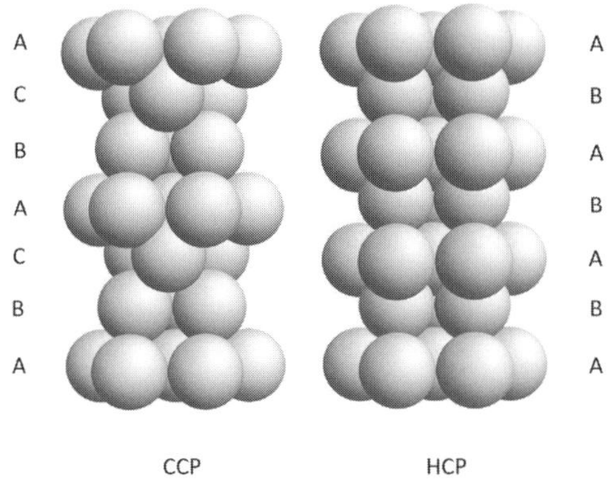

Figure 3.2 The cubic closest packing (CCP) and hexagonal closest packing (HCP) structures. (Credit: WOLFRAM Demonstrations Project) (A black-and-white version of this figure will appear in some formats. For the colour version, please refer to the plate section.)

the voids point in the opposite direction ("down"); these are the C voids. Each sphere in layer **A** is surrounded by six other, equally sized spheres.

We can stack a second layer **B** of spheres atop layer **A** so that the center of each sphere in layer **B** lies over a B void from layer **A**. If we stack a third layer atop layer **B** so that its spheres are in the same positions as those in layer **A**, and a fourth layer above the third layer so that the spheres in the fourth layer are in same positions as those in layer **B**, etc., the stacking sequence will be **ABAB**... This type of structure exhibits *hexagonal closest packing* (HCP) and contains 26 percent void space (Figure 3.2); the coordination number is 12 and there are six atoms per unit cell. Metals with this structure include Co, Cd, and Zn.

A trigonal hole created in layer **B** that is centered on an underlying sphere from layer **A** has the approximate shape of a tetrahedron and is known as a *tetrahedral hole*. The hole is surrounded by four spheres. A void between spheres in layer **B** that is positioned directly above a void in layer **A** has the approximate shape of an octahedron and is known as an *octahedral hole*. This hole is larger than a tetrahedral hole and is surrounded by six spheres. Both tetrahedral and octahedral holes can be occupied by atoms or ions in real crystals.

There is another way to stack the layers. Just as before, we start with layer **A** and position layer **B** so that the center of each sphere in layer **B** overlies a B void in layer **A**. This time, however, we place the third layer (layer **C**) atop layer **B** so that each sphere in layer **C** is above a C void in layer **A**. The fourth layer is positioned identically to layer **A**, the fifth layer is identical to layer **B**, and the sixth layer is identical to layer **C**, etc. This makes a stacking sequence of **ABCABC**... This structure exhibits *cubic closest packing* (CCP) and, like the HCP structure, contains 26 percent void space (Figure 3.2). It is equivalent to the face-centered cubic (FCC) structure. The coordination number is 12 and there are four atoms per unit cell. A prominent meteoritic mineral with FCC is taenite. The unapproved phase, antitaenite, has the same FCC structure and same Fe/Ni ratio range in composition as taenite but is paramagnetic and has a lower *magnetic moment* (a quantity representing the magnetic intensity and orientation of a magnet). Some metal particles extracted from bulk powdered samples of the Carancas

H4–5 chondrite contain intergrowths of tetrataenite and antitaenite (Munayco et al. 2013). (The meteorite formed a 14.2-m-wide crater in unconsolidated, water-saturated soil and sand near Lake Titicaca in Peru on September 15, 2007.]

3.5 Pauling's Rules

The principles governing the way atoms and ions are arranged in crystal structures are described by *Pauling's Rules* (first published in 1929 by American chemist Linus Pauling). These rules are empirical observations showing how the nature of a crystal structure is determined by the sizes, charges, and types of bonding of its constituent ions. Following is a simplified version of Pauling's Rules:

Rule 1: *Radius ratio rule.* For most ionic compounds, cations are smaller than anions, each cation is surrounded by coordinated anions in the shape of a polyhedron, and the cation/anion radius ratio determines the coordination number and the form of the polyhedron.

Rule 2: *Electrostatic valency rule.* Ionically bonded compounds are electrically neutral. The total strength of the electrostatic bonds reaching an anion is equivalent to the charge of the anion.

Rule 3: *Stability adversely affected by the sharing of polyhedral edges and faces.* The stability of ionically bonded compounds is diminished by the sharing of edges and (particularly) faces by adjacent polyhedra. This is because, the greater the sharing of polyhedral elements, the closer are the cations in the adjacent polyhedra, and the greater the electrostatic repulsion between these cations.

Rule 4: *Avoidance of sharing polyhedral elements in crystals with different cations.* For crystals that contain different cations, those cations that are highly charged (i.e., have a high valence) tend not to occur in adjacent polyhedra. This is because (1) like charges repel and (2) small, highly charged cations have low coordination numbers (i.e., they are surrounded by few anions and commonly require more than half the total charge of the anions to achieve charge balance), thus hindering the sharing of these anions with other cations.

Rule 5: *Principle of Parsimony.* The number of essentially different kinds of constituents in a crystal tends to be small. In most crystals, there are only a few different kinds of coordination polyhedra.

These principles underlie the basis for crystal stability. The sizes and charges of atoms and ions determine their packing arrangements within crystals and account for the myriad physical, chemical and optical properties of minerals. Nevertheless, a recent study of about 5,000 oxides showed that only ~13 percent simultaneously obeyed all of the last four rules (George et al. 2020). As stated by George et al., new and improved rules need to be developed.

4
Properties of Minerals
Explanations and Applications

One broad-brush approach to systematizing minerals is to divide them into two categories: (1) native elements and their alloys (e.g., Au, Ag, Cu, C, S, Ni_3Fe, Fe_5C_2) and (2) compounds (e.g., sulfides, sulfates, oxides, halides, silicates, carbides, carbonates, phosphides, phosphates, tungstates, nitrides). Compounds are grouped together based on their monatomic or polyatomic anions. This is a natural classification system (first adopted by Jacob Berzelius) because the chemical properties of minerals within these groups tend to be similar. For example, the properties of anhydrite ($CaSO_4$) more closely resemble those of other sulfates such as barite ($BaSO_4$) and celestine ($SrSO_4$) than those of other Ca-rich minerals, e.g., calcite ($CaCO_3$), oldhamite (CaS), fluorite (CaF_2), antarcticite ($CaCl_2 \cdot 6(H_2O)$), lime (CaO), apatite ($Ca_5(PO_4)_3(F,Cl,OH)$), scheelite ($CaWO_4$), portlandite ($Ca(OH)_2$), powellite ($CaMoO_4$), and wollastonite ($CaSiO_3$).

As chemically and structurally diverse solids, minerals exhibit a large array of properties; some properties are manifest in essentially all mineral species (e.g., crystal symmetry, hardness, habit, specific gravity), others are present to a significant degree in only a limited number (e.g., phosphorescence, magnetism, radioactivity, water solubility). Each mineral has a unique set of properties enabling its identification. However, superficial resemblances between mineral species can lead to confusion (e.g., gold and pyrite (FeS_2), a.k.a. "fool's gold"; or hypersthene ($(Mg,Fe)SiO_3$) (Fs_{30-50}) and hornblende, ($NaCa_2(Mg,Fe)_4AlSi_6Al_2O_{22}(OH,F)_2$). Some related minerals with similar optical properties can be definitively distinguished only by their chemical composition, e.g., niningerite ((Mg,Fe)S) and keilite ((Fe,Mg)S); hornblende and kaersutite ($NaCa_2(Mg,Fe)_4(Ti,Fe)Al_2Si_6O_{22}(OH,F)_2$); arfvedsonite ($Na_3(Mg,Fe)_4AlSi_8O_{22}(OH,F)_2$) and riebeckite ($Na_2(Fe^{2+}{}_3,Fe^{3+}{}_2)Si_8O_{22}(OH)_2$). Several recently described minerals are so rare and so small (e.g., wassonite (TiS); panguite ($(Ti,Al,Sc,Mg,Zr,Ca)_{1.8}O_3$); nuwaite ($Ni_6GeS_2$)), that physical and optical properties such as color, hardness, and specific gravity cannot be determined directly; in such cases, some properties can be calculated or inferred from coarse examples of the minerals' synthetic analogs.

4.1 Major Physical Properties

Crystal Symmetry. Every mineral crystal[1] corresponds to one of 230 space groups. The atoms within these crystals are arranged in a few basic ways determined by their sizes and bonding.

[1] Although the element mercury is considered a mineral, it is a liquid at room temperature and thus lacks such properties as crystal symmetry and hardness. However, mercury freezes at −38.83°C, forming crystals with a rhombohedral structure; temperatures lower than this occur in wintertime at high latitudes on Earth.

Even though two or more crystalline phases may have the same chemical composition, they have distinct crystal structures, different physical properties and nonoverlapping equilibrium temperature-pressure stability ranges. Distinguishable phases with the same composition are known as *polymorphs*. A classic example is SiO_2; approved mineral phases include quartz, cristobalite, tridymite, coesite, stishovite, and seifertite. Each is present in meteorites. (In fact, there are two crystal forms of quartz, two of cristobalite, and seven of tridymite.) The first natural occurrences of coesite and stishovite were discovered in sheared Coconino sandstone from Meteor Crater, Arizona. Seifertite has been found so far only in martian and lunar meteorites and was produced by shock metamorphism during impact-excavation and launch.

Habit. Euhedral crystals display ideal external forms with characteristic interfacial angles (e.g., cubic, octahedral, dodecahedral), signifying their internal three-dimensional geometric structures. Crystals with nonideal morphology may have impinged on adjacent grains during growth, suppressing their ideal external forms. The habit of a crystal is the actual external form, controlled by its immediate growth environment. Numerous terms describe these crystal shapes such as equant (roughly equal in three dimensions), fibrous (thin and elongated prisms), acicular (needlelike), amygdaloidal (almond shaped), dendritic (branching and treelike), reniform (kidney shaped), tabular (book shaped), and botryoidal (resembling a bunch of grapes). Different crystals of the same mineral species may exhibit a wide variety of habits. For example, magnetite in CI carbonaceous chondrites may exhibit framboidal, spherulitic, plaquette, and platelet habits in the same meteorite.

Cleavage. Cleavage describes the manner in which a mineral breaks along specific planes parallel to crystal faces, reflecting the internal atomic arrangement. Cleavage has two fundamental characteristics: quality and crystallographic direction. *Quality*: Some minerals with "perfect cleavage" (e.g., micas, graphite) consist of parallel atomic planes with strong intraplanar bonds and weak bonds between adjacent planes; these minerals break along easily identifiable planes. Minerals with "good cleavage" (e.g., enstatite, millerite) break along imperfect, but distinguishable planes; minerals with "poor cleavage" (e.g., apatite, zircon) break along surfaces that are difficult to recognize as planar. Some minerals (e.g., pyrite, magnetite, quartz) do not show any cleavage. Minerals can exhibit different quality cleavages along different planes, e.g., anhydrite (perfect parallel to {010}, very good parallel to {100}, and good parallel to {001}). *Crystallographic direction*: Minerals cleave along planes that have specific angular relations to each other so that characteristic forms are produced: cubic (three planes intersecting at 90° angles, e.g., halite), octahedral (four planes, e.g., fluorite), rhombohedral (three planes intersecting at angles that are not 90°; e.g., calcite), dodecahedral (six planes; e.g., sphalerite), prismatic (two planes, e.g., orthoclase), and pinacoidal or basal (one plane, e.g., graphite).

Parting. Parting is the breakage of minerals along planes of structural weakness that result from twinning, planar exsolution lamellae, or pressure-induced crystal deformation (due to tectonic, isostatic, or shock forces). Minerals can continue to grow after they part, healing the parting fractures. Within individual mineral grains, parting may produce striations or planes with high concentrations of inclusions. Whereas cleavage is exhibited by all grains of a mineral species, parting is dependent on the particular environmental conditions experienced by an individual grain. Minerals that exhibit parting include magnetite (octahedral parting) and corundum (rhombohedral parting).

Fracture. Every brittle substance can break. Fracture is the type of mineral breakage that does not adhere to parting surfaces or cleavage planes. The major types of fracture include: (a) *conchoidal fracture* (characteristic of glass, opal, quartz, and diamond), leaving smooth curved surfaces resembling the interior of a clam shell, (b) *hackly (jagged) fracture* (characteristic of native metals such as Cu and Ag), resulting in jagged, uneven surfaces, (c) *splintery (fibrous) fracture* (characteristic of chrysotile and kyanite), resulting in sharp, spearlike textures, and (d) *irregular (uneven) fracture* (characteristic of pyrite and magnetite), producing uneven rough surfaces.

Twinning. Most minerals are present as single crystals adjacent to other crystals; their mutual boundaries are randomly oriented. In contrast, some mineral species occur commonly as intergrowths of two separate crystals sharing lattice points along a smooth boundary. This assemblage is a contact twin; its two crystal components are related to each other by symmetry operations, e.g., reflection by a mirror plane, rotation about an axis shared by the crystals, or inversion through a point. The boundary between these two crystals is known as a composition surface; in most cases, it is also a twin plane.

Many *contact twins* consist of numerous parallel crystals, easily discernable under the microscope and related to each other by reflection through a mirror plane. This type of twinning is called "albite twinning" or "polysynthetic twinning" and is common in albite and clinoenstatite. Large crystals of plagioclase commonly exhibit striations on crystal faces, a manifestation of polysynthetic twinning. The polysynthetic twinning in clinoenstatite forms through inversion from protopyroxene during rapid cooling; it is nearly ubiquitous in pyroxene-rich chondrules in unequilibrated chondrites. Multiple twins that are not mutually parallel are called cyclic twins; these occur commonly in rutile and chrysoberyl.

Penetration twins consist of two separate crystals of the same mineral phase that appear to be passing through each other; they are common in orthoclase, pyrite, staurolite, and cinnabar, and are typically related to each other by rotation through a twin axis.

Deformation twins, also known as mechanical twins or gliding twins, form when a crystal is stressed and rows of atoms slide past each other. These occur commonly in quartz and calcite on Earth during regional metamorphism and shock metamorphism. Neumann lines (a.k.a. Neumann bands) are fine parallel lines in low-Ni metallic Fe (i.e., kamacite); they are plate-shaped lamellae produced as mechanical twins during shock processes and occur mainly in iron meteorites.

Minerals from different crystal systems tend to exhibit different types of twinning. For example, monoclinic minerals (e.g., clinoenstatite) commonly have twins formed by reflection on the {001) or (100} planes; isometric minerals (e.g., spinel) commonly have twins formed by reflection on one of the octahedron faces.

Hardness. The hardness of a mineral phase is a measure of how well it resists scratching on a smooth, fresh surface. It is a function of the binding forces among atoms within the crystal and can vary with crystallographic direction. The most commonly used hardness measurement is the Mohs scale (Table 4.1), developed by Friedrich Mohs in 1824. It is a relative scale, based on the ability of minerals to scratch each other, a procedure first articulated by Pliny the Elder in the first century CE.

Weaker bonds in the softer mineral are broken by the harder mineral (which has stronger chemical bonds). The scale ranges in increments of ½ units from 1 (talc, the softest mineral,

Table 4.1 *Mohs hardness of common minerals*

Mohs Hardness	Type Example
1	Talc – $Mg_3Si_4O_{10}(OH)_2$
2	Gypsum – $CaSO_4{\cdot}2H_2O$
3	Calcite – $CaCO_3$
4	Fluorite – CaF_2
5	Apatite – $Ca_5(PO_4)_3(F,Cl,OH)$
6	Orthoclase – $KAlSi_3O_8$
7	Quartz – SiO_2
8	Topaz – $Al_2SiO_4(F,OH)_2$
9	Corundum – Al_2O_3
10	Diamond – C

with weakly bonded planes attached to one another by van der Waals bonds) to 10 (diamond, the hardest mineral, with strong covalent bonds).

Zircon ($ZrSiO_4$), which can scratch quartz and be scratched by topaz, has a hardness of 7½. Chromite ($FeCr_2O_4$), which can scratch apatite and be scratched by orthoclase, has a hardness of 5½. Halite (NaCl), which can scratch gypsum and be scratched by calcite, has a hardness of 2½. Molybdenite (MoS_2) and montmorillonite ($(Na,Ca)_{0.3}(Al,Mg)_2Si_4O_{10}(OH)_2{\cdot}nH_2O$) can be scratched by gypsum and are assigned hardness values of 1–1½. Some sources list a few native elements as having Mohs hardness values <1: e.g., Cs (0.2), Rb (0.3), K (0.4), Na (0.5), and Li (0.6).

Another useful scale is Vickers hardness, developed by Robert Smith and George Sandland at the Vickers Ltd. engineering firm in 1921. It is based on indentation hardness, i.e., how well a given material can resist plastic deformation when a standard load is applied by a pyramid-shaped indenter (with a square cross section) made of diamond or another hard substance. The indentation in the target is examined microscopically in reflected light and a Vickers hardness number (VHN) is assigned, expressed in units of kg/mm^2.

Vickers hardness can be used to gauge shock effects in iron meteorites. For example, unshocked kamacite with 7.0 wt% Ni has a VHN of ~155 (using a 100 g load); if Neumann bands are present, indicative of relatively mild shock, the VHN increases to 170–250; if the kamacite is shocked beyond 13 GPa and exhibits a crosshatched ε structure (a high-density hexagonal close-packed structure), its VHN increases to ~320.

Tenacity. The degree of resistance of a mineral to breaking, crushing, bending, or shredding is called *tenacity*; this property is generally described by one or more of six qualitative descriptions.

1. *Brittle* – readily broken or powdered. Most minerals with ionic bonding (e.g., halite) are brittle.
2. *Malleable* – can be hammered into thin sheets. Many malleable minerals (e.g., gold, metallic Fe) have metallic bonds. Native metals tend to be soft, malleable, ductile, and sectile.
3. *Sectile* – readily cut into thin shavings with a sharp knife. Examples of sectile minerals are talc and native copper.

4. *Ductile* – readily drawn into a wire. Many phases with metallic bonds are ductile; these include copper, metallic Fe, and silver.
5. *Flexible* – bends in response to applied stress but does not resume its original shape when the stress is removed. Flexible minerals include graphite, talc, molybdenite, and brucite.
6. *Elastic* – bends in response to applied stress, but resumes its original shape when the stress is removed. Elastic minerals include the micas – muscovite, phlogopite, and biotite.

Although natural metallic Fe-Ni is soft and ductile, metallurgists recently developed an Fe-Ni-Co alloy (containing intermetallic Al-Ti nanoparticles) that is both ultra-strong *and* ductile (Yang et al. 2018). This material is potentially of great use in automobiles, high-speed trains, cryogenic systems, and aircraft.

Iron meteorites, rare though they are, were essentially the only sources of metallic iron available to technologically primitive peoples. The malleability of iron allowed the ancient Egyptians to fashion decorative beads from iron meteorites for necklaces, bracelets and anklets; Tutankhamun (ruled circa 1332–1323 BCE) had a dagger made of meteoritic iron-nickel (Comelli et al., 2016). Other examples of Bronze Age artifacts made of meteoritic iron include a Ugarit axe blade (Syria, 1400 BCE) and several Zhou dynasty axes (China, 1000 BCE). The Inuit people of Greenland harvested metal from the Cape York iron meteorite to make arrowheads, harpoon blades, and knives. (Local Inuit people from Disko Island, Greenland, forged native terrestrial iron into cutting edges for knives.) Native Americans in what is now Argentina made metal-tipped weapons from the Campo del Cielo iron. Present-day artisans offer iron-meteorite knives, rings, pendants, tie clips, bolo tie slides, and wristwatches for sale on the internet.

Specific gravity. The specific gravity (G) of a mineral is a unitless number defined as the ratio of its density (ρ) to the density of pure water at 4°C (which is 1 g cm^{-3}). Alternatively, G is the weight ratio of a mineral to an equal volume of water at 4°C. Specific gravity is a function of two parameters:

(1) *The elements that compose the mineral.* For example, galena (PbS) is isostructural with alabandite (MnS); both are members of the same crystal system (cubic), same crystal class (hexoctahedral), and same space group (Fm3m), but the measured specific gravity of galena (7.58) is much higher than that of alabandite (4.0). This is because Pb has a much greater atomic mass than Mn.

(2) *The way elements are packed together.* For example, stishovite and tridymite both have the same chemical composition (SiO_2), but the measured specific gravity of stishovite (4.28) is much higher than that of tridymite (2.26). This is because stishovite has a more compact structure in which each Si cation is in octahedral coordination, bound to six oxygen atoms; in tridymite, each Si cation is in tetrahedral coordination, bound to only four oxygen atoms.

The specific gravity of a particular mineral varies as a function of composition. In the olivine solid-solution series, the specific gravity increases linearly from 3.26 for forsterite (Mg_2SiO_4) to 4.39 for fayalite (Fe_2SiO_4). This is because Fe has a much greater atomic mass than Mg. Plagioclase (albite ($NaAlSi_3O_8$) to anorthite ($CaAl_2Si_2O_8$)) undergoes a coupled substitution in which the exchange of Si^{4+} for Al^{3+} is charge balanced by the exchange of Ca^{2+} for Na^+. The measured density of plagioclase increases modestly from 2.63 in albite to 2.76 in anorthite. This is due mainly to Ca having an appreciably greater atomic mass than Na.

The densest terrestrial minerals include chengdeite (Ir_3Fe; $\rho = 19.3$ g cm^{-3}) and unalloyed native iridium ($\rho = 22.65$ g cm^{-3}), both found in alluvial placer deposits. The densest natural crystalline substance ($\rho = \sim 5 \times 10^{13}$ g cm^{-3}) occurs in the outer crusts of neutron stars where neutrons sit side by side within a crystal lattice. A cubic millimeter of this matter (the size of a poppy seed) would have about the same mass as 17 fully fueled Saturn V rockets (only 13 were ever launched).

Luster. Materials reflect, transmit, and absorb different amounts of light. Opaque phases (many of which have metallic bonding) reflect a significant fraction of incident light; transparent phases (which commonly have ionic or covalent bonds) transmit most incident light. One artificial material called Vantablack (made from vertically aligned carbon nanotubes) absorbs 99.965 percent of incident visible light; it reflects 0.035 percent. A newer "superblack" coating of carbon nanotubes absorbs 99.995 percent of visible light. Minerals are routinely divided into those with a *metallic luster* (which typically reflect ~20–50 percent of incident light) and those with a *nonmetallic luster* (which typically reflect ~5–10 percent).

Light can be characterized as an electromagnetic wave with electric and magnetic fields that are perpendicular to one another and also perpendicular to the direction of motion. When light reaches the surface of a metal, the delocalized electrons cancel out its electric field, resulting in the production of a new light wave that propagates away from the metal (i.e., it is reflected).

Minerals with metallic luster include precious metals (e.g., gold, silver, iridium), base metals (e.g., copper, iron), some sulfides (e.g., pyrrhotite, pyrite, marcasite, chalcopyrite, bornite, pentlandite, galena), some oxides (e.g., chromite, magnetite), and some hydroxides (e.g., manganite). Gold ultimately derives its name from the Proto-Indo-European word meaning "to shine or glisten."

Some phases are described as exhibiting a *submetallic luster*; they are typically opaque to semiopaque and reflect ~15–20 percent of the incident light. Examples include hematite, ilmenite, cinnabar, cuprite, and psilomelane.

Nonmetallic minerals exhibit a range of lusters although the categories are imprecise and vary somewhat between texts. Different texts may also characterize the same mineral as having different lusters. The problem is compounded by the fact that different specimens of a particular mineral may appear to have somewhat different lusters; a prime example is malachite ($Cu_2CO_3(OH)_2$), with specimens ranging widely in luster – adamantine, vitreous, silky, and earthy. Many minerals are characterized as having an intermediate luster, e.g., adamantine to vitreous; vitreous to pearly; vitreous to greasy. The nonmetallic lusters include the following:

Vitreous: A moderately brilliant appearance, resembling glass (e.g., quartz, plagioclase, orthopyroxene, hornblende, calcite, siderite, smithsonite, topaz, beryl, chrysoberyl, tourmaline, scapolite, fluorite, spinel, anhydrite, ice, sylvite, niter, borax, azurite). Approximately 70 percent of known minerals have a vitreous luster (including almost all silicates).

Subvitreous: A duller, less-brilliant appearance than glass (e.g., howiite).

Adamantine: Brilliant appearance, resembling a diamond (e.g., diamond, cerussite, crocoite, cubic zirconia, pyrargyrite). Minerals with an adamantine luster tend to have high indices of refraction.

Subadamantine: A less-brilliant appearance than diamond (e.g., garnet, corundum, zincite).

Resinous: The appearance of resin or amber (e.g., sphalerite, sulfur, monazite, pyromorphite).

Earthy (Dull): Resembling a clump of dirt, commonly found in materials with a rough granular surface that randomly scatters light (e.g., kaolinite, glauconite, carnotite). Bauxite, the principle ore of aluminum, is a sedimentary rock consisting of a mixture of diaspore, gibbsite, and böhmite; it is commonly described as having an earthy or dull luster.

Pitchy: Resembles tar (e.g., uraninite, cyrtolite – a form of zircon). Many minerals with this luster are radioactive and have undergone partial metamiction (destruction of the crystal structure via radioactive decay).

Greasy: Resembles a coating of oil, fat, or grease (e.g., cordierite, jadeite).

Waxy: Resembles paraffin or wax (e.g., nephrite (jade), chalcedony, cerargyrite, turquoise).

Silky: Appears fibrous and resembles silk (e.g., asbestos, ulexite).

Pearly: Appears iridescent, resembling a pearl (e.g., stilbite, talc, brucite).

Color. Many minerals have characteristic colors that aid identification, e.g., gold nuggets are bright golden yellow; copper is reddish orange; silver is grayish white; chalcopyrite ($CuFeS_2$) is brassy yellow; malachite ($Cu_2CO_3(OH)_2$) is green. Some minerals are even named after their characteristic colors: e.g., olivine ($(Mg,Fe)_2SiO_4$) for the olive-green color of magnesian samples; albite ($NaAlSi_3O_8$) from the Latin word for "white"; azurite ($Cu_3(CO_3)_2(OH)_2$) from the Persian word for "blue"; celestine ($SrSO_4$) from the Latin word for "sky"; chlorite ($(Fe,Mg,Al)_6(Si,Al)_4O_{10}(OH)_8$) from the Greek word for "green"; rhodonite ($(Mn,Fe,Mg,Ca)(SiO_3)$) from the Greek word for "rose"; and zorite ($Na_2Ti(Si,Al)_3O_9{\cdot}nH_2O$) from the Russian word for "the rosy radiance of the sky at dawn." These minerals have a characteristic color and are called *idiochromatic*.

A related example is cryolite (Na_3AlF_6), many specimens of which are colorless or white. With a refractive index similar to that of water, colorless cryolite seems to disappear when placed in water. The name comes from the Greek *kryos* (ice) and *lithos* (stone). It was initially recovered on the west coast of Greenland; the indigenous people referred to it as "the ice that never melts."

Some ancient cultures attributed spiritual significance to precious metals because of their color. The Incas esteemed gold as the sweat of the Sun and silver as the tears of the Moon. In Greek mythology, there is King Midas with the golden touch; Chrysus (Khrysos), who is the spirit or daimon of gold; and Danaë (mother of Perseus), who was impregnated by Zeus after he came to her as a shower of gold.

Different specimens of most mineral phases can display a variety of colors due to major compositional variations, minor chemical impurities, tarnishing, crystallographic defects, small inclusions, or light absorption by electronic bands. These species are *allochromatic*.

Color changes caused by major compositional variations. There are several well-known examples of mineral composition affecting color (e.g., Hurlbut and Klein 1977). The color of olivine ($(Mg,Fe)_2SiO_4$) changes from pale yellow green or olive green in both forsterite (Mg_2SiO_4) and magnesian olivine to dark, brownish green in fayalite (Fe_2SiO_4); pure forsterite grains from CM2 Murchison are white. The color of sphalerite ($(Zn,Fe)S$) changes from white to yellow to brown to black (a variety known as marmatite) with increasing amounts of Fe^{2+} substituting for Zn^{2+}. The color of tremolite ($Ca_2Mg_5(Si_8O_{22})(OH)_2$) changes from white to green to black with increasing amounts of Fe^{2+} substituting for Mg^{2+}. The color of electrum (a naturally occurring Au-Ag alloy containing 20–80 percent gold and 20–80 percent silver) changes from bright golden yellow to pale gold with increasing proportions of Ag. The ancient Greeks called electrum with a high Ag content "white gold." Electrum that contains substantial amounts of alloyed Cu appears brassy.

Color changes caused by minor chemical impurities. The impurities responsible for color in minerals are generally small amounts of the transition metals Ti, V, Cr, Mn, Fe, Co, Ni, and Cu. These elements have partly filled 3*d* orbitals containing unpaired electrons. It takes much less energy to bump an unpaired electron to a higher (i.e., more excited) energy state than to bump up a paired electron; visible light (with wavelengths between ~400 and 800 nm) has enough energy to excite unpaired electrons, but not paired ones. When a particular wavelength of visible light excites an unpaired electron, that wavelength is absorbed by the electron and removed from the spectrum. The light that passes through the crystal is missing that particular wavelength. The ability of small amounts of transition metals to produce colors in minerals has earned them the nickname "chromophores."

The most well-known example of a particular mineral exhibiting different colors caused by chemical impurities is corundum (Al_2O_3). Pure nonporous crystals of corundum are transparent and colorless. Ruby (from the Latin word for "red") is a corundum crystal with a blood red color; the most sought-after rubies are those described as "pigeon's blood," which display a secondary purple hue. Sapphire is any corundum crystal that displays a color other than red, prototypically blue, but in some cases, violet, green, yellow, orange, black, gray, white, or colorless. Rubies contain small amounts of Cr^{3+} (up to ~1 wt%), substituting for Al^{3+} in octahedral coordination. When photons with wavelengths corresponding to yellow-green and violet light strike the crystal, they are absorbed by the unpaired electrons in the outer shells of the Cr atoms, thereby bumping electrons up to more-excited energy states. The wavelengths of visible light that are not absorbed (and, hence, pass through the crystal) are dominated by red.

Minor Cr^{3+} impurities are also responsible for the green color of emerald (a variety of beryl ($Be_3Al_2SiO_6$)). Pure colorless beryl is sometimes called goshenite. The crystal structure and composition of beryl differ from those of corundum; less energy is required to bump unpaired electrons in the outer shells of Cr atoms in beryl up to higher energy states. These electrons absorb photons with wavelengths corresponding mainly to violet and red light, allowing blue-green wavelengths to pass through the crystal, thus coloring emeralds green. Other gemstone varieties of beryl include aquamarine (which is a shade of blue and contains Fe^{2+} impurities), heliodor (greenish yellow with Fe^{3+}), golden beryl (yellow with Fe^{3+}), morganite (pink with Mn^{2+}), and red beryl (red with Mn^{3+}).

Color changes caused by tarnishing. A tarnish is a thin film that forms at the surface of a material due to corrosion, usually caused by oxidation. Metallic Cu can change from coppery red when fresh to green because a patina (verdigris) of basic copper carbonate develops during tarnishing. Man-made alloys of brass (Cu,Zn) and bronze (Cu,Sn) also develop a green patina as they tarnish. Another natural example is bornite (Cu_5FeS_4), commonly known as peacock ore; fresh specimens (which are bronzy, brown, or copper-red) acquire a blue-violet-purple iridescent film as they tarnish.

Color changes caused by crystallographic defects. Frenkel defects are locations within a crystal structure where an ion is missing that would normally be present. In a minority of cases, the missing ion is an anion that was replaced by an electron, thereby maintaining charge balance. This is an *electron color center*. The electron is available to absorb incident photons of specific wavelengths and get bumped up to higher energy states. The absorbed photons are subtracted from the continuous spectrum, adding color to the mineral.

A *hole color center* derives its label from the absence of an electron from its regular location. The examples often presented in texts involve varieties of quartz (SiO_2). In smoky quartz, some Si^{4+} cations are replaced with Al^{3+}; for charge balance to be maintained, there will commonly be a monovalent cation (e.g., H^+ or Na^+) close by in the lattice. This weakens the forces on the electrons in the outer shells of the O anions bonded to the Al^{3+}. Intense radiation can dislodge one of these electrons (which settles elsewhere in the lattice), resulting in the O having an unpaired electron in its outer shell. Because an electron is missing, there are available energy levels into which other electrons around the O could get bumped up if they absorb a photon of the right energy. That particular wavelength is subtracted from the continuous spectrum, lending a brownish color to smoky quartz. Amethyst (purple-colored quartz) derives its color in an analogous way except that, in this case, Fe^{3+} is substituted for Si^{4+}. If smoky quartz or amethyst is heated to ~400°C or ~450°C, respectively, the dislodged electron returns to its original location and the crystal resumes its original color. This restorative process is known as *bleaching*.

Color changes caused by small inclusions. Many individual crystals can incorporate inclusions as they grow. Opaque inclusions can reflect light; translucent inclusions can refract light. Jasper (a fine-grained reddish-brown variety of quartz) contains small inclusions of hematite (Fe_2O_3), a red mineral of the same composition as rust. Milky quartz contains small pockets of fluid, mainly water. Rose quartz may derive its pink color from microscopic inclusions of a fibrous phase, possibly dumortierite ($Al_7BO_3(SiO_4)_3O_3$) (Figure 4.1) (Goreva et al. 2001; Ma et al. 2002). Rutilated quartz contains acicular inclusions of rutile (TiO_2) and can appear golden, silver, red, or black; the rutile needles can be randomly distributed throughout the crystal or can occur together in bundles.

Color changes caused by light absorption by electronic bands. When atoms form molecules, their electron orbitals overlap. Because the *Pauli exclusion principle* forbids electrons in the same atom from having the same quantum numbers, these orbitals split into molecular orbitals wherein each electron assumes a different energy state. The number of atoms in a macroscopic

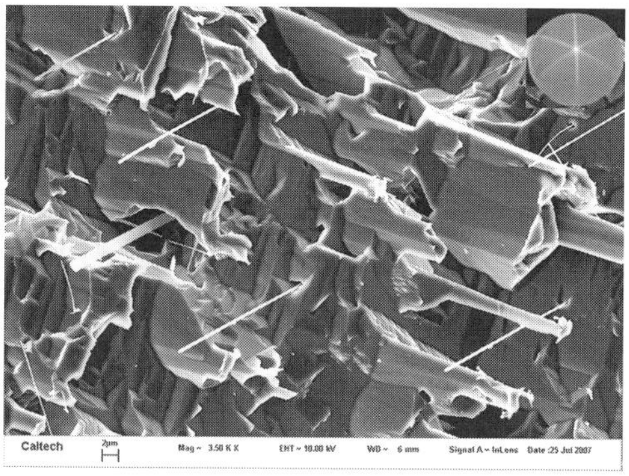

Figure 4.1 Rose quartz. Color caused by inclusions of pink dumortierite-related fibers. After Ma et al. (2002). BSE image. Inserted optical photo courtesy of George Rossman, Caltech. (A black-and-white version of this figure will appear in some formats. For the colour version, please refer to the plate section.)

crystal are very large (~10^{22}); the number of orbitals is also large and the individual energy levels are very closely spaced. The spacing is so close that the energy levels can be regarded as a continuous band. The electrons can become delocalized and move freely from the highest fully occupied electronic band to the lowest unoccupied (but still higher-energy) band. If the energy gap between these bands is lower than the energy of violet light, then blue-violet light is absorbed by the crystal and subtracted from the continuous spectrum; red light passes through. This mechanism is responsible for the red color of cinnabar (HgS). If the energy gap between these bands is lower than the energy of all visible light (as is common in metals), then visible light is more or less completely absorbed and the crystal appears opaque.

Streak. The true color of a mineral can often be determined in its powdered form. Streak is the trail of mineral powder left on an unglazed white porcelain tile (a streak plate) by scratching. The tile has a hardness of ~7, so it cannot be used for harder minerals. Powders of transparent minerals (including many with ionic or covalent bonds) reflect most of the incident light, so their streaks are light in color. Powders of opaque minerals (including many metals) absorb most of the incident light and thus make dark or black streaks. Minerals that are often mistaken for meteorites include the opaque iron oxides hematite (Fe_2O_3, red streak, nonmagnetic) and magnetite (Fe_3O_4, black streak, magnetic) or intergrowths of the two (brownish streak, moderately magnetic).

Diaphaneity (transparency). The extent to which light is transmitted through a substance is its transparency. If light is transmitted through the entirety of a substance and an underlying image can be readily discerned, the substance is *transparent*; if the light exits in such a manner that an underlying image is distorted or blurry, the substance is *translucent*; if light is completely absorbed or reflected, the substance is *opaque*. No real substance is completely transparent. Ordinary window glass transmits only 92 percent of the visible light that strikes it perpendicular to the surface; about 4 percent of the incident light is reflected at each surface (where there is an abrupt change in index of refraction).

For many minerals, the amount of energy available in visible light is insufficient to excite valence electrons to jump to the lowest unoccupied energy band; that failure leaves visible light unabsorbed and the crystal transparent. However, in metals, the lowest occupied energy bands are only partly full; this allows electrons to jump to excited states at relatively low photon energies. Hence, metals characteristically absorb light and are opaque.

For translucent minerals that absorb some of the light, their degree of transparency will be, in part, a function of their thickness. If 10 percent of the incident light is absorbed by a millimeter-thick specimen, then a 3-cm-thick specimen will essentially be opaque.

An important biological use of mineral diaphaneity involves Iceland spar (a transparent variety of calcite). Many species of trilobites had compound eyes made of tiny lenses of calcite. The c-axis was perpendicular to the lens; radially varying Mg/Ca corrected for chromatic aberration. The only modern animal that has calcite eyes is the brittle star (an echinoderm related to starfish).

4.2 Other Physical Properties

Play of colors. Some minerals appear to change color as incident light strikes them at different angles. These minerals are *pseudochromatic*; the variegated effect results from reflection and refraction of light as it encounters internal boundaries between layers with different indices of refraction. The regularly spaced spherules of silica gel within precious opal are all nearly the same size. Because this size is about the same as the wavelength of visible light, the spherules serve as a diffraction grating, splitting polychromatic white light into its ROYGBIV rainbow components (i.e., red, orange, yellow, green, blue, indigo, violet).

Internal iridescence in a mineral occurs when light encounters discontinuities inside a crystal (e.g., cleavage planes, fractures, twins, exsolution lamellae) and is reflected or diffracted. Prime examples occur in the feldspars: "moonstone" (which consists of stacked, alternating layers of orthoclase and albite); labradorite (some specimens of which consist of very fine exsolution lamellae of $An_{47–58}$); and albite (some specimens of which consist of exsolution lamellae of An_0 and An_{25}). Each of these minerals has a unique label for its internal iridescence: moonstone – *adularescence*; labradorite – *labradorescence*; albite – *peristerite*.

External iridescence (or surface iridescence) is produced by light interference as it is reflected by films on an object's surface: e.g., soap bubbles, oil slick on water. Iridescence is common in the organic realm: e.g., butterflies, abalone shells, catfish, peacock feathers, rainbow boas, begonia leaves. It even occurs in beef. Minerals and manufactured products that commonly exhibit surface iridescence include many opaque phases that undergo surface oxidation and tarnishing. Examples include metals (e.g., bismuth, titanium), oxides (e.g., limonite, hematite, goethite), and sulfides (e.g., sphalerite, bornite, rainbow pyrite, chalcopyrite). Rainbow hematite displays a variety of beautiful interference colors due to a thin-film coating of Al-phosphate nanocrystals (Figure 4.2) (Ma and Rossman 2006).

Chatoyancy. Chatoyancy is the silky appearance of a mineral specimen that reflects a band of light perpendicular to closely packed stacks of parallel fibers that either make up the mineral (e.g., satin spar gypsum ($CaSO_4{\cdot}2H_2O$); tiger's eye quartz (SiO_2)) or occur as inclusions (e.g., rutile (TiO_2) within cat's eye chrysoberyl ($BeAl_2O_4$)).

Asterism. Within some cabochons or spheres of hexagonal mineral specimens, inclusions occur in bundles oriented 120° from each other. Each set can exhibit a chatoyancy, combining to form three bands of light (also oriented 120° from each other), producing a six-pointed star. This feature is called an asterism[2] and is most commonly seen in ruby and sapphire (which are varieties of corundum). Some samples of rose quartz display an asterism due to acicular fibrous inclusions of dumortierite oriented 120° from each other (Figure 4.1). Other minerals that can display asterism include diopside, feldspar (moonstone), and garnet.

The most famous star sapphire is the *Star of India* with a weight of 563.35 carats; it was mined in Sri Lanka and sits on display at the American Museum of Natural History in New York. Among the most famous star rubies is the 138.7-carat *Rosser Reeves Ruby*, also mined in Sri Lanka. It is on display at the Smithsonian Institution in Washington, DC. (One carat weighs 200 mg.)

Dispersion (fire). Polychromatic white light is dispersed when it enters a gemstone, each ROYGBIV component traveling at a different speed depending on its wavelength; blue light (short wavelength) travels more slowly than red light (long wavelength). The resultant rainbow of colors is known as dispersion or fire. Gems with higher indices of refraction tend to display

[2] "Asterism" is derived from the Greek word for "star." The term also denotes simple star patterns that do not form an entire single constellation, e.g., the Big Dipper, a.k.a. the Plough (part of Ursa Major); the Summer Triangle (the three brightest stars from three different constellations – Altair from Aquila, Deneb from Cygnus, Vega from Lyra); Orion's Belt, a.k.a. the Three Kings or the Three Sisters (from our vantage point, they are three aligned, evenly spaced bright stars near the center of Orion).

Figure 4.2 Color effect of rainbow hematite due to thin-film interference by coating of Al-phosphate nanorods on the hematite surface. After Ma and Rossman (2006). Upper row: optical photos under reflected light; lower row: BSE images. Optical photo courtesy of George Rossman, Caltech. (A black-and-white version of this figure will appear in some formats. For the colour version, please refer to the plate section.)

greater fire; diamond has the highest index of refraction and the greatest fire of natural gemstones.

Gemstones can be cut to increase the dispersion, so incident light undergoes numerous internal reflections. The aptly named "brilliant cut" was designed by jewelers to maximize gemstone fire – the brilliant is shaped like two cones placed base to base. The top half (the *crown*) is a cone with 33 facets. It is truncated relatively near the base; the flat portion at the top is the *table*. The bottom half of the stone (the *pavilion*) has 25 facets; the apex is usually cut off parallel to the table, forming the small *culet*. The crown and pavilion are separated by the *girdle*, which commonly is also faceted.

Incandescence. A substance that emits light (including visible light) because it has reached a high temperature is incandescent. Nearly all solids and liquids begin to emit dull red light when heated to ~798 K (525°C), i.e., the so-called *Draper point*. As the temperature rises, the brightness increases and the color shifts toward white and blue. The terms "red hot" and "white hot" were first used by blacksmiths as relative indicators of the temperature of ferrous metal.

Thermoluminescence (TL). Some minerals emit small amounts of visible light when heated to ~50 to 100°C, far below the temperature required for incandescence. These minerals exhibit *thermoluminescence* or TL. Minerals that exhibit TL include fluorite, calcite, apatite, and feldspar. British meteor-astronomer Alexander Herschel, grandson of William Herschel, discovered meteorite TL in 1899; he sprinkled dust grains from L6 Middlesbrough on a hot plate and observed a distinct yellowish-white glow.

In a series of papers (e.g., Sears et al., 1980, 1982, 1991a, 1991b), Derek Sears and colleagues showed that TL sensitivity increases with metamorphic grade in ordinary and carbonaceous chondrites. (TL sensitivity was defined as the amount of TL induced in a sample by a standard dose of radiation after any accumulated TL was drained away by annealing the sample at ≥500°C.] Within the type-3 ordinary chondrites (OCs), TL sensitivity increases by three orders of magnitude, enabling their division into subtypes (3.0–3.9). The TL sensitivity in these meteorites correlates with other metamorphic parameters, e.g., decreases in olivine compositional heterogeneity, decreases in bulk volatile contents, increases in the modal abundance of feldspar, and increases in the degree of matrix recrystallization. Increases in TL sensitivity are caused by the development of secondary feldspar (which serves as a TL phosphor at moderate temperatures), mainly through the devitrification of chondrule glass. A detailed discussion of thermoluminescence in meteorites and lunar samples can be found in Sears et al. (2018).

Cathodoluminescence (CL). Cathodoluminescence is produced by the emission of visible-light photons from a luminescent sample after electrons collide with a phosphor. In many cases, the electron source is a scanning electron microscope (SEM) or electron microprobe.

Because the CL and TL properties of type-3 OCs correlate, CL can potentially be used to study low-temperature metamorphic processes. The least-equilibrated type-3 OCs (e.g., LL3.00 Semarkona) contain low-FeO chondrules with primary glassy mesostases[3] that exhibit bright yellow CL; by type 3.4, chondrule mesostases exhibit blue CL. This color change is due to the growth of secondary albite that luminesces blue. Coarse forsterite grains in the least-equilibrated OCs exhibit red CL; by type 3.4–3.6, the CL disappears as olivine becomes more ferroan (due to metamorphic equilibration with small FeO-rich olivine grains within the fine-grained matrix). Numerous CL images of meteorites and lunar rocks were published by Sears (2018).

Luminescence. An object that is not incandescent, but still emits light is *luminescent*; this occurs after the object has absorbed energy from an external source such as electromagnetic radiation (e.g., ultraviolet light), mechanical energy (e.g., high-speed collision), or thermal energy (e.g., emplacement in a high-temperature environment where it may be heated by conduction).

The phenomenon of an object emitting light after having absorbed electromagnetic radiation, is called *photoluminescence*. The emitted light is generally of lower energy (i.e., longer wavelength) than the incident radiation. Minerals can absorb electromagnetic radiation of different wavelengths (e.g., ultraviolet, visible, X-rays), exciting electrons to higher energy states. When electrons drop to unoccupied lower energy levels, light is emitted. If the light is in the visible range, the phenomenon is called **fluorescence** (if the glow ceases almost immediately after the mineral stops receiving incident electromagnetic radiation) or **phosphorescence** (if the glow persists, in some cases for several hours, after the cessation of incident radiation). It takes longer for excited electrons in phosphorescent minerals than in fluorescent minerals to reach their ground states.

Fluorescence (which derives its name from fluorite) is commonly caused by impurities in the minerals. Different specimens exhibit different degrees of fluorescence; individual specimens may not fluoresce at all, particularly if they contain Fe or Cu. Some examples of minerals that fluoresce in ultraviolet light are in Table 4.2.

[3] The mesostasis is the last-formed fine-grained or glassy interstitial material in an igneous rock.

Table 4.2 *Common fluorescent minerals*

Mineral	Formula	Impurity	Fluorescence
Fluorite	CaF_2	Eu^{2+}	blue
Calcite	$CaCO_3$	Mn^{2+}	red, orange
Willemite	Zn_2SiO_4	Mn^{2+}	green
Wollastonite	$CaSiO_3$	Mn^{2+}	orange
Ruby	Al_2O_3	Cr^{3+}	red
Diamond	C	N, N_2, B	blue, yellow

Triboluminescence. Some nonmetallic minerals become faintly luminous after chemical bonds are broken during mechanical abrasion (e.g., striking, crushing, ripping, rubbing, scratching). It is likely caused by charge separation during fracturing, followed by charge recombination that causes an electrical discharge that ionizes the air, resulting in flashes of light. Minerals that exhibit triboluminescence include quartz (white or pale yellow), diamond (blue or red) and opal (orange).

Triboluminescence has been exhibited by certain materials (e.g., Mn-doped Zn sulfide) in hypervelocity impact experiments, but the phenomenon occurs most commonly in the mouths of people chewing wintergreen-flavored candy in the dark.

Electrical properties. The *electrical conductivity* of a mineral is a function of the type of chemical bonding. Minerals with metallic bonding contain delocalized electrons that move freely through the crystal allowing strong electrical conductivity if an external voltage is applied or an ambient magnetic field changes direction. Some sulfide minerals have partial metallic bonding and are electrical semiconductors. Minerals with covalent or ionic bonds are nonconductors (insulators) because their electrons are tightly bound to the atoms and are not free to move around in response to an applied voltage. Graphite is covalently bonded parallel to the *c* axis but contains delocalized electrons that can facilitate electrical conductivity perpendicular to the *c* axis. For anisotropic minerals (including graphite and hematite), electrical conductivity is a function of crystallographic direction.

Piezoelectric minerals produce a voltage if they are deformed by applied mechanical stress: they become positively charged at one end of the crystal and negatively charged at the other end. The phenomenon is essentially confined to minerals that lack a center of symmetry; they possess at least one polar axis and have different crystal forms at each end. Pressure directed at one end of the axis induces a flow of electrons toward the other end, producing charge separation. Piezoelectric minerals include quartz, tourmaline, topaz, berlinite ($AlPO_4$), and macedonite ($PbTiO_3$).

Pyroelectric minerals become positively charged at one end and negatively charged at the other (i.e., a temporary voltage is created) due to changes in temperature. These changes cause slight movement of the atoms within the lattice, inducing changes in the polarization of the crystal and the development of an electrical potential. The phenomenon is confined to minerals with a single polar axis and was first observed in tourmaline.

Magnetism. Electron orbitals around an atom can contain up to two electrons, but because of the Pauli exclusion principle, the electrons must have opposite spins (+½ or −½). Although each of the spinning electrons in a filled orbital produces a magnetic field, the field is canceled out: +½ + −½ = 0. Minerals that do not contain elements with unpaired electron orbitals cannot produce a magnetic field; in contrast, minerals that contain elements with unpaired electron orbitals *can* produce a magnetic field.

Diamagnetic minerals (e.g., quartz) contain no elements with unpaired electrons; i.e., they have a net-zero magnetic moment. However, when placed in proximity to an external magnet, the electrons in diamagnetic minerals will move slightly, producing in the mineral a weak magnetic field opposite in polarity to that of the external magnet. As a consequence, diamagnetic minerals are weakly repelled by a magnet.

Paramagnetic minerals (e.g., olivine) contain elements or ions (e.g., Fe^{2+}) with unpaired electrons. Although each Fe^{2+} ion in olivine has four unpaired electrons in its 3*d* orbitals (each with spin +½) and has a nonzero magnetic moment, the Fe^{2+} ions are randomly oriented and the magnetic moments cancel each other out. However, when placed in proximity to an external magnet, the magnetic moments tend to align subparallel to the external field. Consequently, paramagnetic minerals are weakly attracted to a magnet.

Ferromagnetic materials include (1) the native metals iron, nickel, and cobalt, (2) alloys of these metals with themselves (e.g., permalloy with ~20 percent iron and ~80 percent nickel) or other elements (e.g., steel – iron and carbon; alnico – iron, aluminum, nickel, cobalt, copper, and, in some instances, titanium), and (3) rare-earth-bearing alloys (e.g., yttrium and cobalt; neodymium, iron, and boron; samarium and cobalt). Although the magnetic moments of adjacent ions in small regions (*magnetic domains*) of a grain tend to be parallel, in the absence of an external magnetic field, the numerous domains typically are randomly oriented, resulting in a net-zero magnetic moment for the grain. When an external magnetic field is applied, the domains tend to align, making the grains strongly attractive to a magnet. If the external field is strong, aligned magnetic domains within ferromagnetic materials expand while askew domains shrink; this results in permanent magnetization.

Heating of a ferromagnetic substance above a certain temperature – the *Curie temperature* (which varies for each substance) – results in randomization of the magnetic domains and loss of ferromagnetism. If the material is later cooled below the Curie temperature, the ferromagnetic properties are spontaneously reacquired.

The principal ferromagnetic minerals in meteorites are the metallic Fe-Ni phases, mainly kamacite (low-Ni iron) and taenite (and, to a lesser extent, tetrataenite), present in iron meteorites, pallasites, mesosiderites, and many chondrites. NWA 6259 is an extremely Ni-rich ungrouped iron meteorite (43 wt% Ni); it consists of 95 vol% tetrataenite. Because tetrataenite has high coercivity (the ability to withstand an applied external magnetic field before becoming demagnetized), NWA 6259 is a natural permanent magnet. However, in order for it to become magnetized, it would need to have its magnetic domains aligned by cooling through the Curie temperature in the presence of a strong magnetic field. This probably did not happen.

In *antiferromagnetic* materials, including hematite, NiO, and FeMn alloys, the magnetic moments of neighboring valence electrons tend to be antiparallel, so that macroscopic grains have net-zero magnetic moments.

In *ferrimagnetic* minerals, ions in adjacent crystallographic sites have magnetic moments that are antiparallel, resulting in a net-zero magnetic moment. However, other ions within the structure have unpaired electrons, producing a nonzero magnetic moment (and the mineral behaves like a ferromagnet).

The most common ferrimagnetic mineral is magnetite, which contains one Fe^{3+} ion in tetrahedral coordination and one in octahedral coordination. The ions have opposite magnetic moments that cancel each other out. Nevertheless, adjacent octahedrally coordinated Fe^{2+} ions are aligned, resulting in magnetite having a net nonzero magnetic moment.

Small grains of magnetite are present in most terrestrial igneous and metamorphic rocks. In meteorites, magnetite occurs primarily in members of some carbonaceous-chondrite groups (e.g., CI, CK, CO, oxidized CV) and in aqueously altered type-3 OC (e.g., LL3.00 Semarkona) and CO3 chondrites (e.g., CO3.00 DOM 08006).

Some specimens of magnetite called *lodestone*[4] are permanently magnetized. Many samples consist of microscopic intergrowths of magnetite and maghemite (Fe_2O_3 containing chemical impurities – Ti, Al, Mn); other samples consist mainly of titanomagnetite ($Fe^{2+}(Fe^{3+},Ti)_2O_4$). Lodestone became permanently magnetized in the presence of a strong external magnetic field. However, because the Earth's field (0.25–0.65 gauss) is too weak, it seems likely that lightning bolts were responsible. These bursts of electrical energy are associated with very intense (albeit fleeting) magnetic fields. The process of producing permanent magnets this way is known as *lightning-induced remanent magnetization* (LIRM). This mechanism is supported by the characteristic occurrence of lodestone at or near the Earth's surface (particularly in exposed, topographically high areas), not inside mines.

Radioactivity. The most common radioactive isotopes that are incorporated into minerals include ^{14}C ($t_{1/2} = 5{,}730$ years), ^{40}K ($t_{1/2} = 1.25 \times 10^9$ years), ^{87}Rb ($t_{1/2} = 4.9 \times 10^{10}$ years), ^{232}Th ($t_{1/2} = 1.405 \times 10^{10}$ years), ^{235}U ($t_{1/2} = 7.038 \times 10^8$ years), and ^{238}U ($t_{1/2} = 4.468 \times 10^9$ years). Each "parent" isotope decays to "daughter" isotopes; if those daughters are also radioactive, they decay in turn. The final set of daughter products differs for each of the common parent isotopes: $^{14}C \rightarrow {}^{14}N$; $^{40}K \rightarrow {}^{40}Ar + {}^{40}Ca$; $^{87}Rb \rightarrow {}^{87}Sr$; $^{232}Th \rightarrow {}^{208}Pb + {}^{4}He$; $^{235}U \rightarrow {}^{207}Pb + {}^{4}He$; $^{238}U \rightarrow {}^{206}Pb + {}^{4}He$.

The half-life ($t_{1/2}$) is the time it takes for half the atoms in a bulk sample of a parent isotope to decay. The half-lives of all known isotopes (natural and artificial) range over nearly 55 orders of magnitude. In general, the shorter the half-life, the greater the atomic instability. The shortest half-life is that of ^{7}H (7.3×10^{-31} years); this hydrogen isotope has one proton and six neutrons. It is extremely unstable. The longest half-life may be that of ^{120}Te (2.2×10^{24} years), about 14 orders of magnitude greater than the age of the universe.

The ^{14}C isotope is formed in the Earth's atmosphere after thermal neutrons produced by cosmic rays react with ^{14}N; ^{14}C then reacts rapidly with oxygen to form ^{14}CO, which itself oxidizes more slowly to form $^{14}CO_2$ (radioactive carbon dioxide). The $^{14}CO_2$ molecule behaves chemically like nonradioactive $^{12}CO_2$ and $^{13}CO_2$ and is dissolved in ocean water and taken up by plants. From there, ^{14}C moves up the food chain. It enters biological tissue and is eventually incorporated into calcite, aragonite, and apatite within shells, teeth, and skeletons.

4 The earliest known reference to lodestone is from the sixth century BCE Greek philosopher Thales of Miletus. The word "lodestone" is derived from the Middle English *lode*, meaning "way," a reference to the stone's use as an aid to navigation. The word "magnet" is probably derived from the ancient city of Magnesia in Anatolia (Asia Minor) where some lodestones were collected.

Radioactive ^{40}K constitutes 0.012 percent of naturally occurring K and is incorporated into any mineral that normally contains K (e.g., orthoclase, sylvite). Radioactive ^{87}Rb constitutes 27.8 percent of naturally occurring Rb and, as an alkali metal, is incorporated in trace amounts into minerals that are rich in Na and K.

Thorium and uranium have no stable isotopes and are incorporated into only a few minerals as major components, mainly thorianite (ThO_2) and uraninite (pitchblende) (UO_2). Monazite ($(Ce,La)PO_4$) can contain up to 2 percent Th and is the most important commercial source of the element. Besides uraninite, U occurs as a major constituent in a few rare minerals (e.g., coffinite, brannerite, davidite) and as a minor constituent in about a dozen other phases including autunite, carnotite, tobernite, uranophane and zeunerite. Both Th and U also commonly enter allanite ($(Ce,Ca,Y,La)_2(Al,Fe^{3+})_3(SiO_4)_3(OH)$) and zircon ($ZrSiO_4$) where their decay causes radiation damage to the crystal structure. This process, known as *metamictization*, can result in the grain becoming amorphous, less dense, softer, and darker. Surrounding grains of other phases may also suffer radiation damage and form pleochroic haloes visible in thin section.

An isotope becomes unstable when its atomic nucleus has too much internal energy; in most instances this is the result of an excessive number of neutrons. In the case of hydrogen, ^{1}H (one proton, no neutrons) and ^{2}H (deuterium (D), one proton, one neutron) are stable, but ^{3}H (tritium (T), one proton, two neutrons) is not. For rubidium, ^{85}Rb (37 protons, 48 neutrons) is stable, but ^{87}Rb (37 protons, 50 neutrons) is not. The slightly higher neutron/proton ratio in ^{87}Rb relative to ^{85}Rb (1.35 vs. 1.30) is sufficient to make the heavier isotope radioactive. Because the strong nuclear force that binds atomic nuclei is very short range, in large nuclei (those with $\geq$210 nucleons), the attractive force is only marginally stronger than the repulsive electromagnetic forces arising from like positive charges being in close proximity.

The principal modes of radioactive decay are:

> **Alpha decay** – emission of an alpha particle (two protons and two neutrons, identical to the nucleus of ^{4}He) plus gamma rays. Because of the strong nuclear force, alpha decay is inhibited. However, because the alpha particle can be considered a wave that is not completely localized, there is a very small probability that the alpha particle/wave will be present outside the nucleus, where the strong force is not a factor. This is an example of quantum tunneling or the tunnel effect, which permits alpha decay.

Beta decay – the emission of a beta ray (an electron or positron) as well as a neutrino from the nucleus by the transformation of one type of nucleon to the other. In beta minus decay, a neutron transforms into a proton in the nucleus, producing an electron and an electron anti-neutrino. In beta plus decay (a.k.a. positron emission), a proton transforms into a neutron, producing a positron and an electron neutrino. Beta decay is a result of the weak force; this allows a quark to change type from up to down or down to up via the exchange of a W boson and the production of an electron+antineutrino or a positron+neutrino pair. The change in the type of quark transforms protons to neutrons or vice versa: a proton consists of one down quark (with an electric charge of $-1/3$) and two up quarks (each with an electric charge of $+2/3$); a neutron consists of two down quarks and one up quark.

Positron emission – the transformation of a proton to a neutron plus a positron and a neutrino.

Electron capture – the capture of an electron by a proton in the nucleus and its transformation into a neutron.

Fission – the splitting of a large nucleus into two nuclei of substantial mass.

Radionuclides with lower neutron/proton ratios tend to undergo alpha decay, positron emission, or electron capture; those with higher neutron/proton ratios tend to undergo beta decay.

Radiometric dating utilizes the known rate of decay of unstable isotopes to their daughter products to obtain the length of time that has elapsed since the grain cooled below its *closing temperature* (the temperature below which there is a cessation of significant diffusion of parent or daughter isotopes out of the crystal). Accurate age determinations depend on gauging the amount of daughter product initially in the grain. It must also be assumed that the grain has not been altered (by heat, water, or mechanical stress) to the point where the concentrations of parent or daughter isotopes were significantly changed; any suspected alteration must be considered when evaluating uncertainties.

The age of the Solar System is often assumed to be the time elapsed since the formation of the first coarse solid materials in the nebula. These ancient materials would be the refractory inclusions (i.e., Ca-Al-rich inclusions or CAIs). The oldest (U-corrected) radiometric ^{207}Pb-^{206}Pb ages of refractory inclusions correspond to a brief interval of formation 4,567.30±0.16 Ma (Connelly et al. 2012).

When ^{238}U fissions, the massive daughter nuclei separate at high speeds, tunneling through the surrounding crystal. These damaged grains can be mounted and chemically etched to make the tracks visible in the optical microscope. The number of tracks per unit area depends on the initial U content and on the time that has elapsed since the grain cooled below its closure temperature. This technique, known as *fission track dating*, is commonly used with U-bearing minerals (e.g., zircon, apatite, mica) to obtain information on such topics as the thermal evolution of the Earth's continental crust, the formation ages of meteorites, tektites, moon rocks, and terrestrial impact craters, as well as the ages of archeological artifacts.

Solubility. Some mineral phases are soluble to one degree or another in water, hydrochloric acid (HCl), sulfuric acid (H_2SO_4), and/or nitric acid (HNO_3). Minerals that are readily soluble in water include chlorides (e.g., halite (NaCl); sylvite (KCl)), carbonates (e.g., nahcolite ($NaHCO_3$); trona ($Na_2CO_3 \cdot NaHCO_3 \cdot 2H_2O$)), sulfates (e.g., chalcanthite ($CuSO_4 \cdot 5H_2O$); halotrichite ($FeAl_2(SO_4)_4 \cdot 22H_2O$)), and nitrates (e.g., niter a.k.a. saltpeter (KNO_3)). Minerals that are readily soluble in HCl include arsenates (e.g., adamite (Zn_2AsO_4OH)), carbonates (e.g., aragonite ($CaCO_3$); calcite ($CaCO_3$); dolomite ($CaMg(CO_3)_2$); magnesite ($MgCO_3$); strontianite ($SrCO_3$)), chlorides (e.g., sylvite), hydroxides (e.g., brucite ($Mg(OH)_2$); inderite ($MgB_3O_3(OH)_5 \cdot 5(H_2O)$)), oxides (e.g., tenorite (CuO)), phosphates (e.g., purpurite ($MnPO_4$)), sulfates (e.g., chalcanthite; halotrichite; melanterite ($FeSO_4 \cdot 7H_2O$)), and vanadates (e.g., carnotite ($K_2(UO_2)_2(VO_4)_2 \cdot 3H_2O$)). Native iron is readily soluble in HCl and HNO_3; native copper is readily soluble in HNO_3 but only slightly soluble in HCl.

Field geologists often carry along a small eyedropper bottle of dilute HCl to verify if a rock is limestone (a sedimentary rock composed mainly of calcium carbonate). Calcium carbonate fizzes as it releases bubbles of CO_2 when HCl is dripped on it; the reaction is $CaCO_3 + 2HCl \rightarrow CO_2 + H_2O + CaCl_2$. Dolomite is less soluble and may fizz only if powdered.

Taste. Water-soluble minerals may dissolve in the mouth, react with taste receptors on the tongue, and produce a discernable taste. The most obvious example is halite (rock salt ($NaCl$)), which tastes salty. Other minerals which have noticeable tastes include hanksite ($Na_{22}K(SO_4)_9(CO_3)_2Cl$) – salty; sylvite ($KCl$) – salty, somewhat bitter; glauberite ($Na_2Ca(SO_4)_2$) – salty, somewhat bitter; epsomite ($MgSO_4{\cdot}7H_2O$) – bitter; ulexite ($NaCaB_5O_6(OH)_6{\cdot}5H_2O$) – alkaline; borax ($Na_2B_4O_7{\cdot}10H_2O$) – sweet, alkaline; melanterite ($FeSO_4{\cdot}7H_2O$) – sweet, astringent, metallic; soda niter ($NaNO_3$) – cooling; chalcanthite ($CuSO_4{\cdot}5H_2O$) – sweet, metallic (but slightly poisonous).

Odor. Minerals may have a perceptible odor caused by the escape of molecules from the surface of a crystal to the surrounding air where they can be wafted to the nose and react with olfactory receptors. Some weakly bonded minerals (i.e., those with van der Waals bonds) can have a noticeable odor at room temperature; some more-strongly bonded phases (e.g., those with ionic, covalent, or metallic bonds) may have discernable odors only if heated, moistened, struck, or powdered. A few prominent examples of odiferous minerals include (1) clays such as kaolinite ($Al_2Si_2O_5(OH)_4$) – earthy (argillaceous) odor, especially when moistened; (2) pyrite (FeS_2) – rotten egg (fetid) odor caused by emanating H_2S when heated or struck; and (3) arsenopyrite ($FeAsS$) – garlic odor, particularly when heated. (Arsenic-bearing minerals can emit toxic fumes when heated.) Odor and taste are the only diagnostic mineral properties that can be evaluated only by their ingestion.

Feel. Minerals that have some van der Waals bonding can break apart when rubbed; this property can be perceived as a greasy feel. Graphite has a greasy feel; its layers are bonded by van der Waals forces. The structurally related phyllosilicates, talc ($Mg_3Si_4O_{10}(OH)_2$) and pyrophyllite ($Al_2Si_4O_{10}(OH)_2$), have a greasy, waxy feel; adjacent layers in these phases are joined together by weak electrostatic and van der Waals forces.

Many fibrous minerals (which tend to have ionic, not van der Waals bonds) have a silky feel. For example, fibrous riebeckite, a variety known as crocidolite or blue asbestos ($Na_2Fe^{2+}{}_3Fe^{3+}{}_2Si_8O_{22}(OH)_2$), has just such a silky feel.

Efflorescence. The process of efflorescence derives its name from the French phrase "to flower out." Water can be lost from the structure of hydrated minerals, particularly under hot or dry conditions. Bound water can dissolve chemical salts (e.g., chlorides, fluorides, sulfates) within the crystal; the solution can migrate to the surface of the grain where the salts can precipitate as a powder. One such primary hydrated phase is chalcanthite ($CuSO_4{\cdot}5H_2O$); under dry conditions, it can lose water and precipitate small grains of the anhydrous mineral, chalcocyanite ($CuSO_4$), on the grain surface.

Related to efflorescence is the process of *transformation*: Gypsum ($CaSO_4{\cdot}2H_2O$) can lose water in the gas phase, transforming affected portions of the crystal into anhydrite ($CaSO_4$). Borax ($Na_2B_4O_7{\cdot}10H_2O$) can undergo dehydration and transform into tincalconite ($Na_2B_4O_7{\cdot}5H_2O$). Spontaneous dehydration can transform torbernite ($Cu(UO_2)_2(PO_4)_2{\cdot}12H_2O$) into metatorbernite ($Cu(UO_2)_2(PO_4)_2{\cdot}8H_2O$). Similarly, when autunite ($Ca(UO_2)_2(PO_4)_2{\cdot}10\text{–}12H_2O$) dries out, it can transform into meta-autunite ($Ca(UO_2)_2(PO_4)_2{\cdot}2\text{–}6H_2O$).

Exposure to visible light can transform realgar (As_4S_4) into pararealgar (As_4S_4). Although both phases are monoclinic, pararealgar occurs as a bright yellow powder and has a bright yellow streak; realgar is red or red-orange and has a red or red-orange streak. Pararealgar also has a somewhat lower specific gravity (3.52 vs. 3.56) and is slightly softer on the Mohs hardness scale (1–1.5 vs. 1.5–2).

Deliquescence. The opposite of efflorescence is deliquescence; this latter process occurs when a crystal absorbs water from moist air and dissolves. Deliquescent minerals are water-soluble phases that have high chemical affinities for water; they are often used as desiccants. Phases exhibiting deliquescence include chlorides and nitrates: tachyhydrite ($CaMg_2Cl_6{\cdot}12H_2O$), hydrophilite ($CaCl_2$), carnallite ($KMgCl_3{\cdot}6(H_2O)$), halite (NaCl), sylvite (KCl), and gwihabaite ($(NH_4,K)(NO_3)$), the ammonium analog of saltpeter (KNO_3).

Related to deliquescence is the process of *decomposition*: The most prominent example involves marcasite (FeS_2). Under conditions of high humidity, some marcasite specimens react with moisture and oxygen in the air to form encrustations of melanterite ($FeSO_4{\cdot}7H_2O$) via a reaction similar to

$$4\ FeS_2 + 13\ O_2 + 30\ H_2O \rightarrow 4\ FeSO_4{\cdot}7H_2O + 2\ H_2SO_4 + 2\ SO_2.$$

The sulfuric acid produced by this reaction can dissolve nearby grains. Mineral collectors have dubbed this process "pyrite decay," "pyrite rot," and "pyrite disease." (Pyrite is a polymorph of marcasite.)

4.3 Optical Properties

Optical properties of minerals are largely confined to microscopical observation and are potentially diagnostic. The topic is addressed in numerous optical mineralogy textbooks such as *Introduction to Optical Mineralogy* (Nesse 2012); *Optical Mineralogy: The Nonopaque Minerals* (Phillips and Griffen 1981); and *Introduction to Ore Microscopy* (Shrivastava and Rani 2012).

Refractive index. The speed of light in a vacuum is a constant, c, equal to 299,792,458 $m\ s^{-1}$. Light travels at a speed v that is less than c in all nonopaque substances, e.g., for visible light, v = ~0.9997c in air, ~0.763c in ice, ~0.752c in water, ~0.6596c in halite, ~0.658c in window glass, ~0.565c in sapphire and ~0.4137c in diamond. These values are only approximate because they can vary with a substance's composition, temperature, pressure, and the proportions of fractures, inclusions, and impurities.

The refractive index n of a material is the ratio of c to v, i.e., $n = c/v$. When light passes from one medium to another with a different value of n, the light bends (i.e., it refracts). The amount of bending is described by Snell's Law, the relationship between the angles of incidence and refraction: $(\sin\theta_2)/(\sin\theta_1) = v_2/v_1 = n_1/n_2$, where θ_1 is the angle of incidence, θ_2 is the angle of refraction, and both angles are measured relative to a line perpendicular to the interface between the two media.

The higher the refractive index of a substance, the greater the amount of bending (i.e., the smaller the angle θ_2). The refractive indices of the common materials mentioned previously are: $n_{air} = 1.000293$, $n_{ice} = 1.31$, $n_{water} = 1.330$, $n_{halite} = 1.516$, $n_{window\text{-}glass} = 1.52$, $n_{sapphire} = {\sim}1.77$, and $n_{diamond} = 2.417$. A twig in a brook that appears to bend at a certain angle because of the differences in n between air and water would appear significantly more bent if it were partly embedded in diamond instead of water.

Because every gemstone has a unique index of refraction, in principle, one need only measure the gemstone's angle of refraction (with a refractometer or with oils of known refractive index) to identify it. The longer the wavelength of light, the greater its velocity within a mineral; thus, red light travels faster than violet light. Because velocity v is inversely related to index of refraction n (Snell's Law), red light has a lower value of n than violet light. Modern laboratory refractometers use monochromatic light to obtain accurate determinations of a mineral's index of refraction.

The specific gravity G of a mineral is related to its refractive index n by the approximation: $[(n - 1)/G] = K$, where K is a constant depending on the mineral's composition. For olivine (a solid solution between forsterite (Mg_2SiO_4) and fayalite (Fe_2SiO_4)), the specific gravity increases linearly from 3.222 for forsterite to 4.392 for fayalite; the indices of refraction also increase from forsterite to fayalite. In the early 1980s, Brian Mason of the Smithsonian Institution used a standard set of refractive index fluids (immersion oils) to determine the olivine composition of Antarctic chondrites (as a rapid and inexpensive alternative to electron microprobe analysis).

The amount of light that is reflected as it passes from one medium to the next depends on the index of refraction and the angle of incidence (as described by the *Fresnel equations*). The higher the index of refraction and the smaller the angle of incidence, the higher the percentage of light that gets reflected. Diamond, with its very high index of refraction, reflects a significant fraction of incident light. The high index of refraction also accounts, in part, for diamond's brilliant adamantine luster.

Interference colors and birefringence. Transparent materials are either isotropic (i.e., uniform in all directions) or anisotropic (nonuniform). (Reflected light microscopy with plane-polarized light can often be used to distinguish isotropic and anisotropic minerals.) The isotropic group includes gases, liquids, glasses, and isometric (cubic) crystals such as diamond (C) and langbeinite ($K_2Mg_2(SO_4)_3$). The anisotropic group includes transparent members of the non-isometric crystal systems: triclinic, monoclinic, orthorhombic, tetragonal, trigonal, and hexagonal.

Isotropic materials have a single refractive index (n); light moving through these materials travels at the same velocity in all directions.

Anisotropic minerals have more than one refractive index; the velocity of light varies with crystallographic direction. A beam of light passing through an anisotropic mineral is divided into two polarized rays that vibrate in planes that are perpendicular to each other. One of the two rays, the *fast ray*, has a lower index of refraction (n_f) and moves faster through the crystal; the other ray, the *slow ray*, has a higher index of refraction (n_s) and moves more slowly.

Every anisotropic crystal has one or two orientations, called *optic axes*, in which a beam of light is not divided into two rays (a phenomenon known as *double refraction*). Those minerals with a single optic axis are tetragonal or hexagonal and are referred to as *uniaxial* crystals. Minerals with two optic axes are triclinic, monoclinic, or orthorhombic and are called *biaxial* crystals.

Thin sections of anisotropic minerals viewed with a petrographic microscope through crossed polarizers (and not along an optic axis) display *interference colors* due to transmitted light being divided into two rays. In the time it takes for the slow ray to pass through a mineral, the fast ray passes through that same distance plus an additional distance; the difference between these distances is called *retardation* (Δ). Once the light rays leave the mineral and reach open air, they travel at the same velocity (because air is isotropic) and the retardation value remains

Table 4.3 *Maximum birefringence values for common minerals*

Mineral	Birefringence	Mineral	Birefringence
Leucite	0.001	Tourmaline	0.025
Nepheline	0.005	Epsomite	0.028
Orthoclase	0.007	Forsterite	0.033
Corundum	0.008	Epidote	0.040
Quartz	0.009	Muscovite	0.045
Albite	0.011	Olivine	0.050
Anorthite	0.013	Biotite	0.060
Hypersthene	0.018	Calcite	0.172

constant. The value of Δ depends on the thickness of the mineral (d) and the difference between the indices of refraction of the slow ray and the fast ray (i.e., $n_s - n_f$) so that: $\Delta = d\,(n_s - n_f)$. Because petrographic thin sections have a ideal standard thickness of 30 μm, the differences in the Δ values among minerals in such sections will depend solely on ($n_s - n_f$), assuming that the mineral grains are not narrower than the thickness of the section.

The unitless quantity ($n_s - n_f$) is known as **birefringence** (δ). Different orientations of the crystal have different values for δ, ranging from 0 for the optic axis to some maximum value that is characteristic of the particular mineral. Maximum birefringence values for some common minerals are listed in Table 4.3; these values can be diagnostic.

For a particular orientation of the grain, the slow and fast rays may be of equal magnitude but vibrating in opposite directions. The waves will cancel each other out and the mineral will appear dark. This occurs when the retardation is a whole multiple of the wavelength: $\Delta = i\,\lambda$, where i is an integer and λ is wavelength.

In contrast, in those mineral orientations where the retardation is equal to one-half wavelength ($\Delta = (i + ½)\,\lambda$), the waves are vibrating in the same direction and constructively interfere. Light will pass through the upper polarizer.

Interference colors are produced when polychromatic light (i.e., white light) passes through the mineral and the crossed polarizers. For a particular retardation, some wavelengths will cancel each other out and others will constructively interfere to different extents. When one wavelength is canceled out, its complementary color becomes visible. First-order interference colors are produced when the path difference between the rays is 1λ; second-order colors are produced at 2λ, third-order colors at 3λ, etc.

Because interference colors in a mineral depend on only three factors (orientation, thickness, and birefringence), if the unknown specimen is of uniform thickness in a standard petrographic thin section and different grains of the specimen are viewed so that the maximum interference color is observed, then that color can be used to help identify the phase. All that is needed is a chart of interference colors (a.k.a. a Michel-Lévy Chart) that can be found online and in any optical mineralogy textbook.

Accessory plates can be inserted in the optical path in a petrographic microscope to help determine the order of interference colors. In these plates, the fast ray is parallel to the long dimension. The most commonly used plate is the gypsum plate (or full-wave plate); it can add or subtract an amount of retardation equivalent to one full order of interference color. It also gives the mineral a distinct magenta hue (the interference color between the first and second order in the chart). Other commonly used accessory plates are the quartz wedge and the mica plate (or quarter-wave plate).

Relief. The index of refraction (n) of an unknown mineral sample can be determined with a set of immersion oils, typically with an uncertainty of 0.003. If n differs between the sample and the immersion oil, the mineral appears to be at a different height from the oil. This appearance is *relief*. Differences in n between the mineral and the oil (Δn) determines the degree of relief: $\Delta n > 0.12$, high relief; $\Delta n = 0.04$–0.12, moderate relief; $\Delta n < 0.04$, low relief.

Positive relief occurs when $n_{\mathrm{mineral}} > n_{\mathrm{oil}}$; negative relief occurs when $n_{\mathrm{mineral}} < n_{\mathrm{oil}}$. An easy way to determine if the relief is positive or negative is to use *Becke lines*; these are rims of light at the perimeter of the grain. They can be observed microscopically with a medium-power objective lens when the image is slightly out of focus; when the distance between the objective lens and the sample is increased, the Becke lines move into the phase with the higher index of refraction.

Chagrin. Chagrin is the fine-scale surface roughness of a specimen, illuminated by reflection and refraction at the boundaries of small grains. It is apparent only in those cases where the mineral has an index of refraction differing significantly from the surrounding medium (be it epoxy or immersion oil). The larger the difference in n between the mineral and the surrounding medium, the more intense the chagrin. This property is best observed microscopically when the iris diaphragm is shrunk (reducing the amount of light in the optical path) and an intermediate-power objective lens is used (e.g., 10× to 25×). Minerals exhibiting strong chagrin appear to have a pitted surface.

Pleochroism. In plane-polarized light, some minerals will change color as the stage is rotated. This is due to the absorption by the mineral of different wavelengths of light at different orientations; as the slow ray and fast ray are differentially absorbed, different colors are produced. Common pleochroic minerals include tourmaline, hypersthene, cordierite, and spodumene. Olivine grains can show a pale green pleochroism. In some shocked meteorites, particularly shergottites, some mosaicized olivine grains have a brown color and are pleochroic (light-brown to clear); the brown color may be due to oxidation of ferrous iron at relatively high Po_2.

Extinction. Isotropic minerals and amorphous materials always appear dark under crossed polarizers. Anisotropic minerals appear dark if viewed along the optic axis; for any other orientation they appear dark once every 90° as the microscope stage is rotated. These are the orientations where light rays traversing the mineral are vibrating parallel to the vibration directions of the two polarizers. This dark appearance is known as *extinction*.

Another diagnostic property is *extinction angle* which can vary between 0° and 89°. This is the angle between the long dimension of the grain (or the principal cleavage direction) and the position where the grain undergoes extinction. The angle can be measured with the stage goniometer. If the grain goes extinct at 0° or 90°, it is said to have *parallel extinction*. Common minerals that exhibit parallel extinction include olivine, orthopyroxene, muscovite, zircon, apatite, tourmaline, and rutile. Grains that have a nonzero extinction angle are said to exhibit *inclined extinction*; clinopyroxene grains exhibit this type of extinction.

Minerals with two cleavages such as spodumene and K-feldspar exhibit two distinct extinction angles. Minerals that lack cleavage (e.g., quartz) or are not elongated do not have an extinction angle (although fractures in quartz are sometimes misidentified as cleavage planes).

Undulose extinction (a.k.a. *undulatory extinction*) occurs when a mineral has been deformed so that it exhibits a wavy, mottled pattern of extinction; different regions of such a grain appear dark at different angles than other regions of the same grain. Deformation can result from tectonic forces or via shock. Terrestrial minerals that commonly exhibit undulose extinction include quartz and feldspar; in meteorites, all coarse silicate phases can exhibit undulose extinction when weakly shocked.

Mosaic extinction occurs in shock-deformed grains (particularly olivine) that have been severely strained; such grains exhibit an extinction pattern resembling a stained glass window. Grains no longer exhibit sharp extinction under crossed polars and show mottled or mosaic extinction as the undeformed domains are smaller than the thickness of the section. The greater the degree of shock above a certain threshold, the greater the degree of misorientation of the domains (Carter et al. 1968; Stöffler et al. 1991).

A major criterion for the determination of the shock stage of chondritic meteorites is the type of extinction exhibited by olivine, low-Ca pyroxene and plagioclase grains (see Table 8.3). Shock stage can be determined by petrographic examination as well as by micro X-ray diffraction (Izawa et al. 2011) and bulk X-ray diffraction (Imae et al. 2019). For example, at higher levels of shock, plagioclase grains undergo increasing plastic deformation and strain-related mosaicity. These quantifiable changes are manifested by (a) increases in the range of the angles of undulatory extinction and (b) increases in the averaged full width at half maximum (FWHM) values of x-ray diffraction peaks (Pickersgill et al. 2015).

Sign of elongation. An additional optical property of minerals, one that is often listed in determinative tables, is the *sign of elongation*. In minerals with a positive elongation, the slow light ray is parallel to the long axis of the grain; these minerals are "length slow." In minerals with a negative elongation, the slow ray is perpendicular to the long axis of the grain; these grains are "length fast." The sign of elongation can be readily determined with an accessory plate. To find the sign of elongation, rotate the stage until the mineral goes extinct, then rotate the stage an additional 45° and introduce the accessory plate. The long axis of the mineral grain will be parallel to that of the accessory plate. If the interference color order decreases, the mineral is length slow and has a positive elongation. If the interference color order increases, the mineral is length fast and has a negative elongation.

Optic sign. The optic sign of an unknown mineral can be compared to those of known minerals in determinative tables to aid identification. To understand this property, it is necessary to examine the concept of the *optical indicatrix*. This is an imaginary three-dimensional geometric figure that illustrates the index of refraction (n) in any direction through a mineral. It is a sphere for isometric phases because n is the same in every direction; it is an ellipsoid for anisotropic phases because n varies with crystallographic direction.

Uniaxial minerals. The uniaxial indicatrix is an ellipsoid centered around the crystal's c axis. The distance from the center of the figure to the edge along the c axis is a function of the value of n parallel to the c axis and is labeled n_ε where ε stands for "extraordinary"; the distance from the center to the edge perpendicular to the c axis is labeled n_ω, where ω stands for "ordinary." For minerals with $n_\varepsilon > n_\omega$, the ellipsoid is prolate; if $n_\varepsilon < n_\omega$, the ellipsoid is oblate. The maximum birefringence is equivalent to $|n_\varepsilon - n_\omega|$.

When light enters an anisotropic mineral along a direction not parallel to the optic axis (i.e., not parallel to the *c* axis), it splits into two rays, vibrating in planes that are perpendicular to each other. One of these rays moves parallel to the optic axis; it vibrates in directions perpendicular to the *c* axis and is called the *ordinary ray* or *omega ray* (ω ray). This ray has a constant velocity through the crystal and has a constant refractive index: n_ω. The second ray travels in a different direction, not along the *c* axis; this ray is called the *extraordinary ray* or *epsilon ray* (ε ray). This light ray has different velocities in different directions as it travels through different angles of the atomic structure. For those extraordinary rays that are perpendicular to the ordinary ray, the associated refractive index is n_ε. Other extraordinary rays travel at intermediate velocities through random angles of the crystal and are associated with intermediate indices of refraction $n_{\varepsilon'}$. The intermediate birefringence associated with randomly oriented sections is equivalent to $|n_{\varepsilon'} - n_\omega|$.

The *optic sign* of a uniaxial mineral determines the shape of the indicatrix. The optic sign is positive for those minerals in which $n_\varepsilon > n_\omega$; the extraordinary rays are the slow rays and the ellipsoid is prolate. The optic sign is negative for those minerals in which $n_\varepsilon < n_\omega$; the extraordinary rays are the fast rays and the ellipsoid is oblate.

The optic sign of a uniaxial mineral can be determined by the following procedure: (1) Select a mineral specimen that shows minimal birefringence. (2) Place the condensing lens into the light path to achieve conoscopic illumination. This yields a convergent light cone that will focus to form an image above the objective. (3) Use a high-power objective lens to bring the mineral into focus. (4) Cross the polarizers. (5) Remove the ocular lens or insert the Bertrand lens. A centered optic-axis figure will appear that looks like a cross; this figure consists of two *isogyres* that flare out toward their outer edges. These figures are produced by the extinction of the conoscopic light as it traverses the upper polarizer. The center of the cross is called the *melatope*; this marks the *c* axis. (6) Insert a gypsum plate into the optical path. Optically positive minerals will show an increase by one full color order in the NE and SW corners of the isogyre pattern because the slow direction of these minerals is parallel to the slow direction of the gypsum plate. Optically negative minerals will show a decrease in color order in the NE and SW corners of the isogyre pattern; in these minerals the slow direction is perpendicular to the slow direction of the gypsum plate. Additional details can be found in any optical mineralogy textbook.

Biaxial minerals. The biaxial indicatrix is a triaxial ellipsoid whose axes can be labeled X, Y, and Z. X is the shortest axis; it is parallel to the vibration direction of the fast light ray and is associated with the minimum index of refraction (n_α). Z is the longest axis; it is parallel to the vibration direction of the slow light ray and is associated with the maximum index of refraction (n_γ). Y is the intermediate axis, associated with an intermediate index of refraction (n_β). The vibration directions of the fast and slow rays reside within the XY plane of the ellipsoid (the *optic plane*); this plane is also associated with the minimum and maximum indices of refraction (n_α and n_γ). Biaxial minerals exhibit maximum birefringence ($|\alpha - \gamma|$) when their optic plane is parallel to the microscope stage.

Triaxial ellipsoids contain two circular cross sections, each with a radius of n_β. Light rays that are perpendicular to the two circular sections travel along the two optic axes. The angle between the optic axes in the optical plane is called the *optic angle* or *2V*. The line that bisects the acute angle between 2V is the *acute bisectrix*; the line that bisects the obtuse angle between

2V is the *obtuse bisectrix*. In biaxial positive minerals, Z is the acute bisectrix and is associated with the vibration direction of the slow light ray. In biaxial negative minerals, X is the acute bisectrix and is associated with the vibration direction of the fast light ray.

The optic sign of a biaxial mineral can be determined by the following procedure: (1) Use a high-power objective lens to focus on the crystal. (2) Insert the upper polarizer. (3) Place the condensing lens into the light path to achieve conoscopic illumination. (4) Insert the Bertrand lens into the light path. An acute bisectrix figure is formed when the acute bisectrix is perpendicular to the stage. As the stage rotates, the isogyres change shape from a cross at the 0° and 180° positions to two separate curved bands that resemble mirror images of one another and achieve maximum separation at the 45° and 135° positions. At 45° and 135° the slow ray is oriented NE–SW and the fast ray is oriented NW–SE. (5) Insert the gypsum plate. For biaxial positive minerals, color order increases in the NE and SW quadrants (on the concave side of the isogyres) because the optic plane containing the slow ray is parallel to the slow-ray direction of the gypsum plate. For biaxial negative minerals, color order decreases in the NE and SW quadrants because the optic plane containing the slow ray is perpendicular to the slow-ray direction of the gypsum plate. Optical mineralogy textbooks can provide more detailed explanations.

Reflected-light color, reflection pleochroism, and bireflectance. Minerals display colors in plane-polarized reflected light: e.g., gold appears yellow-gold; copper appears pale pink to reddish copper; kamacite appears white or light blue-gray; silicates appear medium to dark gray. Opaque isometric minerals (e.g., gold, silver, copper, pyrite, galena) retain the same color as the stage is rotated; some non-isometric phases (e.g., covellite, CuS) display minor to moderate color changes (*reflection pleochroism*) as the stage is rotated.

Anisotropic minerals can vary in reflectivity as the stage is rotated; this is due to the angle changing between the light path through the microscope and the crystallographic directions in the mineral. Such changes in brightness are known as *bireflectance*. Minerals that exhibit reflection pleochroism are also bireflectant. Strongly bireflectant minerals include graphite (C), covellite (CuS), molybdenite (MoS_2), and stibnite (Sb_2S_3); moderately bireflectant minerals include marcasite (FeS_2), hematite (Fe_2O_3), pyrrhotite ($Fe_{1-x}S$), and cubanite ($CuFe_2S_3$).

Internal reflections. In minerals that are partly transparent, incident light can penetrate below the surface before being reflected by a fracture, small inclusion, cleavage plane, or internal grain boundary. Such internal reflections can have characteristic colors in reflected light and appear as brighter, diffuse regions of the grain. They are observable in plane-polarized light but are best seen at high magnification with the upper polarizer engaged. Many silicates and carbonates display white or colorless internal reflections combined with low reflectivity. Some opaque phases display internal reflections of different colors depending on grain composition or the abundance of fractures and inclusions. For example, only those sphalerite grains with low Fe display internal reflections. Opaque minerals with internal reflections include anatase (blue), azurite (blue), cassiterite (yellow, brown, white), cinnabar (red), chromite (brown), cuprite (red), goethite (yellow, red), hematite (red), malachite (green), orpiment (yellow, white), realgar (yellow, brown), rutile (yellow, red, brown, white), scheelite (white), sphalerite (yellow, red, brown, white), stibnite (red), and wolframite (brown).

5

Identification of Meteoritic Minerals in Reflected Light, by Backscattered Electron Imaging, and by Energy Dispersive X-Ray Spectroscopy, Wavelength-Dispersive X-Ray Spectroscopy, and Electron Backscatter Diffraction Analysis

5.1 Reflected Light

Many meteoritic minerals can be identified microscopically in reflected light; these include both opaque and nonopaque phases. The appearance of numerous opaque minerals is discussed in detail in two comprehensive books by Paul Ramdohr (1969, 1973).

Silicates. Although silicate phases are transparent (Figure 5.1), they often can be distinguished in reflected light. They all appear gray, but in general, the higher the silica content, the darker the shade of gray. For example, cristobalite and albite are darker gray than olivine or pyroxene. Phases with higher SiO_2 contents tend to have lower indices of refraction and lower specific-gravity values: e.g., cristobalite (100 wt% SiO_2; n = 1.484 to 1.487; G = 2.33), albite (68.74 wt% SiO_2; n = 1.528 to 1.542; G = 2.63), enstatite (59.85 wt% SiO_2; n = 1.650 to 1.679; G = 3.209), forsterite (42.71 wt% SiO_2; n = 1.636 to 1.772; G = 3.222). (The overall range in n given here is the maximum difference shown among the crystallographic directions in the biaxial minerals albite, enstatite, and forsterite; the range in n in uniaxial cristobalite reflects the difference between the ordinary and extraordinary rays.) In phases with lower n, more light is transmitted through the mineral and less is reflected; by definition, a substance with lower reflectivity is less bright, i.e., darker.

Minerals with high silica contents tend to have open structures with low densities (and relatively low coordination numbers). There are fewer electrons per unit volume in these low-density minerals to slow incident electromagnetic radiation; a higher proportion of incident light is transmitted (less is absorbed and reradiated), thus lowering the index of refraction.

The overlap in n between olivine and low-Ca pyroxene makes it difficult to distinguish these minerals from each other in reflected light by their color alone. Other petrographic features such as grain shape, grain size, and associated phases can be used to aid identification. Of course, the least ambiguous way to identify minerals is to analyze them by energy dispersive X-ray spectroscopy (EDS) used along with scanning electron microscopy (SEM) (i.e., SEM-EDS) or by electron probe microanalysis (EPMA); once the identity of a phase in a particular meteorite is established by SEM-EDS/EPMA, other similarly sized instances of it in the same meteorite can be identified with some confidence by reflected-light microscopy alone (using shade, shape, and associated phases).

Amphibole has been reported in a few iron meteorites, enstatite chondrites, and R6 chondrites. The hornblende grains in R6 chondrites are pleochroic in transmitted light and medium dark gray in reflected light. They are readily distinguished from anhydrous silicates because

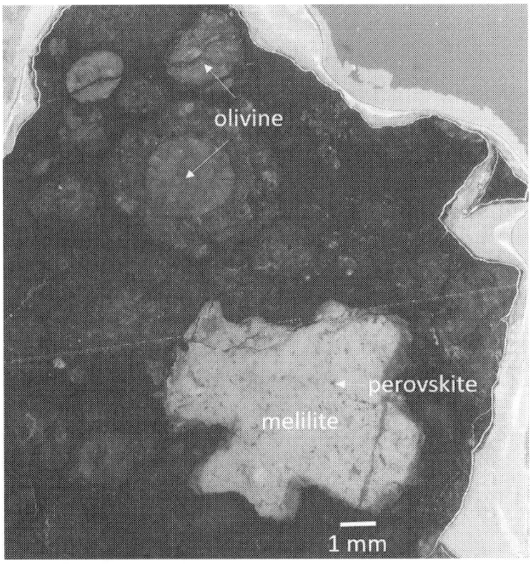

Figure 5.1 Olivine in chondrules and melilite in a CAI from the Efremovka CV3 chondrite. Reflected light. (A black-and-white version of this figure will appear in some formats. For the colour version, please refer to the plate section.)

Figure 5.2 Phenomenon of "silicate darkening" (a.k.a. "shock darkening)" observed in the Tenham L6 ordinary chondrite. Reflected light. (A black-and-white version of this figure will appear in some formats. For the colour version, please refer to the plate section.)

many of the hornblende grains contain irregular dehydrated patches consisting of 5–20-μm-long pyroxene blades flanked by dark pore spaces (Rubin 2014).

Shock effects. Some shocked chondrites appear dark or black, a phenomenon commonly called "shock darkening" or "silicate darkening" (Figure 5.2). In these rocks, metallic Fe-Ni and sulfide have been shock melted and injected into fractures in adjacent silicate grains where they form curvilinear trails of tiny opaque blebs. In some cases, the trails crosscut one another. In transmitted light, the affected silicate grains can appear very dark or as a pastiche of transparent areas traversed by dark trails of opaque phases. In reflected light, these silicate grains display trails of bright blebs; however, only those opaque grains situated at the surface will be visible in reflected light. This feature is a prominent indicator of shock.

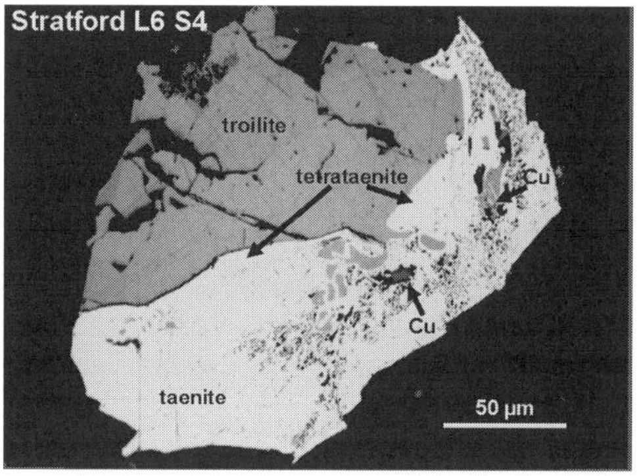

Figure 5.3 Metallic Cu in the shock stage S4 Stratford L6 chondrite. Also evident is a thin band of tetrataenite at the interface between troilite and taenite. Reflected light. (A black-and-white version of this figure will appear in some formats. For the colour version, please refer to the plate section.)

Metallic Fe-Ni. The main metallic Fe-Ni phases in meteorites are kamacite[1] (α-(Fe,Ni)), taenite (γ-(Fe,Ni)), tetrataenite (FeNi), and awaruite (Ni_3Fe). All are whitish, but kamacite is whiter (or a lighter blue-gray) and brighter than taenite; taenite appears creamier with a very slight yellow tinge. Kamacite and taenite are more readily distinguished when they are in contact. Tetrataenite is slightly yellower than kamacite and has a somewhat creamy color; it can be distinguished from taenite and kamacite because it is optically anisotropic. It occurs mainly at the outside boundary of taenite grains adjacent to silicate and troilite (Figure 5.3). Awaruite, present in some oxidized chondrites, particularly CV_{OxA} and R chondrites, appears white.

In lightly carbon-coated sections, metallic Fe-Ni grains, irrespective of Ni content, appear tan in reflected light; they appear brown under heavier carbon coats.

Metallic Fe-Ni grains in chondrites vary significantly in size and shape. Important examples include:

(a) ~5-μm-sized rounded blebs inside chondrules that crystallized during chondrule cooling,
(b) Submicrometer-sized blebs of kamacite with very low Ni (commonly <1 wt%) that formed via reduction of FeO in relict olivine phenocrysts during a chondrule-remelting event,
(c) ~200-μm-sized irregular grains in equilibrated OC that coarsened during thermal metamorphism,
(d) ~2–50-μm-thick shock veins (also commonly containing troilite) that can be of submillimeter to centimeter length in shocked chondrites,
(e) large nodules (up to 3 cm), in some cases associated with troilite, in shocked OC that may have formed by melting, vaporization, and fractional condensation,
(f) rare euhedral awaruite grains in chondrules in CV_{OxA} chondrites such as Allende,
(g) an occurrence of a 100-μm-sized euhedral tetrataenite grain (attached to an anhedral troilite grain of similar size) in an impact-melt-rock clast in the Jelica LL6 chondrite breccia (Rubin 1994a), and
(h) "swathing kamacite," i.e., 400–1,500-μm-thick rims of kamacite surrounding troilite, silicate, and schreibersite-rich inclusions in iron meteorites, pallasites and mesosiderites.

[1] Kamacite is not an IMA-approved name. The formal name for body-centered cubic low-Ni metallic Fe is "iron," but the name kamacite is so entrenched in the meteorite literature, it is used throughout this book.

When etched in Nital (a 2 percent solution of nitric acid in ethanol), the compositionally zoned, border regions of taenite grains adjacent to kamacite appear dark and are often called "cloudy zoned taenite." These regions generally contain ~30–40 wt% Ni and consist of a fine-grained, foamy intergrowth of kamacite and taenite. In many instances, there are clear, unetched regions on both sides of the cloudy zoned taenite. The thin band adjacent to kamacite is tetrataenite.

Martensite is metallic Fe-Ni, typically with ~8–15 wt% Ni, that has a distorted body-centered cubic structure (α_2). It forms from taenite during rapid cooling (i.e., $\gamma \rightarrow \alpha_2$), usually associated with a shock-heating event. When etched with Nital, martensite generally appears riddled with dark brown to black needles or thin plates; in different occurrences, these plates can be packed close together or spread somewhat apart.

Plessite[2] is a mixture of kamacite and taenite and/or tetrataenite. Buchwald (1975) described several varieties of plessite (e.g., spheroidized, acicular, comb, net, cellular) that vary in grain size and grain shape and depend on the Ni content of the original martensite. Wood (1964) concluded that plessite forms by the decomposition of martensite (i.e., $\alpha_2 \rightarrow \alpha + \gamma$). However, Goldstein and Michael (2006) found that parent taenite crystals were retained at the boundaries of martensite laths during cooling after martensite formation; they proposed a more-complex reaction: $\gamma \rightarrow \alpha_2 + \gamma \rightarrow \alpha + \gamma$. Comb plessite does not form from martensite at all; it is just a micro-Widmanstätten pattern that nucleates at lower temperatures.

Shock can melt troilite and cause it to mix with and melt adjacent kamacite grains; shock can also shatter individual grains of kamacite and troilite, transforming them into polycrystalline assemblages. The polycrystallinity of the troilite can best be detected in reflected light after inserting the upper polarizer and rotating the microscope stage. The percentage of polycrystalline kamacite and troilite grains varies in shocked ordinary chondrites from <5 percent to >95 percent.

Moderately shocked kamacite grains that are at least 20 μm across can develop Neumann lines (a.k.a. Neumann bands); they are evident in etched sections. These bands are plate-shaped mechanical twin lamellae with white interiors and thin dark boundaries; they range from 1 to 10 μm in width and can extend to several centimeters in length in hexahedrites. A single kamacite crystal can include Neumann lines oriented in up to 12 different directions.

Kamacite grains that experience shock pressures exceeding 130 kb can develop an hexagonal (HCP) ε-iron structure that inverts to deformed kamacite on cooling. In etched sections, this structure has a moderately dark crosshatched pattern. It typically occurs as subparallel striations oriented in several directions.

Copper. Copper appears pale pink in freshly polished sections and darker reddish-copper in older sections that have somewhat tarnished or oxidized surfaces. Many metallic Cu occurrences are in shocked (or shocked and annealed) equilibrated ordinary chondrites (e.g., Rubin 1994b); the Cu typically occurs inside metallic Fe-Ni grains in association with small irregularly shaped grains of troilite. In L/LL6 Trysil (shock stage S3), metallic Cu is present as an elongated band between kamacite and troilite within a shock vein. In L6 Sixiangkou (shock stage S5), metallic Cu occurs as rare 0.5-μm-thick rinds at the outside edges of small horseshoe-shaped metallic Fe-Ni cells within coarse grains of shock-melted troilite.

Phosphides. The main phosphide in meteorites is schreibersite ($(Fe,Ni)_3P$). It is a white phase, slightly grayer (with a pale creamy rose tint) than kamacite and browner than cohenite.

[2] Martensite and plessite are not IMA-approved names, but both terms are entrenched in the meteorite literature (and, in the case, of martensite, the metallurgical literature as well).

Schreibersite can be most readily identified when it is in contact with metallic Fe-Ni (as is most often the case). It appears in different forms, including patches adjacent to kamacite in enstatite chondrites, in association with sulfide in many irons, as rhabdites (prismatic or plate-shaped crystals) surrounded by metal in many hexahedrites and octahedrites, and as round nodules within metallic Fe-Ni in a few impact-melt clasts in ordinary chondrites. (Octahedrites and hexahedrites are structural types of iron meteorites discussed in Section 8.5).

Sulfides. Troilite (FeS) and pyrrhotite ($Fe_{1-x}S$) appear tan to yellow-brown; some pyrrhotite occurrences are a creamy color with a pink tint. Pyrite (FeS_2) is a light yellow; chalcopyrite ($CuFeS_2$) is light golden yellow. Pentlandite ($(Fe,Ni)_9S_8$) is a light creamy yellow. Unfortunately, daubréelite ($FeCr_2S_4$), alabandite (MnS), ferroan alabandite ((Mn,Fe)S), niningerite ((Mg,Fe)S), and keilite ((Fe,Mn,Mg,Ca,Cr)S) in enstatite meteorites can have similar gray appearances and are usually difficult to distinguish microscopically. Nevertheless, daubréelite in many cases occurs as light to medium gray exsolution lamellae and as thick patches within troilite grains.

Joegoldsteinite ($MnCr_2S_4$), the Mn analog of daubréelite, is a rare phase, so far identified only as small grains in the Social Circle IVA iron and in EH4 Indarch (Isa et al. 2016). Its color and reflectivity are very similar to those of daubréelite. The two phases can best be distinguished by electron-microprobe analysis.

Oldhamite (CaS) also appears gray to pink in freshly polished sections, but over time, it often becomes blemished, tarnished, and somewhat brown or greenish as it reacts with oxygen and moisture in the air. Oldhamite is also a transparent phase (verifiable in transmitted light), unlike other meteoritic sulfides.

In lightly carbon-coated sections, troilite and pyrrhotite appear dark brown or dark blue in reflected light.

Oxides. Chromite ($FeCr_2O_4$) appears steel gray with a slight yellow tint; spinel ($MgAl_2O_4$) is commonly pink (Figure 5.4), ulvöspinel (Fe_2TiO_4), brownish gray; ilmenite ($FeTiO_3$), gray with a pinkish tinge. Magnetite (Fe_3O_4) has a bluish gray color, in some cases with a brown tint;

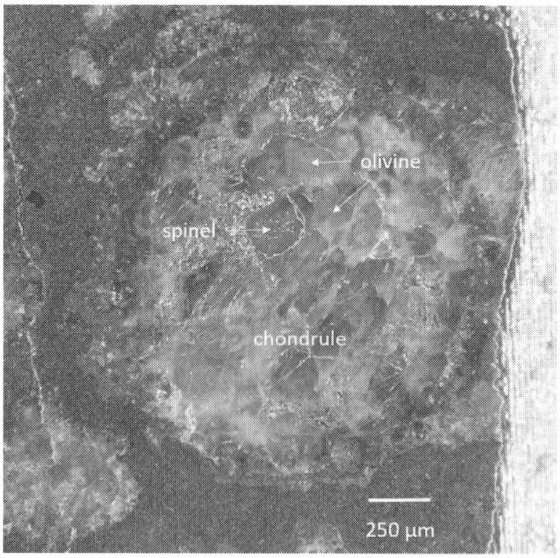

Figure 5.4 Pink spinel in a chondrule from the Allende CV3 chondrite. Reflected light. (A black-and-white version of this figure will appear in some formats. For the colour version, please refer to the plate section.)

rutile (TiO_2) appears bluish-white and is commonly associated with chromite and ilmenite. A Mg-Al-chromite clast in the Raguli H3.8 chondrite contains a couple of light-gray grains of pyrophanite ($MnTiO_3$) (Krot et al. 1993).

Weathering products consisting mainly of limonite, goethite (α-FeO(OH)), or lepidocrocite (γ-FeO(OH)) are light gray with a bluish tint; they generally appear smooth with no fractures. In mildly weathered ordinary chondrites, these phases occur at the edges of metallic Fe-Ni grains and form thin veinlets traversing silicate grains. In heavily weathered samples, there can be complete replacement of metallic Fe-Ni and sulfide grains as well as the formation of millimeter-thick limonite veins up to several centimeters in length. Wlotzka (1993) devised a weathering scale for OC based mainly on the oxidation of metallic Fe-Ni and the formation of limonite.

Chromite is present mainly as (1) coarse isolated grains; (2) blocky intergrowths with metallic Fe-Ni, troilite, and/or silicate; (3) wormy, blocky, and irregular grains in chromite-plagioclase assemblages (formed by shock); and (4) thin laths (with and without plagioclase or feldspathic glass) in narrow veinlets (also a shock product) within olivine grains.

Magnetite in aqueously altered carbonaceous chondrites occurs as platelets, plaquettes, spheroids, and framboids, either as individual grains or in clusters. In CV_{OxA} chondrites such as Allende, magnetite is present as both small and large grains; some grains are anhedral, others occur within ellipsoidal-shaped inclusions. In CK4 and CK5 chondrites, magnetite occurs (a) in large submillimeter-sized nodules intergrown with sulfide (pentlandite, pyrite, chalcopyrite, and mackinawite), chlorapatite and, in some cases, ilmenite, and (b) as small grains, a few micrometers in size, associated with sulfide (mainly pentlandite) and dispersed through the recrystallized matrix.

Graphite. Graphite grains typically display strong reflection pleochroism. In meteorites, graphite occurs most commonly inside grains of kamacite as black polygonal or spherulitic grains, V-shaped connected laths, or irregular aggregates of randomly oriented laths. There are dark rounded nodules in some irons that consist of intergrown graphite and troilite. In some cases, the nodules have troilite cores surrounded by graphite; in others, the troilite and graphite are intergrown and the nodules appear black with a yellow tinge.

Elongated euhedral graphite grains (some with pyramidal terminations) occur in enstatite-chondrite impact-melt breccias, a few relatively unshocked ureilites (carbon-rich olivine-pyroxene achondrites) and a few shock-melted clasts in ordinary chondrites. Many of the euhedral graphite grains in the enstatite-chondrite impact-melt breccias are completely enclosed within silicate; these grains appear medium gray and some have a yellow tinge. A few graphite grains surrounded by silicate occur in assemblages of two or three laths, many with V-shaped morphologies.

In some iron meteorites and enstatite chondrites, graphite occurs within kamacite as compact cubic or pseudohexagonal forms, dubbed *cliftonite*. A few cliftonite grains exhibit sector zoning. Buchwald (1975) described some graphite inclusions in anomalous irons as "basket weave" graphite consisting of mutually perpendicular stubby graphite grains.

Carbides. Cohenite ($(Fe,Ni)_3C$) appears white; it is orthorhombic and moderately anisotropic. Haxonite ($(Fe,Ni)_{23}C_6$) is a similar color but may appear slightly darker; it is cubic and isotropic. Edscottite (Fe_5C_2), which has been found so far only as small lathlike grains in the Wedderburn IAB-Complex iron meteorite, is white.

Nitrides. Osbornite (TiN) is a rare phase with a gold-yellow color; it is similar in brightness to gold. Most grains are very small and have been identified mainly in shock-melted portions of enstatite meteorites. Carlsbergite (CrN) is present in some low-Ni IIIAB irons as small oriented laths and irregular grains. It is white to gray with a rose tint, and, because of its hardness, stands in prominent relief over adjacent metallic Fe-Ni grains.

5.2 Backscattered-Electron Imaging with Energy-Dispersive X-Ray Spectroscopy, Wavelength-Dispersive X-Ray Spectroscopy, and Electron Backscatter Diffraction Analyses

Minerals can be distinguished in backscattered-electron (BSE) images and by fast elemental analysis using a scanning electron microscope or electron probe with an energy-dispersive X-ray spectroscopy (EDS) system, and, if needed, by structural analysis using electron backscatter diffraction (EBSD) in the SEM. Quantitative elemental analysis of minerals is normally carried out using the electron probe with the wavelength-dispersive X-ray spectroscopy (WDS) system. Nowadays, field emission scanning electron microscopy (FESEM) is commonly used for high-resolution electron imaging down to 1 nm resolution.

The mean atomic number Z of a substance is positively correlated with increasingly lighter gray levels. Differences of 0.1 Z are discernable. Iron has a higher atomic number ($Z = 26$) than the other abundant elements in the most common silicate and oxide minerals in meteorites. Consequently, BSE images and Fe X-ray maps appear very similar (Figure 5.5). Minerals that contain even small amounts of FeO will have a moderately high mean atomic number and appear brighter in BSE images. Whereas the mean atomic number of enstatite (Fs_0) is 10.6, that of Fs_5 low-Ca pyroxene is 10.9. The mean atomic number of forsterite (Fa_0) is also 10.6; that of Fa_5 olivine is 11.0. Albite (10.7) and silica (10.8) both appear darker than Fa_5 olivine or Fs_5 low-Ca pyroxene in BSE images. Anorthite (11.9) is as bright as Fa_{16} olivine or Fs_{21} low-Ca pyroxene. Ferrosilite (16.9) and fayalite (18.7) are much brighter.

Chromite (19.9) and troilite (22.4) are substantially brighter than the common silicate phases. However, the orthosilicate zircon ($ZrSiO_4$) is brighter still (24.8). Kamacite with 6.8 wt% Ni

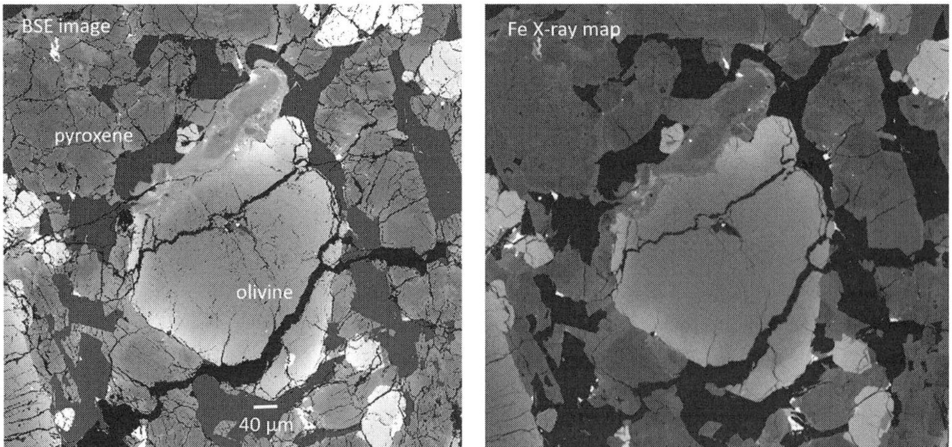

Figure 5.5 BSE and Fe X-ray map showing an olivine phenocryst in the Tissint shergottite.

and 0.5 wt% Co has a mean atomic number of 26.1. Taenite with 40.0 wt% Ni and 0.3 wt% Co has a mean atomic number of 26.8 and can readily be distinguished from kamacite in BSE images (if brightness and contrast controls are properly adjusted).

Graphite, carbide, and phosphide grains are commonly associated with metallic Fe-Ni. Their mean atomic numbers are substantially lower than kamacite and taenite and are distinguishably darker in BSE images: graphite ($Z = 6$), Ni-free cohenite (24.7), Ni-free haxonite (24.9), Ni-free schreibersite (24.3). In contrast, metallic Cu (29), which is commonly associated with small, irregular troilite grains in shocked ordinary chondrites, appears brighter than kamacite, taenite, and troilite in BSE images (and is readily identifiable in reflected light).

The meteoritic phase with the brightest BSE characteristics is metallic Hg ($Z = 80$). Caillet Komorowski et al. (2012) published FESEM and BSE images of nanometer-sized grains of native Hg associated with cinnabar (HgS), metallic Cu, and Cu-sulfide in the Tieschitz H/L3.6 ordinary chondrite.

Elemental analysis by EDS in the SEM or electron probe at micrometer-to-submicrometer scales along with SEM imaging (Figure 5.6) are relatively easy and fast techniques for mineral identification, revealing the identity of most phases along with their chemical formulas. However, for minerals of the same composition, crystal structure analysis by diffraction methods is required.

EBSD in the SEM is an excellent method for *in situ*, nondestructive structure analysis and confirmation. It requires well-polished sections. (EBSD is a technique that provides crystallographic information about a crystal's structure, orientation or strain.) For example, EBSD revealed that there are actually three silica phases (quartz, tridymite, and cristobalite) in the Zakłodzie achondrite (Ma et al. 2012a, 2012b). EBSD has become a critical tool for new mineral investigations in meteorites since 2007 (Ma 2015). Figure 5.7 shows the EBSD pattern of the recently approved mineral davisite from CV_{OxA} Allende with indexing to the diopside-type C2/c structure.

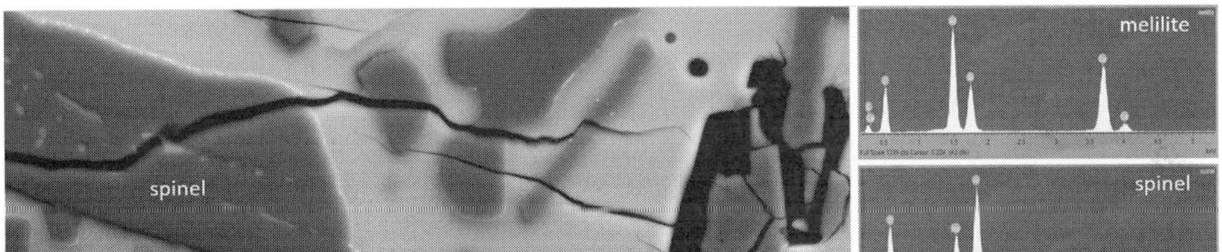

Figure 5.6 BSE image showing melilite, spinel, grossite, and grossmanite in a CAI from Allende. EDS analysis was used for phase identification. (A black-and-white version of this figure will appear in some formats. For the colour version, please refer to the plate section.)

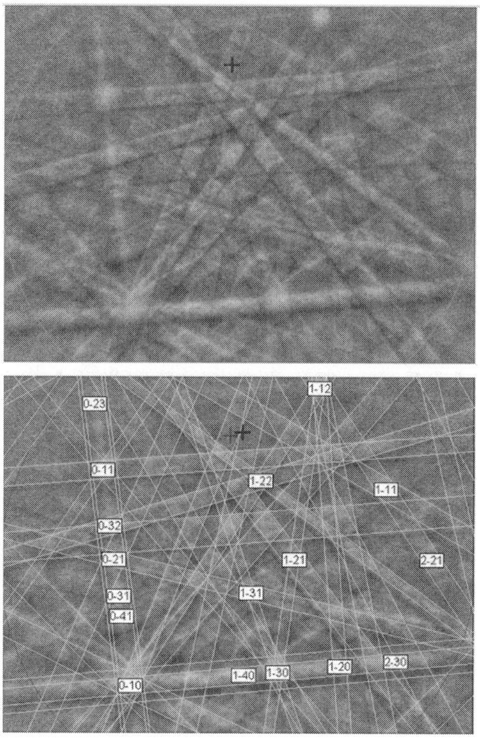

Figure 5.7 EBSD pattern of davisite ($CaScAlSiO_6$) from Allende, indexed to the *C*2/*c* diopside structure.

Over the past two decades, with advanced electron-beam and X-ray techniques, particularly high-resolution analytical scanning electron microscopy, ~170 new meteoritic minerals have been found. Since 2007, a nanomineralogy investigation of meteorites led by Chi Ma at Caltech has revealed more than 40 new minerals. Each newly identified extraterrestrial mineral represents a formation environment with a distinct set of intensive variables (e.g., composition, temperature, pressure, fo_2). Its discovery provides new insights into nebular, parent-body, or shock processes in the early Solar System.

Nanomineralogy is the study of Earth and planetary materials at nanoscales. It focuses on the characterization of nanofeatures (e.g., inclusions, exsolution, zonation, coatings, pores) in minerals and rocks, revealing nanominerals and nanoparticles (Ma 2015). The advanced high-resolution analytical scanning electron microscope (FESEM with EDS and EBSD) and electron probe microanalyzer, allows easy and rapid characterization of solid materials down to nanoscales. Nanofeatures have been discovered in many common minerals and rocks, revealing their physical properties and providing insights into their genesis (Ma and Rossman 2006). New minerals and new materials of important geological significance have been discovered at micro- to nanoscales (e.g., Ma et al. 2012c, 2013c, 2014a). Nanomineralogy has come to play a unique role in Earth and planetary science research.

6
Meteorite Classification and Taxonomy

6.1 History

The first significant step in meteorite classification was taken in 1802 by British chemist Edward Howard who showed that the meteoritic stones and irons he analyzed all contained Ni. This criterion was used to distinguish meteorites from terrestrial rocks, but the reliability of this distinction was soon undermined by two essentially metal-free falls: the Stannern eucrite in the Czech region of the Austrian Empire in 1808 and the Luotolax howardite in Finland in 1813. It became apparent to researchers that, although most stony meteorites contained metallic Fe-Ni, a few did not. This realization somewhat hampered subsequent classification efforts because a few large collections included a handful of metal-free terrestrial rocks posing as meteorites. Some of these "meteorwrongs" were covered with dark crusts, probably desert varnish or weathering rinds.

German chemist Martin Klaproth formally divided stones from irons in 1807. The next major advance was taken in 1843 by Paul Partsch, curator of the Vienna Imperial Collection. He also distinguished stones and irons, but went further by subdividing each group. Irons were classified as "compact" irons (i.e., normal iron meteorites) or as "spongy" or "knotty" irons that contained stone (i.e., stony-iron meteorites). The two principal types of stones were classified as "normal" or "anomalous." There were two varieties of normal stones: Mg-rich rocks (mainly ordinary chondrites) and Mg-poor rocks (such as Stannern and Luotolax). The anomalous stones included Chassigny and the two carbonaceous chondrites known at the time (Alais and Cold Bokkeveld).

In 1862, Gustave Rose at the Mineralogical Museum of the University of Berlin created a detailed classification system, inventing many of the terms and categories used today. He divided irons from stones. Irons were placed into three categories: (1) almost pure iron, (2) pallasites, and (3) mesosiderites. Rose divided stones into seven categories: (1) chondrites (specimens containing the well-known spherulitic objects he dubbed "chondrules"), (2) howardites, (3) chassignites (with only a single member at the time – Chassigny), (4) chladnites (now known as *aubrites* or *enstatite achondrites*, also with only a single member – Bishopville), (5) shalkites (now known as *diogenites*, again with only one member – Shalka), (6) carbonaceous chondrites, and (7) eucrites (from the Greek for "clearly determinable" because their principle minerals – augite and anorthite – were readily identified). Rose also subdivided the chondrites into groups on the basis of color or structure.

In 1863, M. H. Nevil Story-Maskelyne, professor of mineralogy at Oxford and Keeper of Minerals at the British Museum, divided meteorites into aerolites (stones), aerosiderolites (stony-irons), and aerosiderites (irons), the three basic structural categories used today.

Gustav Tschermak at the University of Vienna modified the Rose classification system in 1872 and 1883. He divided the irons by structure into octahedrites, hexahedrites, and ataxites. He subdivided the octahedrites by Widmanstätten pattern into fine, medium, coarse, and curvilinear varieties. (This pattern had been discovered independently by William Thomson in 1804 and Alois von Widmanstätten in 1808.)

Between 1885 and 1904, Aristides Brezina, also at the University of Vienna, made changes to Tshermak's classification system. He was a splitter, creating more than 70 meteorite categories, based mainly on minor distinctions. Secondary characteristics (e.g., those resulting from metamorphism or shock) were used to separate related samples from one another. This led to considerable confusion. For example, in 1929, George Merrill, head curator of geology at the Smithsonian Institution, used the Rose-Tschermak-Brezina classification system to list 11 distinct groups of chondrites: howarditic, white, intermediate, gray, black, spherulitic, crystalline, carbonaceous, orvinite, tadjerite, and "urelite." Many of these groups were then subdivided on the basis of texture.

Needed refinements to the classification scheme were made in the early and mid-twenthieth century by Oliver Farrington at the Field Museum (Farrington 1915), George Prior at the British Museum (Prior 1929, 1953), and Brian Mason at the American Museum (later at the Smithsonian Institution) (Mason 1962).

In 1953, Harold Urey and Harmon Craig at the University of Chicago compiled superior chemical analyses of chondrites from the literature and showed that the data could be arranged into two distinct groups that differed in total iron content and oxidation state. They called the chondrite groups H (high total iron) and L (low total iron) and concluded that they likely originated on different parent bodies. In 1962, Klaus Keil, then at the University of California, San Diego, reproduced the two groups based on specific gravity and modal abundance of metallic Fe-Ni. Two years later, Keil and Kurt Fredriksson (UCSD) analyzed olivine and orthopyroxene in 86 ordinary chondrites by electron microprobe, found distinct differences between the H and L groups and identified a third OC group – LL (low total iron, low metallic Fe). [Since that time, electron microprobe analysis (EPMA) has been the main method for characterizing meteorite samples.]

As meteorites began to be analyzed by advanced analytical techniques and many more samples were recovered from cold and hot deserts in the mid- to late-twentieth century, new meteorite groups were identified by the collective efforts of researchers around the world. These workers established new carbonaceous chondrite groups, split enstatite chondrites into two groups (EH and EL), and identified a new class of chondrite (R), apparently related to OC. Lunar meteorites were first identified in the early 1980s, and about the same time, a consensus was building that shergottites, nakhlites, and Chassigny (SNC samples) were from Mars.[1]

[1] Lunar meteorites resemble Apollo and Luna samples in mineralogy, mineral chemistry, texture, and bulk chemical and isotopic compositions. Analyses of the first recognized lunar meteorite (ALH 81005 – the last Antarctic meteorite collected in the 1981–1982 field season) left little doubt that this rock (e.g., Mason 1982; Marvin 1983), and its subsequently discovered congeners, were lunar. The discovery of lunar meteorites rendered more plausible the idea that some other achondrites could have found their way to Earth from Mars. Meteorites now accepted as martian are all igneous rocks, indicating that they are

The fall of Angra dos Reis in Brazil in 1869 and the recovery of a few related meteorites in Antarctica in the mid- to late-1980s helped establish angrites as a distinct achondrite grouplet (and within a few years, a full-fledged group).

In the last two decades of the twenthieth century and the first two decades of the twenty-first, John Wasson of UCLA and his colleagues analyzed about 750 iron meteorites by instrumental neutron activation analysis (INAA). They placed the irons into more than a dozen distinct groups based on their bulk elemental compositions. Most groups were deemed "magmatic irons"; their compositional trends on element/element diagrams (e.g., Ir vs. Ni) are consistent with formation in asteroidal cores by fractional crystallization. Two iron-meteorite groups (IAB Complex; IIE) have element/element trends inconsistent with fractional crystallization and were dubbed "nonmagmatic irons." They may not have formed in the cores of differentiated asteroids.

The classification scheme used below is the modern one, based on the examination of more than 70,000 meteorites in the world's collections. These samples are all cataloged in the online Meteoritical Bulletin Database: www.lpi.usra.edu/meteor/metbull.php. The list is frequently updated to reflect the latest additions.

6.2 Taxonomy

Meteorites are derived from several kinds of bodies (Figure 6.1): (1) primitive, undifferentiated (i.e., chondritic) asteroids; (2) evolved, differentiated, or partially differentiated asteroids; (3) the Moon; and (4) Mars. Although not usually included in lists of meteorite sources, some micrometeorites and interplanetary dust particles are probably from comets. There is no doubt that samples designated as lunar meteorites are from the Moon and very little doubt that those categorized as martian meteorites hail from Mars.

There is a broad consensus that typical howardite, eucrite, diogenite (HED) samples are from asteroid 4 Vesta. This view is based on several factors: (1) There are spectral similarities of HED meteorites to (a) Vesta (e.g., McCord et al. 1970; Consolmagno and Drake 1977; McSween et al. 2013), (b) a large family of spectrally related small asteroids, i.e., the vestoids (Binzel and Xu 1993; Binzel 2001; Burbine et al. 2001), and (c) a few near-Earth asteroids such as 3551 Verenia, 3908 Nyx, and 4055 Magellan (e.g., Cruikshank et al. 1991). (2) The 505-km-diameter Rheasilvia crater in the southern hemisphere of Vesta is a plausible source for the vestoids; it is estimated that about 1 percent of the volume of Vesta was excavated by the impact. (3) The orbits of the vestoids extend between the orbit of Vesta and gravitational escape hatches (such as the Jovian 3:1 resonance at ~2.5 AU and the ν_6 resonance) that can perturb vestoids toward the inner Solar System. (4) The computed orbit of the Bunburra Rockhole eucrite fall is similar to that of Earth-crossing asteroids.

from a differentiated body. Most have relatively young crystallization ages (Chassigny and the nakhlites formed ~1.3 Ga ago; the shergottite QUE 94201 formed only 0.3 Ga ago), suggesting that these rocks came from a body far larger than any asteroid (and thus one that cooled much more slowly). A plausible candidate is Mars, which has large, relatively young volcanoes. One of these enigmatic achondrites is EETA79001; when this sample was launched off its parent body, portions of the rock were melted, trapping small amounts of gas from the surrounding atmosphere. The melted portions of the rock quenched into veins and pockets of dark glass. Analysis of the gas trapped in the glass showed it to be identical to the martian atmosphere (in CO_2, N_2, ^{40}Ar, ^{36}Ar, ^{20}Ne, ^{84}Kr, and ^{132}Xe) as measured at the martian surface by the Viking landers in 1976 (Bogard and Johnson 1983; Pepin 1985). Similar measurements of gas trapped in impact-produced glass were made later in other martian meteorites.

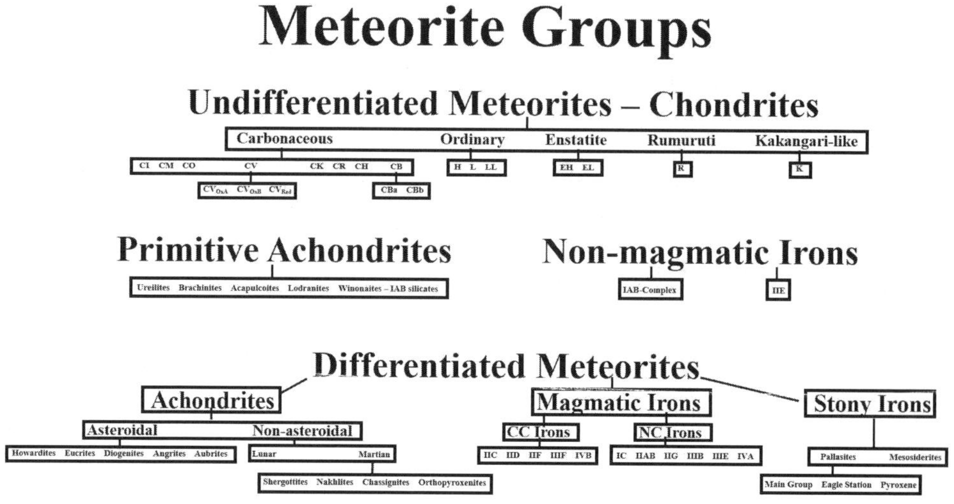

Figure 6.1 Classification of meteorite groups.

It is also clear which parent body provided samples of the Almahata Sitta polymict ureilite that were recovered from the Sudan after the crash of asteroid 2008 TC_3 on October 7, 2008 (Jenniskens et al. 2009). Nearly two dozen howardite specimens were recovered in Botswana after the crash of asteroid 2018 LA on June 2, 2018.

No other individual meteorites or meteorite groups can be assigned to a specific parent body. No meteorites in our collections are known to have been derived from Mercury, Venus, the moons of Mars, or the moons of the giant planets. No meteorites are suspected of originating outside the Solar System, although many primitive chondrites contain presolar grains with just such an origin.

Caveats: (1) Calculations suggest the possibility that meteorites derived from Mercury (Love and Keil 1995; Gladman and Coffey 2009), Venus (Dones et al. 2018), or Phobos and Deimos (Wiegert and Galiazzo 2017) could reach Earth. (2) It is also possible that terrestrial rocks were blasted off Earth and landed on the Moon; some could be present as small clasts among Apollo samples.[2] There might be as much as 200 kg of terrestrial material per km^2 on the lunar surface delivered as meteorites (Armstrong et al. 2002). (3) The recognition that the elongated dark object 1I/2017 U1 Oumuamua is an interstellar interloper on a hyperbolic trajectory traveling at one point in excess of 87 km s^{-1} raises the possibility that collisions of such bodies with meteoroids in the Solar System could ultimately produce meteorites composed of extra–Solar System material. A second interstellar interloper (2I/Borisov), a roughly kilometer-sized comet

[2] Warren et al. (1983) described a 1.8-g felsite clast with major quartz, K-feldspar, and shock-melted glass, and traces of fayalite, ferrohedenbergite, and ilmenite in Apollo 14 breccia 14321g as a lunar granite. Bellucci et al. (2019) later found zircon in the clast and suggested that it was part of a terrestrial meteorite blasted off Earth ~4 Ga ago. However, the sample is not strongly shocked, contrary to what would be expected from a rock launched from Earth (e.g., Rubin 2015b). In addition, the siderophile and chalcophile element concentrations in the rock are typical of Apollo 14 samples and far lower than those in terrestrial rocks. Volatile trace metals Zn, Ga, and Ge are depleted in 14321g below terrestrial granite averages by factors of 27, 2.0, and 18, respectively (Warren and Rubin 2020). It seems very likely that the sample is lunar in origin.

Figure 6.2 Comet 2I/Borisov passing through the solar system. These two images, taken by NASA's Hubble Space Telescope, show the comet appearing near a background galaxy (left) and shortly after perihelion (right). Credits: NASA, ESA, and D. Jewitt (UCLA). (A black-and-white version of this figure will appear in some formats. For the colour version, please refer to the plate section.)

(Figure 6.2), was discovered in 2019; it is on a hyperbolic trajectory (e = 3.08), traveling appreciably faster than Solar System escape velocity. Its tail was ~160,000 km long in late November 2019; it reached perihelion on December 8, 2019. In March 2020, Borisov suffered two consecutive outbursts, consistent with fragmentation of the nucleus. Statistical arguments suggest that, at any one time, there may be ~0.1–0.2 such interlopers (with $r > 100$ m) per cubic astronomical unit within the orbit of Neptune. Calculations imply that several times a century, submeter-sized interstellar meteoroids will encounter the Earth. One candidate is a 500-kg meteoroid that produced a fireball near Papua New Guinea on January 8, 2014; it appears to have entered the Earth's atmosphere on a hyperbolic trajectory at about 44 km s^{-1}, exceeding the escape velocity of the Solar System at 1 AU. (However, the uncertainty in the meteoroid's velocity casts some doubt on its interstellar credentials.)

A *meteorite group* (e.g., H chondrites; angrites; IIIAB irons) comprises samples with a limited range of compositional, textural, and mineralogical characteristics, thought to be consistent with origin from a single parent object. *Bulk chemical composition*: The difference between the highest and lowest values for the CI-normalized Ir/Ni ratios in H-chondrite falls is 0.17, whereas that for all OC falls is 0.56 (Kallemeyn et al. 1989). *Bulk isotopic composition*: The difference between the highest and lowest values of bulk $\Delta^{17}O$ among equilibrated chondrites in each OC group is ~0.3–0.5‰, whereas that among all equilibrated H, L, and LL chondrites is ~0.9‰ (Clayton et al. 1991; Greenwood et al. 2017). *Texture*: The difference between the highest and lowest values in the mean apparent diameter of chondrules (excluding microchondrules) in different LL3.0-3.4 chondrites is ~90 μm (Nelson and Rubin 2002), whereas that among all chondrite groups is ~750 μm (excluding Kakangari-like chondrites and the possibly non-nebular CH chondrites for which only limited data are available) (Friedrich et al. 2015). *Mineral chemistry*: The difference between the highest and lowest values for the mean olivine Fa content among equilibrated chondrites in individual OC groups is ~3–6 mol%, whereas that among all equilibrated H, L, and LL chondrites is ~15 mol% (Rubin 1990). Nevertheless, there are anomalous members of many meteorite groups that could have been derived from separate parent bodies. This is consistent with the spectral characteristics of asteroids that indicate there may be several bodies that are plausible parents of individual meteorite groups (e.g., Burbine 2014; Vernazza et al. 2015).

By convention, five members are the minimum necessary to form a *meteorite group*, but this number is not sacrosanct. Many researchers consider the Kakangari-like (K) chondrites a proper group even though, as of this writing, there are only four members. Unique meteorite specimens are considered *ungrouped* (e.g., the Acfer 094, Adelaide, El Médano 100, and NWA 7821 carbonaceous chondrites; the Al Huwaysah 010, Divnoe, NWA 5400, and NWA 7325 achondrites; the Alikatnima, Guin, Lazarev, and Tucson irons). Related sets of two to four meteorites are often considered *grouplets*;[3] these include the Coolidge and Loongana 001 carbonaceous-chondrite grouplet (Kallemeyn and Rubin 1995), and metal-rich chondrites (sometimes labeled G chondrites; Weisberg et al. 2015) such as NWA 5492, GRO 95551, and NWA 12379 (and its paired specimen, NWA 12273) (e.g., Ivanova et al. 2019a). In time, grouplets tend to turn into groups as new specimens come to light.

Two carbonaceous-chondrite groups have recognized *subgroups* with distinctive textural and mineralogical characteristics: CV chondrites (CV_{OxA}, CV_{OxB}, CV_{Red}) and Bencubbin-like CB chondrites (CB_a, CB_b). All members of a group, irrespective of subgroup, are generally considered to be from the same parent body; the differences among the subgroups are attributed to parent-body processes. Subgroups are roughly analogous to varieties in botany (e.g., the Bisbee spinystar, Desert spinystar, and Kaibab spinystar varieties of pincushion cactus) or to breeds of dog (e.g., Chihuahua, Bulldog, Beagle) or cattle (e.g., Holstein, Hereford, Guernsey).

CV_{OxA} chondrites include Allende, Axtell, ALH 84028, and Tibooburra; secondary phases in CV_{OxA} chondrules include magnetite, Ni-rich sulfide, ferroan olivine (Fa_{30-60}), nepheline, and sodalite. CV_{OxB} chondrites include Bali, Grosnaja, Kaba, and Mokoia; secondary phases in CV_{OxB} chondrules include phyllosilicate, magnetite, Ni-rich sulfide, fayalitic olivine (Fa_{95-100}), and hedenbergite. CV_{Red} chondrites include Arch, Efremovka, Leoville, Vigarano, and QUE 93429. (Although Vigarano is a breccia containing oxidized and reduced lithologies, most of the rock appears to be reduced.) Metallic Fe-Ni is moderately abundant and secondary phases are rare to absent in CV_{Red} chondrites.

CB_a chondrites include Bencubbin, Weatherford, Gujba, Fountain Hills, NWA 1814, and Quebrada Chimborazo 001; they are metal-rich objects containing abundant kamacite globules ranging from several millimeters to about a centermeter in size, large light-colored silicate globules with cryptocrystalline, barred pyroxene, and barred olivine textures, and rare Ca-Al-rich inclusions (CAIs). CB_b chondrites include Hammadah al Hamra 237, MAC 02675 and QUE 94411 and paired samples; these rocks are metal rich and contain ferromagnesian chondrules ranging from less than a millimeter to about a millimeter in size and moderately high abundances of CAIs.

A *meteorite clan* is a higher order than a group; most clans consist of two or more groups originating from separate asteroids that formed in close proximity in the solar nebula (in the case of chondrites) or from separate regions of the same differentiated or partially differentiated body

[3] Nevertheless, some iron meteorites are listed as "ungrouped" even if it is clear they are from the same parent asteroid as one or two other ungrouped irons (e.g., Butler and NWA 859) and should thus be called a grouplet.

(in the case of achondrites, irons, and primitive achondrites). Clan members share numerous compositional, mineralogic, and textural characteristics. The chondrite clans include four non-carbonaceous sets: (1) H, L, and LL ordinary chondrites; (2) Rumuruti chondrites (R chondrites); (3) EH and EL enstatite chondrites; and (4) Kakangari-like chondrites (K chondrites). Chondrite clans also include four carbonaceous sets: (5) CM-CO, (6) CV-CK, (7) CR-CH-CB, and (8) CI. There are ungrouped chondrites associated with most clans. [Some workers (e.g. King et al., 2019; Suttle et al., 2020) have suggested that some ungrouped carbonaceous chondrites constitute a new group labeled CY. These meteorites, mainly from Antarctica, contain abundant Fe sulfides, large grains of periclase, and dehydrated phyllosilcicates.] Primitive achondrite clans include: (1) ureilites, (2) acapulcoites and lodranites, (3) brachinites and related ungrouped meteorites (NWA 6112, MIL 090206, Divnoe), and (4) winonaites and IAB silicate inclusions. The achondrite clans include: (1) shergottites and related samples, nakhlites, chassignites, and orthopyroxenites [martian meteorites, formerly SNCs (shergottites, nakhlites, and Chassigny)], (2) HED samples, (3) lunar meteorites, (4) aubrites, and (5) angrites. Each magmatic iron group is its own clan except for the single clan of IIAB and IIG (which probably originated in the same asteroidal core; Wasson and Choe 2009). The non-magmatic irons consist of two clans: (1) IAB Complex (which might represent more than one parent asteroid; Wasson and Kallemeyn 2002) and (2) IIE.

A *meteorite class* consists of one or more clans that formed in the same general region of the Solar System. The members of a class may be related by one or more whole-rock properties including texture, mineralogy, inclusion abundance, oxidation state, bulk chemistry, and isotopic composition. There are four major chondrite classes: (1) ordinary chondrites (including chondritic inclusions from the IIE-an irons Netschaëvo and Mont Dieu and a unique chondritic clast in the NWA 10214 LL3 chondrite) and R chondrites, (2) enstatite chondrites, (3) carbonaceous chondrites, and (4) Kakangari-like chondrites. There are two major iron-meteorite classes: (1) carbonaceous-chondrite-derived irons (CC irons) – IIC, IID, IIF, IIIF, IVB and (2) noncarbonaceous-chondrite-derived irons (NC irons) – IAB, IC, IIAB, IIE, IIG, IIIAB, IIIE, IVA. [Iron-meteorite groups can also be divided into "magmatic" irons (all groups other than IAB and IIE) and "non-magmatic" irons (IAB and IIE).] Pallasites comprise a class consisting of (1) main-group pallasites, (2) Eagle Station pallasites, and (3) pyroxene pallasites. Mesosiderites form their own class.

There are two separate sets of meteorites of a higher order than class that could be called a "superclass," a term introduced here. Their group members formed on different asteroids, but presumably in the same general locale of the Solar System.

The first superclass is that of enstatite meteorites, which includes (1) EH and EL chondrites, (2) ungrouped and anomalous enstatite chondrites, (3) aubrites (members of which may have originated on more than one parent body), (4) a few reduced metal-rich meteorites, possibly derived from the mantle of the principal aubrite parent asteroid, (5) various reduced impact-melt breccias and impact-melt rocks, and (6) a few meteorites that could be partial-melt residues ultimately derived from enstatite chondrites (Rubin 2015a). It is unclear why some of these bodies melted and others did not. It is possible that the differentiated bodies accreted earlier and acquired higher concentrations of ^{26}Al; alternatively, they could have suffered more intense collisional events and melted.

The second superclass contains (1) HED achondrites, (2) mesosiderites, (3) main-group pallasites, and (4) IIIAB irons. These samples are all differentiated meteorites. They have similar O-isotopic compositions and ultimately formed from material derived from similar (possibly neighboring) O-isotopic reservoirs in the solar nebula. Mesosiderites contain eucrite-like and diogenite-like clasts admixed with abundant metal, but they are unlikely to be from the same parent asteroid as the HEDs. This is because about 40 percent of the large silicate clasts in mesosiderites are gabbros with extremely fractionated rare-earth element (REE) abundances; this rock type is apparently absent among HED samples (Rubin and Mittlefehldt 1992, 1993). [Nevertheless, Haba et al. (2019) argued that mesosiderite silicates, like HEDs, are from Vesta. In contrast, Peplowski et al. (2013) found no clear evidence from *Dawn* spacecraft data that mesosiderite material is present on Vesta.] If HED meteorites are indeed from Vesta, a body that is still intact, then IIIAB irons and main-group pallasites must be from one or two different parent asteroids that disrupted hundreds of millions of years ago (based on their cosmic ray exposure ages; e.g., Eugster 2003). [If HEDs are not from Vesta, it is possible they were derived from the same (disrupted) parent body as IIIABs and/or main-group pallasites.]

There are several meteorites, variously listed as eucrites, anomalous eucrites, and ungrouped achondrites (e.g., NWA 011, Ibitira and NWA 2824, Asuka 881394, NWA 5721) that may be from different parent asteroids than the HEDs. These unusual rocks differ from normal eucrites in texture, bulk chemical composition, and/or bulk O-isotopic composition. Some may hail from unrelated differentiated asteroids and hence not belong to the same superclass as the HED meteorites, mesosiderites, main-group pallasites, and IIIAB irons.

The only higher-order classification division is that which distinguishes differentiated from undifferentiated meteorites. It does not seem to be the case that differentiated meteorites are all from bodies that formed closer to the Sun than the parent asteroids of undifferentiated meteorites. For example, CC iron meteorites may have formed beyond the orbit of Jupiter (Warren 2011a, 2011b; Scott et al. 2018), whereas aubrites probably formed near enstatite chondrites, possibly at $\leq$1.4 AU (Nakashima et al. 2006) or even $\leq$1.0 AU (e.g., Sears 1980; Wasson 1985, 1988; Scott 2007; Rubin 2010a). This indicates that asteroid differentiation occurred in different regions of the Solar System. On the other hand, the distribution of taxonomic types of large asteroids (those with diameters exceeding 50 km) shows that at distances greater than ~3.6 AU, the undifferentiated asteroid types P and D dominate (Gradie and Tedesco 1982). These bodies may have been scattered into the asteroid belt from the Kuiper Belt when the giant planets went through a dynamical instability (Vokrouhlicky et al. 2016). The P and D asteroids have low albedos and featureless reddish spectra. They are modeled as having surfaces composed of organic-rich silicates, anhydrous silicates, and carbon; they may contain water ice in their interiors. Tagish Lake (an ungrouped C2 carbonaceous chondrite) has a reflectance spectrum resembling those of D asteroids. Brown et al. (2002) suggested it might be derived from a cometary meteoroid.

7

Mineralogy of Major Physical Components of Chondrites

7.1 Chondrules

Chondrules are among the most abundant components of chondrites, constituting, for example, 55 vol% of CR carbonaceous chondrites and 65–75 vol% of type-3 OC (Table 7.1). Their average apparent diameters range from ~150 μm in CO3 chondrites to ~900 μm in CV3 chondrites; the overall range in diameter is 0.25 μm to ~5 cm. The most common chondrules in all chondrite groups are ferromagnesian objects, rich in olivine and/or low-Ca pyroxene. There are five common textural types of ferromagnesian chondrules.

(1) Porphyritic, including porphyritic olivine (PO); porphyritic pyroxene (PP); and porphyritic olivine-pyroxene (POP). Porphyritic textures feature large crystals (phenocrysts) surrounded by fine-grained material or glass; these textures are common in igneous rocks. Porphyritic chondrules are the most abundant chondrule textural types (Figure 7.1). PP chondrules contain <10 vol% olivine; PO chondrules contain <10 vol% pyroxene; POP chondrules contain ≥10 vol% olivine and ≥10 vol% pyroxene. In many cases, the mafic silicate grains in porphyritic chondrules are approximately equant and show crystal faces; some grains have skeletal morphologies indicative of rapid cooling. In some finer-grained chondrules, the olivine ± low-Ca-pyroxene grains are rounded. Opaque phases (mainly kamacite) occur commonly as small blebs or droplets within the mesostasis.

(2) Barred olivine (BO). These chondrules are characterized by subparallel continuous or discontinuous plates of olivine separated by patches of mesostasis. Many BO chondrules are surrounded by a thin olivine spherical shell (a rim in two dimensions) approximately perpendicular to the plates (Figure 7.2). In some cases, there are two or more sets of plates oriented in different directions; some BO chondrules have pentagonal outlines. Barred olivine chondrules are sometimes classified as a variety of porphyritic chondrule. Some BO chondrules contain significant pyroxene and are called barred olivine-pyroxene (BOP) chondrules.

(3) Radial Pyroxene (RP). The chondrules consist of laths of low-Ca pyroxene arranged in fan-like arrays radiating from a point at or near the chondrule surface (Figure 7.3). In some cases, there are two or more radiation points. A few RP chondrules contain non-radiating subparallel laths and are texturally analogous to barred olivine chondrules; these objects are sometimes called barred pyroxene (BP) chondrules. In RP and BP chondrules in unrecrystallized chondrites, mesostasis with its included opaque phases is present between the laths and bars. These are nonporphyritic chondrules.

Table 7.1 *Average properties of chondrules, CAIs plus AOIs, and matrix material in different chondrite groups*

Group	Matrix (vol%)	CAIs (vol%)	Chondrules (vol%)	Chondrule Apparent Diameter (μm)	Chondrule Textural Types (number percent relative to the set of all chondrules in each group)						
					POP	PP	PO	BO	RP/BP	C	GOP/GP
CI	99	<0.01	0	–	–	–	–	–	–	–	–
CM	60	1.2	20	270	~95*				0.6	1	
CO	34	1.0	40	150	69	18	8	2	2	1	<0.1
CV	35	3.0	45	900	~94*			6	0.2	0.1	<0.1
CR	42	0.12	55	700	96–98*				0.3	0.4	
CK	50	4	15	900	>99*				<1	<1	
R	42	0.04	40	400	~92*				0.4	1.2	
H	12	0.02	70	450	47	21	16	6	3	4	3
L	12	0.02	70	500	47	21	16	6	3	4	3
LL	12	0.02	70	550	47	21	16	6	3	4	3
EH	8	0.01	70	230	4	77	0.1	0.1	13	5	1
EL	10	0.01	70	500	~87*			~0.1	6	7	

Matrix abundance data from Rubin (2011) and Rubin and Kallemeyn (1993); CAI data modified from Rubin (2011) and Scott and Krot (2014); chondrule abundance data modified from Scott and Krot (2005); chondrule size data from Friedrich et al. (2015); chondrule textural types modified from Rubin (2000, 2010a) and Scott and Krot (2014). Unfilled boxes = not determined. Apparent diameter is the diameter measured in thin section; it is less than the true diameter. Among ordinary chondrites, chondrule textural types were measured only for the LL group (Nelson and Rubin, 2002); values listed for H and L chondrites are assumed to be the same as those of LL chondrites.

Chondrule textural types: POP = porphyritic olivine-pyroxene; PP = porphyritic pyroxene; PO = porphyritic olivine; BO = barred olivine; RP = radial pyroxene; BP = barred pyroxene; C = cryptocrystalline; GOP = granular olivine-pyroxene; GP = granular pyroxene.

* Value is the total for all textural types of porphyritic chondrules.

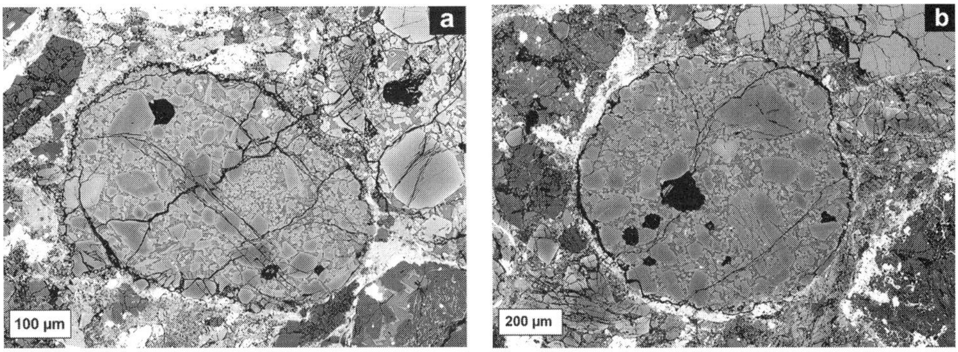

Figure 7.1 Porphyritic olivine chondrules in the Semarkona LL3.00 chondrite. BSE images.

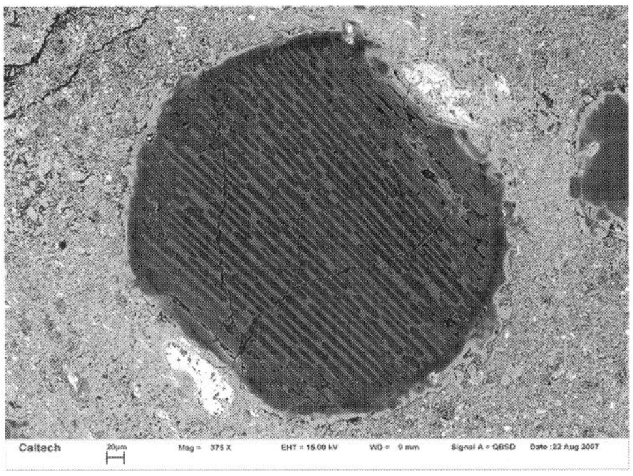

Figure 7.2 A barred olivine chondrule from CV3 Allende. BSE image.

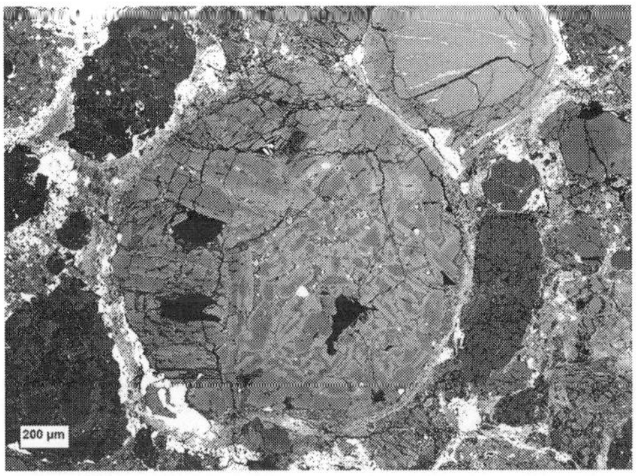

Figure 7.3 Porphyritic pyroxene chondrule (center) and radial pyroxene chondrule (upper right) from the Semarkona LL3.00 chondrite. BSE image.

(4) Cryptocrystalline (C). Cryptocrystalline chondrules consist of numerous patchy domains containing $\leq$2-μm-sized low-Ca pyroxene grains and associated mesostasis. Due to their fine grain size, they typically appear darker in plane-polarized transmitted light than RP chondrules. These also are nonporphyritic chondrules.
(5) Granular olivine-pyroxene (GOP). Granular olivine-pyroxene chondrules consist of densely packed anhedral olivine and/or low-Ca pyroxene grains and small patches of associated mesostasis (with included opaque phases). Although some of these objects have microporphyritic textures, they are generally considered nonporphyritic chondrules.

There are two structural/chemical types of porphyritic chondrules: Type I chondrules (Low-FeO) and Type II chondrules (High-FeO). They differ in the FeO/(FeO+MgO) ratios in their mafic silicates; the cutoff between Type I and Type II was initially set at 10 mol% Fa or Fs, but reset by Grossman and Brearley (2005) to 2 mol%.

Type I chondrules tend to have smaller phenocrysts and glassy mesostases with blebs of kamacite. Some have poikilitic textures (in which large grains of one mineral contain at least several smaller grains of one or more other minerals). The coarse pyroxene grains tend to be clinoenstatite exhibiting polysynthetic twinning. Some Type I POP chondrules are enriched in low-Ca pyroxene phenocrysts at their inside margin; the long axes of these pyroxene grains tend to be subparallel to the chondrule surface. (In the literature, the following nomenclature is often used: Type I PO = Type IA; Type I PP = Type IB; Type I POP = Type IAB).

Type II chondrules generally have larger euhedral phenocrysts surrounded by a darker, glassy to microcrystalline mesostasis containing few (or no) blebs of metallic Fe-Ni. Additional opaque phases in the mesostasis can include troilite and chromite. The olivine grains are characterized by higher P concentrations as well as higher FeO/(FeO+MgO) ratios; many grains show strong compositional zoning with core-to-rim increases in the concentrations of FeO, Cr_2O_3, MnO, and CaO. (Type II PO = Type IIA; Type II PP = Type IIB; Type II POP = Type IIAB).

Synthetic-chondrule crystallization experiments have shown that porphyritic chondrules crystallized from incomplete melts that included nuclei and relict mafic silicate grains. In contrast, the BO, RP, and C chondrules (which are sometimes called "droplet chondrules" because of their typical near-spherical shapes) appear to have formed by quenching of totally molten silicate droplets (although there are exceptions; e.g., Rubin 2013a).

In nearly every case, chondrule mesostases are feldspathic or silico-feldspathic, i.e., they are rich in normative plagioclase or plagioclase plus silica. Overgrowths of Ca pyroxene flank low-Ca pyroxene grains in many PP and POP chondrules; crystallites of Ca pyroxene also occur as isolated or clumped, quasi-acicular grains in the mesostasis. In Type IIAB and IIB chondrules, there are commonly four pyroxene phases: (1) low-Ca clinopyroxene, (2) orthopyroxene, (3) pigeonite, and (4) Ca-pyroxene (typically augite). Tiny grains of merrillite are present in some porphyritic chondrules.

Porphyritic chondrules are more friable than droplet chondrules, breaking readily along grain boundaries when mechanical stress is applied. The proportion of porphyritic chondrules decreases with increasing chondrule size interval, suggesting that large porphyritic chondrules are more easily fragmented by parent-body processes than droplet chondrules. Many of the chondrule fragments in ordinary chondrites have recognizable common textures. It is very likely that, in OC, isolated coarse silicate grains in the matrix, including mono- and polycrystalline assemblages, are derived from mechanically disrupted porphyritic chondrules.

A few percent of the chondrules in ordinary chondrites are compound objects consisting of two or more conjoined chondrules. There are three principal varieties: (1) *enveloping compound chondrules*, in which the inner primary chondrule is surrounded by a spherical shell (typically with a porphyritic texture); (2) *sibling compound chondrules*, bilobate objects consisting of two partly merged chondrules of similar size, and (3) *adhering compound chondrules*, consisting of a large primary chondrule and one or more (in rare cases, more than a dozen) small droplet chondrules that are attached to (and commonly spread out along) the surface, more closely resembling mounds than spheroids. Many BO chondrules have two or more nested olivine rims, presumably representing chondrule surfaces that were later coated with fine-grained material that melted during subsequent chondrule-heating events.

Microchondrules are defined arbitrarily as chondrules <40 μm in diameter. They occur in several principal settings: (1) as abundant objects within FeO-rich fine-grained silicate-rich rims around normal-sized low-FeO chondrules in ordinary chondrites, (2) as the most common chondrules in rare clasts in type-3 ordinary chondrites and OC regolith breccias, (3) as rare objects in clasts in some primitive carbonaceous chondrites, (4) as rare individuals in the fine-grained matrices of many type-3 ordinary and carbonaceous chondrites, (5) as common objects in CH chondrites such as ALH 85085, and (6) as rare objects within fine-grained micrometeorites. Chondrule rims contain two types of microchondrules: (a) abundant low-FeO RP and C objects rich in low-Ca pyroxene and (b) relatively rare high-FeO-olivine-rich objects (e.g., Krot et al. 1997a). Rubin et al. (1982) estimated that microchondrules constitute 0.02 vol% of H3, 0.006 vol% of L3, 0.0005 vol% of LL3, 0.04 vol% of CO3, and 0.001 vol% of CV3 chondrites. Microchondrules are most abundant in the two chondrite groups (H and CO) with the smallest mean diameters of normal-sized chondrules.

A small proportion of normal-sized chondrules have unusual mineralogy. These include silica-rich chondrules (some containing merrihueite or roedderite), chromite-rich chondrules, Al-rich chondrules, and glass-rich chondrules (a variety of Al-rich chondrule).

Silica-rich chondrules in ordinary chondrites consist mainly of ferroan low-Ca pyroxene and silica (probably cristobalite) with minor to accessory amounts of Ca-pyroxene, glass, merrillite, metallic Fe-Ni and troilite (Brigham et al. 1986). In one anomalous porphyritic chondrule in H/L3.9 Bremervörde, silica (~40 vol%) occurs as long laths in the chondrule interior (mainly on one side of the chondrule) and as small, more-equant grains at the chondrule margin. Low-Ca pyroxene (~60 vol%) is compositionally zoned; it is magnesian ($Fs_{15}Wo_{0.1}$) near the chondrule margin and appreciably more ferroan ($Fs_{65}Wo_2$) toward the center. A radial pyroxene chondrule in this same meteorite contains much less silica (~10 vol%), occurring as randomly oriented dendrites as well as lamellae located between laths of low-Ca pyroxene. Several ordinary chondrites also contain silica-pyroxene-fayalite clasts that may be silica-rich chondrule fragments.

Merrihueite ($(K,Na)_2(Fe,Mg)_5Si_{12}O_{30}$) and roedderite ($(Na,K)_2Mg_5Si_{12}O_{30}$) are the Al-poor members of the osumilite group of minerals; these phases have a framework structure consisting of SiO_4 tetrahedra forming hexagonal double rings. Some silica-bearing porphyritic chondrules and clasts (likely chondrule fragments) in ordinary chondrites contain merrihueite or roedderite (Krot and Wasson 1994). One chondrule consists of coarse low-Ca pyroxene and silica phenocrysts and accessory amounts of merrihueite, fayalitic olivine, ferroan low-Ca pyroxene, troilite, and feldspathic mesostasis. Another chondrule has large low-Ca pyroxene phenocrysts poikilitically enclosing small grains of silica and roedderite.

Chromium-rich chondrules are rare objects, constituting <0.1 percent of OC chondrules (Krot and Rubin 1993). They exhibit a wide variety of textures – porphyritic, granular, barred, radial, and cryptocrystalline. There are two types of Cr-rich chondrules: (1) Chromian-spinel chondrules consist of chromian-spinel ($(Mg,Fe^{2+})(Cr,Al,Fe^{3+})_2O_4$) grains surrounded by plagioclase or feldspathic mesostasis of plagioclase composition; also present in many of these chondrules are accessory ilmenite, Ca-pyroxene, merrillite, and olivine. Some of these chondrules contain ~50 vol% chromite and/or chromian spinel. (2) Chromium-rich mafic silicate chondrules consist of chromite, olivine, plagioclase and accessory merrillite, troilite, metallic Fe-Ni, Ca-pyroxene, and ilmenite. The chromite grains are typically of micrometer size and are surrounded by mesostasis, rendering it nearly opaque. Chromium-rich chondrules are not known from highly unequilibrated OC; they have been found only in OC of type 3.7 or 3.8 and higher. These chondrules are thus unlikely to be nebular products. They have been modeled as having formed by (1) parent-body thermal metamorphism of spinel-bearing Al-rich chondrules (see the next paragraph) (Wlotzka 2005) or (2) impact melting of chromite-rich precursors among equilibrated OC (Krot and Rubin 1993; Rubin 2003).

Bischoff and Keil (1984) defined Al-rich chondrules as round objects with igneous textures that contain $\geq$10 wt% bulk Al_2O_3; most are a few hundred micrometers in apparent diameter. They divided Al-rich chondrules into four principal groups on the basis of bulk chemical composition: (1) Ca-Al-rich chondrules ($\leq$5.0 wt% Na_2O), (2) (Ca,Na)-Al-rich chondrules (>5.0 wt% Na_2O), (3) Na-Al-rich chondrules ($\leq$2.6 wt% CaO, >4.1 wt% Na_2O), and (4) Na-Al-Cr-rich chondrules (<4.8 wt% CaO, >4.7 wt% Na_2O, $\geq$2.4 wt% Cr_2O_3). The Ca-Al-rich chondrules contain spinel, Al-Ti diopside (fassaite), olivine, low-Ca pyroxene, Ca-pyroxene, plagioclase, and glass. The (Ca,Na)-Al-rich chondrules are composed of spinel, Al-Ti diopside, low-Ca pyroxene, Ca-pyroxene, olivine, plagioclase, nepheline, and glass. The Na-Al-rich chondrules consist of skeletal olivine, minor Ca-pyroxene and spinel, and abundant sodic glass. The Na-Al-Cr-rich chondrules are fine-grained, nearly opaque objects containing abundant oxides (chromite, chromian-hercynian spinel, ilmenite), minor skeletal olivine, plagioclase and low-Ca pyroxene. An image of an Al-rich chondrule is shown in Figure 7.4.

Glass-rich chondrules are a variety of Al-rich chondrule; they have similar mineralogy and bulk chemical compositions (Krot and Rubin 1994). The glass-rich chondrules contain abundant Si-, Al-, and Na-rich glass or Ca-Al-rich microcrystalline mesostasis surrounding dendritic and skeletal crystals of magnesian olivine, Al-rich low-Ca pyroxene and Al-Ti diopside (fassaite); some contain Mg-Al spinel. There are two varieties: porphyritic glass-rich chondrules with 55–85 vol% glass and nonporphyritic chondrules with 90–99 vol% glass.

Rubin (2004) found that one glass-rich chondrule in LL3.00 Semarkona consists of 73 vol% glass, 22 vol% skeletal forsterite, 3 vol% Al-Ti diopside (with 17.8 wt% Al_2O_3) and 2 vol% low-Ca pyroxene (with up to 16.3 wt% Al_2O_3). Calculations suggest that 20 percent of the tetrahedral sites and 13 percent of the octahedral sites in the low-Ca pyroxene are occupied by Al. The grains exhibit approximately parallel optical extinction, implying they are probably orthorhombic. This is consistent with literature data showing that alumina stabilizes the orthoenstatite structure relative to protoenstatite at low pressure.

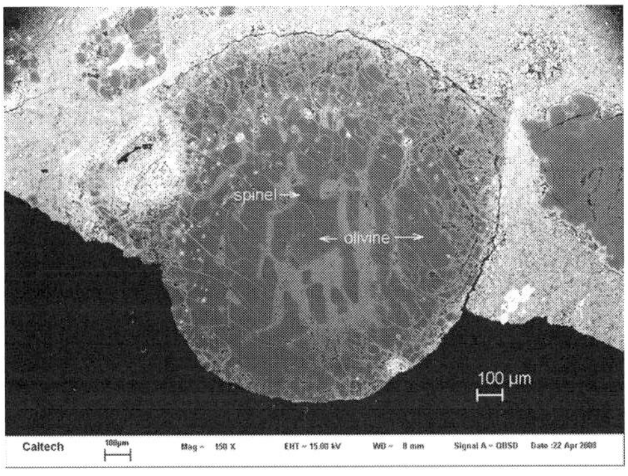

Figure 7.4 An Al-rich chondrule from Allende. BSE image.

Anorthite-rich chondrules (Krot and Keil 2002) in CR and CH chondrites are another variety of Al-rich chondrule. They consist of MgO-rich low-Ca pyroxene and olivine phenocrysts, metallic Fe-Ni nodules, interstitial anorthite, Al-Ti-Cr-rich low-Ca and Ca-pyroxene, and microcrystalline mesostasis containing silica, anorthite, and Ca-pyroxene. A few have regions resembling ferromagnesian chondrules and a few contain relict CAIs.

Chondrules in type-3 enstatite chondrites contain clinoenstatite±forsterite embedded in halogen-bearing glass. Minor phases present in some chondrules include calcic pyroxene, anorthite, silica, kamacite, troilite, niningerite ((Mg,Fe)S) or ferroan alabandite ((Mn,Fe)S), daubréelite ($FeCr_2S_4$), oldhamite (CaS), caswellsilverite ($NaCrS_2$) (and other Cr-rich sulfides), schreibersite, and perryite ($(Ni,Fe)_8(Si,P)_3$). Many BOP chondrules contain orthopyroxene. In EH3 Y-691 and Qingzhen, PP and RP chondrule mesostases consist mainly of glass, but some also contain fine-grained enstatite, diopside, plagioclase, and silica. A small BO chondrule in Y-691 contains the only known grains of wassonite (TiS) (Nakamura-Messenger et al. 2012). (BO chondrules are very rare in enstatite chondrites.)

Parent-body aqueous alteration has affected chondrules in many chondrites, producing numerous secondary phases. In CV_{OxA} chondrites such as Allende, iron-alkali-halogen metasomatism has transformed glassy or microcrystalline mesostases into polymineralic assemblages containing major nepheline and sodalite and minor andradite, grossular, hedenbergite, kirschsteinite and wollastonite. Low-Ca pyroxene phenocrysts have been altered to ferroan olivine (Fa_{25-50}).

In CV_{OxB} chondrites, chondrule mesostases have been aqueously altered to phyllosilicate (e.g., Fe-bearing saponite and Na-rich phlogopite intergrown with accessory Al-rich serpentine in Mokoia). Also present in Mokoia chondrules are assemblages of fayalite, magnetite, troilite, and pentlandite.

Aqueous alteration in type-3 OC has produced secondary phases inside chondrules including Ca-pyroxene and plagioclase (from glassy mesostasis) and carbide-magnetite assemblages from kamacite and troilite. These assemblages contain carbides (including cohenite and haxonite) of variable Ni (1–14 wt%) and Co (0–1.3 wt%) contents, magnetite, troilite, Ni-rich sulfide (including pentlandite), metallic Fe-Ni with 25–70 wt% Ni (including taenite, tetrataenite and awaruite), and Co-rich kamacite (with up to 35 wt% Co).

7.2 Calcium-Aluminum-Rich Inclusions

These refractory objects were recently reviewed by Krot (2019). Calcium-aluminum-rich inclusions (CAIs) are common constituents of most carbonaceous-chondrite groups and are also present as very rare components in ordinary (0.02 vol%), enstatite (0.01 vol%) and R (0.04 vol%) chondrites (Table 7.1). The most abundant primary minerals in CAIs are Al-Ti diopside (fassaite), anorthite, corundum, forsterite, grossite, hibonite, melilite, perovskite and spinel. A number of other refractory and ultrarefractory phases have been discovered as small grains within CAIs in recent years including addibischoffite ($Ca_2Al_6Al_6O_{20}$), allendeite ($Sc_4Zr_3O_{12}$), burnettite ($CaVAlSiO_6$), davisite ($CaScAlSiO_6$), grossmanite ($CaTi^{3+}AlSiO_6$), hexamolybdenum (Mo,Ru,Fe), kangite ($(Sc,Ti,Al,Zr,Mg,Ca,\square)_2O_3$), krotite ($CaAl_2O_4$), kaitianite ($Ti^{3+}{}_2Ti^{4+}O_5$), kushiroite ($CaAlAlSiO_6$), machiite ($Al_2Ti_3O_9$), panguite ($(Ti^{4+},Sc,Al,Mg,Zr,Ca)_{1.8}O_3$), paqueite ($Ca_3TiSi_2(Al,Ti,Si)_3O_{14}$), rubinite ($Ca_3Ti^{3+}{}_2Si_3O_{12}$), tistarite ($Ti_2O_3$), and warkite ($Ca_2Sc_6Al_6O_{20}$), as listed in Table 7.2 and shown in Figures 7.5 to 7.15.

Table 7.2 *Refractory and ultrarefractory minerals in Ca-Al-rich inclusions and amoeboid olivine inclusions*

Name	Formula
Addibischoffite	$Ca_2Al_6Al_6O_{20}$
Akermanite	$Ca_2MgSi_2O_7$
Al,Ti-diopside	$Ca(Mg,Ti,Al)(Si,Al)_2O_6$
Allendeite	$Sc_4Zr_3O_{12}$
Anorthite	$CaAl_2Si_2O_8$
Anosovite (not approved)	$(Ti^{4+},Ti^{3+},Mg,Sc,Al)_3O_5$
Baddeleyite	ZrO_2
Baghdadite	$Ca_3(Zr,Ti)Si_2O_9$
Burnettite	$CaV^{3+}AlSiO_6$
Calzirtite	$Ca_2Zr_5Ti_2O_{16}$
Corundum	Al_2O_3
Davisite	$CaScAlSiO_6$
Dmisteinbergite	$CaAl_2Si_2O_8$
Dmisteinbergite Ba-rich	$(Ca,Ba)Al_2Si_2O_8$
Eringaite	$Ca_3Sc_2Si_3O_{12}$
Forsterite	Mg_2SiO_4
Gehlenite	$Ca_2Al(SiAl)O_7$
Grossite	$CaAl_4O_7$
Grossmanite	$CaTi^{3+}AlSiO_6$
Hexaferrum	(Fe,Os,Ir,Mo)
Hexamolybdenum	(Mo,Ru,Fe)
Hibonite	$CaAl_{12}O_{19}$
Ir-dominant alloy	(Ir,Ru,Os)
Kaitianite	$Ti^{3+}{}_2Ti^{4+}O_5$
Kangite	$(Sc,Ti,Al,Zr,Mg,Ca,\square)_2O_3$

Table 7.2 (*cont.*)

Name	Formula
Khmrabaevite	TiC
Krotite	$CaAl_2O_4$
Kushiroite	$CaAlAlSiO_6$
Lakargiite	$CaZrO_3$
Machiite	$Al_2Ti_3O_9$
Mullite	$Al_6Si_2O_{13}$
Os-dominant alloy	(Os,Ir,Ru)
Osbornite	TiN
Panguite	$(Ti,Al,Sc,Mg,Zr,Ca)_{1.8}O_3$
Panguite Zr-rich	$(Ti,Zr,Al,Sc,Mg,Ca)_{1.8}O_3$
Paqueite	$Ca_3TiSi_2(Al,Ti,Si)_3O_{14}$
Perovskite	$CaTiO_3$
Pt-dominant alloy	(Pt,Ir)
Quartz	SiO_2
Rhenium (not approved)	Re
Rhonite	$Ca_2(Mg,Al,Ti)_6(Si,Al)_6O_{20}$
Rubinite	$Ca_3Ti_2Si_3O_{12}$
Ru-dominant alloy	(Ru,Ir,Os)
Rutile	TiO_2
Spinel	$MgAl_2O_4$
Tazheranite	$(Zr,Ti,Ca,Y)O_{1.75}$
Thortveitite	$Sc_2Si_2O_7$
Ti^{3+},Al,Zr-oxide	$(Ti^{3+},Al,Zr,Si,Mg)_{1.95}O_3$
Tistarite	Ti_2O_3
Warkite	$Ca_2Sc_6Al_6O_{20}$
Zirconolite	$CaZrTi_2O_7$
Zirkelite	$(Ti,Ca,Zr)O_{2-x}$

Many CAIs experienced aqueous alteration on their parent asteroids, a process that produced such secondary minerals as anorthite, Ca-pyroxene, grossular garnet, nepheline, and sodalite. (Both anorthite and Ca-pyroxene are primary *and* secondary phases.) Minor and accessory secondary minerals include monticellite ($CaMgSiO_4$), wadalite ($Ca_6Al_5Si_2O_{16}Cl_3$), and larnite (Ca_2SiO_4), as well as the recently discovered adrianite ($Ca_{12}(Al_4Mg_3Si_7)O_{32}Cl_6$) (Figure 7.16; Ma and Krot 2018), beckettite ($Ca_2V_6Al_6O_{20}$) (Figure 7.17; Ma et al., 2016a), chlormayenite (renamed from brearleyite ($Ca_{12}Al_{14}O_{32}Cl_2$)) (Figure 7.10; Ma et al., 2011c), hibonite-(Fe) ($(Fe,Mg)Al_{12}O_{19}$) (Figure 7.18; Ma 2010), and hutcheonite ($Ca_3Ti_2(SiAl_2)O_{12}$) (Figure 7.16; Ma and Krot 2014). There are also micrometer-sized crystals of two chalcogenide minerals (nuwaite (Ni_6GeS_2) and butianite (Ni_6SnS_2)) within veins and fractures in igneous diopside grains in an Allende Type B1 CAI (Figure 7.19; Ma and Beckett 2018); these phases formed by vapor deposition during late-stage alteration.

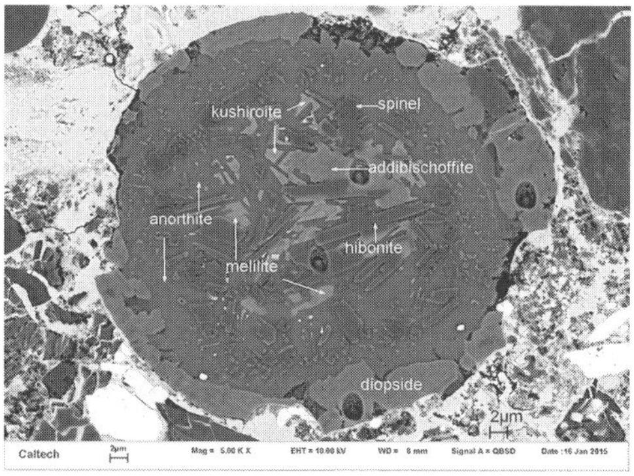

Figure 7.5 Addibischoffite ($Ca_2Al_6Al_6O_{20}$) in a CAI from the Acfer 214 CH chondrite (Ma et al. 2017a). BSE image.

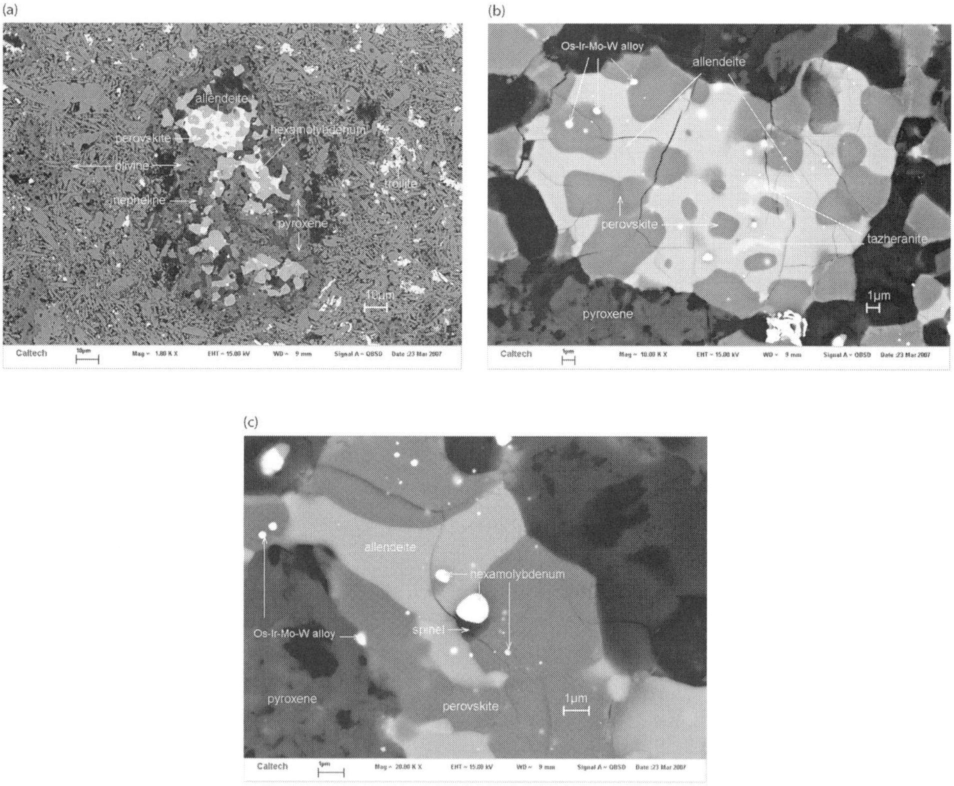

Figure 7.6 Allendeite ($Sc_4Zr_3O_{12}$) and hexamolybdenum (Mo,Ru,Fe) in an ultrarefractory inclusion from Allende (Ma et al. 2014a). BSE images.

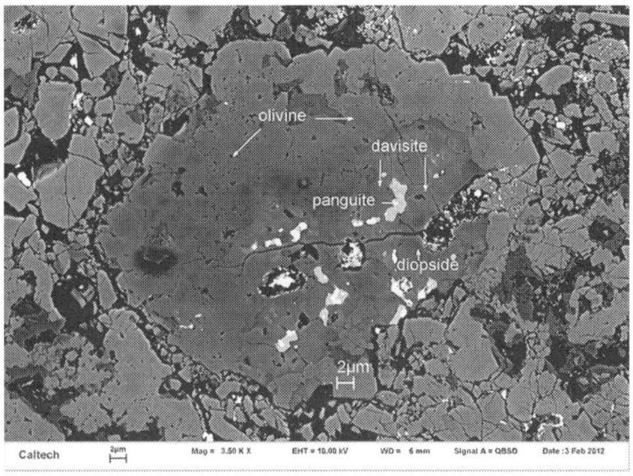

Figure 7.7 Davisite ($CaScAlSiO_6$) and panguite ($(Ti^{4+},Sc,Al,Mg,Zr,Ca)_{1.8}O_3$) in an ultrarefractory inclusion within an amoeboid olivine inclusion from Allende (Ma et al. 2012c). BSE image.

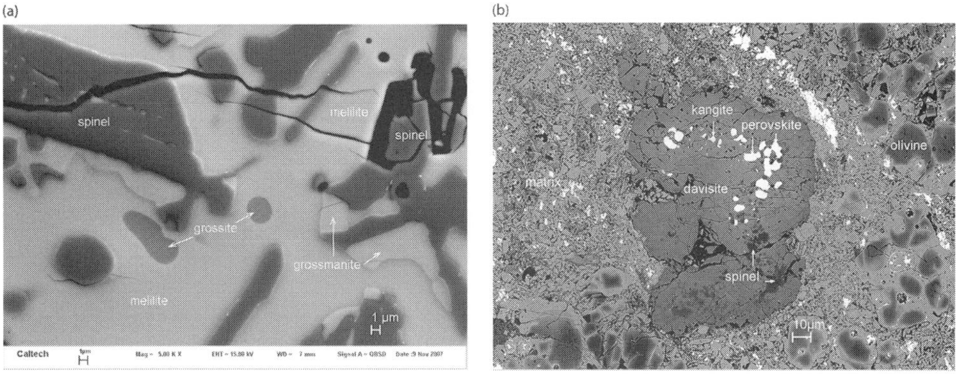

Figure 7.8 (a) Grossmanite ($CaTi^{3+}AlSiO_6$) in a Type B CAI, (b) kangite ($(Sc,Ti,Al,Zr,Mg,Ca,\square)_2O_3$) and davisite in an ultrarefractory inclusion from Allende (Ma and Rossman 2009c; Ma et al. 2013c). BSE images.

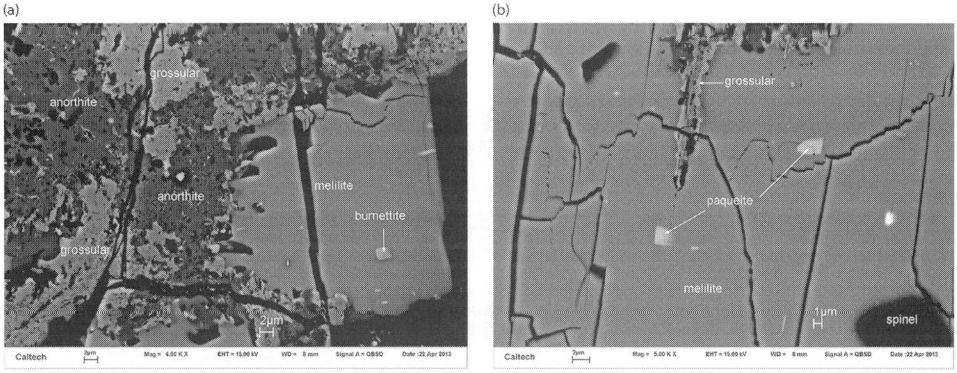

Figure 7.9 (a) Paqueite ($Ca_3TiSi_2(Al,Ti,Si)_3O_{14}$) and (b) burnettite ($CaVAlSiO_6$) in a V-rich Fluffy Type A CAI from Allende (Ma and Beckett 2016a). BSE images.

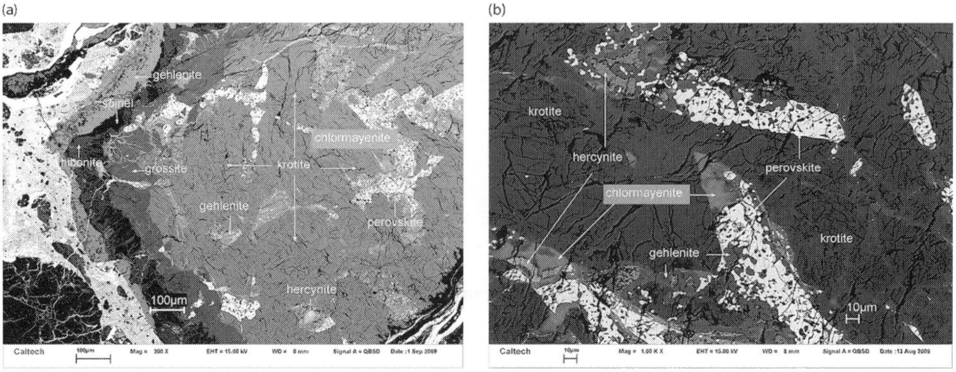

Figure 7.10 Krotite ($CaAl_2O_4$) and chlormayenite (renamed from brearleyite; $Ca_{12}Al_{14}O_{32}Cl_2$) in a CAI dubbed *Cracked Egg* from the NWA 1934 CV3 chondrite (Ma et al. 2011c, 2011d). BSE images.

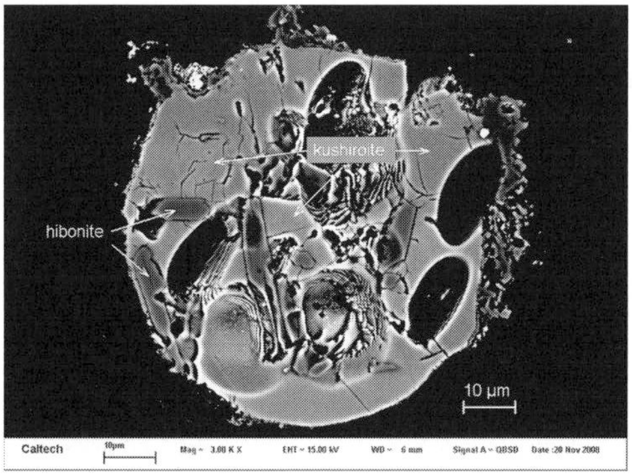

Figure 7.11 Kushiroite ($CaAlAlSiO_6$; Calcium Tschermak's pyroxene) with hibonite in the Murray CM2 chondrite (Ma et al. 2009). (Calcium Tschermak's pyroxene is well known as an important component in pyroxene; it is a member of the Ca-clinopyroxene group in which Al dominates in the M1 site.) BSE image.

Figure 7.12 Machiite ($Al_2Ti_3O_9$) with corundum in the matrix of the Murchison CM2 chondrite (Krot et al. 2020). BSE image.

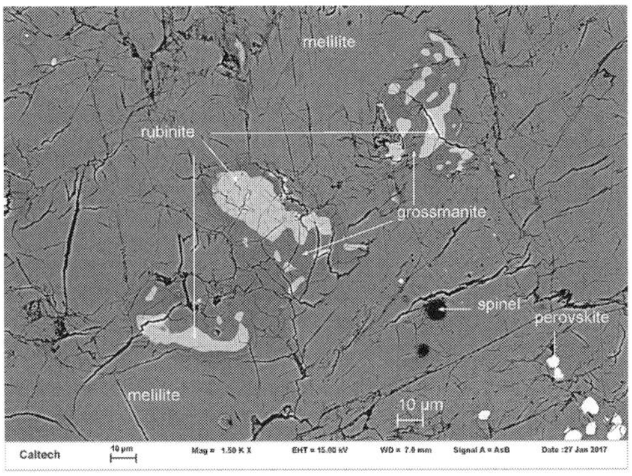

Figure 7.13 Rubinite ($Ca_3Ti^{3+}{}_2Si_3O_{12}$) and grossmanite in a compact Type A CAI from the Efremovka CV3 chondrite. After Ma et al. (2017d). BSE image.

Figure 7.14 Tistarite (Ti_2O_3) in a cluster of refractory minerals, along with corundum and khamrabaevite (TiC) grains within a chondrule from Allende (Ma and Rossman 2009a). BSE image.

Figure 7.15 Warkite ($Ca_2Sc_6Al_6O_{20}$) in an ultrarefractory inclusion from the Vigarano CV3 chondrite (Ma et al. 2020a). BSE image.

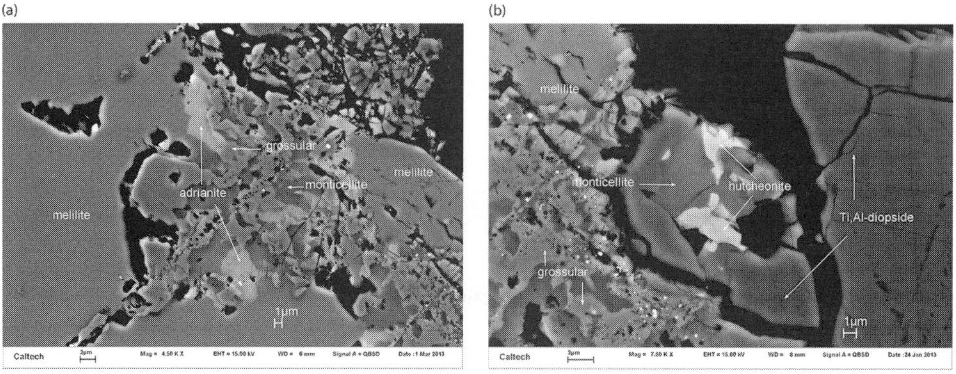

Figure 7.16 (a) Adrianite ($Ca_{12}(Al_4Mg_3Si_7)O_{32}Cl_6$) and (b) hutcheonite ($Ca_3Ti_2(SiAl_2)O_{12}$), two new secondary minerals, in a FUN CAI from Allende (Ma and Krot 2014, 2018). BSE images.

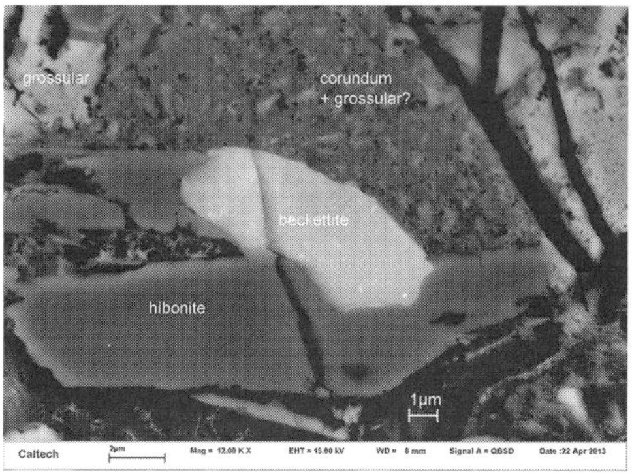

Figure 7.17 Beckettite ($Ca_2V_6Al_6O_{20}$), a secondary mineral in in a fluffy Type A V-rich CAI from Allende (Ma et al. 2016a). BSE image.

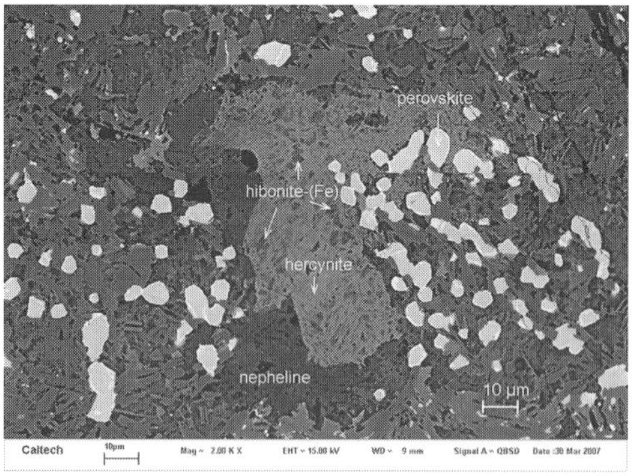

Figure 7.18 Hibonite-(Fe) ($(Fe,Mg)Al_{12}O_{19}$), a secondary mineral from Allende (Ma 2010). BSE image.

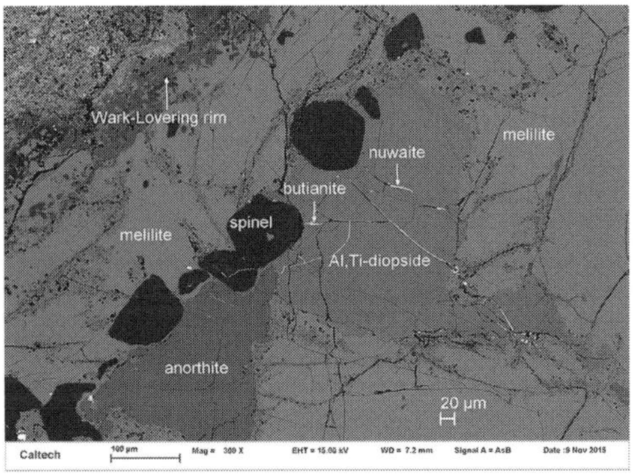

Figure 7.19 Nuwaite (Ni_6GeS_2) and butianite (Ni_6SnS_2), two new secondary minerals, in a Type B CAI from Allende (Ma and Beckett 2018). BSE image.

Figure 7.20 Monipite (MoNiP) in a Type B CAI from Allende (Ma et al. 2014b). BSE image.

Also produced during secondary alteration are opaque assemblages dubbed "Fremdlinge" (German for "foreigners") by El Goresy et al. (1977, 1978). These assemblages consist of refractory metal nuggets, enriched in Pt-group elements (PGEs), and varying in diameter from 1 to 1000 µm. The nuggets are surrounded by metallic Fe-Ni, V-rich magnetite, and Fe-Ni sulfides; these phases are associated with molybdenite (MoS_2), kamiokite ($Fe_2Mo_3O_8$), tungstate, phosphate, and the recently discovered monipite (MoNiP) (Figure 7.20; Ma et al. 2014b), and majindeite ($Mg_2Mo_3O_8$) (Figure 7.21; Ma and Beckett 2016b). Although most refractory metal nuggets occur in CAIs, a few also occur in chondrules and matrix material (e.g., Daly et al. 2017). They have also been found in several presolar graphite grains (Croat et al. 2013), indicating that they are not all parent-body alteration products.

A significant proportion of the CAIs can be divided into different categories (A, B, C) that differ moderately in shape, size, mineralogy, and mineral chemistry. The individual groups can themselves be subdivided. The largest CAIs are 2–3 cm in size and occur exclusively in CV3 chondrites; most CAIs outside the CV group are <1 mm.

Compact Type A CAIs (CTAs) have spheroidal shapes and were apparently molded by surface tension during moderate melting. CTAs consist mainly of zoned melilite enclosing

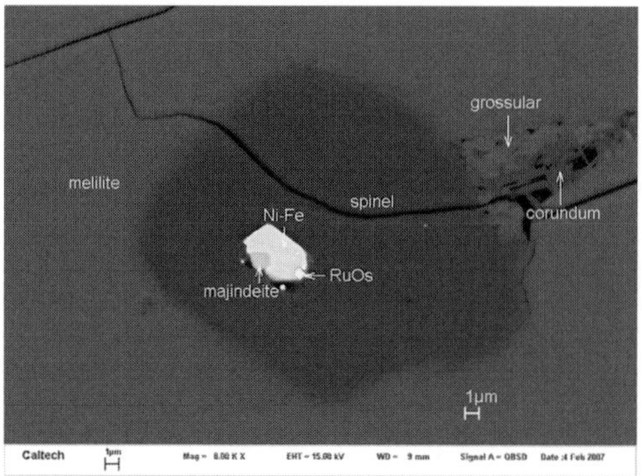

Figure 7.21 Majindeite ($Mg_2Mo_3O_8$) in a Type B CAI from Allende (Ma and Beckett, 2016b). BSE image.

spinel, hibonite, perovskite, and noble metal nuggets; some CTAs consist almost entirely of monomineralic melilite. Other CTA inclusions contain small amounts of Al-Ti-rich diopside (fassaite) and relatively rare grains of rhönite ($Ca_2(Mg,Al,Ti)_6(Si,Al)_6O_{20}$).

The CTAs are typically surrounded by Wark–Lovering rims. These rims are a few tens of micrometers thick and generally consist of a series of mono- or bimineralic layers of melilite, spinel and diopside along with some hibonite and perovskite (Wark and Lovering,1977).

Fluffy Type A CAIs (FTAs) have irregular shapes and appear not to have experienced significant melting as individual objects. The FTAs in CV3 chondrites are ~0.3–3 cm in size. Most of the centimeter-sized FTAs are aggregates of numerous melilite-rich nodules; many nodules resemble individual CTAs with Wark–Lovering rims.

Type B CAIs formed very early in the history of the Solar System – about 4.567 Ga ago. They occur exclusively in CV chondrites as multimillimeter- to multicentimeter-sized coarse-grained spheroidal igneous objects that consist mainly of melilite, Al- and Ti-rich diopside, spinel, and minor anorthite.

There are three principal varieties of Type B CAIs: Type B1 inclusions consist of a coarse-grained melilite mantle ($Åk_{\sim 10-72}$) [Åk = mol% åkermanite] surrounding a core of pyroxene, spinel, anorthite, metallic Fe-Ni, and noble metal nuggets. Type B2 inclusions have no melilite mantle and are mineralogically similar to B1 cores. Type B3 inclusions (the least-common variety) contain coarse forsterite grains along with pyroxene, spinel, and melilite; they are also known as forsterite-bearing Type B inclusions or FoB CAIs. Some FoBs also contain minor anorthite.

Type C CAIs are igneous objects that consist primarily of spinel, calcic pyroxene and anorthite exhibiting poikilitic or ophitic textures. (In igneous rocks with ophitic textures, plagioclase crystals are surrounded completely by pyroxene grains.)

One special group of CAIs is dubbed "FUN" inclusions because these bodies exhibit Fractionation and Unidentified Nuclear effects; they are characterized by unusual isotopic compositions of Mg, O, Ba, Ti, Ca, and other elements (e.g., Wasserburg et al. 1977). They are mineralogically and texturally indistinguishable from normal (non-FUN) CAIs.

Spinel-rich inclusions are fine-grained CAIs consisting predominantly of spinel, melilite, anorthite, Al- and Ti-rich diopside with minor hibonite and perovskite and rare forsterite. The spinel-rich inclusions in reduced CV3 chondrites have been little affected by parent-body aqueous alteration: (1) they contain only rare grains of secondary nepheline and sodalite and (2) all spinel and almost all Al-Ti-diopside grains are ^{16}O-rich.

Hibonite-silicate spherules are small objects (≲170 μm) of varying textures. Some inclusions consist of hibonite laths surrounded by glass enriched in an Al-pyroxene component, others of hibonite laths within Al-pyroxene crystals. Some of the spherules are texturally complex objects that contain hibonite, Al-pyroxene, melilite, and grossite.

Different chondrite groups have different populations of CAIs. For example, Type B CAIs occur only in CV chondrites. The main types of CAIs in CM chondrites include spinel-pyroxene inclusions, spinel inclusions, spinel-pyroxene-olivine inclusions, pyroxene inclusions, pyroxene-olivine inclusions, and hibonite-bearing inclusions. CM CAIs come in both simple and complex varieties; some of the inclusions have nodular, banded, or distended structures. Platy hibonite grains (PLACs) are largely confined to CM chondrites. Melilite (which is highly susceptible to aqueous alteration) is rare in CM CAIs, presumably due to such alteration. Although CAIs are rare in enstatite chondrites, those that are present tend to be rich in spinel; hibonite is less abundant and perovskite occurs in accessory amounts. Grossite-rich spherules, hibonite-rich spherules, and Al-diopside spherules are most common in CH chondrites.

Rare grains of the refractory mineral osbornite (TiN) have been found in CAIs and in CAI fragments in the Isheyevo CB/CH-like carbonaceous chondrite and in the ALH 85085 CH3 chondrite. The mineral was also identified amidst *Stardust* samples (presumably CAI fragments) collected from comet 81P/Wild.

A subset of CAIs has been called ultrarefractory inclusions. These objects are relatively rare and tend to be small. They have been found in CM, CO, CR, and CV chondrites – some are individual objects; others occur as enclaves within AOIs or as individual nodules within large fine-grained CAIs. Many of the inclusions are enriched in the most refractory rare-earth elements (REEs) – Lu, Er, Ho, and Dy. The inclusions contain one or more ultrarefractory phases such as tazheranite, zirconolite, Y-rich perovskite, Zr- and Y-rich perovskite, davisite, warkite, Sc-bearing Al-Ti diopside, and very rare baddeleyite (e.g., Krot et al. 2019) (Figure 7.22). Some inclusions are associated with TiO_2-, Sc_2O_3-, Y_2O_3-, and ZrO_2-bearing

Figure 7.22 An ultrarefractory inclusion in Allende containing zirconolite and tazheranite (cubic zirconia; $(Zr,Ti,Ca,Y)O_{1.75}$) (Ma and Rossman 2008). BSE image.

silicate glass. Many of the ultrarefractory inclusions also contain more common refractory phases such as melilite, hibonite, and spinel.

7.3 Amoeboid Olivine Inclusions

The irregularly shaped bodies known as amoeboid olivine inclusions (AOIs) or amoeboid olivine aggregates (AOAs) are similar in abundance to CAIs in carbonaceous chondrites (Figure 7.23). The AOIs are similar in mineralogy, texture, and isotopic composition in all carbonaceous-chondrite groups. AOIs are fine-grained (typically 1–20 μm) and consist of major olivine, major to minor calcic pyroxene, minor anorthite, accessory metallic Fe-Ni and perovskite, and in some minimally altered carbonaceous chondrites, accessory spinel ± low-Ca pyroxene ± melilite. Some CV3 AOIs also contain refractory and ultrarefractory oxides such as grossite ($CaAl_4O_7$), davisite ($CaScAlSiO_6$), and panguite ($(Ti,Al,Sc,Mg,Zr,Ca)_{1.8}O_3$) (Ma et al. 2012c) (Figure 7.7). Many AOIs have porous interiors. Some pristine AOIs consist mainly of fine-grained forsterite surrounding small spheroidal CAIs (with Al-Ti diopside, spinel, and anorthite along with accessory melilite and perovskite); this texture is consistent with forsterite condensing from nebular gas onto small previously formed CAI nuggets.

Very rare AOIs have also been reported in the Semarkona LL3.00 ordinary chondrite. These are fine-grained objects that characteristically contain forsterite (Fa_{0-1}), Al-Ti diopside (with 0.2–2.0 wt% TiO_2 and 4–10 wt% Al_2O_3) and anorthite (Itoh et al. 2007). Many of the forsterite grains enclose 2–10-μm-sized grains of metallic Fe-Ni. No alteration phases are present.

In CV_{OxA} chondrites, AOIs contain nepheline, sodalite and hedenbergite formed by secondary aqueous alteration. In CV_{OxB} chondrites, AOIs have a lower abundance of feldspathoids and instead contain low-Al phyllosilicate between olivine grains, and veinlets and blebs of magnetite, pyrrhotite ± pentlandite. In metamorphosed CO chondrites, AOIs contain ferroan olivine; the higher the degree of metamorphism, the larger the proportion of ferroan olivine and the greater the hercynite/spinel ($FeAl_2O_4/MgAl_2O_4$) ratio.

AOI mean size varies among chondrite groups: LL3, 200 μm; CM2, 225 μm; CO3, 260 μm; CR2, 600 μm; CK3, 900 μm; CV3, ~3,000 μm (Rubin 2013b). The three groups with the largest

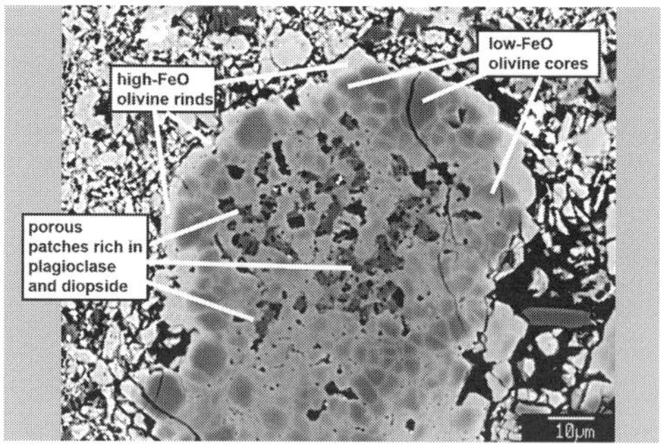

Figure 7.23 Amoeboid olivine inclusion in Allende. The inclusion has a very thick olivine rim enclosing small isolated interior patches of plagioclase, diopside and large pores. Olivine in the rim consists of low-FeO cores surrounded by ferroan olivine; the lobate periphery of the rim is composed of ferroan olivine. BSE image.

AOIs also have the largest mean chondrule diameters: CR2, 700 μm; CK3, 900 μm; CV3, 900 μm (Friedrich et al. 2015).

7.4 Plagioclase-Olivine Inclusions

Plagioclase-olivine inclusions (POIs) are 0.5–5-mm-diameter, mostly spheroidal, chondrule-like objects in carbonaceous chondrites. They have been reported in CV chondrites (Allende, Leoville, Vigarano) and ungrouped carbonaceous chondrites (Adelaide, Ningqiang) (Sheng et al. 1991; Lin and Kimura 1996, 1997). These objects have igneous textures and consist of abundant plagioclase (Ab_{2-18} with no K_2O) and olivine (Fa_{1-37}), with variable amounts of Al- and Ti-bearing fassaite, aluminous enstatite (with up to 7.8 wt% Al_2O_3), diopside, FeO-poor pigeonite, and spinel. Rare phases include sapphirine ($(Mg,Al)_7(Mg,Al)O_2(Al,Si)_6O_{18}$) as well as numerous Ti-rich oxides: armalcolite ($(Mg,Fe)Ti_2O_5$), Ca-armalcolite ($CaTi_2O_5$), zirconolite ($(Ca,Ce)Zr(Ti,Nb,Fe^{3+})_2O_7$), rutile ($TiO_2$), ilmenite ($FeTiO_3$), perovskite ($CaTiO_3$), geikielite ($MgTiO_3$), and members of a poorly characterized Ti-rich mineral series (dubbed Phase T). These latter phases contain (in wt%): TiO_2 (77.9–84.0%), CaO (0.62–10.3%), MgO (0.61–11.7%), and FeO (0.34–5.2%); the stoichiometric formula ranges from $R^{2+}R^{4+}{}_3O_7$ to $R^{3+}{}_2R^{4+}{}_2O_7$ (where R includes Ti, Mg, Ca, Al, Cr, Fe, V, Si, Na, Zr, Sc, and Mn). Secondary phases formed by parent-body aqueous alteration (e.g., nepheline, sodalite, hedenbergite, awaruite) constitute <5 vol% of the inclusions.

POIs lack melilite and refractory-siderophile-rich opaque assemblages. None possesses a Wark–Lovering rim.

There are several textural varieties of plagioclase-olivine inclusions: Group 1 POIs have subophitic textures and contain abundant plagioclase (20–66 vol%) and spinel (5–30 vol%). Group 2 POIs have subophitic textures and contain abundant plagioclase (30–60 vol%) and minor spinel (<5 vol%). Group 3 POIs have porphyritic, granular, and barred textures; they contain the highest modal abundances of olivine (>40 vol%), lesser amounts of plagioclase (<25 vol%), and rare spinel.

The POIs are intermediate in bulk composition between Type C CAIs and normal ferro-magnesian chondrules. About half contain isotopically fractionated Mg, located mainly in spinel. The isotopic heterogeneity of these igneous objects indicates that melting was incomplete and relict phases were preserved.

7.5 Dark Inclusions

Two very different types of objects in chondritic meteorites have been called "dark inclusions" – one set includes oldhamite- and REE-rich clasts in the Abee EH impact-melt breccia; the other includes aqueously altered clasts in carbonaceous chondrites.

7.5.1 Dark Inclusions in the Abee EH Chondrite

These irregularly shaped, fine-grained objects are dark in color with a yellow tinge; they constitute ~0.2 vol% of the Abee breccia and range in maximum dimension from ~2–13 mm.

They were described by Rubin and Keil (1983), Sears et al. (1983), and Rubin and Scott (1997). Relative to average EH chondrites, the dark inclusions are enriched in troilite, oldhamite, and graphite; they are depleted in metallic Fe-Ni and plagioclase. The fine grain size and high abundance of troilite in the dark inclusions probably account for their color.

Most of the metallic Fe-Ni in the dark inclusions is of martensitic composition (with 9–17 wt% Ni); in contrast, martensite is virtually absent from the Abee matrix. The enstatite and plagioclase in the dark inclusions are richer in CaO and compositionally more heterogeneous than in the rest of Abee. Schreibersite in the dark inclusions is richer in Ni than schreibersite in other Abee components.

The dark inclusions are enriched in REE (~4 × EH) except for Eu, reflecting the relatively high modal abundance (up to 9 wt%) of oldhamite (a major carrier of REEs in enstatite chondrites). The inclusions are also depleted in siderophile elements (0.25 × EH), reflecting the low modal abundance of metallic Fe-Ni (as low as 2 wt%). Dark inclusions are also depleted in Na, Al, and K, reflecting a low modal abundance of plagioclase (as low as 3 wt%). The depletion in plagioclase is also responsible for the negative Eu anomaly in the REE patterns of the dark inclusions (Sears et al. 1983).

The dark inclusions are probably rapidly cooled impact-melt products; they may be related to a 16-mm-long oldhamite-rich, chondrule-free vein in EL6 Jajh deh Kot Lalu (Rubin et al. 1997).

7.5.2 Dark Inclusions in Carbonaceous Chondrites

Krot et al. (1995, 1999) defined these dark inclusions as black-to-light-gray lithic clasts, typically angular in shape and ranging in size from a few millimeters to a few centimeters, that are made of chondritic components including chondrules, chondrule fragments, CAIs, CAI fragments, and matrix material.

Although dark inclusions in reduced CV3 chondrites, such as Efremovka, have largely escaped the oxidation processes that affected the oxidized CV3 chondrites (e.g., Allende and Bali), they have experienced significant preterrestrial aqueous alteration that modified their initial textural, mineralogical, and compositional characteristics. Dark inclusions in Efremovka consist of pseudomorphs of chondrules and CAIs that have been partly replaced by secondary phases: abundant fine-grained (<0.2 μm) ferroan olivine ($Fa_{35–42}$) containing numerous voids; other tiny ferroan olivine grains are highly strained with numerous dislocations and few voids. Also present in the altered chondrules are minor chromite, chlorite, and interstitial fine-grained phyllosilicates and poorly crystalline Si-Al-rich material (Krot et al. 1999). A few Efremovka dark inclusions contain some of the same secondary phases present in oxidized CV chondrites: TiO_2-bearing andradite, kirschsteinite, and FeO-bearing diopside. Opaque nodules in some chondrules within dark inclusions are composed of grains of Ni- and Co-rich taenite, Co-rich kamacite, wairauite (CoFe), and very rare sulfide.

Altered CAIs in some of the dark inclusions in Efremovka contain abundant FeO-rich material, minor phosphate and metallic Fe-Ni, and rare relict grains of primary CAI phases – melilite, Cr-free spinel, and perovskite.

Dark inclusions in the Allende CV_{OxA} chondrite resemble the Allende matrix in texture, mineralogy, bulk chemistry, and bulk O-isotopic composition. They appear to have undergone oxidation, metasomatism, and mild metamorphism along with the Allende whole rock, but the inclusions exhibit a wider range of alteration. Altered dark inclusions in Allende are traversed by veins (up to 20 μm thick and 3 mm long) containing major hedenbergite and minor andradite and sulfide. The inclusions are surrounded by Ca-rich rims (up to 150 μm thick) that have similar mineral assemblages to the rims around many Allende CAIs. Secondary minerals in these dark inclusions and rims include ferroan olivine, magnetite, Ni- and Co-rich metal, Fe-Ni sulfides, nepheline, sodalite, hedenbergite, andradite, wollastonite, and kirschsteinite (e.g., Krot et al. 1999, 2000a). Phyllosilicates are rare.

CV dark inclusions can be divided into two main groups (e.g., Krot et al. 1995): Type A dark inclusions are similar to their CV hosts, but with smaller chondrules and CAIs. Type B dark inclusions are free of chondrules and CAIs; they are composed mainly of round aggregates of small grains of ferroan olivine set in a matrix of even finer-grained olivine. A third variety of dark inclusion contains no chondrules, CAIs or their pseudomorphs; a centimeter-sized dark inclusion in CV_{Red} Vigarano consists nearly entirely of <5 μm grains of ferroan olivine (Fa_{43-51}) arranged in a set of parallel arcuate bands (Tomeoka and Kojima, 1998). The bands are akin to "dish structures" in terrestrial siltstones and sandstones and likely formed by compaction and dehydration of unconsolidated sediment.

Irregularly shaped *dark clasts* in CR2 chondrites are xenoliths that may be related to Allende dark inclusions. These clasts contain chondrule and mineral fragments, phyllosilicate, magnetite, sulfide, accessory Cr-spinel, metallic Fe-Ni within relict olivine, breunnerite (a.k.a. ferroan magnesite) ($(Mg,Fe)CO_3$), Ca-phosphate, ilmenite, schreibersite, and very rare refractory metal particles rich in Os, Mo, and Ir (Endreß et al. 1994).

Dark-colored CI-, CM- or CR-like clasts occur in some ordinary-chondrite regolith breccias (e.g., Abbott), type-3 ordinary chondrites (e.g., Sahara 987645), CH3 chondrites (e.g., ALH 85085), CR chondrites (e.g., Acfer 097), CB chondrites (e.g., Hammadah al Hamra 237), HED samples (e.g., Kapoeta), and polymict ureilites (e.g., Nilpena). These dark clasts are probably fragments of carbonaceous-chondrite projectiles that collided at low relative velocities with their present hosts. They are not closely related to dark inclusions in CV3 chondrites.

7.6 Carbon-Rich Aggregates

Carbon-rich aggregates occur as isolated inclusions (~5 to 1,000 μm in size) that constitute 0.7–3.2 vol% of a few type-3 ordinary chondrites (e.g., H3.4 Sharps, L3.5 Allan Hills A77011, LL3.5 Allan Hills A78119, and H3.6 Allan Hills A81024). The aggregates were originally characterized as being composed of graphite and magnetite (Scott et al. 1981a, 1981b; McKinley et al. 1981), but subsequent petrographic, geochemical, isotopic, and TEM (transmission electron microscopy) studies (Scott et al. 1988a; Brearley 1990) showed the aggregates to consist of irregularly shaped grains of poorly graphitized carbon along with amorphous carbon, small (<3 μm) well-disseminated grains of metallic Fe-Ni, and minor chromite. (Some aggregates in Sharps contain >50 vol% metallic Fe-Ni.)

The C-rich aggregates also constitute 14–36 vol% of four ~1- to 5-mm-sized chondritic clasts in solar-gas-rich H-chondrite regolith breccias (DT1 and DT2 from Dimmitt, PV1 from Plainview, and WN1 from Weston). In addition, these clasts contain 16–54 vol% chondrules,

18–54 vol% isolated silicate grains, 2–6 vol% metallic Fe-Ni, and 1–16 vol% troilite. The C-rich aggregates occur between chondrules in the clasts, serving the function of matrix material. Olivine and low-Ca pyroxene in the chondrules are compositionally very heterogeneous; e.g., for PV1, mean olivine Fa = 17±7.7 mol%; n = 275 (range Fa 3–43 mol%), and mean low-Ca pyroxene Fs = 10.6±6.3 mol%; n = 241 (range Fs 1–38 mol%) (Scott et al. 1981b).

Although Clast PV1 contains 13 wt% C (compared to 0.1–0.6 wt% C in most type-3 OC), it closely resembles normal H chondrites in chondrule size (~300 μm), bulk Mg-normalized mean refractory lithophile abundance ratio (1.00 ± 0.09 × H) and bulk O-isotopic composition ($\Delta^{17}O$ = 0.66‰). The clast apparently formed from H-chondrite material by shock heating, fracturing, faulting, recrystallization, hydrothermal alteration, and C enrichment from an unknown source (Rubin et al. 2005). (Carbon-rich aggregates in Sharps contain organic material that experienced different thermal histories and may thus have been derived from several sources.)

7.7 Matrix Material

Matrix material in primitive chondrites is fine-grained, mainly silicate material occurring between chondrules, refractory and mafic inclusions, chondrule and inclusion fragments, as well as isolated grains and polycrystalline assemblages of silicate, metal, sulfide, oxide, and other phases (definition modified from Scott et al. 1988b). This material constitutes a significant fraction of type-1 to type-3 carbonaceous, ordinary, enstatite, and R chondrites (Table 7.1). It occurs (a) between chondrules and other components, (b) as fine-grained rims around chondrules, refractory inclusions, metal-sulfide assemblages, and chondritic clasts, and (c) as separate matrix lumps up to 2 mm across. Presolar grains are present in the matrices of primitive meteorites; the least-metamorphosed chondrites tend to contain higher modal abundances of presolar grains.

Because matrix material is fine grained and porous, it is quite susceptible to aqueous alteration, a process that transformed the primary matrix minerals in many carbonaceous chondrites into phyllosilicates and other hydrous and anhydrous phases.

7.7.1 CI Chondrites

Bulk CI chondrites contain ~17 wt% H_2O, a reflection of the high modal abundance of constituent minerals with chemically bound water or hydroxyl. The principal components of CI matrices are silicates (50–60 vol%): Fe-Mg serpentines interlayered with saponite. The two major oxide phases are magnetite (~8–11 wt%) and tiny grains (<8 nm) of ferrihydrite. Carbonate phases (5 vol%) include dolomite, breunnerite (up to 3 mm in size), minor siderite and rare calcite (e.g., Scott and Krot 2014). Sulfides also occur: 10–100-μm-sized laths and plates of pyrrhotite associated with rare elemental S; minor pentlandite (isolated ~20 μm grains and 15-μm-sized inclusions within pyrrhotite); and minor cubanite. Sulfates (gypsum, epsomite, blödite, Ni-rich blödite) are present in CI chondrites, mainly in veins; these phases, however, may have formed by terrestrial alteration. Additional accessory phases reported in CI-chondrite matrices include merrillite, limonite, periclase, brucite, Ti_3O_5, magnesiochromite, and eskolaite.

Thermally metamorphosed CI-like chondrites are discussed in Chapter 11, Section 11.2.3.

7.7.2 CM Chondrites

CM chondrites span a large range of aqueous alteration, from type 2.0 (highly altered) to 2.7 to 2.9 (somewhat to minimally altered), with corresponding changes in matrix mineralogy and mineral chemistry. The most widely used system for classifying CM chondrites by degree of aqueous alteration is that of Rubin et al. (2007), extended by Rubin (2015c). All CM matrices contain abundant Fe-Mg serpentines of variable composition (mainly Fe-bearing chrysotile); their MgO/FeO ratios tend to increase with increasing degrees of alteration. Among the minimally altered CM chondrites, amorphous silicate material appears to be more abundant than phyllosilicates.

Tochilinite-cronstedtite intergrowths (formerly called "PCP" for "poorly characterized phases") occur as isolated clumps with irregular shapes and, in some cases, as multimillimeter-long lenses (Figure 7.24); the "FeO"/SiO_2 composition of these intergrowths decreases with increasing alteration. ("FeO" includes FeO in mafic silicates, oxides, and phyllosilicates; Fe^{3+} in cronstedtite and Fe-oxide; Fe^{2+} in sulfides; and small grains of metallic Fe, all combined and recalculated as FeO.) There are two types of these intergrowths (Tomeoka and Buseck 1985): Type-I intergrowths occur as compact objects rimming metallic Fe-Ni; Type-II intergrowths are compositionally zoned and appear pseudomorphic after olivine and pyroxene. Large clumps of tochilinite-cronstedtite intergrowths constitute 15–40 vol% of CM2.2–2.6 chondrites. The clumps break down with increasing degrees of alteration as S tends to leave the intergrowths to form relatively coarse sulfide grains. Consequently, highly altered CM2.0–2.1 chondrites contain only 2–5 vol% of the intergrowths. At the other end of the alteration scale, tochilinite-cronstedtite intergrowths should be absent in unaltered CM3.0 rocks. In fact, Asuka 12236, perhaps the least-altered CM chondrite, contains no tochilinite-cronstedtite intergrowths (Nittler et al. 2020). The bulk composition of the tochilinite-cronstedtite intergrowths changes with increasing degrees of alteration, reflecting an increase in the whole-rock phyllosilicate/sulfide ratio, an increase in the MgO/"FeO" ratio of the intergrowths, and the loss of S from the intergrowths to form relatively coarse sulfide grains.

The modal abundance of metallic Fe-Ni in CM chondrites decreases during aqueous alteration from about 1.2 to $\leq$0.02 vol%. Although sulfide in Paris (minimally altered) appears to consist only of pyrrhotite and pentlandite, the more-altered CM chondrites contain many sulfide grains

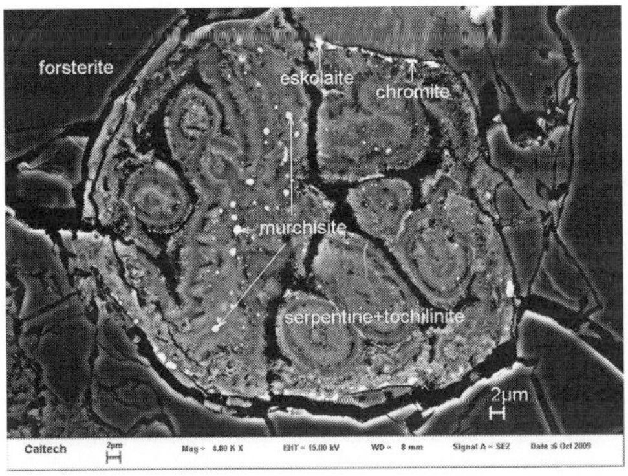

Figure 7.24 Murchisite (Cr_5S_6) in fine-grained tochilinite and serpentine in the altered area of an isolated olivine grain from the Murchison CM2 chondrite (Ma et al. 2011a). BSE image.

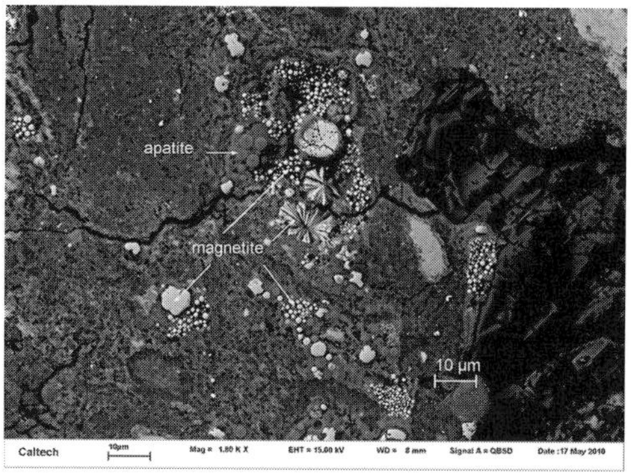

Figure 7.25 Magnetite framboids, spherulites and quasi-equant crystals in Murchison. BSE image.

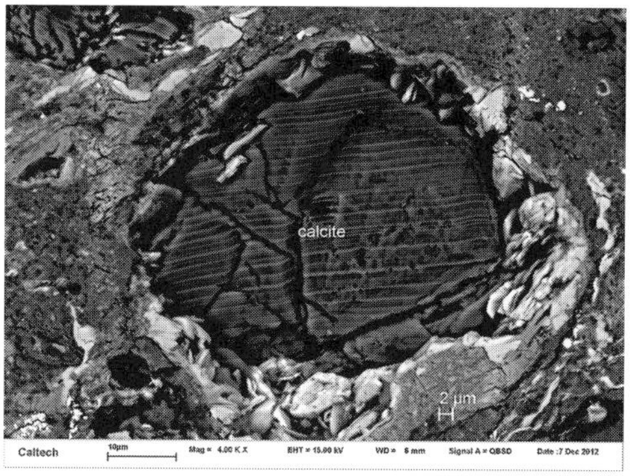

Figure 7.26 Calcite in the matrix of Murchison. BSE image.

with "intermediate" compositions, i.e., with atomic Ni/(Fe+Ni) ratios of 0.10 to 0.40. Magnetite occurs as massive clumps, spheres, and equant grains in Murchison (Figure 7.25). Hematite was detected by X-ray diffraction (XRD) in Nogoya; even though this meteorite is an observed fall, some researchers view the hematite as a terrestrial weathering product.

The carbonate grains in Murchison and Paris are Ca carbonate, as they are in all but the most highly altered CM chondrites (subtype 2.0 and 2.1) (Figure 7.26). In addition to Ca carbonate (calcite and rare aragonite), highly altered CM chondrites contain aggregates of complex carbonates containing Ca, Mg, Fe, Mn, and Ni.

Accessory phases within CM matrices include forsterite, Cl-free apatite, orthoenstatite, Ca-pyroxene, halite, and sylvite. Sulfates (some of which may have formed during terrestrial alteration) include gypsum, bassanite, anhydrite, and thenardite.

The recent discovery of sodalite within a 0.6 mm CAI in CM MET 01075 may indicate that the CM parent body underwent minor alkali-halogen metasomatism similar to that of oxidized CV chondrites. Alternatively, this particular CAI may be an exotic clast incorporated into the CM regolith from another carbonaceous-chondrite parent body.

Thermally metamorphosed CM-like chondrites are discussed in Chapter 11, Section 11.2.3.

7.7.3 CR Chondrites

Like CM chondrites, CR chondrites exhibit various degrees of aqueous alteration (e.g., Harju et al. 2014), ranging from CR1 (a.k.a. subtype 2.0) GRO 95577 (Weisberg and Huber 2007) to subtype 2.8 or 3 (MET 00426 and QUE 99177) (Abreu and Brearley 2010; Harju et al. 2014).

The least-altered CR chondrites contain matrices with submicrometer-sized amorphous ferroan silicate grains admixed with nanometer-scale phyllosilicates (Le Guillou et al. 2015). Some of the amorphous silicates encircle tiny particles of troilite and Fe-Ni sulfide. Only rare coarse grains of phyllosilicates, magnetite and Ca-carbonate are present. Some minimally altered CR chondrites contain very rare nanometer-sized grains of metallic Fe-Ni. The matrices of CR chondrites that have experienced intermediate levels of alteration contain Fe-Mg serpentines, saponite, calcite, magnetite, pyrrhotite, and pentlandite. The phyllosilicate-rich mesostases of the most-altered CR chondrites contain a few elongated aggregates of magnetite framboids.

A detailed petrologic study of the matrix of a minimally altered CR chondrite (LAP 02342) showed that different matrix regions differ in bulk Fe, S, Al, Si, Mg, K, and Na (Wasson and Rubin 2009). Some regions have relatively high bulk Ca, inferred to reflect Ca-carbonate modal abundances on the order of 1 wt%. A few regions contain sulfide laths (3×7 to 14×110 μm) arranged in clusters. About half of the sulfide laths are flanked on one or two of their long sides by 2–5-μm-wide sheaths of phyllosilicate. [A similar sulfide-and-sheath structure in the Kaidun polymict breccia was termed a "sausage in a bun" morphology by Zolensky et al. (1996).] The differences in matrix composition in different regions of LAP 02342 probably mainly reflect heterogeneous parent-body aqueous alteration.

7.7.4 CO Chondrites

CO chondrites were subjected to different degrees of aqueous/hydrothermal alteration; their petrologic subtypes range from type 3.0 for minimally altered rocks (e.g., DOM 08006, Y-81020, Colony, ALHA77307) to type 3.7 and 3.8 for appreciably altered, moderately metamorphosed samples (e.g., Warrenton, Isna, NWA 2131, NWA 5202).

The matrix of CO3.00 ALHA77307 (Brearley 1993a; Brearley and Jones 1998) contains unequilibrated Si- and Fe-rich amorphous silicate with embedded submicrometer sized particles of olivine (Fa_{0-97}), low-Ca pyroxene, metallic Fe-Ni, magnetite, pyrrhotite, pentlandite, fibrous anhydrite, and mixed-layer phyllosilicate. Also present are Fe-Mg serpentine, chlorite, isolated 0.2–4-μm-sized grains of forsterite and isolated grains and grain-clusters of Mg- and Mn-rich olivine (sometimes termed LIME – Low-Iron, Manganese-Enriched – olivine).

Matrix material within CO chondrites of higher petrologic subtype contains abundant ferroan olivine along with minor pyroxene and accessory hercynite and chromite. Warrenton matrix also contains accessory kamacite, taenite, awaruite, and hydrous ferric oxide (but no phyllosilicates).

7.7.5 CV Chondrites

Materials in CV-chondrite matrices include comminuted chondrule and CAI debris (including mineral grains and mineral fragments), minerals formed by secondary processes involving aqueous/hydrothermal alteration and Fe-alkali-halogen metasomatism (particularly in oxidized CV chondrites) and some primary phases (particularly in reduced CV chondrites). Phases that are probably derived from chondrule and CAI fragments include troilite, olivine, low-Ca pyroxene, spinel, anorthite, Ca-, Mg-rich pyroxene (e.g., diopside), Al–Ti diopside (fassaite), and grossular. Phases most likely formed by secondary processes include fayalite and ferroan olivine, hedenbergite, wollastonite, nepheline, sodalite, andradite, monticellite, kirschsteinite, tetrataenite, awaruite, magnetite (many grains with framboidal morphologies), chromite, ilmenite, pyrrhotite, pentlandite, wadalite, adrianite, poorly graphitized carbon and amorphous carbon, apatite, saponite, Fe-bearing saponite, Na phlogopite (aspidolite), Al-rich serpentine, Na-K mica, ferrihydrite, and calcite. Additional secondary phases reported in CAIs and chondrules that may also be present in CV matrices include phlogopite, montmorillonite, clintonite, margarite, chlorite, talc, carbonates, amphibole, and a biopyribole (jimthompsonite). Phases that may be primary include kamacite, troilite, olivine, and low-Ca pyroxene. Some minerals in CV matrices may have both primary and secondary origins.

Oxidized CV chondrites of the Allende subgroup (CV_{OxA}) have very little bulk H_2O (e.g., <0.2 wt% in the Allende fall) and only rare occurrences of phyllosilicate in their matrices; nevertheless, several different phyllosilicate phases are present: phlogopite, montmorillonite, clintonite, margarite, saponite, and chlorite-group minerals. Grains of other secondary phases also occur including Ni-rich metallic Fe (tetrataenite and awaruite), pentlandite, magnetite, nepheline, sodalite, and hedenbergite. The most abundant phase in the Allende matrix is ferroan olivine (Fa_{45-55}) with elongate or “match-stick” morphologies; many of these matrix olivine grains contain small inclusions of magnetite, pentlandite and/or metallic Fe-Ni; some are rimmed by poorly graphitized carbon. This type of rim also surrounds small voids in the matrix.

Matrix material in the Bali-like oxidized subgroup ($CV3_{OxB}$) contains several phyllosilicates: Fe-bearing saponite, saponite, Na phlogopite, Al-rich serpentine, and Na-K mica. Also present are grains of such secondary phases as magnetite, carbonates, secondary Ca-phosphates, Ni-rich sulfides, fayalite, Ca-Fe-rich pyroxene, and andradite. Olivine has a large compositional range (Fa_{16-48}) with distinct peaks at Fa_{19-20} and Fa_{34-38}.

Members of the reduced subgroup ($CV3_{Red}$) have matrices with major olivine and minor to accessory pyrrhotite, low-Ca pyroxene, metallic Fe-Ni, and hydroxylapatite. Portions of the matrices were oxidized and contain magnetite, hedenbergite, andradite, kirschsteinite, ferrihydrite, and saponite. Small olivine grains in the matrix appear to have a flat compositional distribution, ranging from approximately Fa_2 to Fa_{60}. Portions of the Vigarano matrix are relatively coarse and porous and are probably contaminated with comminuted chondrule debris.

7.7.6 CK Chondrites

Kallemeyn et al. (1991) defined the CK carbonaceous-chondrite group and noted that grain sizes in the matrix increased with metamorphic grade: type 3 ($\leq$0.1–10 μm); type 4 (5–50 μm); type 5 (50–200 μm); type 6 (50–300 μm). Matrix material in Ningqiang (classified as either C3-ungrouped or CK3-an) consists of poorly sorted angular, anhedral grains of olivine with minor awaruite, pentlandite, troilite, and magnetite. Matrix in CK4 Karoonda contains 10–25-μm-sized olivine grains with minor low-Ca pyroxene and plagioclase.

One of the least-equilibrated CK chondrites is NWA 5343 (type 3.6 or 3.7; Dunn et al. 2018). Unweathered matrix regions are relatively dark in backscattered electron (BSE) images. They consist of silicate grains (subhedral to anhedral olivine, euhedral Ca-pyroxene and euhedral plagioclase, averaging 15–20 μm) and minor micrometer-sized magnetite grains separated by 10–14 vol% pores. Submicrometer-sized silicate and oxide grains occupy some of the pores.

The matrices of most CK chondrites also contain small angular silicate grains that appear to have been derived from crushed chondrules. Some coarser chondrule fragments are also apparent. The angular silicate grains are separated by irregular pore spaces. Silicate grains in CK3.8 NWA 1559 (one of the least-equilibrated CK chondrites) average 11 μm in size.

7.7.7 Ungrouped Carbonaceous Chondrites

There are many ungrouped carbonaceous chondrites with abundant fine-grained matrix material. A few unique individuals are described in the following paragraphs.

Acfer 094 is a primitive, ungrouped carbonaceous chondrite with unequilibrated silicates; it has the highest concentration of presolar grains (mainly SiC and diamond) in any analyzed meteorite. Approximately 60 percent of Acfer 094 consists of 10- to 50-μm-sized grains of kamacite, pentlandite, troilite, olivine (Fa_{25-60}), and low-Ca pyroxene (Fs_{1-16}) that are probably chondrule fragments. TEM studies (Newton et al. 1995; Greshake 1997) show the fine-grained component of the matrix to consist of 200–300 nm grains of forsterite (Fa_{0-2}), 300–400 nm grains of enstatite (Fs_{0-3}), 1.0–1.5 μm grains of Ca pyroxene, Ni-bearing pyrrhotite, rare pentlandite, and Si-, Mg-, and Fe-rich amorphous material. There are also rare grains of ferrihydrite and phyllosilicates (probably chlorite and serpentine, the latter possibly a product of terrestrial weathering). The matrix contains 11-μm-diameter ultraporous regions consisting of amorphous Fe-Mg silicates, forsterite, enstatite, Fe-Ni sulfide, organics, and ~40 vol% pores. Matsumoto et al. (2019) concluded that the pores formed after ice melted and flowed away; they interpreted these structures as "fossil ice."

Acfer 094 contains widely distributed Fe-rich intergrowths of magnetite and Fe-Ni sulfides with a very heavy bulk O-isotopic composition ($\delta^{17,18}O \approx +180‰$) (Sakamoto et al. 2007; Seto et al. 2008; Abe et al. 2017). This "cosmic symplectite (COS)" material contains three distinct layers: (1) an inner layer of 100–200 nm magnetite and pyrrhotite grains, (2) a symplectic intergrowth of nanometer-sized magnetite and Fe-Ni sulfide grains, and (3) an outer Fe-hydroxide layer. Also present in this material is a single grain of Na-sulfate (Na_2SO_4) (Matsumoto et al. 2020).

Adelaide is an unequilibrated type-3 carbonaceous chondrite with a very fine-grained (typically 0.1–0.5 μm) matrix containing major olivine (Fa_{31-84}) and minor clusters of orthoenstatite (Brearley 1991). Also present are minor amounts of 1–2-μm-sized grains of magnetite, an amorphous Fe-rich phase, rare spherical aggregates of 30–100-nm-sized grains of pentlandite, and rare coarser (2–15 μm) olivine and low-Ca pyroxene grains (possibly chondrule fragments). Metallic Fe–Ni is absent from the matrix, probably due to terrestrial weathering.

Bench Crater is a 1.5 × 3 mm fragment of a carbonaceous chondrite found among Apollo 12 samples. Although currently classified as ungrouped, Bench Crater may be a shocked CM chondrite. Its precursor projectile struck the Moon at a sufficiently low relative velocity to avoid large-scale melting. The principal component in the rock is fine-grained matrix material consisting of pyrrhotite laths, magnetite platelets and framboids, minor pentlandite, Mn-rich ilmenite, carbonate (calcite and dolomite), and rare chalcopyrite. Polycrystalline assemblages of mafic silicates within the matrix were probably derived from fragmented chondrules. Saponite mixed with decomposed serpentine also occurs. The decomposition may have resulted from shock heating when Bench Crater hit the Moon (Zolensky 1997).

LEW 85332 is a unique, highly unequilibrated (type 3.0–3.1) carbonaceous-chondrite breccia (Rubin and Kallemeyn 1990). It has a hydrated matrix containing fine-grained interlayered saponite and serpentine (both with highly variable FeO/(FeO+MgO) ratios) associated with polycrystalline ferrihydrite (probably a product of terrestrial weathering) (Brearley 1997). The phyllosilicates occur in sheaves and as randomly oriented interlocking grains; elongated grains of troilite (0.5–2.0 μm long) occur subparallel to some of the phyllosilicates. Also present are magnesian olivine and low-Ca pyroxene grains; rare examples of pyroxene are intergrowths of monoclinic and orthorhombic phases. There are numerous primitive carbonaceous-chondrite clasts in the breccia with uncharacterized greenish-brown to yellow-brown phyllosilicate-rich matrices containing magnetite (framboids, platelets, round and angular grains), small angular to euhedral troilite grains (some with pentlandite exsolution lamellae), small isolated silicate grains, rare microchondrules, and fragments of normal-sized chondrules. Other clasts contain abundant phyllosilicates (65–70 vol%), major troilite (25–30 vol%), minor mafic silicate grains (1–2 vol%), and rare pentlandite. A unique clast consists mainly of fine-grained silicate-rich matrix material (97 vol%) and 3 vol% of numerous randomly oriented needles (<1–3×2–8 μm in size) of schreibersite with ~12 wt% Ni.

Tagish Lake, an ungrouped C2 chondrite fall, has a high porosity and an extremely low bulk density (1.66 g cm^{-3}); it has the highest bulk C content (5.81 wt%) of any known chondrite whole-rock. Nanodiamonds in the matrix account for 0.4 wt% of the bulk C. The meteorite consists of abundant carbonate-poor and minor carbonate-rich lithologies. The matrix of the carbonate-poor lithology consists of major phyllosilicates (saponite and serpentine) and lesser amounts of magnetite, Fe,Ni-sulfide, and olivine (peaking at Fa_1, but ranging to Fa_{39}) and rare grains of low-Ca pyroxene (peaking at Fs_2). The fine-grained textures of the phyllosilicates closely resemble those of CI chondrites (Brearley 2006). The fine-grained carbonate-rich lithology contains abundant calcite and minor dolomite and siderite. Figure 7.25 shows abundant magnetite in the matrix.

7.7.8 Ordinary Chondrites

In the most unequilibrated type-3 ordinary chondrites, fine-grained opaque silicate matrix material occurs (1) around chondrules, opaque aggregates, and chondritic clasts, (2) as discrete lumps up to 2 mm in size in the whole rock, and (3) as discrete 100–400-μm-sized lumps enclosed within a few chondrules (e.g., Scott et al. 1984). More-or-less in order of decreasing abundance, fine-grained OC matrix consists of ~0.1-μm-sized grains of olivine ($Fa_{\sim1-91}$), and somewhat coarser grains of magnesian to moderately ferroan low-Ca pyroxene ($Fs_{\sim1-30}$), Ca-pyroxene, albitic plagioclase, metallic Fe-Ni (kamacite, taenite, awaruite), troilite, magnetite, spinel, chromite, anorthitic plagioclase, pentlandite, carbide (cohenite, haxonite), merrillite, and twinned calcite crystals. Many type-3 OC do not contain all these phases; the proportions of phases vary from chondrite to chondrite. Chondrule fragments, isolated silicate grains and rare microchondrules are also present within matrix material. The principal bulk compositional difference between various matrix occurrences in the same meteorite is in FeO/(FeO+MgO).

Semarkona (LL3.00) has been moderately aqueously altered; in addition to fine-grained olivine, the matrix contains twinned monoclinic low-Ca pyroxene, abundant Na-bearing, Fe-rich smectite as well as maghemite, Ni-rich pyrrhotite, and numerous grains of Ni-rich metal (taenite and awaruite), pentlandite, nickel phosphide, cohenite, edscottite, and wüstite. Some olivine and pyroxene grains have been altered; their alteration products contain voids. A few moderately altered sulfide grains have been partially replaced by Fe-rich phyllosilicate (Fe-rich chlorite). The presence of pristine amorphous silicates in the matrix indicate that the alteration of the Semarkona whole rock was heterogeneous (Dobrică and Brearley 2020).

Although most olivine grains in Semarkona matrix have low MnO concentrations, a few magnesian grains (LIME olivines) contain up to 1.4 wt% MnO. Matrix in several OC contains amorphous material that is rich in normative albite (e.g., LL3.1 Bishunpur) or has a bulk composition intermediate between that of albite and ferroan olivine (e.g., LL3.3 ALHA76004 and LL3.3 ALHA77015).

In addition to normal fine-grained matrix material, Tieschitz (H/L3.6) contains "white matrix" with clastic grains of albite and nepheline (or an intergrowth of the two) as well as olivine, Ca-pyroxene, sulfide, and metallic Fe-Ni.

With increasing degrees of metamorphism, the proportion of low-FeO olivines tends to decrease, the range of olivine compositions (i.e., σ_{Fa}) decreases significantly, and the concentrations of minor elements in olivine change – mean Al_2O_3, Cr_2O_3, CaO, and NiO tend to decrease and mean MnO tends to increase.

Recrystallized matrix, composed mainly of somewhat coarser mafic silicate grains, occurs in type-3.4–4 OC. It is translucent when viewed microscopically in transmitted light, unlike the (opaque) fine-grained matrix in highly unequilibrated OC. Recrystallized matrix also contains lower bulk S, CaO, and FeO, and higher MgO than opaque matrix. Although Semarkona contains no recrystallized matrix, more-equilibrated OCs contain progressively higher proportions of recrystallized matrix (Huss et al. 1981).

7.7.9 R Chondrites

The fine-grained opaque matrix in R3 chondrites and type-3 clasts in R-chondrite breccias has a clastic texture and contains abundant 2–3-µm-sized ferroan olivine ($Fa_{35–60}$; occurring as isolated anhedral to moderately elongated grains) and minor low-Ca pyroxene (e.g., Kallemeyn et al. 1996; Bischoff et al. 2011; Isa et al. 2014). Chondrule fragments and coarse isolated silicate grains also occur. More-equilibrated R chondrites contain translucent, recrystallized matrix with somewhat coarser olivine grains with narrower compositional ranges.

In most cases, the dominant opaque phase in R-chondrite matrix material is sulfide (typically pyrrhotite); oxide is generally far less abundant. Sulfides include pyrrhotite, troilite, pentlandite, chalcopyrite, and pyrite; oxides include magnetite, chromite, Al-rich chromite, magnetite-chromite solid solution, and ilmenite. There are also rare grains of metallic Fe-Ni (kamacite, martensite, and, more commonly, awaruite) and very rare, small grains of unusual sulfides (erlichmanite (OsS_2), laurite (RuS_2), and a Os,Ru,Fe-rich sulfide), arsenides (sperrylite ($PtAs_2$)), sulfarsenides (irarsite (IrAsS), and a phase close to (IrPt)AsS), and tellurides (the $PtTe_2$ polymorphs chengbolite[1] and moncheite). Phyllosilicates have not been reported.

7.7.10 Enstatite Chondrites

Type-3 enstatite chondrites were first recognized by Prinz et al. (1984), but not examined thoroughly for many years. Enstatite-chondrite matrix material was long thought not to exist, but the discovery of presolar grains in these meteorites (e.g., Ebata et al. 2006, 2007) suggested otherwise.

EH3 chondrites. Rubin et al. (2009) described matrix material in two EH3 chondrites as consisting of three principal components: opaque grains, silicate grains and an inferred fine-gained nebular component (the site of presolar grains). Weisberg et al. (2014) characterized the matrix in ALH 81189 as consisting of ~1–3-μm-sized rounded grains of silica (with both amorphous and crystalline occurrences), forsterite, enstatite and albitic plagioclase admixed with Cr-bearing troilite, oldhamite, daubréelite, schreibersite, small amounts of metallic Fe-Ni and lithic fragments. Rubin et al. (2009) also reported relatively coarse grains of perryite. The matrix in Y-691 contains a similar mineral assemblage, but appears slightly coarser and somewhat porous.

Type-3 enstatite-chondrite whole rocks consist mainly of chondrules and coarse silicates and opaque grains forming an interchondrule clastic matrix. This interchondrule material may have formed mainly from shattered chondrules. The lithic fragments and coarse opaque and silicate grains in the matrix patches are probably fragments of chondrules and grains from these interchondrule regions.

EL3 chondrites. Rubin (2010a) described matrix material in the MAC 88136 EL3 chondrite occurring in patches around chondrules and adjacent to chondrule fragments and metal-sulfide assemblages. Matrix material has a clastic texture and consists of ~25 vol% irregular 4–25-μm-sized chips of metallic Fe-Ni and sulfide amid ~75 vol% silicate grains. These silicates are present as 4–10-μm-sized crushed chondrule fragments along with a fine-grained (presumably nebular) component. Weisberg et al. (2014) examined matrix in MAC 88136 and EL3 QUE 93351 and found it to consist mainly of 1–2-μm-sized irregular-to-angular grains of enstatite and albitic plagioclase along with minor silica. Weisberg et al. (2019, 2020) found that EL3 NWA 8785 contains 33.7 vol% fine-grained FeO-rich matrix consisting of mineral and chondrule fragments along with submicrometer aqueously altered FeO-rich material with variable amounts of FeO, MgO, and SiO_2.

7.7.11 Kakangari

Kakangari, an unequilibrated member (and prototype) of the K-chondrite grouplet, contains 30 vol% fine-grained matrix material with 0.2–1.5-μm-sized grains of major $Fs_{2.7\text{-}5}$ enstatite

[1] Although the mineral name chengbolite is used in this book, it is a recently discredited Bi-poor variety of moncheite.

(disordered intergrowths of ortho and clino) and forsterite (Fa_{0-3}) and minor albite, anorthite, Cr-spinel, troilite, metallic Fe-Ni and ferrihydrite (e.g., Brearley 1989). These phases typically occur in 2–8-μm-sized aggregates. About 5–10 vol% of the matrix consists of a clastic component with 5–20 μm angular mineral fragments (mainly twinned clinoenstatite with minor olivine and Ca-pyroxene), probably derived from broken chondrules.

7.8 Carbide-Magnetite Assemblages

About 10 percent of type-3 H, L, and LL chondrites contain abundant assemblages consisting of carbides (cohenite and haxonite), magnetite, troilite, and high-Ni metallic iron (tetrataenite and awaruite). Cobalt-rich kamacite (with up to 35 wt% Co) occurs only in L and LL assemblages; Ni-rich sulfide (possibly an intergrowth of troilite and pentlandite) occurs only in LL assemblages. The modal abundances of the constituent phases vary from assemblage to assemblage.

Carbide-magnetite assemblages occur as rounded, ellipsoidal, or irregularly shaped polymineralic nodules, 200–600 μm in size. They are present in the fine-grained silicate-rich matrix, in chondrule rims, and within Type I chondrules. The assemblages were described originally by Taylor et al. (1981) and characterized in more detail by Krot et al. (1997b). Primitive CO3.00 chondrites (e.g., DOM 08006) contain carbide-magnetite veins and assemblages wherein metallic Fe-Ni has been partially altered to magnetite.

These objects probably formed from metal-troilite nodules by hydrothermal alteration involving C–O–H-bearing fluids (Krot et al. 1997b) on OC and CO parent bodies. Some Type I chondrules in LL3.00 Semarkona that contain internal kamacite blebs surrounded by mesostasis or olivine also contain carbide grains that extend to the chondrule surface. It appears that fluids circulating through the fine-grained chondrite matrix incompletely penetrated many chondrules. This is consistent with observations that some Semarkona chondrules have altered and unaltered mesostasis regions with a sharp boundary between them, marking the farthest penetration depth of the fluid (figure 6 of Rubin 2013a). The source of the C–O–H-bearing fluids may be ices, hydrated minerals, and/or adsorbed gases, released and mobilized during radiogenic or impact heating.

7.9 Opaque Phases and Opaque Assemblages

7.9.1 Ordinary Chondrites

The principal opaque mineral groups in unweathered equilibrated ordinary chondrites include metallic Fe-Ni (kamacite, taenite, tetrataenite, and, in some highly oxidized LL chondrites, awaruite), sulfide (mainly troilite, but also, in some cases, minor to accessory pentlandite and mackinawite), and oxides (mainly chromite with minor to accessory ilmenite and rutile, and in a few highly oxidized LL chondrites, magnetite). With increasing oxidation state ($H < L < LL$), the modal abundance of metallic Fe-Ni decreases, the taenite/kamacite ratio increases, the average kamacite grain size decreases, the kamacite Co content increases, and the kamacite Ni content decreases. [These changes are accompanied by systematic variations in the silicate and oxide phases: increases in FeO/(FeO+MgO) in olivine, low-Ca pyroxene, chromite, and ilmenite, and increases in the olivine/low-Ca-pyroxene modal ratio.] A few very oxidized LL chondrites (e.g., LL6 Appley Bridge and LL5 Parambu) contain rare metallic Fe grains with 37 wt% Co and 2.5 wt% Ni.

Many shocked OC contain small grains of metallic Cu, in many cases at the interface between metallic Fe-Ni and small, irregularly shaped grains of troilite. Metallic Cu also occurs in rare cases between metallic Fe-Ni and troilite in elongated shock veins and at the interface between metal blebs and surrounding troilite in rapidly cooled metal-sulfide assemblages. The Cu contains ~1.5–2.0 wt% Ni and has a modal abundance in many shocked type 4–6 OC of about $1.2\text{–}1.5 \times 10^{-4}$ vol%.

Some shocked OC contain grains of martensite (α_2, a distorted body-centered cubic structure), formed from high-temperature taenite during quenching. A few OC melt-rock clasts (e.g., in the Dimmitt H-chondrite regolith breccia) contain opaque blebs consisting mainly of schreibersite intergrown with metallic Fe-Ni. The matrices of aqueously altered type-3 OC contain magnetite, carbides (cohenite and haxonite), Ni-rich metal (including tetrataenite and awaruite), and pentlandite.

Many opaque assemblages in OC consist of metallic Fe-Ni associated with troilite. Although many large metallic Fe-Ni grains observed in polished sections appear free of sulfide, it is possible that many (most?) of these grains are attached to troilite in the third dimension. In contrast, many chromite grains do seem to be monomineralic; others are intergrown with kamacite and/or troilite.

About 1 percent of chondrules have concave indentations at their surfaces (sometimes called "craters"); a few of these are partially filled with opaque phases (generally metallic Fe-Ni and troilite) that appear to have been expelled via centrifugal force while the chondrules were spinning and partly molten (e.g., figure 8 of Grossman and Wasson 1985).

Gooding and Keil (1981) described rare metallic chondrules (~0.3 percent of chondrules) that consist mainly of metallic Fe-Ni with accessory sulfide, phosphate, and/or phosphide, and, in some cases, magnetite or silicate grain fragments. Some may be droplets that were expelled and lost from larger silicate chondrules during chondrule formation; others may be parent-body shock products.

Some ordinary chondrites contain large metal nodules, large troilite nodules, or large metal-troilite nodules that range up to ~3 cm in maximum dimension. The metal in these nodules is typically depleted in refractory siderophile elements (e.g., Ir, Pt, Re, Os, Ru, Rh) and Cu by large factors (up to 240) compared to fine-grained metal in the same meteorite. Nevertheless, the nodules are not depleted in W or Mo (both normally refractory), nor in common (Fe, Ni, Co) or volatile (Au, As, Ga, Sb) siderophile elements. These chemically fractionated nodules probably formed by a complicated shock-metamorphic process involving impact melting and vaporization; oxidation of W and Mo to form volatile oxides, concomitant partial oxidation of Re and Os; rapid fractional condensation of refractory siderophiles; transport of the residual vapor; condensation of Cu, S, and Se into vugs and fractures to form sulfide; and condensation of the remaining siderophiles to form metal nodules and veins. During vapor cooling and condensation (or during subsequent parent-body annealing) W, Mo, Re, and Os oxides were reduced to their elemental states (e.g., Widom et al. 1986; Rubin 1995, 1999).

Metal-sulfide melts and silicate melts are immiscible. During impact melting of ordinary chondrites, the fine metal grains tend to coagulate and separate from the silicate melt. A variety of features can be produced: (1) Molten metal can form channels through unmelted or partially melted silicate clasts as in the Portales Valley (e.g., Rubin et al. 2001; Ruzicka et al. 2005) and Rose City (e.g., Mason and Wiik 1966; Rubin 1995) H-chondrite impact-melt breccias. (2) Metal and sulfide can solidify into scattered centimeter-sized rounded nodules as in the PAT 91501 L-chondrite impact-melt rock (e.g., Mittlefehldt and Lindstrom 2001). (3) In the unusual case of the rapidly cooled LL-chondrite impact-melt rock LAR 06299 (Rubin and Moore 2011), metal and sulfide droplets responded to the gravitational field of the asteroid and aligned themselves with gas bubbles (vesicles) in the melt according to density – bubbles at the top end, troilite in the middle, and metallic Fe-Ni at the bottom:

Gas bubble (ρ < 0.001 g cm^{-3})	**↑ UP**
troilite (ρ = ~4.7 g cm^{-3})	
metallic-Fe-Ni (ρ = ~8.0 g cm^{-3})	**↓ DOWN**

Shock veins consisting of metallic Fe-Ni ± troilite occur in many shocked ordinary chondrites; these veins vary in thickness from <1 to >100 μm (Figure 7.27). Some veins consist mainly of small blebs of metal and troilite that form curvilinear trails within silicate grains. This feature leads to the phenomenon called "shock darkening" or "silicate darkening"; it can significantly increase the opacity of silicates, producing "black chondrites" such as L5 Farmington, LL4 Albareto, LL5 Guidder, and LL6 Dongtai. There are also shock-darkened chondritic clasts in a few OC regolith breccias such as H6 Nulles.

Many shocked OC also contain thin veinlets of chromite. There are three varieties: (1) thin, continuous elongated chromite bands; (2) discontinuous rounded and oblong tiny chromite grains; and (3) discontinuous rounded and oblong tiny chromite grains with interspersed plagioclase or glass of plagioclase composition.

A few OC impact-melt breccias (e.g., Smyer; Rubin 2002b) contain troilite-rich regions consisting of unmelted, crushed silicate grains surrounded by troilite and transected by thin troilite veins (some as thin as 0.1 μm). Few grains of metallic Fe-Ni occur, leading to FeS/Fe

Figure 7.27 Thin section of a small individual Chelyabinsk stone showing an extensive network of dark shock veins. Image courtesy of Rhian Jones. Reflected light.

modal ratios ranging up to ~500:1, far higher than the eutectic weight ratio of 7.5:1. The texture of these troilite-rich regions indicates that the sulfide formed from a fluid of very low viscosity, probably an S_2 vapor produced during shock-induced evaporation of troilite.

There are additional opaque phases in the fine-grained matrices of some primitive type-3 ordinary chondrites such as LL3.00 Semarkona. These include Fe-oxide (magnetite), Ni-rich metallic Fe (mainly awaruite), Fe-carbides (cohenite, haxonite, edscottite), and Ni-rich sulfide (pentlandite). Type I chondrules in type-3 OC contain blebs of kamacite and metal of plessitic composition as well as blebs of troilite. Rare troilite grains have daubréelite exsolution lamellae.

7.9.2 R Chondrites

The major sulfides in R chondrites include low-Ni sulfide (primarily pyrrhotite, but troilite in ALH 85151) and pentlandite. Monosulfide solid solution ($Fe_{1-x}S$-$Ni_{1-x}S$) is present in equilibrated R chondrites. Sulfides occur mainly as isolated pyrrhotite grains or as pyrrhotite-pentlandite intergrowths; some pyrrhotite grains have flamelike exsolution lamellae of pentlandite. A few sulfide assemblages contain pyrite, chalcopyrite, and metallic Cu.

Metallic Fe-Ni is absent from most R chondrites. Kamacite is rare, but several R chondrites contain awaruite (Ni_3Fe) with 60–70 wt% Ni; ALH 85151 (R3.6) also contains a cellular troilite-martensite mixture (wherein the metal has 9.9 wt% Ni) and low-Ni metallic Fe (presumably kamacite) with 37 wt% Co. Some R3 chondrites have rare grains of kamacite and tetrataenite.

Although magnetite is absent in most R chondrites, ~6 vol% magnetite (with 0.42 wt% Cr_2O_3) occurs in an unequilibrated 1.8 × 2.4 mm R-chondrite clast in R3.8 PCA 91241. Magnetite is also present in the matrix of R3.5 MET 01149. Up to several percent of Cr_2O_3-rich magnetite occurs in the amphibole- and mica-bearing R6 chondrite LAP 04840 as well as in R4-an Sierra Gorda 010 (R4-an) and R3.6 Dhofar 1514. Other oxides in R chondrites include chromite, magnetite-chromite solid solution, chromite-hercynite-ulvöspinel solid solution, and ilmenite. Some CAIs contain perovskite.

Tiny grains of various noble-metal–rich sulfides, arsenides, sulfarsenides, and tellurides have been reported including erlichmanite (OsS_2), laurite (RuS_2), sperrylite ($PtAs_2$), irarsite (IrAsS), chengbolite ($PtTe_2$), moncheite ($PtTe_2$), and related phases. Noble-metal–rich metallic phases include PtFe(Ir,Ni) alloy, Os with minor Re, Pt, Os-Ir-Ru-Pt alloy, native Au, Au-Ag alloy (electrum), ruthenosmiridim (RuOsIr), rustenburgite (Pt_3Sn), and PdPtSn.

7.9.3 Enstatite Chondrites

Enstatite chondrites contain Si-bearing metallic Fe-Ni (nearly exclusively kamacite although rare grains have higher Ni contents), phosphide (schreibersite), numerous sulfides (mainly troilite, daubréelite, oldhamite, djerfisherite, and caswellsilverite, plus niningerite in EH chondrites or ferroan alabandite in EL chondrites), carbide (mainly cohenite), and graphite. Rare sulfide phases include heideite, wassonite (Figure 7.28), sphalerite, and Mn-bearing pentlandite. Rare grains of silicon nitride (Si_3N_4) of Solar System origin have been found within grains of metallic Fe-Ni, schreibersite, and sulfide (as well as silicate) in EL3, EH3, and

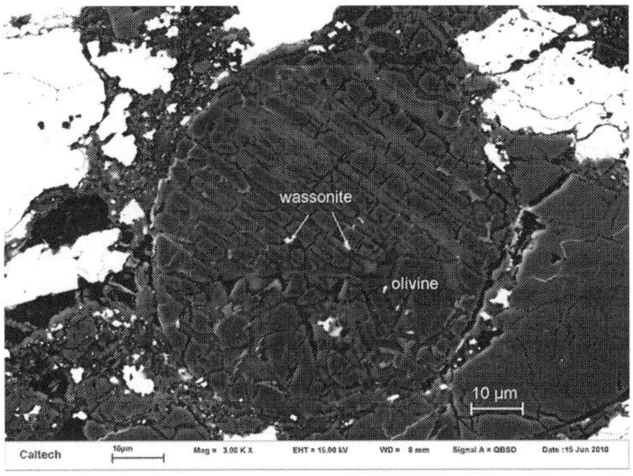

Figure 7.28 Wassonite (TiS) in a barred olivine chondrule in the Y-691 EH3 enstatite chondrite. BSE image.

EH4 chondrites. Type-3 enstatite chondrites also contain silicide (perryite). Some impact-melted EH chondrites contain keilite instead of niningerite. Rare grains of osbornite occur in a few impact-melted enstatite chondrites.

Several enstatite chondrites contain opaque shock veins (detailed in the review paper by Rubin 2015a): (1) Small sulfide veins fill fractures in some euhedral enstatite grains in the Abee EH impact-melt breccia. (2) Several EL3 and EL4 chondrites (e.g., Asuka 881314, MAC 02635, EET 90299, Dar al Gani 1031) have millimeter-long metallic Fe-Ni–rich veins, some of which also contain sulfide and silicate. (3) The Atlanta EL6 breccia contains a 9.4-mm-long kamacite vein, the Blithfield EL impact-melt breccia contains a metal vein several centimeters long, and EL6 Jajh deh Kot Lalu has a 16-mm-long oldhamite-rich vein.

Abee contains 0.1–1-mm-sized kamacite globules as well as kamacite-rich nodules up to 11 mm in maximum dimension; these metal-rich objects commonly contain euhedral enstatite grains. Blithfield has coarse, irregularly shaped metal nodules that lack euhedral enstatite. The Hadley Rille EH chondrite was shocked and partly melted when it struck the Moon (Rubin 1997c); it also contains kamacite globules. It seems likely that all these globules and nodules formed from molten metal that partly separated from a silicate-rich, metal-poor impact melt.

Rounded and subrounded, ~50–300-µm-sized metal-sulfide nodules constitute up to ~6 vol % of some EH3, EL3 and EL4 chondrites (e.g., Van Niekerk and Keil 2011; Weisberg and Kimura 2012). The sulfide phases include troilite, oldhamite, djerfisherite, and, in EH chondrites, niningerite, or in EL chondrites, ferroan alabandite. In addition to kamacite, some nodules contain schreibersite and/or perryite. Suggested models of their formation include condensation (Weisberg et al. 2006; Lehner et al. 2010) or localized impact melting (Van Niekerk and Keil 2011).

Graphite occurs in two principal forms: (1) irregular or euhedral inclusions in kamacite (probably formed by exsolution from C-bearing metallic Fe-Ni during thermal metamorphism) and (2) elongated laths, some with pyramidal terminations, surrounded by kamacite, sulfide or silicate (probably formed by crystallization from impact melts).

Carbide occurs as thin exsolution lamellae and micrometer-sized laths of cohenite within kamacite.

7.9.4 Kakangari-like Chondrites

The meteorites in this grouplet have simple opaque mineral assemblages. All of the meteorites have relatively coarse grains of kamacite, taenite, and troilite. Matrix regions in Kakangari contain small grains of Cr-spinel, troilite, and metallic Fe-Ni along with enstatite, olivine, albite, anorthite, and ferrihydrite. An unusual, possibly foreign, clast in Lea County 002 consists of normally zoned olivine, laths of clinoenstatite, Ca-pyroxene, plagioclase, and small euhedral chromite grains. Opaque phases within Kakangari refractory inclusions include spinel and perovskite. The NWA 10085 K4 chondrite contains several opaque phases – kamacite, taenite, troilite, schreibersite, and euhedral chromite.

7.9.5 Carbonaceous Chondrites

Different carbonaceous-chondrite groups contain different sets of opaque minerals.

CI chondrites contain ~10 wt% magnetite (occurring as euhedral crystals, framboids, spherulites, plaquettes, and platelets in isolated grains or in clusters), sulfides (pyrrhotite with minor pentlandite and cubanite), rare grains of elemental S and very rare 2–5 μm rounded blebs of kamacite within scarce olivine grains. Opaque phases in the matrix include limonite, Ti_3O_5 (possibly kaitianite), eskolaite, and ferrihydrite. Also present are tiny presolar grains – mainly graphite and silicon carbide. A unique 170-μm-diameter CAI found in Ivuna has small grains of spinel and metallic Fe-Ni as well as Fremdlinge enriched in Pt-group metals.

CM chondrites contain metallic Fe-Ni (kamacite) that decreases in modal abundance during aqueous alteration. Paris (type 2.7–2.9) contains 1.2 vol% metallic Fe-Ni; the CM2.0–2.1 chondrites (e.g., MET 01070, QUE 93005) generally contain $\leq$0.02 vol% metallic Fe-Ni. Significantly altered CM chondrites contain many sulfide grains with "intermediate" compositions, i.e., with atomic Ni/(Fe+Ni) ratios of 0.10–0.40. In contrast, sulfide in Paris appears to consist exclusively of pyrrhotite and pentlandite. A few CM sulfide grains have enough Fe to place them in the compositional range of troilite. Some CM chondrites also contain Fe,Ni sulfide grains with substantial Cr and P; these grains may be intergrowths of sulfide and phosphide.

Oxides include magnetite (which occurs as massive clumps, spheres, plaquettes, and equant grains) and very rare hematite (detected by XRD in Nogoya, but it might be terrestrial). Small grains of chromite, hercynite, and spinel have also been reported.

CO chondrites come in two principal varieties: magnetite rich and magnetite poor. Magnetite-rich CO chondrites such as DOM 10104, MIL 090543, and Cargo Muchacho Mountains contain 2–7 vol% magnetite, 0.4–1 vol% metallic Fe-Ni, 1–3 vol% troilite, and minor Fe-carbide (Rubin and Li 2019; Simon et al. 2019). Accessory pentlandite occurs in some of the meteorites. ALH 77307 contains haxonite (with ~5–6 wt% Ni) and cohenite (with 2–4 wt % Ni); both phases have minor concentrations of Co and small but variable amounts of Cr, Si, and P.

Chromite, when present, occurs as small (2–30 μm) euhedral to subhedral grains within FeO-rich Type II chondrules or in association with ferroan olivine grains (presumably chondrule fragments) in the matrix. In DOM 10104 and MIL 11213, small (<4 μm) chromite grains also occur in association with magnetite, both in chondrules and in the matrix.

Kamacite is generally more abundant than Ni-rich metal; tetrataenite is typically more abundant than taenite. Many opaque assemblages consist of cores of metallic Fe-Ni (kamacite, taenite, tetrataenite) and/or carbide (cohenite, haxonite) mantled by magnetite. Carbide grains within chondrules and the matrix of Cargo Muchacho Mountains and DOM 08351 are generally irregular in shape and 10–80 μm in size. They occur as isolated grains and in association with other opaque minerals (metallic Fe-Ni, troilite, and magnetite).

Ornans contains taenite, but little or no kamacite. Many Type I chondrules in magnetite-rich CO chondrites contain abundant opaques. These phases include 10–25-μm-sized spheroidal or irregular masses of kamacite, taenite, and troilite; in some cases, magnetite and pentlandite occur along with or instead of troilite and metallic Fe-Ni. In Ornans chondrules, many of the magnetite grains are intergrown with tiny grains of Ca-phosphate. Type II chondrules are generally free of opaques.

Magnetite-poor CO chondrites such as ALH 83108, Kainsaz, and Moss contain troilite (~0.6–6.2 vol%), metallic Fe-Ni (~0.4–5.0 vol%), rare Fe-carbide, and oxide. Kainsaz and Isna contain chromite and no magnetite, but most magnetite-poor CO chondrites contain both phases. Metallic Fe-Ni occurs as kamacite, taenite, and tetrataenite. Kamacite is by far the most abundant metallic phase in these rocks; it is present as intergrowths and isolated grains up to 300 μm in size. There are rare, 1–20-μm-long rounded to irregular grains of metallic Cu associated with metal and troilite in some magnetite-poor CO chondrites (e.g., Moss, Kainsaz, Lancé, Warrenton, Felix).

Kamacite compositions show systematic increases in Ni and Co and decreases in Cr with increasing subtype; taenite shows corresponding decreases in Ni content (McSween 1977; Scott and Jones 1990). These trends result from parent-body processes involving metamorphism, oxidation and partitioning of elements among silicates, metal phases and carbides. In primitive CO chondrites, most kamacite grains contain ~0.4–0.6 wt% Co, but one grain was reported in CO3.05 Colony with 4.7 wt% Co.

The modal abundance of magnetite correlates with CO3 subtype (Rubin and Li 2019): CO3.0–CO3.1 chondrites all have significant amounts of magnetite (>1.7 vol%); CO3 chondrites of subtype 3.2 and higher have little or no magnetite (0–0.6 vol%). There is also an inverse correlation between the abundances of metallic Fe-Ni and magnetite: magnetite-rich CO chondrites tend to contain little or no metallic Fe-Ni ($\lesssim$1.2 vol%), whereas metal-rich CO chondrites tend to contain little or no magnetite ($\lesssim$0.6 vol%).

CV chondrites are divided into three subgroups: (1) the reduced subgroup ($CV3_{Red}$), (2) the Allende-like oxidized subgroup ($CV3_{OxA}$), and (3) the Bali-like oxidized subgroup ($CV3_{OxB}$). All contain CAIs (formed in the nebula) – the CAIs include such opaque phases as perovskite and spinel.

$CV3_{Red}$ chondrites have escaped major parent-body alteration and, among the CV subgroups, best retain the mineral assemblage they inherited from the solar nebula. These rocks contain moderate amounts of kamacite, taenite, and troilite (as in ordinary chondrites) and only small amounts of magnetite (probably a secondary phase formed by parent-body oxidation of metallic Fe-Ni). Also present are rare carbide-metal-magnetite particles. A few millimeter-sized metallic Fe-Ni grains in Leoville contain hundreds of micrometer-sized grains of silicon nitride (Si_3N_4) as well as 50–500-nm-sized grains of Cr-V nitrides (Leitner et al. 2020).

$CV3_{OxA}$ chondrites contain Type I (low-FeO porphyritic) chondrules with spheroidal to ellipsoidal opaque nodules with major awaruite and magnetite and minor pentlandite and merrillite. [One chondrule contains a unique 290 × 510 μm magnetite-rich ellipsoidal nodule with 9 vol% euhedral awaruite grains (up to 65 μm in size) averaging 71.2 wt% Ni and 2.1 wt% Co (Rubin 1991); see Figure 11.4(a).] Chromite is present in the mesostases of some Type IIA (high-FeO porphyritic olivine) and high-FeO BO chondrules in $CV3_{OxA}$ chondrites. Chromian spinel has been reported in some BO chondrules in Allende. There are rare occurrences of spinel and corundum in chondrules; in some cases, these may be associated with small included CAI fragments. Fremdlinge, which formed during secondary alteration, contain refractory metal nuggets with associated metallic Fe-Ni, V-rich magnetite, and Fe-Ni sulfides.

$CV3_{OxB}$-chondrite matrices contain magnetite, pentlandite, tetrataenite, and awaruite. Bali-like oxidized clasts in Vigarano ($CV3_{Red}$) contain chondrules with opaque nodules consisting of Cr-poor and Cr-rich magnetite, Ni- and Co-rich metal, Ni-rich and Ni-poor sulfide and very rare kamacite (Krot et al. 2000b).

CK chondrites contain a myriad of opaque phases. Many chondrules contain 60–500-μm-sized spheroidal to ellipsoidal concentrically layered opaque nodules composed of major magnetite and pentlandite with minor-to-accessory pyrite, Ni-bearing pyrrhotite, and chalcopyrite. Similar objects (some as large as 1 mm) occur as independent aggregates outside chondrules; some contain accessory chalcopyrite and ilmenite. Magnetite grains in some aggregates have ilmenite and spinel exsolution lamellae. The aggregates probably formed from opaque nodules that were expelled from molten chondrules during chondrule formation. The oxidized mineral assemblages in these objects probably result from parent-body alteration.

Geiger and Bischoff (1995) surveyed the opaque phases in CK chondrites and found that magnetite was the most abundant opaque, averaging ~4 vol% (and containing ~4 wt% Cr_2O_3). Magnetite is the most abundant phase in opaque aggregates; it also occurs as micrometer-sized grains throughout the matrix. The principal sulfide phases in the matrix include pentlandite, pyrite, pyrrhotite, monosulfide solid solution ($(Fe,Ni)_{1-x}S$), troilite, millerite, and chalcopyrite. There are also rare micrometer-sized grains of noble metal–rich sulfides (e.g., $(Os,Ru,Ir)S_2$, PtS, $(Au,Fe,Ag)_2S$, $(Fe,Au,Co)_2S_3$), tellurides (chengbolite ($PtTe_2$)), and arsenides ($(Ir,Pt,Fe,Os)(As,S)_2$).

Metallic Fe-Ni is very rare in CK chondrites. It has been reported in Y-693, Y-82104, and Maralinga. (A single grain of kamacite was identified in Maralinga.)

CR chondrites are distinctive because about 15–20 percent of their chondrules contain one or more ~70–900-μm-sized masses of kamacite within the interior and/or at the margin. Coarse metal grains at chondrule margins tend to be compositionally zoned; they have their lowest Ni contents at the interface with the surrounding silicate matrix. A few grains of martensitic composition (8–12.5 wt% Ni) are present along with kamacite inside chondrules. (Taenite appears to be absent in CR2 chondrites.) Large kamacite masses are also present in the matrix. Metal grains exhibit a positive Co-Ni trend and have solar Co/Ni ratios. Opaque nodules in several CR chondrites contain submicrometer-sized grains of metallic Fe-Ni, Fe sulfides, Fe-oxides, and nanometer-sized grains of metallic Cu and CuFe alloys.

Many chondrules are surrounded by continuous or discontinuous troilite-free rings of numerous ~30–50-μm-sized kamacite blebs (that have relatively low concentrations of refractory siderophile elements). In individual rings, metal blebs are rather uniform in composition, but the Ni contents of metal within different rings can vary by a factor of 2. The Ni contents of coarse metal in chondrule interiors tend to be higher than those of metal within the rings.

Enveloping compound chondrules consist typically of a primary PO or BO chondrule, enclosed by a metal-bleb-laden ring, which is itself surrounded by a spherical shell of PO-chondrule material. In some cases, this shell is encircled by another metal-bleb ring. Igneous rims (which surround ~26 percent of CR chondrules) contain abundant ~5–40-μm-sized metal blebs. These grains tend to be smaller than the larger metal grains in the rings or in chondrule interiors. About 20 percent of the metallic Fe-Ni in CR chondrites occurs in the silicate matrix as 20–100-μm-sized grains.

Associated with a few percent of metallic Fe-Ni grains are the two principal sulfides: pyrrhotite (with ~2 wt% Ni and <0.05 wt% Co) and pentlandite (with ~35 wt% Ni and 0.9 wt% Co).

Magnetite is absent in the least-altered CR chondrites, but becomes increasingly abundant in more-altered specimens. The iron oxides in NWA 5028 have lower analytical totals and contain more Al_2O_3 than magnetite in Renazzo, Al Rais, or NWA 2196 (0.22–0.48 vs. $\leq$0.07 wt% Al_2O_3). The low analytical totals in NWA 5028 iron-oxide grains probably reflect higher concentrations of ferric iron and a greater hematite compositional component. The highly altered CR1 (or CR2.0) chondrite GRO 95577 contains small magnetite grains in the matrix as well as a few kamacite grains with thick magnetite rinds; coarse grains of magnetite occur within the interiors and at the margins of chondrule pseudomorphs in this meteorite.

After taking pairing into account, there are currently three separate meteorites classified as CR6 and one as CR7. Opaque phases in these rocks include kamacite, taenite, chromite, troilite, and, in the case of NWA 6901 (thought to be paired with CR6 NWA 6921 and NWA 7317), rare tetrataenite. None contains magnetite.

CH chondrites are typified by ALH 85085. These rocks contain abundant metallic Fe-Ni (20±5 vol%); the metal grains exhibit a positive Co-Ni trend and have solar Co/Ni ratios. About 75–80 percent of the metal is kamacite, 15–20 percent is plessite and ~2 percent is taenite; rare grains of tetrataenite also occur. The mean size of metal grains is 5–10 μm, but grains range up to ~700 μm. About 15–20 percent of the metal grains larger than 20 μm in ALH 85085 are approximately equant or spheroidal. A significant proportion of metal grains wrap around chondrules and CAIs. Plessite in ALH 85085 contains 8–18 wt% Ni and consists of an intergrowth of kamacite and elongated grains of taenite.

Sulfide constitutes only ~0.5 vol% of ALH 85085. The predominant phase is troilite with ~0.8 wt% Ni and 0.13 wt% Cr; rare grains of troilite within metallic Fe-Ni contain up to 7.5 wt % Cr. Also present in ALH 85085 are rare isolated grains of pentlandite (with 20 wt% Ni) and rare heazlewoodite.

There are two oxide phases in ALH 85085: Mg-Al spinel and ferroan chromian spinel (with 48 wt% Al) together constitute ~0.2 vol% of this meteorite. Dmitryivonvite ($CaAl_2O_4$) was found within a CAI in CH3 NWA 470; this new mineral is a high-pressure phase produced by shock metamorphism (Mikouchi et al. 2009).

ALH 85085 contains ~4 vol% dark C1 clasts. Opaque phases in these clasts include troilite, pentlandite, and magnetite (which occurs as framboids, platelets, spherulites, and isolated <1–5-μm-sized grains).

Bencubbinites (a.k.a. CB chondrites) are divided into coarse-grained CB_a chondrites (e.g., Gujba, Bencubbin, Weatherford, Fountain Hills, NWA 1814) and finer-grained CB_b chondrites (e.g., Hammadah al Hamra 237, QUE 94411, MAC 02675). Metal grains exhibit a positive Co-Ni trend and have solar Co/Ni ratios.

CB_a chondrites contain 40–60 vol% metallic Fe-Ni (with 5–8 wt% Ni), mainly occurring as centimeter-sized ellipsoidal nodules (or nodule fragments) with variable amounts (0.005–5 vol%) of included Cr-bearing troilite. The metal in these nodules contains 0.1–1 wt% Cr and P and is depleted by factors of up to 1,000 relative to CI chondrites in moderately volatile siderophile elements (e.g., Ga, Ge, As, Sb).

CB_b chondrites contain >70 vol% metallic Fe-Ni with 4–15 wt% Ni. Small metal grains in this subgroup exhibit symmetrical compositional zoning; the concentrations of Pt-group elements progressively decrease from grain cores to rims. Also present in the QUE 94411 CB_b chondrite are rare coarse metal nodules like those in CB_a chondrites.

7.10 Presolar Grains

Presolar grains are the oldest meteoritic minerals (e.g., Clayton and Nittler 2004; Lodders and Amari 2005), many predating the Solar System by hundreds of millions of years, some by more than a billion (e.g., Gyngard et al. 2014; Heck et al. 2020). They range from 2 nm to 25 μm (e.g., Gyngard et al. 2018); some of the larger grains form aggregates >200 μm in size. Presolar phases include C polymorphs (diamond, graphite, and lonsdaleite[2]); moissanite (SiC); cohenite ($(Fe,Ni)_3C$); Ti-, Mo-, Zr-, and Ru-rich carbides; metallic iron; Ni-bearing kamacite; nierite (Si_3N_4); corundum (Al_2O_3); spinel ($MgAl_2O_4$); hibonite ($CaAl_{12}O_{19}$); TiO_2; magnesiochromite ($Mg(Cr,Al)_2O_4$); and silicates (predominantly olivine, low-Ca pyroxene, and silicaceous glass), as listed in Table 7.3. Organic compounds also occur. A 2018 report of silica grains within a dust shell around a supernova remnant suggests that presolar silica might also be found in the matrices of primitive meteorites.

Diamonds are the most abundant presolar grains, constituting ~0.04–0.05 percent of primitive chondrites. They are typically euhedral and ~2 nm in size.

Presolar graphite grains range from 1 to 20 μm. Many contain 20–500-nm-sized grains of TiC (khamrabaevite); while some TiC-bearing graphite grains also contain Zr- and Mo-rich carbides, other TiC-bearing grains instead contain kamacite, cohenite, and metallic Fe.

Presolar SiC grains have been found in primitive ordinary, carbonaceous, and enstatite chondrites. Many grains are euhedral; others are anhedral and blocky.

Presolar oxide grains include micrometer- and submicrometer-sized corundum (Al_2O_3), spinel ($MgAl_2O_4$), hibonite ($CaAl_{12}O_{19}$), and TiO_2 (probably rutile).

Presolar silicate grains have been identified in interplanetary dust particles (IDPs), micrometeorites, primitive carbonaceous chondrites, LL3 chondrites, and EH3 chondrites mainly by their O-isotopic anomalies. These presolar silicates include olivine, low-Ca pyroxene,

[2] Lonsdaleite, often called hexagonal diamond, appears to be faulted and twinned cubic diamond and not a discrete phase (Németh et al., 2014).

Table 7.3 *Presolar minerals and phases*

Name	Formula
Amorphous Carbon	C
Amorphous Silicate	Mg-Fe silicate
Chromite	$FeCr_2O_4$
Cohenite	Fe_3C
Corundum	Al_2O_3
Diamond	C
Enstatite	$(Mg,Fe)SiO_3$
Eskolaite	Cr_2O_3
Fe-Sulfide	FeS
Forsterite	$(Mg,Fe)_2SiO_4$
Graphite	C
Hibonite	$CaAl_{12}O_{19}$
Iron	Fe
Khamrabaevite	TiC
Magnesiochromite	$Mg(Cr,Al)_2O_4$
Magnetite	Fe_3O_4
Mo,Zr,Ru-Carbide	(Mo,Zr,Ru)C
Moissanite	SiC
Nierite	Si_3N_4
Oldhamite	CaS
Osbornite	TiN
Rutile	TiO_2
Spinel	$MgAl_2O_4$

Al-rich silicate, and amorphous material; whereas most presolar silicates are magnesian, a few contain more FeO than MgO.

7.11 Calcium-Plagioclase-Bearing Clasts

The major silicate phases in equilibrated ordinary chondrites are olivine (ol), orthopyroxene (opx), Ca-rich clinopyroxene (cpx, typically augite or diopside), and sodic plagioclase (plag); pigeonite (pig) also occurs. These same phases are the principal minerals in common terrestrial volcanic and plutonic rocks, although the plagioclase in these rocks is typically calcic: basalt (plag + cpx ± opx or pig); norite (plag + opx); troctolite (plag + ol); orthopyroxenite (opx); harzburgite (ol + opx); lherzolite (ol + cpx + opx); wehrlite (ol + cpx). Rounded-to-angular chondrule-sized clasts, mineralogically similar to (although finer-grained than) these terrestrial igneous rocks (particularly troctolites, basalts, norites, and orthopyroxenites) have been reported in several ordinary chondrites (e.g., Hutchison and Bevan 1983; Mittlefehldt et al. 1995); following are descriptions of a few of them.

Hutchison (1982) reported two basaltic clasts in H/L3.6 Tieschitz: one contains a matrix of twinned calcic plagioclase (An_{83}) surrounding zoned subcalcic augite-pigeonite grains with small olivine blebs; the other clast has abundant magnesian low-Ca pyroxene ($Fs_{14}Wo_4$), minor magnesian olivine (Fa_{15}), and minor plagioclase (An_{60-80}).

Grossman and Rubin (1986) studied an ellipsoidal 2.5 × 3.5 mm, holocrystalline, lherzolitic/noritic clast in LL3.4 Manych that consists of 61 vol% olivine ($Fa_{14.9}$), 19 vol% low-Ca pyroxene ($Fs_{12.0}Wo_{4.0}$), 11 vol% pigeonite ($Fs_{15.1}Wo_{14.5}$), and 9 vol% calcic plagioclase ($Ab_{18.8}Or_{1.3}$); only trace amounts of metallic Fe-Ni and sulfide are present.

As summarized by Mittlefehldt et al. (1998), troctolitic clasts in L5 Barwell, L6 Y-75097, and L6 Y-793241 have H-chondrite-like O-isotopic compositions and thus cannot have been derived from their L-chondrite hosts by impact melting. Subhedral-to-euhedral olivine grains constitute ~70–80 vol% of the clasts; the remainder consists of crystalline plagioclase (or maskelynite), minor chromite, and minor-to-accessory merrillite or chlorapatite. Pyroxene appears to be absent. Olivine in the clasts has the same composition as olivine in the hosts. Plagioclase is generally more calcic in the clasts than the hosts (e.g., An_{44-61} vs. An_{19} for Y-793241); the Barwell clast is compositionally zoned: plagioclase near its margin is less calcic (An_{20}) than its core (An_{70-74}).

Suggested origins for the clasts include impact melting of host or projectile material, igneous activity on differentiated asteroids (although the clasts differ in texture and mineral chemistry from known achondrites), and formation in the solar nebula by flash heating and extensive devolatilization. Different clasts may have formed by different processes.

The Manych clast studied by Grossman and Rubin (1986) shows a smooth decrease in elemental abundances with increasing volatility, consistent with melting of normal chondritic material accompanied by devolatilization. The bulk chemical composition of the clast could have formed by devolatilization of small amounts of Si, intermediate amounts of Cr and Mn, and major amounts of Na and K. If such a process occurred, the chondritic precursors of the clast must have remained at high temperatures for longer periods than normal chondrules; the clast could contain fractionated K isotopes. It is not known if the clast formed in the nebula (perhaps in a dust-rich environment) or by impact melting on its parent asteroid.

The troctolitic clasts in the L chondrites could not have formed by impact melting host material because the clasts have H-chondrite-like O-isotopic compositions. Although the clasts' REE patterns generally appear to be fractionated, the reported data may not be representative.

7.12 Eclogitic Clasts

Terrestrial eclogites are mafic metamorphic rocks rich in omphacite $(Ca,Na)(Mg,Fe^{+2},Al)Si_2O_6$) and pyrope garnet ($Mg_3Al_{22}(SiO_4)_3$). Kimura et al. (2013) reported 1–3-mm-sized, granular-textured, eclogitic clasts in CR2 NWA 801. Constituent minerals in the clasts include omphacitic pyroxene, pyrope-rich garnet, olivine, orthopyroxene, and minor graphite, phlogopite, chlorapatite, metallic Fe-Ni, troilite, and pentlandite. No plagioclase or spinel-group minerals were observed. The omphacite has a disordered *C*2/*c* structure; this phase contains 37 mol% diopside-hedenbergite, 34 mol% jadeite, and 19 mol% enstatite-ferrosilite. The clasts appear to have formed at moderately high temperatures and pressures (940–1,080°C and 2.8–4.2 GPa) and may be achondrite projectile fragments derived from a large differentiated parent body (Hiyagon et al., 2016).

7.13 Impact-Melt-Rock Clasts

Ordinary chondrite regolith and fragmental breccias contain up to ~3 vol% lithic clasts with igneous textures (typically porphyritic, poikilitic, skeletal, or glassy). The clasts range in size from ~40 μm to ~4 cm and vary in shape from spheroidal to angular and blocky. Many are light colored and depleted in metallic Fe-Ni and sulfide relative to their hosts. Some form sharp boundaries with their hosts; others have blurred, recrystallized interfaces. The impact-melt-rock clasts generally did not form *in situ*; a typical clast was probably derived from elsewhere on the same parent body, but it is unclear if the distance from the point of origin to the present host is centimeters or tens of kilometers. Those with recrystallized boundaries appear to have been annealed after incorporation into their present location.

These igneous clasts have been widely interpreted as having formed by impact melting (e.g., Fodor and Keil 1975, 1976, 1978; Keil 1982; Rubin et al. 1983; Rubin 1985). In most cases the progenitor of a clast was the host meteorite; these clasts have bulk chemical and O-isotopic compositions as well as olivine and low-Ca pyroxene compositions similar to those of their hosts. Some of these clasts contain relict, unmelted ferromagnesian chondrules, and/or chondrule fragments. A minority of clasts differ from their hosts in mineral chemistry, bulk chemistry, and/or bulk O-isotopic composition and presumably formed from projectiles.

Poikilitic impact-melt-rock clasts consist commonly of large monocrystalline pyroxene grains (oikocrysts) containing a few to many smaller olivine grains (chadacrysts). In many cases the pyroxene grains are orthopyroxene; in a few cases, they are low-Ca clinopyroxene.

Porphyritic clasts contain numerous olivine and/or low-Ca pyroxene phenocrysts of various sizes; the clasts consist mainly of subhedral grains separated by a glassy or microcrystalline mesostasis. Such textures tend to be different than those of common Type I and II porphyritic chondrules.

Some clasts cooled quickly and contain skeletal olivine and/or low-Ca pyroxene grains surrounded by a glassy or microcrystalline mesostasis. A particular clast in the Dimmitt H-chondrite regolith breccia experienced a cooling gradient: one side is glassy with small skeletal olivines; the grains become coarser and the proportion of glass decreases progressively toward the opposite side of the clast. Other clasts quenched to become glass-rich spherules.

Lunning et al. (2020) reported several porphyritic impact-melt-rock clasts in R chondrites; they contain normally zoned olivine phenocrysts within glassy mesostasis. They are described in more detail in Section 11.5.

A few impact-melt-rock clasts have been described from enstatite chondrites. A ~2 × 4-cm-sized clast in EH3 Parsa has a porphyritic texture comprising euhedral, subhedral, and skeletal enstatite crystals, accessory forsterite, and small blebs of kamacite and sulfide surrounded by a cryptocrystalline mesostasis (Bhandari et al. 1980; Rubin 1985). A 2.5 cm clast in EL6 Hvittis has a fine-grained hypidiomorphic granular texture and consists of interlocking 50–100-μm-sized grains of enstatite and, to a lesser extent, plagioclase with minor to accessory amounts of troilite, ferroan alabandite, daubréelite, oldhamite, graphite, kamacite, martensite, low-Ni schreibersite, and sinoite (Rubin 1983a). There are relict inclusions in the clast that contain trapped host material consisting of coarse grains of silicate, sulfide, and kamacite, and minor graphite and sinoite. A 110-μm-diameter albitic-composition glass spherule in EH3 Adhi Kot contains a few small rounded grains of troilite (Rubin 1983b).

Few impact-melt-rock clasts have been reported in carbonaceous chondrites. Rare exceptions include those in $CV3_{Red}$ RBT 04143 and $CV3_{OxA}$ LAR 06317, which consist of small olivine grains surrounded by vesicular glass (Lunning et al. 2016).

7.14 Impact-Melt Pockets

In contrast to impact-melt-rock clasts, impact-melt pockets were produced *in situ* from host material, just like shock veins. The melt pockets in L-group ordinary chondrites are typically <500 μm across and consist of dark-colored glass or a cryptocrystalline groundmass with bulk compositions that are typically enriched in the plagioclase chemical component [SiO_2 (40–58 wt%), Al_2O_3 (2–11 wt%), Na_2O (0.6–4.9 wt%), K_2O (0.1–1.2 wt%)] and depleted in the mafic chemical component [FeO (2–16 wt%), MgO (9–33 wt%), Cr_2O_3 (0.1–1.8 wt%)] relative to their host (Dodd and Jarosewich 1979, 1982). For comparison, average L-chondrite falls contain: 39.7 wt% SiO_2; 2.2 wt% Al_2O_3; 0.95 wt% Na_2O; 0.11 wt% K_2O; 14.5 wt% FeO; 24.7 wt% MgO; and 0.53 wt% Cr_2O_3 (Jarosewich 1990). Most melt pockets contain relict olivine and/or low-Ca pyroxene grains derived from their host; some also contain rounded or irregular grains of metallic Fe-Ni and troilite. A few contain chromite. Some melt pockets are rounded and quasi-equant; others are elongated or somewhat sinuous.

Impact-melt pockets are not restricted to ordinary chondrites:

(1) The RKPA80259 EH5 impact-melt breccia contains melt pockets of feldspathic glass adjacent to coarse grains of enstatite. In many cases, small spheroidal blebs of metallic Fe-Ni and/or troilite also occur (Fagan et al. 2000).
(2) There are also numerous impact-melt pockets in contact with CAIs in the $CV3_{Red}$ chondrites Leoville and Efremovka (Acquadro et al. 2019). The pockets contain glass, normally zoned olivine grains with branching, skeletal and so-called Christmas-tree morphologies, and opaque globules consisting of sulfide and metallic Fe-Ni.

8

Petrologic and Mineralogical Characteristics of Meteorite Groups

In this book, minerals in chondrites are considered primary if they formed in the solar nebula (even if they formed by reaction of a different precursor phase with nebular gas) and secondary if they formed on their parent body. Minerals in differentiated meteorites are considered primary if they formed by igneous processes or upon initial cooling, and secondary if they formed during subsequent alteration.

8.1 Chondrites: Undifferentiated Meteorites

Chondrites are undifferentiated meteorites; they were derived from asteroids that did not experience wide-scale or global melting. Metallic Fe-Ni and sulfide did not quantitatively separate from silicate as immiscible liquids in chondrite parent bodies. Chondrites tend to retain features inherited from the solar nebula; nevertheless, most individual meteorites experienced mild to severe secondary processing on their parent asteroids due to aqueous alteration, thermal metamorphism, and/or shock metamorphism. Subsequent terrestrial weathering of these meteorites can obliterate primitive features, oxidize minerals, and affect bulk chemical and isotopic compositions.

8.1.1 Ordinary Chondrites

Often abbreviated "OC," ordinary chondrites comprise the clan with the most individual meteorites; ~80 percent of observed falls are ordinary chondrites. There are three major groups: H (high total iron), L (low total iron), and LL (low total iron, low metallic iron) that vary systematically in numerous mineralogical, chemical, physical, and isotopic properties (Table 8.1). The principal differences among the groups result from their oxidation state. These variations were recognized in the early twentieth century and enshrined as "Prior's Rules." George T. Prior, Keeper of Mineralogy at the British Museum of Natural History from 1909 to 1927, generalized that the lower the amount of metallic iron in a chondrite, the greater the amount of FeO in its silicate minerals. Table 8.1 shows that the average amount of metallic Fe-Ni-Co decreases (in wt%) from H (17.8) to L (8.33) to LL (3.56) as iron becomes progressively more oxidized, with increasing proportions of iron changing from Fe^0 to Fe^{2+}. Concomitant oxidation processes are responsible for (a) increasing FeO in olivine, low-Ca pyroxene, chromite, and ilmenite; (b) increasing Co and Ni in kamacite; (c) decreasing

Table 8.1 *Systematic variations in mean mineralogical, textural, bulk chemical, physical and bulk isotopic properties of ordinary-chondrite groups*

	Olivine Fa (mol%)	Low-Ca-pyx Fs (mol%)	Plagioclase Ab (mol%)	Chromite FeO (wt%)	Ilmenite FeO (wt%)	Kamacite Co (mg/g)	Kamacite Ni (mg/g)	Kamacite/taenite Modal Ratio	Modal (vol%/vol%) Olivine/low-Ca-pyx ratio
Source of Data	a	b	b	b	b	a	a	c	d
Petrologic Type	4–6	4–6	4–6	4–6	4–6	4–6	4–6	3–6	4–6
H	18.8	17.2	81.9	31.2	40.9	4.7	69.4	5	1.4
L	24.7	21.3	84.2	33.0	41.9	8.0	65.3	1	2.0
LL	29.4	24.1	85.9	34.5	44.2	26.0	54.9	0.4	3.6

	Metallic Fe-Ni–Co (wt%)	Chondrule Diameter (μm)	Bulk Fe (wt%)	Bulk Ni/Mg	Bulk Ir/Ni ($\times 10^{-5}$)	Bulk Ir/Au	Bulk density (g cm^{-3})	Bulk $\delta^{18}O$ (‰)	Bulk $\Delta^{17}O$ (‰)
Source of Data	e	f	e	g	g	g	h	i	i
Petrologic Type	3–6	3	3–6	3–6	3–6	3–6	4–6	4–6	4–6
H	17.8	400	27.45	0.115	4.85	3.63	3.53	4.08	0.73
L	8.33	500	21.93	0.083	4.16	3.26	3.42	4.70	1.07
LL	3.56	550	19.63	0.067	3.54	2.91	3.28	5.04	1.26

Source of data: a = Rubin (1990); b = Gomes and Keil (1980); c = Afiattalab and Wasson (1980); d = Hutchison (2004); e = Jarosewich (1990); f = Friedrich et al. (2015); g = Kallemeyn et al. (1989); h = Consolmagno et al. (2006); i = Greenwood et al. (2017). pyx = pyroxene. Data for metallic Fe-Ni-Co, bulk Fe, and bulk density are for observed falls only; all other data are for both falls and finds.

kamacite/taenite modal ratios, because less Fe is available to produce kamacite; and (d) higher olivine/low-Ca-pyroxene ratios (because in olivine, the divalent-cation/Si ratio is twice that in low-Ca pyroxene: 2:1 vs.1:1, as more divalent Fe becomes available).

Other systematic variations from H to L to LL include (e) decreasing total Fe; (f) decreasing siderophile/lithophile element ratios (e.g., Ni/Mg); (g) decreasing bulk density; (h) lower refractory-siderophile/common-siderophile ratios (e.g., Ir/Ni); (i) lower refractory-siderophile/ volatile-siderophile ratios (e.g., Ir/Au); (j) heavier O isotopes; and (k) increasing chondrule size.

In addition to the three major OC groups, there are some OC with intermediate properties that have been designated H/L and L/LL. These samples may simply be anomalous H, L, or LL chondrites, or they might represent the first members of new OC groups (from otherwise-unsampled asteroids) with intermediate mineralogical and chemical properties.

The anomalous IIE-iron Netschaëvo contains highly recrystallized OC clasts informally designated HH (very high total iron, very high metallic Fe). These clasts are more reduced than H chondrites: they contain modally more abundant metallic Fe-Ni, olivine with lower Fa contents, and kamacite with lower Co concentrations (e.g., Olsen and Jarosewich 1971; Rubin 1990). Their bulk O-isotopic composition is lighter than that of H chondrites, and their bulk siderophile-element/Ni ratios tend to lie along extensions of established H-L-LL trends. Chondritic clasts in the Mont Dieu IIE iron are intermediate in their properties between those in H chondrites and the Netschaëvo clasts (Van Roosbroek et al. 2015). A few clasts in meteorite breccias (e.g., in LL3–6 NWA 10214) appear to be ungrouped OC that also lie along the OC mineralogic and geochemical trends (Rubin et al. 2017).

Ordinary chondrites have undergone thermal metamorphism and have been ordered into a sequence of "petrologic types," ranging from type 3 (the least metamorphosed) to type 6 (the most metamorphosed) (Van Schmus and Wood 1967). The type-3 chondrites have been further divided into subtypes 3.0 to 3.9 (e.g., Sears et al. 1980). Grossman and Brearley (2005) found that the least-metamorphosed chondrites show systematic variations in (a) Cr concentration in ferroan olivine, (b) alkalis in the glassy mesostases of low-FeO (Type I) and high-FeO (Type II) chondrules, (c) the extent of albite crystallization from the glass of Type II chondrules, (d) concentrations of S and Na in matrix material, and (e) distribution of troilite in the whole rock. These systematic differences were used to define subtypes 3.00 to 3.15 at the low end of the metamorphic scale.

The least-metamorphosed type-3 ordinary chondrites consist of a few basic components: (a) chondrules, chondrule fragments, and coarse isolated monocrystalline and polycrystalline grains, (b) opaque phases – principally metallic Fe-Ni (kamacite and taenite) and sulfide (troilite with very rare examples containing daubréelite exsolution lamellae), (c) fine-grained silicate matrix material (described in Chapter 7) including organic components and tiny presolar grains, (d) rare refractory inclusions, and (e) rare ameboid olivine inclusions. Some type-3 OC breccias also include foreign clasts derived from different asteroids. Matrices in aqueously altered type-3 OC such as LL3.00 Semarkona contain additional opaque minerals: magnetite, Ni-rich metal (mainly awaruite), Ni-rich sulfide (pentlandite), Fe carbide (cohenite, haxonite, edscottite), and wüstite.

Numerous chemical, textural, and mineralogical properties vary with increasing metamorphic grade from type 3.00 to type 6 (Table 8.2) including (a) loss of volatile components, (b) increasing recrystallization, (c) systematic changes in TL and CL characteristics, (d) destruction of chondrule glassy mesostases and crystallization of secondary feldspar, (e) decreasing grain-to-grain heterogeneity in mineral composition, (f) transformation of

Table 8.2 *Variations of properties with petrologic type in ordinary chondrites*

Category	Parameter	Petrologic Type												
		3.0	3.1	3.2	3.3	3.4	3.5	3.6	3.7	3.8	3.9	4	5	6
Bulk Composition	C (wt%)	0.3–0.6					0.2–0.5			<0.3				
	H_2O (wt%)	1.0–2.5							<1.0					
	Indium (ng/g)	8–100										0.1–8		
	$^{36}Ar_p$ (10^{-8} g^{-1} STP)	45–60			15–50				5–15			0.5–5		0.2–2
Texture	Chondrule textural integration	None						Incipient		Discernable		Minor	Moderate	Extensive
	Chondrule appearance	Very sharply defined										Well-defined	Readily delineated	Poorly defined
Thermoluminescence (TL)	Sensitivity (×1,000, relative to the Dhajala H3.8 chondrite)	<1		1–2.2	2.2–4.6	4.6–10	10–22	22–46	46–100	100–220	220–460	300–600	600–2,000	
	Glow curve shape	Irregular		Sharp peak										
	Peak temperature (°C)	170–200		<140			>140							
Cathodoluminescence (CL)	Yellow CL in chondrule meso	Common		Rare			Absent							
	Blue CL in chondrule meso	Present		Common			Dominant							
	Red CL in chondrule olivine	Present					Rare to absent					Absent		
	Matrix CL	Red	Some red areas	Low CL			Increasingly blue →						Matrix absent	
Matrix	Hydrated phases	Abundant	Present		Absent								Matrix absent	
	Presolar graphite	Present			Absent								Matrix absent	
	Presolar diamonds	Decreasing →								Absent			Matrix absent	
	FeO/(FeO+MgO) in matrix normalized to whole rock	>1.4					1.1–1.4			<1.1			Matrix absent	
	% matrix recrystallization	<20				>20						100	Matrix absent	
	Sulfur (wt%)	>1.0	0.5–1.0		<0.5								Matrix absent	

Table 8.2 (*cont.*)

Category	Parameter	Petrologic Type 3.0	3.1	3.2	3.3	3.4	3.5	3.6	3.7	3.8	3.9	4	5	6
Feldspathic Material	Texture of chondrule meso	Increasing number density of crystallites →										Increasingly recrystallized →		
	Isotropic glass	Common				Less common		Rare				Absent		
	Type I chondrule meso	Modal albite uncommon				Modal albite present						Mesostasis absent		
		Decreasing normative anorthite →										Mesostasis absent		
	Type II chondrule meso	Albite rare	Increasing modal albite →			Modal albite present						Mesostasis absent		
	Secondary feldspar (μm)	Submicrometer size										2	2–25	50–150
Olivine	PMD of FeO	>33					15–33			5–15		<5		
	Type I chondrules	FeO<2 wt% CaO>0.3 wt% common			FeO increases to equilibrium value CaO decreases to equilibrium value					FeO near equilibrium CaO near equilibrium		FeO uniform CaO uniform		
	Type II chondrules	FeO converges to equilibrium value								FeO near equilibrium		FeO uniform		
		Cr_2O_3>0.3 wt%	Cr_2O_3 heterogeneous	Cr_2O_3<0.1 wt%										
Pyroxene	PMD of FeO low-Ca pyx	>30							20–30			5–20	<5	
	Structure of low-Ca pyx	Mainly monoclinic										Mixed	Orthorhombic	
	Size of Ca-pyx grains (μm)	Mainly submicrometer grains in chondrule mesostasis										<1	2–5	5–30
Nonsilicates	Phosphates	Some Na- and Fe-rich		Ca-rich, increasing modal abundance and grain size →								Ca-rich, large grains		
	Chromite	Tiny grains in chondrule mesostasis and within olivine phenocrysts								Moderate-sized grains		Coarse grains		
	Kamacite composition	Co heterogeneity high			Co heterogeneity moderate		Co heterogeneity low							
		P-, Cr-, Si-bearing	P-, Cr-, Si-free											
	Mean kamacite grain size (μm)	~20–80										~50–100	~100–300	~100–300
	Mean troilite grain size (μm)	~20–80										~20–100	~20–150	~20–150
	Taenite	Unzoned			Zoned									
	Carbides and magnetite	Common			Present		Rare to absent							

Modified from table 1 of Huss et al. (2006), which contains additional references. Metal and troilite grain size data modified from Rubin et al. (2001). ng/g = nanograms per gram, equivalent to ppb; $^{36}Ar_p$ = primordial ^{36}Ar; meso = mesostasis; pyx = pyroxene; PMD = percent mean deviation, modified from Dodd et al. (1967) – it is determined in the following way: Obtain the wt% FeO in each grain of a particular mineral, calculate the mean of these values, determine the absolute value of the deviation of each analysis from that mean, obtain the average of these deviations and divide that value by the average FeO content, then multiply the result by 100.

monoclinic to orthorhombic low-Ca pyroxene, (g) crystallization of secondary Ca-pyroxene grains, (h) increasing modal abundance and coarsening of phosphate, (i) crystallization of secondary chromite grains, (j) coarsening of kamacite and troilite grains, (k) development of compositional zoning in taenite, and (l) destruction of fine-grained carbides and magnetite produced earlier during parent-body aqueous alteration.

The primary minerals in unaltered portions of the least-equilibrated OC include olivine, low-Ca pyroxene (major clinopyroxene and rare orthopyroxene), kamacite, taenite, troilite, and, nearly exclusively within chondrules, crystallites of Ca-pyroxene, pigeonite, and rare tiny grains of merrillite. About 18 percent of chondrules in LL3.00 Semarkona contain primary igneous plagioclase laths (1–8 μm wide, up to 60 μm long); they are highly variable in composition ($Ab_{95.7}Or_{2.4}$ – $Ab_{0.3}Or_{0.2}$), but all are low in K_2O ($Or_{<3}$) (Lewis and Jones 2019). In addition, rare small grains of primary chromite occur in some Type II chondrules. Rare Ca-Al-rich inclusions (CAIs), amoeboid olivine inclusions (AOIs), and presolar grains in these rocks each have their characteristic allotment of minerals (discussed in Chapter 7).

Mineral phases that appear in OC during thermal metamorphism include orthopyroxene, Ca-pyroxene, plagioclase, phosphate, chromite, ilmenite, and rutile:

Low-Ca pyroxene. Type IB porphyritic pyroxene chondrules in unequilibrated type-3 OC (e.g., type 3.0–3.5) contain low-Ca clinopyroxene phenocrysts exhibiting polysynthetic twinning on (100), inclined extinction, and shrinkage cracks perpendicular to the twinning planes. These phenocrysts formed as a metastable phase by inversion from protoenstatite (the stable polymorph between 1,000°C and 1,557°C) with an accompanying change in volume during chondrule quenching. During thermal metamorphism, the concentrations in low-Ca pyroxene of FeO and MnO increase while those of Cr_2O_3 and Al_2O_3 decrease. At a temperature of ~630°C (probably exceeded in type-5 OC), clinoenstatite transforms into (ortho)enstatite, accounting for the absence of low-Ca clinopyroxene in unshocked type-5 and -6 OC. (At shock stage ≥S3, polysynthetically twinned low-Ca clinopyroxene can form from orthopyroxene by shock heating and quenching, accounting for its presence in many shocked OC of all petrologic types.)

Ca-pyroxene. In type 3.0–3.5 OC, Ca-pyroxene is present mainly in chondrules, where it occurs as small crystallites within the mesostasis and as overgrowths on low-Ca pyroxene phenocrysts. (It also occurs in rare CAIs.) Ca-pyroxene is essentially absent within fine-gained silicate matrix material in type-3 OC. With increasing metamorphic grade, Ca-pyroxene increases in size from submicrometer grains in type-4 OC, to 2–5 μm grains in type-5 OC, and to grains tens of micrometers in size in type-6 OC. The precursor material for Ca-pyroxene in equilibrated ordinary chondrites is chondrule mesostasis in type-3 OC: the two most abundant CIPW-normative constituents in the mesostases of Type I and Type II chondrules in LL3.00 Semarkona are plagioclase (66 and 45 wt%, respectively) and diopside (22 and 14 wt%). Coarse crystalline plagioclase and pyroxene are evident between the olivine bars in some barred olivine (BO) chondrules in type 5–6 OC.

Plagioclase. Relatively coarse grains of plagioclase are very rare in type 3.0–3.5 OC except for anorthite phenocrysts in rare Al-rich chondrules, small oligoclase grains within igneous rims around ~10 percent of chondrules, and in very rare CAIs and AOIs. During thermal metamorphism, chondrule mesostases devitrify (evident in type 3.7–3.9 OC) and albite crystals nucleate and grow. Type-5 OC contain 2–10-μm-sized plagioclase grains and type-6 OC contain many plagioclase grains exceeding 50 μm. In H chondrites, plagioclase compositions are albitic and homogeneous in petrologic types 4–6; in contrast, plagioclase in L and LL chondrites is compositionally heterogeneous in type 4 rocks, but becomes albitic and homogeneous by type 6 (Lewis and Jones 2016).

Phosphate. The mesostases in Type II chondrules in LL3.00 Semarkona average ~1.3 wt% P_2O_5 and contain accessory 2–10 μm grains of merrillite. Although the mesostases in Type I chondrules have lower concentrations of P_2O_5 (<0.02 wt%), some of these chondrules contain similarly sized merrillite grains. These small phosphate grains crystallized within the mesostasis during chondrule cooling. Type-3 and -4 OC contain fine-grained merrillite and/or chlorapatite within troilite-metal assemblages; chlorapatite also occurs as 2–15-μm-thick rinds around some kamacite and taenite grains. In these cases, the source of the P is probably adjacent metallic Fe-Ni. There are similar variations in the Cl/F ratio of apatite in OCs of all petrologic types; suggesting that apatite composition is controlled by local parent-body conditions (Jones et al. 2016). Phosphate grains in the matrix are tens of micrometers in size; some grains are as large as 300 μm. The Ca, Fe, Mg, and Na in phosphate grains in equilibrated OC could have been derived from chondrules or olivine, pyroxene, and plagioclase grains in the matrix. The source of the halogens is likely chondrule mesostasis; the low abundances of Cl and F in some analyses of mesostasis may indicate that these elements are mobilized at very low levels of metamorphism or aqueous alteration.

Chromite. Chondrules in type-3 OC contain some small chromite grains up to ~10 μm in size. Coarse secondary chromite grains (up to 200 μm) within unshocked type 4–6 OC occur as isolated grains and in association with metallic Fe-Ni and troilite. With increasing metamorphic grade, chromite tends to become richer in FeO, MnO, and TiO_2 (although MnO may decrease between type 5 and 6). The source of the Fe and Cr in chromite could be mafic silicate grains; olivine and low-Ca pyroxene in type-3 OC have variable FeO/(FeO+MgO) ratios, and some low-Ca pyroxene grains contain >1 wt% Cr_2O_3. Chromite-bearing silicate chondrules are rare objects, occurring only in metamorphosed OC where they constitute ~0.1 percent of the chondrules. Phenocrysts and subhedral chromite grains in these chondrules range up to 200 μm in size. Their apparent absence in type 3.0–3.7 chondrites suggests they are not nebular objects. Krot and Rubin (1993) proposed they formed by impact melting, but much more work needs to be done to evaluate this possibility.

Ilmenite and rutile. Type-3 OC do not contain coarse grains of ilmenite or rutile, but such grains are present in some type-5 and -6 OC. In Farmington (a shock-darkened L5 chondrite), ilmenite occurs in ~500-μm-sized clusters that also include chromite, metallic Fe-Ni, and troilite. Some of these ilmenite grains contain rutile exsolution lamellae; others have exsolution lamellae of both rutile and chromite. Allegan (H5) contains discrete grains of rutile in the matrix, not associated with ilmenite or chromite.

In each OC group, there is an apparent systematic increase in oxidation state (exemplified by mean olivine Fa content) with petrologic type (Rubin 1990):

H3 – $Fa_{17.6}$	L3 – $Fa_{23.5}$	LL3 – $Fa_{27.7}$
H4 – $Fa_{18.3}$	L4 – $Fa_{24.0}$	LL4 – $Fa_{28.5}$
H5 – $Fa_{18.6}$	L5 – $Fa_{24.7}$	LL5 – $Fa_{29.5}$
H6 – $Fa_{19.4}$	L6 – $Fa_{25.1}$	LL6 – $Fa_{29.7}$

There are additional petrologic indicators of progressive oxidation that correlate with petrologic type, most clearly illustrated by comparing type-3 or type-4 OC with type-6 OC. These include (1) increasing kamacite Co content (wt%): H3 – 0.46, H6 – 0.48; L3 – 0.78, L6 – 0.82; LL3 – 1.74, LL6 – 2.74; (2) increasing kamacite Ni content (wt%): H3 – 6.67, H6 – 6.87; L3 – 5.71, L6 – 6.57; LL3 – 5.35, LL6 – 5.49; (3) decreasing modal metallic Fe-Ni abundance (wt%): H4 – 18.5, H6 – 17.8; L4 – 8.9, L6 – 8.2; LL4 – 4.3, LL6 – 2.8; and (4) increasing normative olivine/pyroxene ratio: H4 – 1.26, H6 – 1.45; L4 – 1.64, L6 – 1.93; LL4 – 2.41, LL6 – 3.03. These correlations are probably due to the mobilization of minor amounts of water, likely bound within phyllosilicates. (Phyllosilicates have been well documented in LL3.00 Semarkona and LL3.1 Bishunpur.) With increasing heat (radiogenic and/or collisional) phyllosilicates were dehydrated; water was driven out of the crystal lattices and became available as an oxidant. The olivine/pyroxene ratio increased due to the replacement of low-Ca pyroxene by ferroan olivine.

One of the most oxidized ordinary chondrites is the Morokweng LL6 breccia, found as a set of fragments within the ~70-km-diameter, 145-Ma-old Morokweng impact crater in South Africa (Maier et al. 2006). The fragments, which range up to 25 cm in size, occur at a depth of ~766 m within an 870-m-thick melt sheet. The samples contain well-preserved, porphyritic, radial pyroxene, and barred olivine chondrules. Olivine ranges from Fa_{30} to Fa_{33}; the latter value is higher than mean Fa in any other LL chondrite. Other phases in the Morokweng chondrite include orthopyroxene, plagioclase, chromite, apatite, pyrrhotite and pentlandite. No grains of troilite or metallic Fe-Ni were identified.

Most ordinary chondrites, including Morokweng, were also affected by shock metamorphism. Individual samples can be assigned shock stages varying from S1 (unshocked) to S6 (very strongly shocked) (Table 8.3). Shock processes affected the type of optical extinction and the development of planar fractures in silicate minerals, the formation of clinoenstatite lamellae in orthopyroxene, whole-rock recrystallization and melting, creation of high-pressure phases, and the transformation of plagioclase into maskelynite (formed either by solid-state shock disordering or by quenching of a dense mineral melt at high shock pressures). Other shock-metamorphic processes commonly involve brecciation, shearing, and impact melting.

The interstitial glass in OC impact-melt rocks (e.g., Ramsdorf, PAT 91501, LAR 06299) tends to be feldspathic; these glasses are typically enriched in SiO_2, Al_2O_3, Na_2O, and K_2O relative to the whole rocks. These glasses are similar in composition to the ~100–200-μm-sized melt-pocket glasses in shocked OC studied by Dodd and Jarosewich (1979). The enrichment in a plagioclase compositional component in these melts is due to the low impedance of crystalline plagioclase to shock compression and the mineral's ready compressibility.

The criteria for shock-stage assignment is given in Table 8.3. Data from the *Catalogue of Meteorites* (Grady 2000) show that the mean shock stage of ordinary chondrites increases with petrologic type: type 3, S2.26 (n = 103); type 4, S2.32 (n = 240); type 5, S2.64 (n = 557); type 6, S3.17 (n = 751). This is also illustrated by the uneven distribution of shock stages among the petrologic types: e.g., type 3 (22.3% S1, 44.7% S2, 20.4% S3, 9.7% S4, 2.9% S5, 0% S6); type 6 (6.8% S1, 18.8% S2, 42.3% S3, 19.3% S4, 8.0% S5, 4.8% S6). These data do not take possible pairings into account, but the trends are robust.

This correlation is consistent with collisional heating being a contributing factor to ordinary-chondrite metamorphism. Although it could be argued that the relatively low mean shock stage

Table 8.3 *Shock-stage determination mainly as a function of type of extinction and fracture in common silicate minerals in chondritic meteorites*

Shock Stage	Description	Olivine	Low-Ca pyroxene	Plagioclase
S1	Unshocked	Sharp optical extinction; irregular fractures	Sharp optical extinction; irregular fractures	Sharp optical extinction; irregular fractures
S2	Very weakly shocked	Undulose extinction; irregular fractures	Undulose extinction; irregular fractures; some planar fractures	Undulose extinction; irregular fractures
S3	Weakly shocked	Undulose extinction; irregular fractures; planar fractures	Undulose extinction; irregular fractures; planar fractures; clinoenstatite lamellae on (100)	Undulose extinction
S4	Moderately shocked	Weak mosaic extinction; planar fractures	Weak mosaic extinction; planar fractures; clinoenstatite lamellae on (100)	Undulose extinction; partly isotropic grains
S5	Strongly shocked	Strong mosaic extinction; planar fractures	Strong mosaic extinction; planar fractures	Maskelynite
S6 (applies to restricted local regions in or near melt zones)	Very strongly shocked	Solid-state recrystallization; yellow-brown staining; ringwoodite; melting	Grain melting; majorite	Shock melting and formation of normal glass
S7	Whole-rock melting			

Adapted from Stöffler et al. (1991, 2018), Rubin (2015a), Rubin et al. (1997), and Izawa et al. (2011).

of type-3 OC reflects the greater friability of these rocks (i.e., that highly shocked type-3 chondrites would tend to fall apart), there is little difference in the rheological strength of type-5 and -6 OC. Because the mean shock stages of these two petrologic types are significantly different (S2.64 and S3.17, respectively), it is plausible that impact heating is partly responsible for OC metamorphism.

8.1.2 R Chondrites (Prototype Rumuruti)

Most R chondrites are of petrologic type $\geq$3.6 and contain abundant olivine (53–77 vol%) with a sharp compositional peak at $Fa_{37–40}$. Nevertheless, the total range in Fa is broad: 0.4–45.4 mol%. Low-Ca pyroxene tends to be much less abundant (~0–13 vol%) and is essentially absent in Y 793575. In R chondrites with low-Ca pyroxene, the phase is monoclinic and commonly exhibits polysynthetic twinning; it also tends to be compositionally heterogeneous: e.g., the compositional distribution in Rumuruti and PCA 91002 is $Fs_{\sim 0–30}$. Other pyroxene phases in R chondrites include pigeonite, augite, and diopside. Plagioclase is mostly sodic, in the compositional range of albite or oligoclase ($An_{6–18}$). More-calcic grains ($An_{35–69}$) are present in Acfer 217; rare potassic grains occur in Acfer 217 ($An_{7.9}Or_{46.4}$) and Rumuruti ($An_{4.2}Or_{87.3}$).

Primary minerals within the rare CAIs include spinel, hibonite, Al-Ti diopside (fassaite), perovskite, anorthite, and accessory olivine. Nepheline, sodalite, and ilmenite were formed in some CAIs as secondary minerals during parent-body aqueous alteration.

Matrix material is abundant in R chondrites. It contains small grains of ferroan olivine ($Fa_{35–60}$) and low-Ca pyroxene, chondrule fragments, and isolated mafic silicate grains. Opaque phases include sulfides (pyrrhotite, troilite, pentlandite, chalcopyrite, pyrite, rare lead sulfide, rare arsenic sulfide); oxides (magnetite, chromite, Al-rich chromite, magnetite-chromite solid solution, ilmenite); phosphates (chlorapatite, hydroxylapatite, merrillite); rare metallic Fe-Ni (awaruite, kamacite, martensite); rare metallic Cu, rare noble metal–rich phases (PtFe(Ir,Ni) alloy, Pt metal, Os with minor Re, Pt, Os-Ir-Ru-Pt alloy, native Au, electrum, ruthenosmiridim, rustenburgite, PdPtSn); and very rare, small grains of noble-metal–rich sulfides (erlichmanite, laurite, Os,Ru,Fe-rich sulfide), arsenides (sperrylite), sulfarsenides (irarsite ((IrPt)AsS)), and tellurides (chengbolite, moncheite).

Three R6 chondrites have OH-bearing phases: MIL 11207 and LAP 04840 contain hornblende and phlogopite; MIL 07440 contains accessory titan-phlogopite but no hornblende.

Very rare Cl-rich phases (in addition to phosphate) have been reported. These include a single grain of bismuth chloride in LAP 04840 and a scapolite-like phase with 6.7 wt% Cl in chondrule mesostases in ALH 85151 and Carlisle Lakes.

8.1.3 Carbonaceous Chondrites

All carbonaceous-chondrite groups except CH (which has been modeled as having a non-nebular origin) have CI- and Mg-normalized refractory-lithophile-element abundances $\geq$1.0 (i.e., at or higher than CI-chondrite levels); in all non-carbonaceous chondrite groups, the abundances are $<$1.0 (i.e., lower than CI-chondrite levels). (The refractory lithophiles are elements that would condense from a cooling solar-composition gas at high ambient temperatures and partition into silicate and oxide phases. They have 50 percent condensation temperatures ≥1,400 K at 10^{-4} bar and include Al, Ca, Sc, Ti, V, Y, Zr, REE, and Hf.) Carbonaceous chondrites also contain appreciably more fine-grained silicate matrix than ordinary or enstatite chondrites. Except for the CI group (which has been extensively altered), carbonaceous chondrites also contain more CAIs than other chondrite groups. (Nevertheless, one altered CAI has been reported in the Ivuna CI chondrite. Also present in CI chondrites are a few small ^{16}O-rich hibonite and spinel grains that were probably derived from mechanically disrupted CAIs.) Even CB chondrites, which also may not have a nebular origin, tend to contain more CAIs than non-carbonaceous chondrites.

Carbonaceous chondrites and non-carbonaceous chondrites plot in widely separated regions in diagrams of O-, Ti-, Cr-, and Ni-isotopic composition (i.e., ε^{50}Ti vs. ε^{54}Cr; Δ^{17}O vs. ε^{54}Cr; ε^{54}Cr vs. ε^{62}Ni) (Warren 2011a, 2011b). Carbonaceous and non-carbonaceous chondrites also vary in Mo-isotopic composition; they plot along distinct *s*-process mixing lines on ε^{95}Mo vs. ε^{94}Mo diagrams (Budde et al., 2016, 2017, 2018). Many researchers have expressed support for the suggestion of UCLA cosmochemist Paul Warren that carbonaceous chondrites formed beyond the orbit of Jupiter and non-carbonaceous chondrites were derived from inner-Solar-System bodies.

Carbonaceous chondrites comprise more individual groups than any other meteorite class. They contain eight major groups, a few grouplets, and more than a dozen unique samples. The groups can be distinguished from one another by numerous physical and chemical properties, including mineralogy, texture, chondrule size, chondrule abundance, bulk chemical composition, and O-isotopic composition.

CI chondrites (prototype Ivuna). These meteorites have a bulk chemical composition closest to that of the solar photosphere (Figure 8.1). This similarity includes elements with different geochemical affinities: refractory lithophiles, common lithophiles, moderately volatile lithophiles, refractory siderophiles, common siderophiles, moderately volatile siderophile/chalcophile elements, and highly volatile siderophile/chalcophile elements (Table 8.4). CI chondrites best represent the bulk chemical signature of the condensed materials from which the bodies in the Solar System formed.

The principal preterrestrial mineral constituents of CI chondrites, discussed in Chapter 7, include (1) Fe-Mg serpentines with interlayered saponite; (2) magnetite and tiny grains of ferrihydrite;[1] (3) carbonates (dolomite, breunnerite, siderite, calcite); (4) sulfides (pyrrhotite, rare elemental S, pentlandite, cubanite); and (5) miscellaneous accessory phases (merrillite, periclase, Ti_3O_5, magnesiochromite, eskolaite). There are also rare isolated grains of olivine (some containing 2–5-µm-sized spherical metallic Fe-Ni blebs, some containing silicate melt inclusions with shrinkage bubbles), low-Ca pyroxene (mainly orthoenstatite), and Ca-pyroxene, all probably derived from fragmented chondrules. Small rare grains of hibonite and spinel were probably broken out of altered CAIs. Presolar grains include graphite, diamond, silicon carbide and corundum.

CM chondrites (prototype Mighei). Some CM chondrites are breccias: e.g., Aguas Zarcas consists of highly altered, chondrule-poor regions and less-altered, chondrule-rich regions; Murray contains scattered millimeter-sized clasts appreciably more altered than the host. The

[1] Some researchers maintain that most (or even all) occurrences of ferrihydrite in meteorites are due to terrestrial alteration.

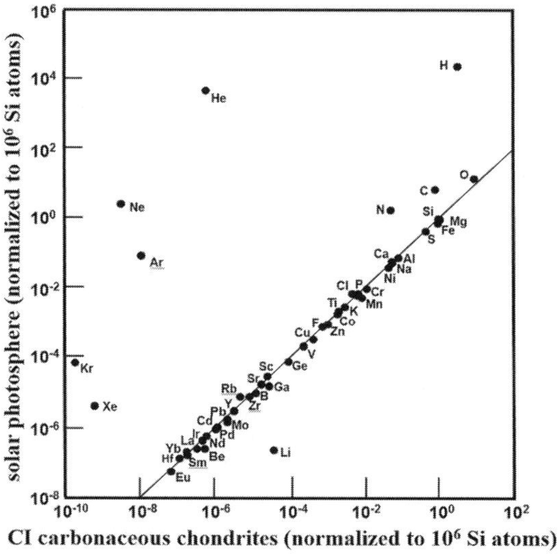

Figure 8.1 Log-log plot of selected elemental abundances in the solar photosphere versus those in CI carbonaceous chondrites. The photosphere/CI-chondrite ratios for most elements are close to unity; these elements include refractory, common and volatile elements with lithophile, siderophile and chalcophile affinities. Noble gas abundances are much higher in the Sun than in CI chondrites because they did not quantitatively condense into solids. Ar, Kr, and Xe are present in the P1 "planetary-type" gas component in primitive chondrites, probably trapped within pores in fine-grained matrix material; the P1 component is depleted in Ne, consistent with Ne being farther from the one-one line. H is much closer to the line than He because some H was incorporated into organic molecules and OH-bearing minerals (e.g., phyllosilicate, amphibole, phosphate). In contrast, He is much more abundant in the Sun than in CI chondrites; He is the most unreactive element, forming rare compounds only at high pressures. N, C and O are higher in the Sun because a portion of each of these elements formed compounds (e.g., CO, CO_2, N_2, HCN, H_2O) that remained in the gas to a significant extent. Lithium is depleted in the Sun, probably due either to Li burning during nuclear fusion or to Li loss to a protoplanetary disk. Diagram modified from McSween and Huss (2010) and Dye et al. (2015).

entire set of CM chondrites averages ~60 vol% fine-grained matrix material. In unweathered CM samples, the matrix consists of abundant Fe-Mg serpentines (mainly Fe-bearing chrysotile); variable amounts of isolated clumps of tochilinite-cronstedtite intergrowths; minor metallic Fe-Ni; some sulfide (pyrrhotite, pentlandite, grains of intermediate composition, and murchisite); oxide (small clumps of magnetite and rare hematite and eskolaite); carbonate (calcite and carbonate phases containing Ca, Mg, Fe, Mn, and Ni); and accessory forsterite, Cl-free apatite, orthoenstatite, Ca-pyroxene, halite, and sylvite.

There are additional components in CM chondrites:

Chondrules. Textural types include porphyritic olivine, porphyritic pyroxene, porphyritic olivine-pyroxene, barred olivine, barred pyroxene, granular olivine-pyroxene, radial pyroxene, and cryptocrystalline. They contain major olivine and low-Ca clinopyroxene with a few blebs of metallic Fe-Ni, schreibersite, and murchisite; some metal-free chondrules contain small grains of sulfide. A few chondrules in the least-altered CM chondrites (e.g., Paris) contain isotropic glassy mesostases.

Table 8.4 *Elements with diverse geochemical affinities that are similar in abundance in the solar photosphere and CI chondrites*

Atomic Number	Element	Geochemical Affinity	Photosphere/ CI Ratio
4	***Be***	***RLE***	***0.98***
11	**Na**	**MVLE**	**1.07**
12	**Mg**	**CLE**	**0.94**
13	***Al***	***RLE***	***1.02***
14	**Si**	**CLE**	**0.99**
15	**P**	**MVLE**	**1.02**
19	**K**	**MVLE**	**1.03**
20	***Ca***	***RLE***	***1.07***
23	***V***	***RLE***	***1.02***
24	**Cr**	**CLE**	**0.99**
26	Fe	CSE	0.92
27	Co	CSE	1.05
28	Ni	CSE	1.05
39	***Y***	***RLE***	***1.06***
40	***Zr***	***RLE***	***1.02***
41	***Nb***	***RLE***	***1.07***
42	*Mo*	*RSE*	*0.93*
44	*Ru*	*RSE*	*1.11*
45	*Rh*	*RSE*	*1.05*
46	Pd	CSE	1.01
48	Cd	HVSCE	1.05
51	Sb	MVSCE	0.99
57	***La***	***RLE***	***0.91***
58	***Ce***	***RLE***	***0.90***
60	***Nd***	***RLE***	***1.04***
62	***Sm***	***RLE***	***0.92***
64	***Gd***	***RLE***	***1.10***
66	***Dy***	***RLE***	***0.96***
68	***Er***	***RLE***	***0.93***
77	*Ir*	*RSE*	*0.97*
82	Pb	HVSCE	0.89

All the listed elements have solar-photosphere/CI-chondrite abundance ratios between 0.89 and 1.11. For refractory elements, the 50 percent condensation temperatures (T_c) from a gas of solar composition at 10^{-4} bar are 1,850–1,400 K; for common elements, T_c = 1,350–1,250 K; for moderately volatile elements, T_c = 1,230–640 K; for highly volatile elements, T_c <640 K. RLE = refractory lithophile element; MVLE = moderately volatile lithophile element; CLE =common lithophile element; CSE = common siderophile element; RSE = refractory siderophile element; HVSCE = highly volatile siderophile/chalcophile element; MVSCE = moderately volatile siderophile/chalcophile element. Lithophile elements in bold; refractory elements in italics. Data from Palme and Jones (2005).

Chondrule fragments and isolated mafic silicate grains. These bodies are derived from disrupted chondrules and contain the same mineral phases.

CAIs. CM refractory inclusions can be classified by their principal minerals: spinel-pyroxene, spinel, spinel-pyroxene-olivine, pyroxene, pyroxene-olivine, hibonite-bearing. Textural types include simple and complex inclusions with nodular, banded or distended structures. The olivine is forsteritic. Pyroxene phases include diopside, Al-rich diopside, Al-Ti diopside, rare enstatite, and davisite. Some CAIs contain small grains of perovskite and hibonite. Only rare CAIs contain gehlenitic melilite. There are also rare ultrarefractory inclusions in CM chondrites containing Y-rich perovskite, davisite, Sc- and Ti-rich clinopyroxene, warkite, kangite, panguite, thortveitite, hibonite, and machiite. Although some workers have suggested that CM CAIs formed in the nebula with little melilite (in contrast to CAIs in CV chondrites), it seems much more likely that the abundant melilite initially present in CM CAIs was mostly destroyed by early-stage parent-body alteration. It was replaced by phyllosilicates, voids, and zeolite-like Na-Al silicates. Even in Paris, one of the least-altered CM chondrites (subtype 2.7–2.9), very few CAIs contain melilite (Rubin 2015c).

Compound chondrule-CAI objects. These objects formed by the low-velocity collisions of CAIs and chondrules in the nebula probably accompanied or followed by minor remelting. Their constituent minerals are the same as those of typical CAIs and chondrules.

AOIs. These objects contain major forsterite enclosing pores and minor wormy patches of diopside and accessory Al-rich diopside.

Presolar grains. Presolar grains in CM chondrites resemble those in other primitive carbonaceous, ordinary and enstatite chondrite groups. Their mineralogy was described in Chapter 7.

CO chondrites (prototype Ornans). Chondrules in primitive CO3 chondrites contain major olivine, clinoenstatite, and feldspathic glass. Many clinoenstatite grains are polysynthetically twinned and have Ca-pyroxene overgrowths; pigeonite also occurs. Type I chondrules contain appreciable metallic Fe-Ni and minor sulfide. Spinel is present in Type I chondrules; chromian hercynite occurs in Type II chondrules. In more metamorphosed CO chondrites, anorthite grains (in many cases intergrown with nepheline) occur within the mesostasis.

Amoeboid olivine inclusions in CO3.0 chondrites contain major forsterite, anorthite, Al-Ti diopside (fassaite) and opaque phases (mainly kamacite). A few AOIs also contain spinel. Troilite is rare to absent. The AOIs also contain 8–20 vol% voids. AOIs in CO 3.3–3.8 chondrites contain forsterite, anorthite, Al-Ti diopside, opaque phases (kamacite, taenite, troilite), spinel, pleonaste (an unapproved ferroan variety of spinel), Ca-phosphate (probably merrillite), and voids.

The primary minerals within CAIs include melilite, spinel, clinoenstatite, diopside, Al-Ti diopside, forsterite, hibonite, anorthite, kamacite, perovskite, and rare corundum and grossite; secondary CAI minerals include hercynite, nepheline, sodalite, ilmenite, monticellite, and troilite. Rare ultrarefractory inclusions contain Y-rich perovskite, davisite, Al-diopside, grossite, warkite, panguite, kangite, eringaite, spinel, and melilite.

The primary phases in CO3 matrix material include Si- and Fe-rich amorphous silicate, olivine, low-Ca pyroxene, and metallic Fe-Ni. Secondary phases include magnetite, pentlandite, pyrrhotite, anhydrite, ferric oxide, Fe-Mg serpentine, and chlorite.

Opaque phases (reviewed in Chapter 7) include metallic Fe-Ni (kamacite, taenite, tetrataenite); sulfide (troilite, pentlandite); carbide (haxonite, cohenite); and oxide (magnetite, chromite). Some Type I chondrules in Ornans contain magnetite intergrown with tiny Ca-phosphate grains.

CV chondrites (prototype Vigarano). The group is divided into three subgroups: the reduced subgroup ($CV3_{Red}$), the Allende-like oxidized subgroup ($CV3_{OxA}$), and the Bali-like oxidized subgroup ($CV3_{OxB}$), each with a distinguishable set of secondary minerals.

The reduced-subgroup members are the most pristine; their chondrules consist predominantly of olivine, low-Ca pyroxene, glass, and droplets of metallic Fe-Ni (kamacite and taenite). Additional primary phases in these rocks include: (1) troilite mainly in chondrules and matrix and (2) anorthite, gehlenite, spinel, diopside, Al-Ti diopside (fassaite), perovskite, and hibonite in CAIs (and, to a lesser, extent AOIs). There are many other primary refractory phases that occur in accessory amounts in CV CAIs including grossite, corundum, addibischoffite, allendeite, baghdadite ($Ca_3(Zr,Ti)Si_2O_9$), burnettite, davisite, dmisteinbergite (from a single FUN inclusion in Allende), eringaite, grossmanite, hexaferrum, hexamolybdenum, kangite, krotite, kushiroite, panguite, paqueite, rubinite, tazheranite, thortveitite, warkite, and zirconolite. Presolar grains in reduced CV chondrites include diamond and silicon carbide. There are some secondary minerals in $CV3_{Red}$ samples, perhaps a result of brecciation: these phases include minor saponite, serpentine, and ferrihydrite in Vigarano.

Secondary phases in CV_{OxA} chondrites include magnetite, Ni-rich sulfide, ferroan olivine (Fa_{30-60}), nepheline, sodalite, tetrataenite, awaruite, merrillite, monticellite, kirschsteinite, hutcheonite, dmisteinbergite, adrianite, andradite, grossular, hedenbergite, wollastonite, butianite, and nuwaite. Secondary phases in CV_{OxB} chondrites include phyllosilicate (Fe-bearing saponite, Mg-rich saponite, Fe-rich serpentine), magnetite, Ni-rich sulfide, fayalitic olivine (Fa_{95-100}), hedenbergite, tetrataenite, and awaruite.

The intrinsic bulk magnetic properties of CV chondrites are functions of the abundance, proportions, and grain sizes of their ferromagnetic minerals: magnetite, pyrrhotite, kamacite, tetrataenite, and awaruite. Saturation magnetization (M_S) is proportional to mineral abundance; saturation remanent magnetization (M_{RS}), and magnetic susceptibility (χ) are functions of mineral abundance and the magnetic microstructures of these phases.

CK chondrites (prototype Karoonda). The CK group lacks highly unequilibrated members; petrologic types range from ~3.6 to 6. The principal primary phases in chondrules include olivine and low-Ca pyroxene; those in AOIs include olivine, anorthite, and diopside; those in CAIs include olivine, fassaite, and anorthite (e.g., MacPherson and Delaney 1985). Rare troilite and very rare kamacite also occur. Secondary phases in CK chondrites include magnetite, pentlandite, pleonaste, Ca-pyroxene, ilmenite, pyrite, Ni-bearing pyrrhotite, chalcopyrite, monosulfide solid solution, millerite, and tiny rare grains of noble-metal–rich sulfides, tellurides, and arsenides.

It has been suggested that CK3 chondrites can be distinguished from CV3 chondrites by magnetite composition. Dunn and Gross (2017) found that CK3 magnetite contains $\geq$3.0 wt% Cr_2O_3, $\geq$0.1 wt% TiO_2, $\geq$0.2 wt% NiO, and $\geq$2.0 wt% Al_2O_3; CV3 magnetite has lower concentrations of these oxide components.

CR chondrites (prototype Renazzo). The distinctive textures of CR chondrites were discussed in Chapter 7. Primary phases in CR chondrules include olivine, low-Ca pyroxene, metallic Fe-Ni (kamacite and some martensite), troilite, and feldspathic glass; free silica occurs within igneous rims around many chondrules. Primary minerals in CAIs include melilite,

Ca-pyroxene, spinel, hibonite, and grossite; rare ultrarefractory inclusions include davisite. Matrix regions in the least-altered CR chondrites contain amorphous ferroan silicate grains associated with tiny particles of troilite and Fe-Ni sulfide. The metallic Fe-Ni grains in CR-chondrite whole rocks have a solar Ni/Co ratio and exhibit a positive correlation between their Ni and Co contents (Weisberg et al. 1993). Secondary phases in CR chondrites include Fe-Mg serpentines, saponite, calcite, magnetite, pyrrhotite, pentlandite, iron-oxide, and rare tetrataenite. In the CR2.0 chondrite GRO 95577, there are chondrule pseudomorphs (aqueously altered chondrules) containing opaque assemblages consisting of kamacite cores surrounded by iron-oxide rinds. One such rind surrounds a core that is compositionally equivalent to a mixture enriched in iron carbonate (siderite) and ferrous sulfate.

CH chondrites (prototype ALH 85085). These rocks are breccias consisting of chondrules, CAIs, abundant opaque phases and a few volume percent C1 clasts and reduced clasts. Chondrules contain major forsterite, enstatite, and glass, and rare spinel, ferroan chromian spinel, ferroan olivine, and ferroan low-Ca pyroxene. Non-chondrule regions of the host contain abundant metallic Fe-Ni (major kamacite, minor plessite, and accessory taenite and tetrataenite), accessory sulfide (mainly Ni- and Cr-bearing troilite along with rare pentlandite and heazlewoodite), and accessory oxide (spinel, ferroan chromian spinel, and magnetite). As in CR chondrites, the metallic Fe-Ni grains in CH chondrites exhibit a positive correlation between their Ni and Co contents; they also have a solar Ni/Co ratio. The principal phases in CAIs are hibonite, grossite, spinel, perovskite, melilite, and Al-rich diopside; some hibonite-rich, melilite-free inclusions contain glass. Rare ultrarefractory inclusions include perovskite, grossite, warkite, allendeite, hibonite, spinel, and melilite. The most common type of C1 clast consists of magnetite (framboids, spherulites, platelets, and isolated grains), pentlandite, merrillite, troilite, olivine, dolomite, and phyllosilicates. The reduced clasts contain blebs of kamacite; adjacent silicate grains exhibit reduction halos.

CB chondrites (prototype Bencubbin). The coarse-grained CB_a chondrites such as Bencubbin and Gujba contain (1) abundant metallic Fe-Ni (present mainly as centimeter-sized ellipsoidal nodules with minor to accessory troilite); (2) abundant light-colored silicate nodules with cryptocrystalline, barred pyroxene, or barred olivine textures (consisting of magnesian monoclinic and orthorhombic low-Ca pyroxene, magnesian olivine, feldspathic glass, and rare Ca-pyroxene); (3) rare refractory inclusions (with spinel, Al-Ti diopside and melilite); (4) rare matrix material consisting of silicate glass and fused droplets of metallic Fe-Ni and (5) some high-pressure phases formed by shock (majorite, wadsleyite, coesite, grossular-pyrope, possibly majorite-pyrope$_{ss}$, associated with a variety of C-rich phases including graphite, amorphous to poorly graphitized carbon, diamond, and bucky diamond) (e.g., Garvie et al. 2011). Also present are amorphous carbonaceous nanoglobules. Not every CB_a chondrite includes all these components. Bencubbin contains rare ordinary-chondrite and carbonaceous-chondrite clasts; Weatherford contains ordinary- and R-chondrite clasts. Unusual V-rich sulfides were identified in CB_a chondrite Sierra Gorda 013, including V-rich daubréelite ($Fe(Cr,V)_2S_4$), V-rich brezinaite ($(Cr,V,Fe)_3S_4$), and a new V-rich sulfide ($(V,Fe,Cr)_4S_5$) (Ivanova et al. 2019b).

The finer-grained CB_b chondrites such as Hammadah al Hamra 237 and QUE 94411 contain (1) very abundant metallic Fe-Ni (including chemically zoned individual grains); (2) common refractory inclusions (with spinel, hibonite, melilite, Al-Ti diopside, and forsterite); and

(3) hydrated lithic clasts containing magnetite framboids and platelets, prismatic sulfide (pentlandite and pyrrhotite), complex Fe-Mn-Mg-Ca carbonate, and phyllosilicate (serpentine and saponite).

Isheyevo. This single 16 kg stone is the only meteorite currently classified as CH/CBb (e.g., Ivanova et al. 2006). Different lithologies in this breccia vary in their modal abundance of metallic Fe-Ni (7–90 vol%); most have 50–60 vol%. Metal-rich lithologies are also rich in CAIs and resemble CBb chondrites; metal-poor lithologies resemble CH chondrites. About 70 to 80 percent of the metal grains in Isheyevo are compositionally homogeneous; they have solar Co/Ni ratios and contain ~4–25 wt% Ni; some of the Ni-poor grains contain round inclusions of troilite with 2.5–13 wt% Cr. About 20 to 30 percent of the metal grains in Isheyevo are compositionally zoned; Cr increases and Ni and Co decrease from grain center to edge. All lithologies in the rock contain metallic Fe-Ni, chondrules, CAIs, AOIs, and heavily hydrated clasts. There are two populations of chondrules: (1) porphyritic chondrules, including Type I (forsterite – Fa_2, low-Ca pyroxene – $Fs_2Wo_{1–2}$, glassy-to-fine-grained mesostasis); Type II (olivine – $Fa_{10–38}$, low-Ca pyroxene – $Fs_{10–12}Wo_{1–2}$, minor chromite, metallic Fe-Ni and troilite, alkali-rich mesostasis); Al-rich (Al-diopside, spinel, Ca-Al-Mg-rich mesostasis; plagioclase-rich Al-rich chondrules contain anorthite, forsterite, low-Ca pyroxene, Ca-pyroxene, metal, and, in some cases, spinel); silica-rich (low-Ca pyroxene, Ca-pyroxene, silica, glassy mesostasis), and metal-rich varieties, and (2) metal-free cryptocrystalline and skeletal olivine-pyroxene chondrules. There are two populations of CAIs: highly refractory spherules (containing hibonite, grossite, Al-rich pyroxene, perovskite, gehlenite, and spinel) and less-refractory objects (containing melilite, Al-Ti diopside, and anorthite; some with forsterite). The hydrated clasts consist of phyllosilicates, lath-shaped Fe-Ni sulfide grains, Ca- and Mg-rich carbonates, and magnetite (framboidal and platy morphologies). Isheyevo displays distinct sedimentary laminations composed of alternating silicate-rich and metallic-Fe-Ni-rich layers of millimeter-to-centimeter thickness (Morris et al. 2015; Garvie et al. 2017). These layers were modeled as having formed by gentle sedimentation of debris generated within an impact plume. (See Section 11.7.)

Coolidge-Loongana 001 grouplet. These two carbonaceous chondrites are significantly weathered. Their primary minerals include kamacite, taenite, troilite, chromian spinel, Ca phosphate, homogeneous olivine, heterogeneous polysynthetically twinned low-Ca clinopyroxene, and minor Ca-pyroxene (e.g., Kallemeyn and Rubin 1995). Coolidge has some chondrules with feldspathic glass containing Ca-pyroxene crystallites and small grains of plagioclase ($An_{45–95}$); a few chondrules contain spinel. Some chondrules consist of augite, plagioclase, and silica. Refractory inclusions contain plagioclase, Ca-rich pyroxene and compositionally zoned spinel grains with aluminous cores and chromian rims. Amoeboid olivine inclusions contain olivine, metallic Fe-Ni, troilite, and spinel. Primary phases in the matrix include olivine, pyroxene, and plagioclase; spinel may also occur. It has been suggested that Hammadah al Hamra 073 and Sahara 00182 may be related to this grouplet (Weber et al., 1996; Choe et al., 2010).

Ungrouped carbonaceous chondrites. There are scores of ungrouped carbonaceous chondrites; some are unique, some are related to another sample or two (and constitute part of a grouplet), and some share affinities with established groups. The list includes Acfer 094 (C2; CM related), Adelaide (C2; possibly CO or CV related), Belgica 7904 (C2; CM related;

thermally altered), Dar al Gani 055 (C3; CV related), Dar al Gani 978 (C3; unique), GRA 98025 (C2; unique), Ningqiang (C3; CK or CV related), NWA 1152 (C3; CV related), QUE 99038 (C2; unique), Sahara 00182 (C3; unique), and Tagish Lake (C2; unique). In general, the mineralogical characteristics of these meteorites are similar to those of other carbonaceous chondrites of the same petrologic type.

8.1.4 Enstatite Chondrites

Enstatite chondrites are highly reduced rocks containing modally abundant clinoenstatite and/or orthoenstatite with very low FeO contents (typically <0.5 wt%). Unweathered samples all contain schreibersite and Si-bearing kamacite;[2] many contain graphite (e.g., Keil 1989; Weisberg and Kimura 2012). Every enstatite chondrite also contains troilite as well as some other sulfides (e.g., niningerite or ferroan alabandite; djerfisherite; daubréelite; oldhamite; caswellsilverite) that have high concentrations of elements that are found in silicates and oxides in more-oxidized rocks; these elements include Na, Mg, K, Ca, Ti, Cr, and Mn. There are two separate groups of enstatite chondrites: EH (high iron) and EL (low iron); both groups have members spanning the full range of petrologic types (3–6). They differ principally in the following parameters:

Chondrule size. The average chondrule diameter in EH3 chondrites is far smaller than that in EL3 chondrites (230 μm vs. 500 μm).

Sulfide mineralogy. In addition to other sulfide phases, EH chondrites contain niningerite; in many shocked EH chondrites, the niningerite has been converted to keilite. EL chondrites contain ferroan alabandite ((Mn,Fe)S) or, in some cases, alabandite (MnS), instead of niningerite. Only very rare grains of ferroan alabandite have been reported in shocked EH chondrites.

Degree of reduction. The most noticeable difference between EH and EL chondrites in the oxidation state is in the concentration of Si in kamacite. It is higher in EH than EL chondrites but generally increases with petrologic type in both groups: ~2 wt% in EH3, 2.6–3.5 wt% in EH4, ~4 wt% in EH6; and ~0.3–0.5 wt% in EL3, 0.9–1.8 wt% in EL5, 1.1–1.7 wt% in EL6.

Bulk abundances of siderophile elements. This is readily illustrated by the higher siderophile-element/Mg ratios in EH compared to EL chondrites: Os/Mg (6.2×10^{-6} vs. 4.2×10^{-6}); Ir/Mg (5.3×10^{-6} vs. 3.7×10^{-6}); Ni/Mg (1.65×10^{-1} vs. 9.22×10^{-2}).

The petrographic and mineralogic characteristics of enstatite chondrites were described in the preceding paragraphs. In summary, the silicate phases in enstatite chondrites include monoclinic and orthorhombic low-Ca pyroxene, forsterite, calcic pyroxene, plagioclase, silica (cristobalite, quartz, and tridymite), roedderite and glass; a few shocked samples contain fluor-richterite or fluor-phlogopite. Fine-grained coesite occurs within a narrow shock vein in EH3 Asuka 10164. Enstatite-chondrite sulfides include troilite, niningerite, keilite, ferroan alabandite or alabandite, daubréelite, oldhamite, djerfisherite, caswellsilverite, and rare heideite, wassonite, sphalerite, and Mn-bearing pentlandite. Metallic Fe (nearly exclusively Si-bearing kamacite) is abundant; a few metallic Fe grains with ~8 to 11 wt% Ni also occur. Lawrencite ($FeCl_2$) is

[2] A very rare exception is the EH-chondrite portion of the Galim polymict breccia; this portion suffered impact melting and oxidation. Its kamacite grains contain very little Si (Rubin 1997c).

present as thin rims around silica grains, as inclusions within kamacite and troilite and as isolated matrix grains in the EH4 Indarch fall. Also occurring in enstatite chondrites are silicides (perryite in type-3 samples), nitrides (osbornite, nierite), oxynitrides (sinoite), phosphides (schreibersite), carbides (cohenite), graphite, and shock-produced diamonds twinned on {111}. Rare CAIs contain spinel, hibonite, and perovskite.

Greenwood et al. (2020) listed several samples of unusual composition, texture, or mineralogy that are either anomalous EH or EL chondrites or ungrouped enstatite chondrites; these meteorites include Y-791510, LEW 87223, and Y-793225.

8.1.5 Kakangari-Like Chondrites (Prototype Kakangari)

The Kakangari grouplet currently has four members: Kakangari (K3), LEW 87232 (K3), Lea County 002 (K3), and NWA 10085 (K4) (e.g., Weisberg et al. 1996; Utas et al. 2017). Most chondrules are porphyritic (Type I) and contain more pyroxene than olivine; BO chondrules are present; RP, C and Type II porphyritic chondrules are very rare or absent. Chondrule olivine and pyroxene grains are MgO-rich; e.g., Kakangari olivine averages $Fa_{2.2}$ and low-Ca pyroxene averages $Fs_{4.4}$. Pyroxene is present as clinoenstatite, Ca-pyroxene (Wo_{30-50}), and pigeonite. In addition to olivine and pyroxene, Kakangari chondrules contain feldspathic mesostasis, silica, chlorapatite, metallic Fe-Ni, and troilite.

Refractory inclusions in Kakangari tend to be spinel-rich; some contain small grains of perovskite. These inclusions are rimmed by sodalite and Al-rich diopside. One reported hibonite-rich inclusion is rimmed by Cr-Al spinel, sodalite, and Al-rich diopside. One fassaite-rich inclusion with minor olivine was reported. Also present in Kakangari are AOIs, some of which contain fassaite rimmed by spinel.

The fine-grained component of matrix material in K3 chondrites consists of major enstatite (disordered intergrowths of ortho and clino), major forsterite, minor albite, anorthite, Cr-spinel, troilite, metallic Fe-Ni, and ferrihydrite. Coarse metallic Fe-Ni grains include kamacite (5–8 wt % Ni, 0.2–0.5 wt% Co) and taenite (24–34 wt% Ni). The matrix in K4 NWA 10085 appears coarser and moderately recrystallized; mafic silicates in the matrix are compositionally uniform ($Fa_{2.8-3.2}$; $Fs_{3.6-4.5}$) and identical to those in chondrules.

Minerals in NWA 10085 include enstatite, forsterite, sodic plagioclase ($An_{13.3}Wo_{3.6}$), kamacite, taenite, troilite, Fe-phosphide, and euhedral chromite. The meteorite also contains ~20 vol% limonite, formed during terrestrial weathering.

8.2 Primitive Achondrites

8.2.1 Ureilites (Prototype Novo-Urei)

Ureilites are carbon-bearing, ultramafic, olivine-pyroxene achondrites (e.g., Goodrich 1992). Their olivine/(olivine+pyroxene) modal ratios vary widely from 0.1 to 1.0, although in most samples, olivine is more abundant than pyroxene. The total modal abundance of (olivine+pyroxene) is typically 80 to 90 vol%. Most ureilites are highly equilibrated rocks with a mean olivine endmember composition somewhere in the range of Fa_{5-25}, with a significant peak at Fa_{21}; the dominant pyroxene phase is uninverted pigeonite with up to 1.26 wt% Cr_2O_3 and a mean endmember composition somewhere in the range of $Fs_{8-20}Wo_{5-11}$. Some ureilites contain pyroxene grains with clinopyroxene lamellae; in different samples, the lamellae range in thickness from ~0.01 μm (LEW 85440) to ~50 μm (LEW 88774).

Olivine grain cores have high concentrations of CaO (~0.30–0.45 wt%) and Cr_2O_3 (~0.56–0.85 wt%) and have FeO/MnO ratios of ~17–53. Olivine grains in contact with carbonaceous matrix material have 10–100-μm-thick reduced rims consisting of forsterite with abundant blebs of low-Ni kamacite. Reduced rims also occur on pyroxene grains (where they consist of enstatite and low-Ni kamacite) and, in LEW 88774, on chromite grains (where they consist primarily of Cr-rich cohenite, brezinaite (Cr_3S_4), and eskolaite (Cr_2O_3)). In some ureilites, the olivine and pyroxene grains contain small opaque spherules consisting mainly of C-bearing metallic Fe-Ni, cohenite, and troilite.

The most abundant phase in the C-rich matrix material between coarse silicate grains is graphite; in addition to graphite, some shocked ureilites contain variable amounts of chaoite, diamond, "lonsdaleite" (actually faulted and twinned cubic diamond), and/or organic carbon. The graphite in most samples is fine-grained, but a few ureilites (e.g., ALH 78091, ALH 83014, Nova 001) contain ~500-μm-long euhedral graphite blades with pyramidal terminations.

Interstitial phases that occur in at least some ureilites include metallic Fe-Ni (with up to ~3 wt% Si, ~1.5 wt% Cr, and ~1 wt% P), schreibersite, troilite, brezinaite, low-Ca pyroxene, augite, Al-Ti diopside, silica, Si-Al-alkali glass, chromite, and corundum. Associated with the metal in the GRA 95205 ureilite, is a Sn- and In-rich component that exhibits superconductivity (Wampler et al. 2020).

Polymict ureilites contain clasts of monomict ureilites along with other lithic clasts (e.g., porphyritic enstatite, enstatite-granular-olivine, basaltic, angritic, phosphate-bearing feldspathic melt-rock, carbonaceous-chondritic matrix) (Goodrich et al. 2004, 2015). Also present are a variety of interstitial phases including carbon, suessite, sulfide, chromite, apatite, halite, and sylvite. Plagioclase in the lithic clasts ranges in composition from An_{0-100}; grains of intermediate composition (An_{30-80}) tend to be free of K_2O. Carbonaceous-chondrite-matrix-like clasts in Nilpena consist of saponite, serpentine, magnetite, pentlandite, pyrrhotite, and ferrihydrite.

8.2.2 Brachinites (Prototype Brachina)

Brachinites consist (in vol%) of 74–98% olivine, 4–15% Ca-pyroxene (Cr-diopside or Cr-augite), 0.5–2% chromite, 3–7% Fe-sulfide (mainly Ni-bearing troilite, but Brachina itself also contains pentlandite), minor phosphate (mainly chlorapatite, but Hughes 026 also contains merrillite), minor Ni-rich metal (taenite and tetrataenite), 0–10% plagioclase, and, in some cases, accessory orthopyroxene (e.g., Mittlefehldt et al. 2003; Keil 2014). Olivine generally forms equigranular textures; interstitial phases include augite, chromite, and, in some cases, plagioclase.

Although individual brachinites contain homogeneous unzoned olivine, the overall compositional range is Fa_{30-35}. Olivine has low Cr_2O_3 (0.01–0.08 wt%), moderate MnO (0.3–0.6 wt%), and moderate CaO (0.05–0.27 wt%).

8.2.3 Acapulcoites and Lodranites (Prototypes Acapulco and Lodran)

These two related groups have OC-like mineralogy and consist of orthopyroxene, olivine, Cr-diopside, sodic plagioclase, metallic Fe-Ni (kamacite and taenite), phosphide (schreibersite and melliniite), troilite, merrillite, chlorapatite, chromite, graphite, and rare metallic Cu (in acapulcoite Dhofar 1222). Acapulcoites are fine-grained rocks (150–230 μm); several contain a few volume percent (up to 6 vol%) relict BO, RP, GOP, PP, PO, and POP chondrules. Lodranites are appreciably coarser grained (540–700 μm) and do not contain relict chondrules (e.g., Rubin 2007; Keil and McCoy 2018).

Acapulcoites and lodranites have similar compositional ranges in olivine and orthopyroxene ($Fs_{3.3-13.3}$, $Fs_{3.7-13.8}$). Many chondrule-free acapulcoites have Ca-pyroxene grains with thin exsolution lamellae. Lodranites contain inverted pigeonite grains with coarse Ca-pyroxene exsolution lamellae.

In volume percent, acapulcoites contain ~74–89 percent silicates, ~4–18 percent kamacite, 0–0.3 percent taenite, 2.7–6.4 percent troilite, 0–1.7 percent chromite, and $\leq$0.1 percent graphite; lodranites contain ~65–94 percent silicates, ~1–28 percent kamacite, 0–0.1 percent taenite, 0.1–3.1 percent troilite, 0–0.7 percent chromite, and $\leq$1.3 percent graphite. Although there are exceptions, lodranites tend to be depleted in troilite and plagioclase relative to acapulcoites. EET 84302 is a coarse-grained rock, depleted in troilite, but not in plagioclase; many workers consider it intermediate between acapulcoites and lodranites.

Members of both groups have suffered reduction: (1) olivine and orthopyroxene have appreciably lower FeO/(FeO+MgO) ratios ($Fa_{4.2-14.5}$, $Fs_{3.1-13.7}$ in acapulcoites; $Fa_{3.3-13.3}$, $Fs_{3.7-13.8}$ in lodranites) than those in OC (e.g., $Fa_{18.8}$, $Fs_{17.2}$ in H4–6); (2) some olivine and orthopyroxene grains exhibit reverse compositional zoning; (3) many samples have lower values of Fa than Fs[3] (e.g., $Fa_{4.6}$, $Fs_{6.9}$ in acapulcoite Superior Valley 014; $Fa_{9.4}$, $Fs_{12.6}$ in lodranite FRO 90011) compared to $Fa_{18.8}$, $Fs_{17.2}$ in mean H4–6 chondrites; (4) olivine grains in acapulcoites and lodranites have low FeO/MnO ratios (16–18 and 8–25) compared to H4–6 olivine (32–38); (5) some acapulcoites have high orthopyroxene/olivine modal ratios (e.g., 1.7 in Monument Draw; 1.4 in Dhofar 312) compared to H chondrites (~0.74); (6) some acapulcoites contain phosphide (rare in OC); (7) both groups have very high kamacite/taenite modal ratios (e.g., 54 in acapulcoite Dhofar 1222; ~280 in lodranite GRA 95209) compared to OC (e.g., 5 in H chondrites), and (8) some samples have high metal/troilite modal ratios (e.g., 3.6 in acapulcoite LEW 86220; 238 in lodranite EET 84302) compared to OC (~2 in H, ~1 in L, ~0.4 in LL). Graphite is the likely reductant.

Reduction probably occurred at elevated temperatures, mainly by one of the following reactions:

$$(\mathrm{Mg,Fe})\mathrm{SiO_4} + \mathrm{C} = \mathrm{MgSiO_3} + \mathrm{Fe} + \mathrm{CO}\uparrow$$

or

$$2(\mathrm{Mg,Fe})\mathrm{SiO_4} + \mathrm{C} = 2\mathrm{MgSiO_3} + 2\mathrm{Fe} + \mathrm{CO_2}\uparrow$$

[3] This is due to faster Fe-Mg diffusion in olivine than in orthopyroxene during reduction.

8.2.4 Winonaites (Prototype Winona) and IAB Silicates

Winonaites and IAB irons were likely derived from the same parent asteroid; individual winonaites could be large IAB silicate inclusions freed from their iron-meteorite hosts. Most winonaites are fine-to-medium-grained equigranular rocks with OC-like minerals plus a few reduced phases (e.g., Hunt et al. 2017). These rocks consist of (in vol%): 15 to 30% forsterite (Fo_{95-100}), 25 to 45% percent orthopyroxene (En_{91-100}), 1 to 7% Cr-diopside, 6 to 14% sodic plagioclase (An_{7-25}), 0.1 to 13% metallic Fe-Ni, 1 to 19% troilite, 0 to 0.6% graphite, 0 to 1% schreibersite, 0 to 0.2% chromite and accessory phosphate (merrillite), daubréelite, and alabandite. Mineral compositions are more reduced than those in H chondrites.

One of the most primitive winonaites is NWA 1463, which contains abundant, readily identifiable relict chondrules. It has a recrystallized chondritic texture, resembling that of a type-5 OC. Its mineral compositions are in the range of typical winonaites.

Hammadah al Hamra 193 is an unusual winonaite consisting mainly of large orthopyroxene grains enclosing olivine and plagioclase. It contains fluorapatite (instead of merrillite) and large grains of the amphibole fluoro-edenite enclosing Ca-pyroxene, orthopyroxene, plagioclase, and olivine. The texture of HaH 193 suggests that fluoro-edenite formed from diopside, olivine, and plagioclase, and replaced clinopyroxene.

IAB Complex irons include meteorites with silicate inclusions similar in mineralogy to those in winonaites. Many inclusions are angular to subrounded; they are typically millimeter to centimeter in size with granoblastic textures. Also present in many IAB irons are small (<100 μm) silicate crystals surrounded by coarse metallic Fe-Ni grains. The most common phases in silicate inclusions are forsterite, enstatite, Cr-diopside, sodic plagioclase, phosphate (mainly chlorapatite and merrillite), chromite, magnesiochromite, metallic Fe-Ni, troilite, schreibersite, and graphite. Rare phases include cohenite, alabandite, ferroan alabandite, daubréelite, sphalerite, metallic Cu, and K-feldspar. San Cristobal inclusions contain antiperthite (exsolution lamellae of K-feldspar within sodic plagioclase), roedderite, haxonite, and brianite.

Carlton and Dayton are Ni-rich IAB Complex irons, previously designated IIICD. Many of their silicate inclusions are rich in phosphate. Carlton inclusions contain abundant chlorapatite and farringtonite-brianite; one inclusion has a single 175 × 975 μm grain of chladniite in association with olivine, orthopyroxene, sodic plagioclase, schreibersite, and troilite (McCoy et al. 1994). Dayton inclusions lack olivine, graphite, and carbide but contain free silica along with brianite and panethite. Plagioclase is less calcic in the Dayton and Carlton inclusions than in those in other IAB irons (e.g., $An_{1.1-4.9}$ vs. $An_{9.2-21.5}$), probably due to the partitioning of Ca into the abundant phosphates in the Dayton and Carlton inclusions.

The IAB silicate inclusions have been divided into distinct groups (Benedix et al. 2000): (1) *Non-chondritic*: e.g., basaltic, gabbroic or Mg-andesitic inclusions (Cr-diopside, sodic plagioclase and minor orthopyroxene, olivine, troilite, and metallic Fe-Ni) in Caddo County; troctolitic inclusions in Ocotillo; and coarse-grained olivine-rich inclusions in Winona and Mt. Morris (Wisconsin). (2) *Rounded*: e.g., inclusions rich in graphite and troilite (some of which contain olivine, orthopyroxene, and plagioclase) in Odessa and Toluca. Toluca inclusions commonly also contain metallic Fe-Ni, schreibersite, and cohenite. (3) *Angular silicate*: inclusions most closely resembling winonaites in texture and mineralogy. The abundances of these inclusions vary significantly: some IAB irons contain a few volume percent (e.g., El Taco mass of Campo del Cielo); others contain ~40 vol% (e.g., specimens of Landes, Lueders, Woodbine). There are fine-grained and medium-grained silicate assemblages as well as coarse monomineralic grains. Schrader et al. (2017) reported quasi-circular, 0.39–1.34-mm-diameter, polymineralic, graphite-free, silicate assemblages in Campo del Cielo that resemble recrystallized POP chondrules. These objects consist of olivine ($Fa_{3.8–4.4}$), orthopyroxene ($Fs_{4.7–6.4}Wo_{0.9–1.5}$), Ca-pyroxene ($Fs_{1.6–3.2}Wo_{43.8–47.3}$), albitic plagioclase and minor pyrrhotite, subhedral chromite, and, in one case, rutile.

NWA 468 is a IAB iron with ~55 vol% silicates including olivine ($Fa_{5.3}$), low-Ca pyroxene (both clinopyroxene with polysynthetic twinning and orthopyroxene; $Fs_{8.9}Wo_{1.9}$), diopside ($Fs_{3.7}Wo_{45.4}$), and plagioclase ($Ab_{78.7}Or_{2.6}$). Also present are chromite, troilite, schreibersite, and metallic Fe-Ni (kamacite, taenite, plessite). A few percent limonite, formed by terrestrial weathering, also occurs. The metal host of the meteorite consists of ~5-mm-wide zones of duplex plessite formed from original 5-mm-sized γ crystals.

8.3 Asteroidal Achondrites

8.3.1 Howardites, Eucrites, Diogenites

Most members of the three related groups of howardites, eucrites, and diogenites (HEDs) were probably derived from the same parent asteroid, widely assumed to be 4 Vesta (e.g., McSween et al. 2013; Mittlefehldt 2015). Spectral reflectivity measurements of the surface of Vesta show that many regions are basaltic and some, particularly in the southern hemisphere, are probably diogenitic, enriched in material dredged up from the lower crust by one or more major collisions. Howardites are surface breccias that are mechanical mixtures of eucrites and diogenites.

Eucrites are igneous volcanic monomict breccias or unbrecciated rocks rich in pyroxene (with <10 vol% orthopyroxene) and calcic plagioclase. Ordinary non-cumulate eucrites were significantly metamorphosed; they contain pigeonite (initially $Fs_{48–58}Wo_{6–15}$) with fine augite exsolution lamellae parallel to (001) and compositionally zoned plagioclase grains (e.g., $An_{80–93}$ in Juvinas). Lava-like eucrites were mildly metamorphosed; they contain zoned pigeonite ($Fs_{30–80}$), zoned plagioclase and minor ilmenite. Cumulate eucrites are coarse-grained gabbros: Binda type samples contain (1) orthopyroxene with blebby augite and (2) inverted low-Ca clinopyroxene (initially $Fs_{32–46}Wo_{7–16}$); Moore County-type samples contain orthopyroxene, inverted pigeonite with thick augite exsolution lamellae parallel to (001), and compositionally unzoned plagioclase (in the range $An_{90–98}$). The cumulate eucrites Moama, Moore County, and Y-980433 all possess monoclinic tridymite. Moama contains orthorhombic tridymite lamellae within monoclinic tridymite grains (indicating moderately rapid cooling at temperatures <400°C); Y-980433 contains quartz associated with an impact melt vein. Polymict eucrites are breccias with <10 vol% primary orthopyroxene. Minor and accessory phases present in some eucrites include olivine, silica (tridymite and quartz), ilmenite, chromite, phosphate (merrillite and apatite), low-Ni kamacite, troilite, zircon, and baddeleyite. A few polymict eucrites contain pyroxferroite; troilite, hedenbergite, and silica are present in NWA 1109. The NWA 5738 eucrite breccia has veins of essentially Ni-free kamacite and veins with major Ca-plagioclase and Fe^{+3}-bearing Cr-spinel. High-pressure minerals produced by impact events on Vesta were recently identified in several eucrites, including tissintite, coesite, stishovite, majorite, and vestaite (e.g., Pang et al. 2018).

Diogenites are orthopyroxenites (monomict breccias or unbrecciated rocks) with <10 vol% plagioclase. They consist largely (~84 to 100 vol%) of orthopyroxene ($Fs_{15-33}Wo_{1-2}$) with minor chromite; some contain major olivine or silica. Metallic Fe-Ni and troilite are present in minor to accessory amounts. A few diogenites have noritic lithologies and consist of significant orthopyroxene, Ca-plagioclase and olivine as well as minor pigeonite and augite. Diopside is present in some diogenites as exsolution products. Rare phosphate also occurs.

Howardites are polymict breccias composed predominantly of eucritic and diogenitic clasts. Some (e.g., Bholghati, Jodzie, Kapoeta, Y793497) also contain a few volume-percent carbonaceous-chondrite clasts (~80% CM2, ~20% CR2), impact-melt-coated breccia clasts, and impact-melt rock clasts. The 2015 Sariçiçek fall contains several types of clasts: (1) plagioclase-rich with inclusions of chromite and ilmenite; (2) silica-rich with blebs of troilite; (3) plagioclase-silica intergrowths, also containing large blebs of troilite; (4) ophitic and subophitic basaltic clasts with laths of plagioclase and zoned clinopyroxene. The matrix of Sariçiçek consists of Ca-pyroxene ($Fs_{50\pm18}Wo_{26\pm20}$) (including many grains with orthopyroxene exsolution lamellae), orthopyroxene ($Fs_{28\pm6}Wo_{2\pm1}$), plagioclase ($An_{83.5-89.6}Or_{0.2-4.0}$), silica, kamacite, troilite, chromite, ilmenite, and rare grains of olivine (Fa_{18-21}); fayalitic olivine is also present.

8.3.2 Angrites (Prototype Angra dos Reis)

Angrites are ancient (~4.56 Ga old) igneous rocks (e.g., Keil 2012). They are silica-undersaturated, alkali-depleted, and moderately Ca- and Ti-rich with normative nepheline and larnite (Ca_2SiO_4, abbreviated La). Most are unbrecciated. There are three textural subgroups: (1) fine-to-coarse-grained, quench-textured, basaltic-diabasic rocks with ophitic, dendritic, or porphyritic textures, (2) coarse-grained plutonic-subvolcanic rocks, and (3) the NWA 8535 dunite. The only observed fall, Angra dos Reis (AdoR) (1.5 kg), is a granular olivine pyroxenite containing 93 vol% homogeneous Al-Ti diopside ($Fs_{12}Wo_{55}$) along with minor Ca-bearing olivine (Fa_{46}; 1.3 wt% CaO), hercynite, and troilite, and accessory Mg-kirschsteinite, celsian, merrillite, titanomagnetite, baddeleyite, and metallic Fe-Ni. There are also a few reports of plagioclase grains among mineral separates.

Other angrites contain 20–24 vol% normally zoned Al-Ti diopside (with $Fs_{21}Wo_{53}$ cores and $Fs_{48}Wo_{52}$ rims and fairly constant CaO), 29–42 vol% normally zoned olivine, 33–36 vol% Ca-plagioclase, accessory spinel-group minerals (including hercynite, spinel, Cr-pleonaste, ulvöspinel and titanomagnetite), glass, metallic Fe-Ni, troilite, and merrillite, and rare carbonates, baddeleyite, celsian, and rhönite. Olivine phenocrysts have Fa_{10-27} cores, Fa_{55-100} rims, and <0.1 wt% CaO; euhedral groundmass olivines have Fa_{34} cores and $Fa_{58}La_{37}$ rims. LEW 86010 contains anorthitic plagioclase and homogeneous olivine (Fa_{63}; 1.5–2.6 wt% CaO) with kirschsteinite exsolution lamellae. The phases identified in NWA 12774 include olivine, Ca-pyroxene (fassaite), troilite, anorthite, Fe-sulfide (possibly pyrrhotite), silico-phosphates, amorphous carbon, graphite, and kamacite.

D'Orbigny, the largest angrite (16.55 kg), has a subophitic texture and contains normally zoned pyroxene (with $Fs_{20}Wo_{52}$ cores and $Fs_{49}Wo_{51}$ rims), normally zoned olivine (with cores of $Fa_{39.7}La_{1.3}$ and rims of $Fa_{75.9}La_{11.5}$), anhedral subcalcic kirschsteinite grains, laths of anorthitic plagioclase, and minor hercynite, skeletal ulvöspinel, troilite, merrillite, and a calcium silico-phosphate phase. Also present are large vugs lined with augite and rare centimeter-sized magnesian olivine crystals (Fa_{9-11}).

NWA 4590 (213 g) has an unusual cumulate texture and consists of 33 vol% Al-Ti diopside with rare pigeonite exsolution lamellae, 28 vol% anorthite (some grains partially enclosing ulvöspinel), 14 vol% calcic olivine with 10–50-μm-wide kirschsteinite exsolution lamellae, 18 vol% ulvöspinel, 5 vol% kirschsteinite with thin olivine exsolution lamellae, and accessory glass, merrillite, Cl-bearing calcium silico-phosphate, troilite, and kamacite.

8.3.3 Aubrites a.k.a. Enstatite Achondrites (Prototype Aubres)

This achondrite group comprises highly reduced differentiated rocks with very little oxidized iron (e.g., Keil 2010). Aubrites contain 75–98 vol% enstatite ($Fs_{0.06}$), 0.3–16 vol% albite ($Ab_{92.7}Or_{3.2}$), 0.2 to 8 vol% diopside ($Fs_{0.08}Wo_{43.5}$), 0.3 to 10 vol% forsterite ($Fa_{0.01}$), and minor to accessory amounts of a large variety of phases: metallic Fe-Ni (Si-bearing kamacite, Si-poor kamacite, Si-bearing taenite), numerous sulfides (Ti-bearing and Cr-bearing troilite, heideite, caswellsilverite, oldhamite, niningerite, djerfisherite, brezinaite, daubréelite, alabandite, ferromagnesian alabandite, sphalerite, Sb- and Ag-sulfides), graphite, silica (cristobalite, tridymite), roedderite, schreibersite, perryite, osbornite, metallic Cu, and feldspathic glass.

Aubritic enstatite is mainly disordered orthopyroxene, but clinoenstatite occurs in some microporphyritic clasts in Norton County. Exsolved diopside constitutes up to ~25 vol% of host enstatite grains in some aubrites, indicating that these pyroxene grains are inverted pigeonite. Enstatite grains in aubrites are generally several millimeters to several centimeters in size; Peña Blanca Spring has enstatite crystals as large as 6×8×10 cm.

Although most aubrite plagioclase is albitic, ALHA78113 and QUE 97289 contain compositionally heterogeneous plagioclase ($Ab_{28.4-97.6}An_{0.0-70.6}Or_{0.2-5.2}$ and An_{4-35}, respectively). Larned, an anomalous, highly shocked aubrite, contains impact-produced plagioclase glass that is much less sodic and much more potassic than the average composition of crystalline plagioclase in aubrites ($Ab_{79}Or_{21}$ vs. $Ab_{92.7}Or_{3.2}$). This rock also contains a phase described as "ballen silica," patches of silica with a fish-scale texture commonly found in impact-melt rocks associated with terrestrial impact craters (e.g., Schmieder et al. 2009).

Most aubrites are breccias. The majority are monomict fragmental breccias; a few are regolith breccias containing solar-wind-implanted rare gases. Cumberland Falls, ALH 78113, and Larned are polymict breccias. Numerous dark clasts in Cumberland Falls were derived from ordinary chondrites; the clasts have all been shocked and some are impact-melt breccias. The ordinary-chondrite clasts have been reduced: olivine is very magnesian ($Fa_{0.4-0.7}$), low-Ca pyroxene grains are reversely zoned, and some reduced phases are present – daubréelite and schreibersite.

Rare phases are present in some clasts in other aubrites. Larned contains a millimeter-sized websterite clast consisting of enstatite, diopside, schreibersite, and sulfides (including greigite). Secondary veinlets of a phase similar to greenalite are also present in Larned. One 400 × 800 μm forsterite-and-enstatite-rich clast in ALH 83009 is rimmed by troilite (with inclusions of geikielite and perovskite), alabandite, djerfisherite, daubréelite, and schreibersite.

Mayo Belwa is an impact-melt breccia containing ~5 vol% vugs; some vugs are lined with granular enstatite, acicular diopside, minor cristobalite, and bundles of albite laths. Other vugs are lined with bundles of acicular grains of fluor-richterite.

Norton County, the largest aubrite (1.1 metric tons), contains diverse igneous clasts including dunites, plutonic orthopyroxenites, plutonic pyroxenites, plagioclase-silica rocks, impact-melt breccias, and impact-melt rocks containing both albitic (An_6) and anorthitic (An_{92}) plagioclase.

The unbrecciated aubrites Shallowater and Mount Egerton appear to be from a different asteroid than the majority of aubrites (Keil et al. 1989). These two rocks contain much more metallic Fe-Ni (8.4 and 21 wt%, respectively) than average aubrites (~0.5 wt%). The metal abundance in Mount Egerton is similar to the average in enstatite chondrites (~22 wt%). Both Shallowater and Mount Egerton have bulk positive Eu anomalies, unlike most aubrites. These two meteorites also have near-identical cosmic ray–exposure ages (27.0 Ma and 26.9 Ma), and it seems possible they are from the same parent body.

Phases produced from primary minerals in aubrites by terrestrial weathering include (a) schöllhornite ($Na_{0.3}(H_2O)CrS_2$), derived from caswellsilverite ($NaCrS_2$); (b) portlandite ($Ca(OH)_2$), vaterite (hexagonal $CaCO_3$), calcite (trigonal $CaCO_3$), bassanite ($CaSO_4{\cdot}½H_2O$), and barite ($BaSO_4$), all likely derived from oldhamite (CaS); and (c) goethite (FeO(OH)), derived from metallic Fe-Ni.

Although most of the reduced phases in aubrites also occur in enstatite chondrites, there are some significant differences. (1) Troilite in enstatite chondrites contains ~0.5 wt% Ti, whereas troilite in aubrites averages 2.8 wt% Ti, with a pyroxene-oldhamite inclusion in Bustee containing troilite with up to ~25 wt% Ti. (2) Diopside in enstatite chondrites is very rare (occurring as small grains in some type-3 samples as well as constituting ~1 vol% of the EL6 impact-melt breccia EET 90102), whereas diopside averages 2.2 vol% in aubrites and ranges up to 20 vol% in some Norton County clasts. (3) The mean sulfide/metal volume ratio in EL6 chondrites (~0.6) is much lower than in aubrites (~2.5). (4) Taenite is a rare phase in enstatite meteorites, but is less abundant in enstatite chondrites than aubrites. (5) Metallic Cu has not been reported in enstatite chondrites, but is a rare phase in aubrites. (6) Sinoite is present in many shocked EL chondrites but has not been reported in aubrites.

There are also petrologic and geochemical differences between aubrites and EL melt rocks. (1) Aubrites are much coarser (1-10 mm-sized grains vs. 10-100 μm-sized grains). (2) Aubrites have much less metallic Fe-Ni (~0.2 vol% vs. ~5–15 vol%). (3) Aubrites contain disordered orthopyroxene; some EL melt rocks contain enstatite with clinoenstatite exsolution lamellae. (4) Most EL melt rocks are not brecciated. (5) Most aubrites are depleted in REE relative to CI and almost all have negative Eu anomalies; at least some EL melt rocks, e.g., Ilafegh 009, do not have Eu anomalies. These differences, plus others discussed by Keil (2010), strongly support the conclusion that aubrites are not derived from either the EL or EH parent asteroid.

8.3.4 Ungrouped Achondrites

About 3 percent of known achondrites are ungrouped (not taking potential pairings into account). Many of these rocks are related to established groups, but differ somewhat in detailed petrologic, mineralogical, or geochemical properties. A few of the more interesting ungrouped asteroidal achondrites are described in this section.

NWA 6693. This FeO-rich 5.1-kg rock has a coarse poikilitic cumulate texture with pigeonite oikocrysts ranging up to 1.4 cm in size (Warren et al. 2013). It contains (in vol%) 70% pyroxene (both low-Ca pyroxene – $Fs_{39.9}Wo_{3.2}$ and Ca-pyroxene – $Fs_{20.0}Wo_{40.3}$); 16% olivine ($Fa_{49.4}$ with 0.77 wt% NiO); 13% feldspar (Ab_{92}, with almost no discernible twin lamellae); 0.6% Cr-spinel; 0.4% awaruite and trace amounts of merrillite and Ni-rich sulfide (probably pentlandite and heazlewoodite). Very rare K-rich feldspar ($Ab_{57.3}Or_{42.5}$) also occurs. The bulk rock is highly depleted in S and chalcophile elements but shows only minor to negligible depletions in common and refractory siderophile elements (Co, Ni, Ir, and Os are 0.7–1.0 × CI). The bulk O-isotopic composition ($\Delta^{17}O = -1.08‰$) is far below that of eucrites on the standard three-isotope diagram and is similar to that of acapulcoites-lodranites. NWA 6693 is unique among achondrites in its high modal abundances of very ferroan mafic silicates (including NiO-bearing olivine) and the occurrence of sodic plagioclase, Ni-rich metal, and Ni-rich sulfide.

NWA 6698 and NWA 11575. These andesitic meteorites were described by Váci et al. (2020a). NWA 6698 is a medium-to-fine-grained igneous rock with unzoned euhedral crystals; it consists (in vol%) of 67% plagioclase (andesine to oligoclase), 16% pigeonite, 10% augite, 5% glassy mesostasis with microlites, and accessory-to-trace amounts of chromite, sulfide, and phosphate. NWA 11575 is a fine-grained igneous rock with subhedral crystals; it consists (in vol %) of 62% plagioclase (mainly oligoclase), 37% zoned pyroxene (pigeonite cores surrounded by magnesian augite mantles and ferrous augite rims), and trace amounts of Fe oxides, sulfide, phosphate, K-feldspar, and silica. This meteorite also contains a "dark" lithology consisting (in vol%) of 65% glassy mesostasis with microlites and 35% zoned pyroxene.

NWA 7325. This 345-g rock was recovered as a group of 35 dark green stones. It is a fine-to-medium-grained unbrecciated olivine gabbro with sparse vesicles; it consists of ~25 to 30 vol% Al-Cr-bearing diopside ($Fs_{0.6-1.3}Wo_{44.8-49.3}$), ~15 vol% lobate olivine ($Fa_{2.7-6.0}$, with undulose extinction), ~55 to 60 vol% untwinned calcic plagioclase (An_{85-93}, associated with the olivine), ~1 vol% Fe-sulfide (probably troilite with 2.2 to 6 wt% Cr), accessory ferrochromite, kamacite, and taenite, and rare eskolaite (associated with Fe-sulfide) (Goodrich et al. 2017). The bulk rock is highly depleted in incompatible trace elements (mostly far below 0.3 × CI); it is enriched in light rare-earth elements (LREE) and has a large positive Eu anomaly. NWA 7325 is unique among achondrites in containing very magnesian mafic silicates along with Ca-plagioclase. The bulk O-isotopic composition of NWA 7325 ($\Delta^{17}O = -0.91‰$) plots on the CCAM line and resembles that of ureilites. The rock has been well dated; its Pb-Pb isotopic age (4563.4±2.6 Ma) is essentially identical to that of its Al-Mg age (4563.09±0.26 Ma) (Koefoed et al. 2016). The similarity of some of the bulk major element ratios (e.g., Al/Si, Mg/Si) to that of the Mercurian crust (measured by remote sensing) led Irving et al. (2013) to suggest that the meteorite might be a sample of Mercury. A similar view was expressed by Jabeen et al. (2014) based on the bulk O-isotopic data. However, the community consensus is that the old age of NWA 7325 precludes derivation from a planetary-sized parent body. It may be an impact melt of an early-formed gabbroic lithology on a differentiated asteroid.

NWA 11119. This single 453-g silica-rich (andesitic to dacitic) extrusive crustal rock consists of 87.8 vol% porphyritic phenocrysts, 11 vol% medium-to-fine-grained groundmass (formed from a quenched melt) and 1.2 vol% millimeter-sized spherical vesicles and irregular cavities.

The whole-rock modal abundances (in vol%) of the major phases include 49.2% subhedral-to-anhedral porphyritic plagioclase ($Ab_{12.6\pm7.1}Or_{0.1\pm0.2}$) grains up to ~4 mm in size; 26.3% tridymite laths and anhedral grains (with ~0.8 wt% Al_2O_3) up to 3.4 mm; and 12.3% subhedral-to-anhedral pyroxene (Ca-pyroxene – $Fs_{6.0\pm2.1}Wo_{38.1\pm1.1}$; enstatite – $Fs_{9.5\pm0.3}Wo_{4.6\pm0.1}$) grains up to 3.7 mm. Also present are submicrometer- to micrometer-sized grains in the groundmass: ulvöspinel, ilmenite, troilite, tranquillityite, zircon, metallic Fe-Ni, fayalite, and tsangpoite ($Ca_5(PO_4)_2(SiO_4)$). (Coarser, compositionally zoned grains are also present in the groundmass.) The whole-rock REE pattern is fairly flat with a small LREE depletion and a small positive Eu anomaly. The mean bulk O-isotopic composition ($\Delta^{17}O = -0.96‰$) resembles that of ureilites, NWA 7325 (described in the previous paragraph), and a trachyandesitic clast in the polymict Almahata Sitta ureilite. NWA 11119 has an Al-Mg age of 4,564.8±0.3 Ma (Srinivasan et al. 2018); this geochemically evolved rock crystallized on a differentiated planetesimal only 2.5 to 3.5 Ma after the first refractory solids formed in the solar nebula and before the terrestrial planets accreted.

Dhofar 778. This 156-g harzburgite was described by Pang et al. (2020). Although initially classified as a diogenite, the rock's bulk O-isotopic composition ($\Delta^{17}O = -0.160\pm0.054‰$; 2σ) differs from that of normal HED samples ($\Delta^{17}O = -0.240\pm0.012‰$; 2σ); it may derive from the lower crust or upper mantle of a previously unsampled differentiated asteroid. The rock has both poikilitic and non-poikilitic lithologies; overall, it consists (in vol%) of ~54% zoned orthopyroxene ($Fs_{13-24}Wo_{0.4-4.5}$; Fe/Mn = 29±4); ~44% compositionally homogeneous olivine (Fa_{26-29}, Fe/Mn = 46±3), and ~2% other phases. The latter include plagioclase (An_{72-93}), pigeonite ($Fs_{23-24}Wo_{6-8}$), augite ($Fs_{11-14}Wo_{36-41}$), chromite, troilite, metallic Fe-Ni (7–18 wt% Ni), and highly silicic glass (87–93 wt% SiO_2). Olivine grains range up to 5 mm across; many contain symplectic intergrowths of orthopyroxene and spinel. The silicic glass occurs only within olivine. Dhofar 778 has been significantly shocked (shock stage S4), probably to ~13–17 GPa (1,200–1,600°C); high-pressure phases include ringwoodite, wadsleyite, and coesite.

GRA 06129. This rock, paired with GRA 06128, has a total mass of 644 g. It has a granoblastic texture and highly variable grain sizes (Shearer et al. 2011). It contains abundant (>70 vol%) plagioclase ($Ab_{84}Or_2$) crystals, ranging from a few hundred micrometers to ~5 mm. There are smaller amounts of olivine (Fa_{59}), orthopyroxene ($Fs_{44}Wo_2$), and Ca-pyroxene ($Fs_{19}Wo_{43}$), and minor amounts of opaque phases: ilmenite, troilite, pentlandite (26 wt% Ni), and spinel (with 13 wt% TiO_2, 35 wt% Cr_2O_3, and 3 wt% Al_2O_3). Awaruite (67 wt% Ni) occurs as tiny grains within olivine. The mean bulk O-isotopic composition of the rock is very close to the terrestrial fractionation line ($\Delta^{17}O = -0.06‰$).

Zakłodzie. This ungrouped, 8.68-kg enstatite-rich impact-melt rock has a granoblastic texture and contains ~60 vol% small, tightly packed equigranular enstatite grains with low MnO (0.02 wt%), ~20 vol% metallic Fe-Ni (kamacite and martensite), ~10 vol% troilite with very high Mn (1.35 wt%) and Cr (4.7 wt%), and ~10 vol% plagioclase (with two populations – $Ab_{59-64}Or_{0-0.5}$ and $Ab_{86-89}Or_{9-12}$) (Przylibski et al. 2005). Accessory phases include 0.5-mm-sized Mn-rich keilite as well as free silica (tridymite, quartz, and cristobalite), sinoite, oldhamite, schreibersite, graphite, and amphibole. Also present are two newly discovered sulfide minerals. (1) Buseckite ((Fe,Zn,Mn)S), the Fe-dominant analog of wurtzite (with a $P6_3mc$ structure), occurs as 4–20-μm-sized single crystals (Figure 8.2; Ma et al. 2012a). The grains are black under diffuse illumination and appear nearly opaque grayish brown in transmitted light. (2) Browneite (MnS),

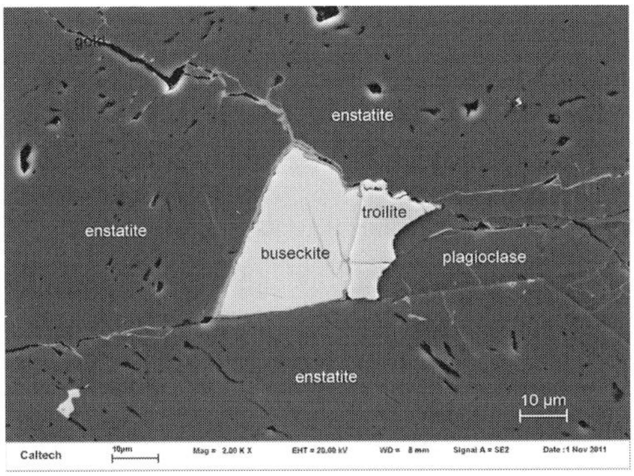

Figure 8.2 Buseckite [(Fe,Zn,Mn)S] occurring with troilite, plagioclase, and enstatite in the Zakłodzie ungrouped enstatite-rich achondrite (Ma et al. 2012a). Secondary electron image.

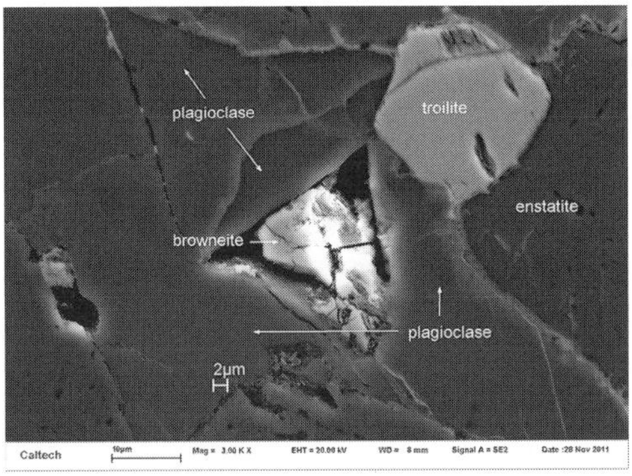

Figure 8.3 Browneite (MnS) occurring in plagioclase with troilite and enstatite in the Zakłodzie ungrouped enstatite-rich achondrite (Ma et al. 2012b). Secondary electron image.

a new member of the sphalerite group with the $F43m$ structure, occurs as a single ~16-µm-sized crystal (Figure 8.3; Ma et al. 2012b). It is yellowish brown, translucent, and brittle. The presence of keilite, sinoite, low-MnO enstatite, and high-Mn troilite are all properties previously suggested to be indicative of an impact-melt origin (e.g., Rubin, 2015a). Zakłodzie is classified as an ungrouped enstatite achondrite.

Itqiy. This 4.72-kg meteorite is a chondrule-free rock comprised mainly of 78 vol% subhedral, equigranular 0.5–4-mm-sized grains of $Fs_{0.2}Wo_{3.0}$ enstatite (exhibiting weak mosaicism), and 22 vol% 0.2–2.0-mm-sized kamacite grains with 3.13 wt% Si (Patzer et al. 2001). There is little sulfide: troilite is very rare; there are a few grains of Mg-Mn-Fe sulfide with compositions intermediate between niningerite and alabandite, and some Fe-Cr sulfide grains. The sulfide grains occur in 0.2–1.0-mm-sized intergranular regions. Plagioclase is also of low abundance relative to enstatite chondrites or aubrites. Itqiy has been modeled as a partial-melt residue ultimately derived from an enstatite chondrite and is currently classified as EH7-an.

8.4 Nonasteroidal Achondrites

8.4.1 Lunar Meteorites

Moon rocks, lunar meteorites, and lunar minerals are discussed in detail in Section 9.4. Here is a brief summary.

About 80 percent of lunar meteorites are broadly anorthositic or feldspathic. These rocks contain varying abundances of anorthite, olivine; orthopyroxene, clinopyroxene (pigeonite, ferropigeonite, diopside, augite, subcalcic augite), silica; oxides (chromite, Ti-chromite, Cr-ulvöspinel, ilmenite, rutile, Cr-pleonaste, baddeleyite), K-bearing glass, and small amounts of kamacite and troilite.

About 15 percent of lunar meteorites are broadly basaltic or gabbroic. They contain major calcic plagioclase and clinopyroxene (augite, subcalcic augite, pigeonite), accessory silica (probably tridymite), rare olivine, and trace amounts of metallic Fe-Ni and troilite. Some of these rocks contain additional phases including hedenbergite, pyroxferroite, ilmenite, chromite, Ti-magnetite, loveringite, monazite, and apatite.

About 5 percent of lunar meteorites are mixed, mingled, or polymict breccias. Their mineral constituents are a combination of those in anorthositic and basaltic samples.

8.4.2 Martian Meteorites

Martian meteoritic minerals are discussed in detail in Section 9.5. Here is a brief summary.

About 82 percent of martian meteorites are shergottites and related rocks containing (1) major pyroxene (augite, subcalcic augite, pigeonite, and/or orthopyroxene), olivine ($Fa_{24–40}$) and maskelynite and/or plagioclase ($Ab_{30–50}$), (2) minor oxides (magnetite, titanomagnetite, ulvöspinel, ilmenite, chromite, hercynite, baddeleyite), phosphates (merrillite, chlorapatite) and sulfide (pyrrhotite, pentlandite), and (3) some late-stage phases (silica, pyroxferroite, fayalite, hercynite, and ferro-kaersutite).

About 9 percent of martian meteorites are nakhlites containing (1) major augite, subcalcic augite, olivine, and mesostasis, and (2) minor-to-accessory pigeonite, orthopyroxene, plagioclase, K-feldspar, silica, titanomagnetite, ulvöspinel, rutile, magnetite, hercynite, chlorapatite, merrillite, pyrrhotite, pyrite, marcasite, and chalcopyrite.

About 1 percent of martian meteorites are chassignites containing $\geq$90 vol% olivine and minor to accessory orthopyroxene ($Fs_{19}Wo_3$), pigeonite, augite, plagioclase, sanidine, chromite, chlorapatite, troilite, pentlandite, ilmenite, rutile, baddeleyite, kaersutitic amphibole, biotite, and phlogopite.

About 7 percent of martian meteorites are classified as polymict breccias. They consist of mixed lithologies and have a variety of the phases present in the other martian-meteorite varieties.

The martian-meteorite suite also includes two pyroxenites: ALH 84001 (with major orthopyroxene and minor to accessory chromite, maskelynite, augite, apatite, pyrite, and carbonate) and NWA 2646 (with major pigeonite, augite, olivine, and maskelynite, and minor oxides, phosphate, and sulfide).

Martian meteorites also contain many high-pressure phases produced during launch off Mars.

8.5 Iron Meteorites

The petrographic/metallographic characteristics of iron meteorites have been expertly reviewed by Buchwald (1975, 1977). Their structural classification is based on the size and shape of kamacite crystals. Hexahedrites (symbol H) have large equiaxial kamacite crystals; in most cases these crystals are larger than the maximum dimension of their host meteorite. Octahedrites characteristically have Widmanstätten patterns consisting of intergrown kamacite and taenite; the kamacite grains occur parallel to the four sets of octahedral planes within the parent taenite crystal (Figure 8.4). The bandwidth of the kamacite lamellae determines the octahedrite class: coarsest (Ogg), >3.3 mm; coarse (Og), 1.3–3.3 mm; medium (Om), 0.5–1.3 mm; fine (Of), 0.2–0.5 mm; finest (Off), <0.2 mm. Plessitic octahedrites (Opl) contain small (<0.2 mm) kamacite spindles scattered in an octahedral arrangement within the precursor taenite grain. Ataxites (D) typically have fine-grained structures difficult to discern without microscopic aid.

Although there are overlaps, there is a tendency (particularly among Ogg to Off octahedrites) for irons with higher bulk Ni contents to have narrower kamacite bandwidths (data from Hutchison 2004). Irons with less Ni generally have lower modal taenite and coarser structures.

H	5.3–5.8 wt% Ni
Ogg	5.6–6.5 wt% Ni
Og	6.1–8.2 wt% Ni
Om	7.1–10.6 wt% Ni
Of	7.3–13.0 wt% Ni
Off	11.3–18.1 wt% Ni
Opl	9.0–14.3 wt% Ni
D	15.8–60.8 wt% Ni

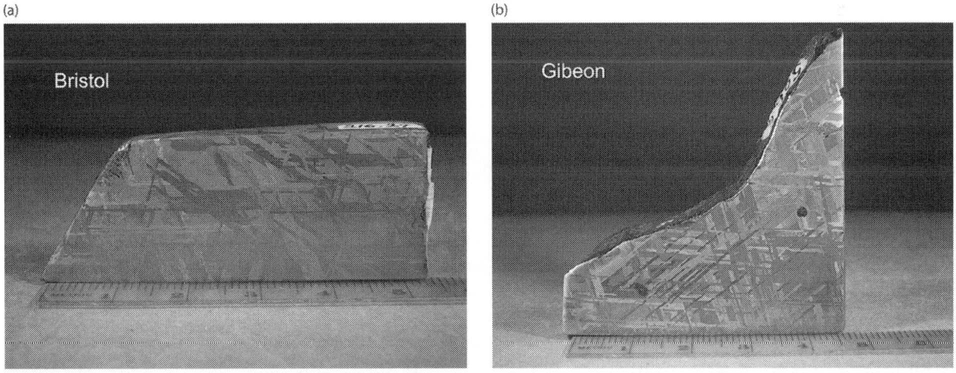

Figure 8.4 The Widmanstätten patterns of the Bristol and Gibeon IVA iron meteorites. Reflected light. (A black-and-white version of this figure will appear in some formats. For the colour version, please refer to the plate section.)

Iron meteorites have also been classified into 13 major chemical groups (each symbolized by a Roman numeral and a one- or two-letter combination, e.g., IIE, IIIAB) based on their bulk concentrations of such siderophile elements as Ni, Co, Au, As, Sb, W, Ga, Ge, Ir, and Cu. A group is defined as having five or more members. The compositions of irons from a single chemical group tend to form tight clusters on element-element diagrams and are assumed to have been derived from a unique asteroid. The sole possible exception is the IIG group that Wasson and Choe (2009) suggested formed from a P-rich magma within the core of the IIAB asteroid. About 10 percent of irons do not belong to a major group and are thus considered "ungrouped." Most ungrouped irons are unique specimens, but some are members of grouplets with two, three, or four members. Altogether, iron meteorites may sample ~60 distinct parent asteroids.

The different iron-meteorite groups have been modeled as being "magmatic" (formed by fractional crystallization from a well-mixed molten metallic core on a differentiated asteroid) or "non-magmatic" (presumed by some researchers to have formed by impact-melting processes on a chondritic asteroid). Most iron groups are magmatic; individual meteorite bulk compositions plot on element-element diagrams (e.g., Ni vs. Au) inside narrow arrays with slopes consistent with fractional crystallization. There are substantial fractionations of refractory siderophile elements (e.g., Ir, Os, Re, Ru, Pt) from common (Ni, Co) and volatile (e.g., Au, Ga, Ge, As, Sb) siderophiles in some of these groups. For example, the concentration ranges of Ir, Re, and Os in group IIAB (currently ~140 members) are ~4,000, ~6,000, and ~10,000, respectively. Fractionations arise because the solid/liquid distribution coefficient D is >1 for refractory siderophile elements and <1 for many common and volatile siderophiles under ideal conditions of core crystallization. However, an important complication is that nonmetallic elements such as S, C, and P within the metallic liquid can affect D significantly.

Groups IAB and IIE are considered non-magmatic; their slopes on element-element concentration diagrams do not conform to formation by fractional crystallization. Many individual IAB and IIE meteorites have silicates with chondritic bulk compositions and a few silicate inclusions contain recrystallized chondrules.

Based on their W-, Mo-, and N-isotopic compositions, iron-meteorite groups can be assigned to carbonaceous or non-carbonaceous isotopic reservoirs that probably were located at different heliocentric distances. Groups that formed from carbonaceous-chondrite-like precursors (CC irons) include IIC, IID, IIF, IIIF, and IVB. Groups that formed from non-carbonaceous-chondrite-like precursors (NC irons) include IAB, IC, IIAB, IIE, IIG, IIIAB, IIIE, and IVA. The magmatic CC-iron groups have higher median refractory siderophile element contents, higher median Ni contents, and higher median Ir/Ni and Ir/Au ratios than magmatic NC-iron groups. Magmatic CC-iron groups also tend to have longer cosmic-ray exposure ages than magmatic NC-iron groups (Rubin 2018).

The most common minerals in iron meteorites are metallic Fe-Ni (kamacite, martensite, taenite, tetrataenite); sulfide (troilite, daubréelite); phosphide (schreibersite, nickelphosphide); carbide (cohenite, haxonite); native carbon (graphite, diamond); native copper; nitride (carlsbergite); oxide (chromite, fusion-crust magnetite); and phosphate (merrillite, chlorapatite). Although Neumann-banded kamacite, martensite, and shock-hatched kamacite are present in many specimens, these are shock products and are not listed below; in general, only primary phases are described. The principal phases in silicate inclusions vary by iron-meteorite group; they were reviewed by Ruzicka (2014) and are described throughout Section 8.5.

8.5.1 Magmatic CC Irons

IIC: All specimens in this small group are plessitic octahedrites (Opl). The samples contain kamacite spindles up to ~800 μm long and ~50–70 μm wide. Some spheroidized plessite (formed via reheating of fine-grained plessite) occurs in Ballinoo. Schreibersite is very common in IIC irons, present as thin lamellae, nodules within kamacite spindles, and discontinuous rims around troilite-rich nodules. Swathing kamacite surrounds some of the schreibersite grains in Perryville. Oriented schreibersite grains in Wiley are arranged as Brezina lamellae (oriented schreibersite crystals) parallel to (110). Troilite-rich nodules in IIC irons range in size from a few millimeters to a few centimeters. The nodules in Ballinoo are fine-grained mixtures of troilite, kamacite, taenite, schreibersite, and rare grains of chromite; those in Perryville contain small rounded droplets of daubréelite. Chromite is also present in Cratheús (1950) and Perryville. Bars of troilite in Unter-Mässing have parallel daubréelite exsolution lamellae. [Although Wiley has been classified as IIC on the basis of its Ir, Ga, Ge, and Ni contents (Wasson 1969), it differs from other IIC irons in Mo- and W-isotopic composition (Kruijer et al. 2017; Bermingham et al. 2018; Hilton et al. 2018) and in its abundances of highly siderophile elements (Tornabene et al. 2019). In addition, unlike other IIC irons, it contains an unusual phosphate assemblage: 20–100-μm-sized inclusions of sarcopside or graftonite (Buchwald 1975).]

IID: Nearly all members of this moderate-sized group are medium octahedrites (Om); the anomalous Arltunga is an ataxite (D) and a couple of specimens (Needles and Wallapai) are fine octahedrites (Of). The IID irons contain elongated (in some cases, swollen) kamacite lamellae and significant plessite (comb, net, martensitic, duplex). Schreibersite is abundant, occurring as Brezina lamellae, blebs, blocks, veinlets, rosettes, rhabdites, and skeletal crystals with swathing kamacite. Troilite is generally a minor phase, but is rare in a few specimens (e.g., Wallapai; Nothing). ("Nothing" is the name of a IID iron from Arizona.) Troilite-rich nodules range from 0.1 to 3 mm; some contain daubréelite exsolution lamellae. One troilite nodule in Carbo contains green glass (presumably formed by shock melting) that devitrified to form cristobalite. Also present in several IID irons (e.g., Arltunga, Carbo, Elbogen, Losttown) are chromite grains. Carbo is unusual in containing 100–800-μm-wide assemblages of haxonite, taenite, and kamacite.

IIF: Most members of this small group are plessitic octahedrites (Opl). Although Del Rio is described as a polycrystalline, Ni-rich ataxite (D), large regions are plessitic. IIF specimens contain scattered long spindles of kamacite (some discontinuous), duplex plessite, and small troilite-rich inclusions. Schreibersite occurs within kamacite spindles and as small particles within plessite. Troilite is minor to rare; it is present as 5–15 mm nodules, some of which contain daubréelite exsolution lamellae. Monahans (1938) has moderately abundant crystals of euhedral chromite. A small graphite inclusion was reported in Del Rio.

IIIF: This small iron group has diverse octahedral structures, including fine (Of), medium (Om), coarse (Og), and coarsest (Ogg) octahedrites. Most specimens contain elongated kamacite lamellae decorated with 0.5–3-μm-sized schreibersite precipitates. Schreibersite also occurs in different samples as small nodules, skeletal grains, millimeter-sized crystals, and rhabdites. Troilite is rare in some samples (e.g., Moonbi, Nelson County). Other specimens have moderately abundant troilite nodules with daubréelite exsolution lamellae; the nodules in St.

Genevieve County are 1–9 mm in diameter and are surrounded by discontinuous 10–50-μm-wide schreibersite rims and 0.5–1.0-mm-thick bands of swathing kamacite. Many IIIF irons have plessite (comb, net, duplex); some coarse plessite fields in St. Genevieve County contain spheroidized 1–15-μm-sized blebs of taenite.

IVB: All members of this moderately small group are ataxites (D). They contain fine intergrowths of kamacite and taenite as well as scattered kamacite spindles up to 500 μm in length. There is abundant duplex plessite. Schreibersite occurs as fine precipitates on kamacite spindles in some meteorites; schreibersite is also present as a few larger crystals (up to 120 μm long), as scattered 1–15 μm grains rimmed by swathing kamacite, as 0.5–2-μm grains within duplex plessite in Hoba (the largest known intact meteorite), and as small grains around a few troilite nodules in Tawallah Valley. The abundance of schreibersite is correlated with bulk Ni content: IVB irons with low Ni (e.g., Tlacotepec, 15.60 wt% Ni) tend to contain relatively little schreibersite, whereas samples with high Ni (e.g., Warburton Range, 18.02 wt% Ni) tend to have more abundant schreibersite (Buchwald 1975; Campbell and Humayun, 2005). Troilite nodules and lenticular bodies in IVB irons tend to be fairly small (0.1–3 mm) and rare; some troilite nodules contain daubréelite exsolution lamellae. Many troilite nodules in Tawallah Valley are surrounded by 10–20-μm-thick bands of swathing kamacite; a few troilite nodules in Santa Clara contain small (<25 μm) grains of silica.

8.5.2 Magmatic NC Irons

IC: This small group comprises mainly coarse octahedrites (Og), although a couple specimens have medium (Om) or medium-to-coarse (Og-Om) structures, and a few are polycrystalline. In addition to kamacite and taenite, several varieties of plessite occur (comb, pearlitic, net, acicular, somewhat spheroidized). Minor phases present in most specimens include troilite with daubréelite exsolution lamellae, schreibersite, and cohenite. Bendegó contains cohenite and scattered grains of haxonite. Several IC irons contain small grains of carlsbergite. Graphite is present as feathery veins in Nocoleche, as intergrowths with troilite in NWA 2743 and as cliftonite crystals in Santa Rosa.

IIAB: This is the second largest magmatic iron group; the members are all hexahedrites (H) or coarsest octahedrites (Ogg), typically containing minor to accessory taenite and little or no plessite. Schreibersite occurs commonly as large (centimeter-sized) masses, skeletal crystals (surrounded by swathing kamacite), lamellar grains, and rhabdites. Old Woman contains centimeter-long chains of schreibersite grains in three distinct orientations 60° apart; in many cases, the individual grains have embayed or skeletal morphologies and are 100–200 μm wide and up to several millimeters long. Also common in IIAB irons are troilite nodules, many with daubréelite exsolution lamellae. Some troilite nodules in Coahuila and Hex River Mountains contain 1–5 mm grains of kosmochlor ($NaCrSi_2O_6$) as well as unidentified silicates. Accessory-to-rare phases in some IIAB samples include graphite, cohenite, carlsbergite, roaldite, and chromite. Sikhote-Alin (the largest meteorite fall in historic times) contains a very large (4 × 8 cm) euhedral chromite grain as well as accessory amounts of millimeter-sized olivine grains associated with troilite and chromite.

IIG: The meteorites in this small, P-rich group are hexahedrites (H) or coarsest octahedrites (Ogg) with very large kamacite crystals. The Ogg specimens (e.g., Auburn) contain small bands of taenite. Schreibersite occurs as massive coarse nodules (up to 4 cm), rosettes, skeletal crystals, lamellar grains, scattered spherules, and thin rhabdite plates with up to 10 different orientations. Troilite is a minor to accessory phase. Some nodules in Auburn are intergrowths of schreibersite, troilite, and daubréelite.

IIIAB: This is the largest magmatic iron group; most members are medium octahedrites (Om); others are fine (Of) or coarse (Og) octahedrites. The principal minerals are kamacite, taenite, and plessite. In many specimens, kamacite contains numerous small rhabdites, platelets of carlsbergite, and rare grains of daubréelite. Low-Ni IIIAB irons contain troilite with small rhabdite inclusions; high-Ni specimens contain elongated oriented troilite inclusions (Reichenbach lamellae) and oriented schreibersite crystals (Brezina lamellae) occurring parallel to (100), (210), (211), and (110). Although 13 distinct phosphate minerals were identified in IIIAB irons (Olsen et al. 1999), only 7 have been characterized: johnsomervilleite, galileiite, graftonite, beusite, sarcopside, buchwaldite, and maricite. Sulfide-rich nodules in Cape York contain troilite, ferroan alabandite, sphalerite, djerfisherite, an uncharacterized K-bearing sulfide, silica, metallic Cu, chromite, and phosphates (e.g., Kracher et al. 1977). Copper-bearing sphalerite and pentlandite are present in Grant. Puente del Zacate is unique in possessing a 7 mm angular silicate inclusion within a 16-mm-diameter troilite nodule. The silicate inclusion has a granoblastic texture and consists (in wt%) of 23% olivine (Fa_4); 14% low-Ca pyroxene (Fs_6Wo_1); 15% Cr-diopside (Fs_3Wo_{47}); 15% plagioclase ($Ab_{82}Or_4$); 1% troilite; 0.5% chromite; 0.5% daubréelite; 4% metallic Fe-Ni; and 27% graphite (or cavities that once contained graphite).

IIIE: Most specimens are coarse (Og) and some are medium (Om) octahedrites. The group was originally described by Scott et al. (1973) and Scott and Wasson (1975). Breen et al. (2016) placed the IIIE irons into four categories reflecting increasing degrees of shock alteration; these categories are discussed in more detail in Chapter 11. The principal mineral phases in IIIE irons are kamacite, taenite, and plessite. In weakly shocked samples, haxonite occurs within plessite, and sulfide inclusions typically consist of polycrystalline troilite with daubréelite exsolution lamellae. Severely shocked samples have sulfide-rich assemblages consisting of fragmental and subhedral daubréelite crystals, 1 to 4 vol% spidery troilite filaments, and 30 to 50 vol% low-Ni kamacite grains (which contain up to 6.0 wt% Co); haxonite in these inclusions has fully decomposed to graphite.

IVA: Most specimens are fine octahedrites (Of) and consist mainly of kamacite, taenite, plessite (comb, net, cellular), abundant troilite, and common daubréelite. Duel Hill (1854) contains sarcopside ($Fe_3(PO_4)_2$). Phosphide is present as rare micrometer-size grains in low-Ni IVA irons; in those with high Ni, phosphide commonly occurs as small grains on kamacite spindles. In addition to metallic Fe-Ni, Social Circle contains troilite, chromite, daubréelite, and rare joegoldsteinite ($MnCr_2S_4$) (Figure 8.5; Isa et al., 2016); the occurrence of brezinaite (Cr_3S_4) is uncertain. Gibeon and Bishop Canyon contain tabular tridymite. Steinbach and São João Nepomuceno (SJN) have clumps of coarse-grained low-Ca pyroxene (up to 3.5 mm long) and tridymite (up to 1×2.6 mm) (e.g., Ruzicka 2014). Most of the low-Ca pyroxene in these two meteorites is orthorhombic ("bronzite") (Steinbach: $Fs_{14.5-15.5}Or_{0.21-1.1}$; SJN: $Fs_{12.6-13.8}Or_{0.45-0.57}$); the remainder is monoclinic ("clinobronzite") with polysynthetic twinning. About half of the orthopyroxene grains in SJN contain small inclusions of troilite, chromite, metallic Fe-Ni or tridymite as isolated grains or in a variety of intergrowths; some of the orthopyroxene grains in Steinbach contain similar inclusions, but at lower modal abundances. Examination of slabs indicates that Steinbach consists (in vol%) of 38.3%

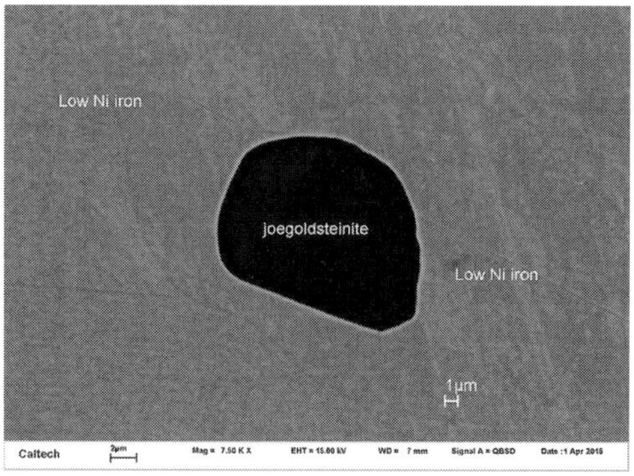

Figure 8.5 Joegoldsteinite ($MnCr_2S_4$) in low-Ni iron (kamacite) from the Social Circle IVA iron meteorite (Isa et al. 2016). BSE image.

orthopyroxene, 1.6% low-Ca clinopyroxene, 21.0% tridymite, 0.0% chromite, 32.5% metallic Fe-Ni, and 6.6% troilite with a modal tridymite/pyroxene ratio of 0.53; SJN consists (in vol%) of 23.9% orthopyroxene, 1.6% low-Ca clinopyroxene, 4.2% tridymite, 0.2% chromite, 65.8% metallic Fe-Ni, and 4.3% troilite with a modal tridymite/pyroxene ratio of 0.16.

8.5.3 Non-Magmatic NC Irons

IAB Complex: This is the largest iron-meteorite group, but it exhibits such compositional diversity that it seems likely that members are from more than a single parent asteroid. Most specimens are coarse (Og), medium (Om), or fine (Of) octahedrites; some are finest octahedrites (Off) and a few are ataxites (D). Wasson and Kallemeyn (2002) divided the IAB Complex into a main group (closely corresponding to the old chemical group IA), five subgroups (sLL: low Au, low Ni; sLM: low Au, medium Ni, originally IIIC; sLH: low Au, high Ni, originally IIID; sHL: high Au, low Ni; sHH: high Au, high Ni), and several grouplets with one to six members.

Besides kamacite, taenite, and plessite (comb, pearlitic, and spheroidized varieties), common phases in IAB Complex irons include troilite, graphite, schreibersite (present both as coarse grains and small rhabdites), nickelphosphide, chromite, and carbide. Whereas cohenite and haxonite are the most common carbides, Wedderburn (sLH) is unique in containing laths of edscottite (Fe_5C_2) (Figure 8.6; Ma and Rubin 2019), minor cohenite, and no haxonite. Canyon Diablo (main group) has two C polymorphs: graphite and diamond. Troilite-rich nodules in Canyon Diablo consist either of pure troilite or of troilite intergrown with different combinations of graphite, daubréelite, chromite, sphalerite, kosmochlor, and rare silicates, rutile, "chalcopyrrhotite" (probably cubanite or isocubanite), mackinawite, and krinovite ($NaMg_2CrSi_3O_{10}$). Cohenite commonly occurs within rims around the troilite nodules. Numerous phosphates are present in Morasko: buchwaldite ($NaCaPO_4$), brianite ($Na_2CaMg(PO_4)_2$), moraskoite ($Na_2Mg(PO_4)F$), czochralskiite ($Na_4Ca_3Mg(PO_4)_4$), merrillite, fluorapatite, and possibly chlorapatite (Karwowski et al. 2016). Associated with the metal in Mundrabilla, is a component rich in Pb, Sn, and In that exhibits superconductivity (Wampler et al. 2020).

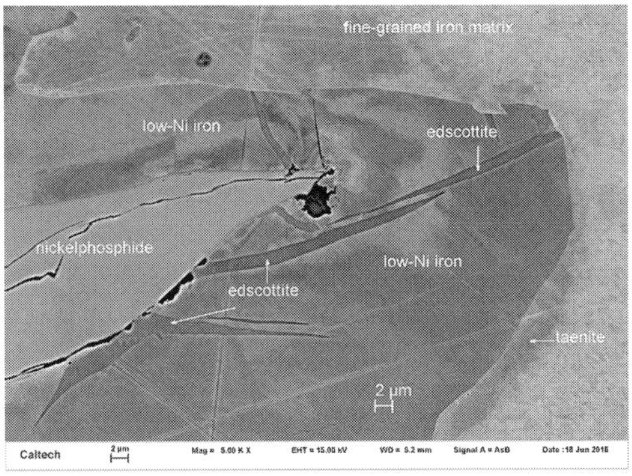

Figure 8.6 Edscottite (Fe_5C_2) in low-Ni iron, with taenite and nickelphosphide in fine-grained iron matrix from the Ni-rich Wedderburn IAB iron meteorite (Ma and Rubin 2019). BSE image.

Silicate-rich inclusions in IAB irons (discussed in more detail in Chapter 11) occur commonly as millimeter- to centimeter-sized nodules. The principal phases in these inclusions are forsterite, enstatite, Cr-diopside, albite, chlorapatite, merrillite, chromite, metallic Fe-Ni, troilite, schreibersite, and graphite. Minor to accessory phases include cohenite, alabandite, daubréelite, sphalerite, and metallic Cu. Toluca (sLL) contains kosmochlor. San Cristobal (unique IAB) contains K-feldspar with antiperthite exsolution lamellae, roedderite, haxonite, and brianite. Various phosphates (chlorapatite, farringtonite, brianite, panethite) are present in Carlton (sLM) and Dayton (sLH). In general, IAB silicate inclusions have undergone prolonged high-temperature metamorphism that recrystallized any chondrules that may have originally been present; mafic silicates in the inclusions were somewhat reduced during metamorphism. Campo del Cielo (main group) appears to contain millimeter-sized, highly recrystallized POP chondrules. It seems likely that similar objects are present in other IAB irons.

IIE: This group is structurally varied, containing coarse (Og), medium (Om), fine (Of), and finest (Off) octahedrites as well as samples best described as polycrystalline kamacite aggregates. The metallography varies among the IIE irons – all contain kamacite and taenite, some have plessite (comb and net), many have schreibersite and troilite, and a few contain chromite and phosphate. There are numerous phosphate phases in Elga: F-rich apatite ($Ca_5(PO_4)_3(F,OH,Cl)$); merrillite ($Ca_9MgNa(PO_4)_7$); panethite ($(Ca,Na)_2(Mg,Fe)_2(PO_4)_2$); chladniite ($Na_2CaMg_7(PO_4)_6$); brianite ($Na_2CaMg(PO_4)_2$); maricite ($NaFePO_4$); czochralskiite ($Na_4Ca_3Mg(PO_4)_4$); tuite ($\gamma$-$Ca_3(PO_4)_2$, a high-pressure phase formed from merrillite); and a new phosphate of composition $Na_2Fe(Mg,Ca)(PO_4)_2$ (Litasov and Podgornykh 2017). The silicate inclusions in IIE irons (e.g., Ruzicka 2014) vary in abundance from 0 vol% to 15 vol %. They are of two general types: (1) Unfractionated inclusions with chondritic bulk compositions are present in Netschaëvo, Techado, Garhi Yasin, Mont Dieu, and Watson 001. Netschaëvo contains angular chondrule-bearing clasts (surrounded by swathing kamacite) that have been recrystallized to petrologic type 6. Chondrule textural types include BO, PO, PP, POP, and RP; constituent minerals include olivine, low-Ca pyroxene, Ca-pyroxene, plagioclase, merrillite, chlorapatite, chromite, troilite, schreibersite, and metallic Fe-Ni. Some quenched impact melt is also present. Highly recrystallized, petrologic-type-6, chondrule-bearing clasts also occur in Techado, Garhi Yasin and Mont Dieu. Watson 001 has a 10-cm-long subrounded-to-subangular inclusion consisting of major coarse orthopyroxene with poikilitically enclosed olivine, Ca-pyroxene, and albite with antiperthite exsolution of alkali feldspar, as well as minor merrillite and chromite. (2) Fractionated inclusions tend to have globular shapes and contain little or no olivine ($\lesssim$2 vol%). They are rich in sodic feldspar (albite or oligoclase), Si-rich glass or tridymite, K-feldspar, antiperthite feldspar, and Ca-pyroxene. Minor to accessory phases include pigeonite, aenigmatite ($Na_2Fe^{+2}{}_5TiSi_6O_{20}$), phosphate (merrillite and apatite – chlorapatite in many specimens, F-rich apatite in Elga), oxide (ilmenite, rutile, armalcolite), pentlandite, and sodalite. Colomera contains an 11-cm-long sanidine crystal as well as yagiite ($(K,Na)_2(Mg,Al)_5(Si,Al)_{12}O_{30}$) and Ca-pyroxene.

8.5.4 Ungrouped Irons

The ungrouped iron meteorites are a diverse lot, both structurally and compositionally. Specimens include ataxites (D) (e.g., ALHA77255; Barbianello); finest octahedrites (Off) (e.g., Prambanan); fine octahedrites (Of) (e.g., ALHA81014; La Garita); medium octahedrites (Om) (e.g., Hammond; Grand Rapids); coarse octahedrites (Og) (e.g., Ban Rong Du; Guin); plessitic octahedrites (Opl) (e.g., NWA 4705; STG 07009); hexahedrites (H) (e.g., Horse Creek), samples with anomalous structures (e.g., Zacatecas (1792); NWA 6163; Nedagolla); and silicate-bearing specimens (e.g., Bocaiuva; NWA 176; Guin). Following are descriptions of a few of the ungrouped irons.

Guin. Guin is a coarse octahedrite (Og) with an unusually high bulk Ni content (9.25 wt%). The meteorite contains ~6 vol% silicate inclusions petrologically and compositionally similar to many of those in IIE irons (Rubin et al. 1986). The largest Guin inclusion (2 × 4 cm) is an ovoid mass consisting mainly of a shock-melted-and-quenched fine-grained plagioclase matrix surrounding coarse partly melted augite grains (some of which are flanked by orthopyroxene). One edge of the inclusion is ringed by 150–600-μm-sized plagioclase grains with polysynthetic twinning; small blebs of troilite occur within the plagioclase grains. The bulk chemical composition of this inclusion is similar to that of some melt pocket glasses in OC produced by the *in situ* preferential shock-melting of plagioclase. Shock veins within the large Guin inclusion consist of fine-grained-to-glassy silicate mesostasis containing small blebs of troilite.

Tishomingo. Tishomingo was described by Buchwald (1975) as "a unique meteorite, unrelated to any other meteorite." It is chemically anomalous with 32.5 wt% Ni (and is among the most Ni-rich irons known); it is highly depleted in volatile siderophile elements (e.g., Ga, Ge, As, Au) and contains very low concentrations ($<$0.05 wt%) of C and P. It is structurally anomalous, consisting of 20 vol% residual taenite and 80 vol% coarse martensite plates; it contains no graphite, carbides, phosphides or silicates (Buchwald 1975; Yang et al. 2014). The martensite plates are 20–50 μm wide and up to 2 mm long. One slab is transected by a 2×40 mm taenite twin (figs. 1742, 1743 of Buchwald 1975); martensite plates within the taenite twin are oriented in two directions: one is parallel to the twin plane, the other to one of the martensite directions outside the taenite twin. Examination at high resolution shows the martensite plates have decomposed; they contain 10–20-nm-wide oriented taenite platelets and 20–100-nm-wide taenite particles. Also present in Tishomingo are numerous "fizzy-textured" metal-sulfide intergrowths; they occur as equant 20–250-μm-sized assemblages and as 10×250 μm elongated masses. The intergrowths consist of irregular troilite grains comingled with metallic Fe-Ni and daubréelite: the cores of the intergrowths comprise taenite (56 wt% Ni; 0.15 wt% Co) and troilite; the rims consist of fine-grained kamacite (2.7 wt% Ni; 2.6 wt% Co) and troilite. Daubréelite occurs mainly in the rims as subrounded 2–5-μm-sized grains; daubréelite is also present as laths up to 50 μm long. Yang et al. (2014) inferred that the metal-sulfide intergrowths formed by shock-melting precursor troilite nodules containing daubréelite lamellae; such nodules occur in many iron meteorites. Corrigan et al. (2005) identified a single grain of stishovite in Tishomingo, indicating a shock pressure between ~10 and 30 GPa. It is possible that a single impact formed the stishovite (from a preexisting silica grain), melted the troilite nodules, and caused thermal decomposition of the martensite (Yang et al. 2014).

Tucson. Tucson is a polycrystalline ataxite (D). The metal phase of Tucson is very reduced (Kurat et al. 2010); it contains 0.8 wt% Si in solid solution (e.g., Miyake and Goldstein, 1974). Troilite, daubréelite, chromite, graphite, and carbides appear to be absent. The meteorite contains ~8 vol% millimeter- and submillimeter-sized elongated silicate inclusions arrayed in curvy subparallel rows; the inclusions consist of major forsterite and enstatite, minor Al-rich diopside and calcic plagioclase (or plagioclase-composition glass) and accessory spinel and brezinaite (Cr_3S_4). They appear to have formed by impact melting and quenching of preexisting silicate inclusions and impact-induced shearing of the host. The presence of brezinaite rather than troilite was ascribed by Buchwald (1975) to the relatively high concentration of Cr in solid solution in the metal. (Although some workers suggested that the silicate was derived from the projectile, the survival of such material after a major impact is unlikely; it is more plausible that all these phases were derived from the target.)

Washington County. The Washington County iron is an ataxite (D) with no hint of a Widmanstätten structure. The meteorite consists of (1) polycrystalline, equiaxed ~50 μm kamacite grains with no Neumann bands; (2) 1–15-μm-sized irregular, amoeba-shaped, cloudy-yellow taenite grains at kamacite grain boundaries; (3) angular 1–15-μm-sized schreibersite grains, commonly intergrown with taenite; and (4) small (1–3 μm) spheroidized schreibersite particles within kamacite grain interiors. No other phases are present – there are no sulfides, silicates, oxides, nitrides, carbides, or graphite. The meteorite is massive; there are no cavities or pores. Washington County contains trapped solar-wind-implanted rare gases (Becker and Pepin 1984), as in stony-meteorite regolith breccias. Although this sulfide-free meteorite is highly depleted in S, it is only moderately depleted in most volatile siderophile elements relative to mean IIIAB irons: e.g., Cu/Ni (1.39×10^{-3} in Washington County vs. 1.87×10^{-3} in mean IIIAB, WC depleted by 26%); Ga/Ni (1.68×10^{-4} vs. 2.38×10^{-4}, depleted by 29%); Ge (2.05×10^{-4} vs. 4.75×10^{-4}, depleted by 57%); As (7.41×10^{-5} vs. 9.65×10^{-5}, depleted by 23%); Au (1.29×10^{-5} vs. 1.21×10^{-5}, enriched by 6%). Washington County may have formed by impact melting, vaporization of sulfide, moderate loss of most volatile siderophile elements, crystallization, and significant annealing.

Willamette. Willamette (the largest intact iron in the United States) is a medium octahedrite (Om). It was previously designated IIIAB, but more recent data show it to be compositionally anomalous; it should be considered an ungrouped iron. There are significant shock effects in Willamette (Rubin et al. 2015). The meteorite contains elongated troilite nodules that were crushed and penetrated by wedges of crushed metal during a collisional event. The troilite nodules formed during initial crystallization of the iron core magma from trapped melt. A major impact caused melting of metal-sulfide assemblages to form lobate taenite masses. Ensuing mechanical damage included impact crushing of the nodules, the jamming of metal wedges into the nodules, and the crushing of metal grains adjacent to sulfide throughout the meteorite. This was followed by post-shock annealing and late-stage shock production of Neumann bands in kamacite.

8.6 Stony-Iron Meteorites

8.6.1 Pallasites

Pallasites can be divided into several categories: main-group pallasites (PMG, ~82%), Eagle Station pallasites (PES, ~7%), and ungrouped pallasites (~11%). A few of the ungrouped pallasites (e.g., Vermillion and Y-8451) are commonly called "pyroxene pallasites," although Wasson and Choi (2003) referred to these two meteorites as "olivine-rich ungrouped irons." The PMG and PES groups differ in isotopic composition, bulk metal composition, and oxidation state.

On such diagrams as ε^{54}Cr vs. ε^{62}Ni, ε^{54}Cr vs. ε^{50}Ti, and ε^{54}Cr vs. Δ^{17}O, PMG pallasites plot within the non-carbonaceous (NC) field along with ordinary chondrites, enstatite chondrites, R chondrites, ureilites, angrites, terrestrial rocks, lunar rocks, martian meteorites, HEDs, mesosiderites; and IIIAB irons. (Other NC iron-meteorite groups include IAB, IC, IIAB, IIE, IIG, IIIE, and IVA.) PES pallasites plot among carbonaceous chondrites on ε^{54}Cr vs. Δ^{17}O diagrams. (CC iron-meteorite groups include IIC, IID, IIF, IIIF, and IVB.) It seems likely that NC and CC bodies formed at different heliocentric distances.

The metal in PES pallasites contains higher concentrations of Ni and refractory siderophile elements than PMG metal. For example, the mean Ir/Ni and Ir/Au ratios in PES (n = 2) are 6.26×10^{-5} and 10.09, respectively, whereas those in PMG (n = 37) are 6.04×10^{-6} and 0.254, respectively. Representative PES/PMG siderophile-element ratios include 10.4 for Ir/Ni and 39.7 for Ir/Au.

PES minerals are more oxidized than corresponding PMG phases. PES olivine has considerably higher FeO/(FeO+MgO) ratios than olivine in most PMG pallasites: $Fa_{\sim 20}$ vs. $Fa_{\sim 12}$. The atomic Fe/Mn ratio in Eagle Station (PES) olivine is 104; the ratio in PMG olivine is 35–60. PES olivine (n = 2) contains 22.6±4.7 µg/g Co and 60.3±21.5 µg/g Ni; PMG olivine (n = 6) contains 7.8±1.2 µg/g Co and 22.3±9.0 µg/g Ni (Mittlefehldt et al. 1998). (PES olivine is also more calcic than PMG olivine: 0.07 vs. ~0.01 wt% CaO.) The atomic Fe/Mn ratio in Eagle Station chromite is 87; the ratio in PMG and PMG-an chromite is 28–34.

As of this writing, among well-characterized pallasites listed in the online Meteoritical Bulletin Database (MBD), the (PMG+PMG-an)/PES ratio is ~12. The small proportion of PES members is probably in part a function of the relatively high oxidation state of its parent asteroid. After melting and differentiation, this oxidized asteroid would likely have had a small iron core encircled by a narrow spherical shell of mixed olivine-metal (i.e., pallasitic) material.

The principal primary minerals in pallasites include metallic Fe-Ni (kamacite, taenite, plessite), olivine, troilite, schreibersite, chromite, phosphate (farringtonite, stanfieldite, merrillite), low-Ca pyroxene (orthopyroxene, clinoenstatite-clinobronzite), and Ca-pyroxene (augite to diopside). Fukang contains silica/felsic inclusions with tridymite, K-rich silicate glass, and an unidentified Ca-Cr silicate phase. (The latter phase also occurs in Pavlodar.) NWA 10019 and Choteau contain minor plagioclase. Vermillion contains cohenite; trace amounts of graphite were reported in Brenham and Krasnojarsk (Mason 1963). Rare or trace phases include (a) pentlandite in Newport (although pentlandite in most pallasites is secondary); (b) mackinawite in Dora, Imilac, Itzawisis, and Thiel Mountains; (c) metallic Cu in Brenham, Glorieta Mountain, Lipovsky, Molong, and Newport; and (d) rutile in Rawlinna 001.

Main-Group Pallasites (PMG). Although pallasites are generally characterized as being half metal and half olivine, the actual modal abundance of olivine in different PMG specimens varies from ~35 to 85 vol% (e.g., Buseck 1977). Olivine grains occur as (1) individual crystals or crystal fragments, ranging from ~1 mm to ~3 cm and (2) as constituents of subrounded-to-subangular polycrystalline aggregates up to 30 cm in maximum dimension. The olivine-rich aggregates typically contain relatively little metal (<10 vol%). Troilite, chromite, schreibersite, and phosphate tend to occur at olivine-metal grain boundaries. In rare cases, small grains of low-Ca pyroxene also occur within symplectic intergrowths with troilite, chromite, phosphate, and/or kamacite at olivine grain boundaries. Troilite also fills fractures within many olivine grains. Some grains of chromite and schreibersite are completely enclosed by metallic Fe-Ni. The abundance of troilite varies from 0.1 to 7.3 vol%; schreibersite varies from 0.3 to 2.8 vol%. Esquel contains (in vol%): ~66% olivine, ~32% metallic Fe-Ni, ~0.5% troilite, ~0.8% schreibersite, ~0.3% chromite, and small amounts of phosphate (on the order of 0.2%) (Ulff-Møller et al. 1998).

A few PMG and PMG-an pallasites contain large olivine-free regions (e.g., up to 7 cm in Esquel, up to 20 cm in Brenham, up to 30–40 cm in Seymchan); olivine is absent from most Glorieta Mountain specimens. These large metal regions commonly display Widmanstätten patterns (Figure 8.7). The majority of PMG pallasites have angular olivine grains (e.g., Ahumada, Esquel, Seymchan, Admire, Huckitta; Imilac, Dora, Albin, Finmarken, Marjalahti, Glorieta Mountain, South Bend, Santa Rosalia); a minority have rounded olivine grains (e.g., Springwater, Thiel Mountains, Pavlodar, Brenham, Krasnojarsk).

In many pallasites, 1–2-mm-thick bands of swathing kamacite surround isolated olivine grains. Plessite commonly occurs in small metal regions between the olivine grains; larger

Figure 8.7 Seymchan pallasite with metal regions displaying Widmanstätten patterns. Reflected light. (A black-and-white version of this figure will appear in some formats. For the colour version, please refer to the plate section.)

regions commonly display a Widmanstätten pattern (e.g., Fukang, Brenham). The bulk chemical composition of PMG metal is similar to that of evolved IIIAB irons; both groups could have been derived from the same differentiated parent body.

PMG olivine ranges in composition from Fa_{11-18} with a large peak at Fa_{12}. In addition to P-free olivine, phosphoran olivine grains (with 3–5 wt% P_2O_5) occur in a few pallasites (e.g., Brahin, Springwater, Rawlinna 001, Zaisho). In these olivines, P^{5+} substitutes for Si^{4+} in tetrahedral sites and there are vacancies in octahedral sites to achieve charge balance.

Merrillite in Fukang contains ~0.4 wt% F but no detectable Cl.

Euhedral chromite grains, up to 5 mm in size, are present at metal-olivine boundaries in Fukang. Sericho (probably a PMG) contains a 5 cm cruciform chromite grain (Boesenberg et al, 2018). In one specimen of Brenham, there is a 10 cm region filled with massive chromite surrounding olivine (Wasson et al. 1999); this texture is analogous to that of metal surrounding olivine in other Brenham regions. The massive chromite is compositionally zoned with the highest FeO contents adjacent to olivine. This chromite region contains more phosphate and more phosphoran olivine than typical Brenham regions. Boesenberg et al. (2018) identified only a single small (~60 μm) chromite grain in Sericho; the grain is reversely zoned with an FeO-rich core (100×Fe/(Fe+Mg) = 92.1; Fe/Mn = 48.7) and a more magnesian rim (100×Fe/(Fe+Mg) = 83.0; Fe/Mn = 36.1). This is the first report of a reversely zoned chromite grain in a pallasite.

Phillips County (PMG-an) is unique in containing abundant troilite. Its troilite/metal ratio is in the range of 5 to 10. Its olivine grains are equiaxed and somewhat rounded; olivine also occurs as numerous tiny angular shards. Olivine and low-Ca pyroxene in Phillips County are more ferroan ($Fa_{18.5}$; $Fs_{16.8}Wo_{0.09}$) than in most PMG members and are similar in composition to those in Springwater ($Fa_{18.0}$; $Fs_{16.6}Wo_{0.07}$) and Zaisho ($Fa_{18.3}$; $Fs_{16.8}Wo_{0.10}$).

Eagle Station Pallasites (PES). There are currently five members of this group: Cold Bay, Eagle Station, Itzawisis, Karavannoe, Oued Bourdim 001. They have textures like those of PMG pallasites in that they consist mainly of olivine enclosed in metallic Fe-Ni. Olivine grains ($Fa_{\sim 20}$) are angular; they tend to be fractured and <1 cm in size. Olivine also occurs as small pointed fragments surrounded by metal. The olivine grains in Karavannoe contain chromite, low-Ca pyroxene, and metallic Fe-Ni. Two pyroxenes occur in Eagle Station: orthopyroxene ($Fs_{17}Wo_1$ with 0.11 wt% Al_2O_3) and Ca-pyroxene (Fs_7Wo_{45} with 0.20 wt% Al_2O_3). Olivine in Oued Bourdim 001 contains rare spherules of troilite up to 1 mm in diameter; olivine grains in this meteorite are surrounded by successive rims of swathing kamacite (up to 1 mm thick) and swathing schreibersite (up to 0.5 mm thick). Phosphates in Karavannoe (mainly merrillite) occur at the boundary between olivine and metal. Both merrillite and stanfieldite are present in Eagle Station.

The metal in Karavannoe is composed of martensite surrounded by 0.05–0.5-mm-thick bands of kamacite. Metal in Oued Bourdim 001 has a martensitic decomposition structure consisting of kamacite platelets (with taenite rims) within a matrix of fine plessite. Also present in the plessite are 10–50-μm-sized schreibersite grains and accessory chromite.

Pyroxene pallasites. The ill-defined pyroxene-pallasite grouplet, consisting initially of only Vermillion and Y-8451, was proposed by Boesenberg et al. (1995). The two meteorites have near-identical O-isotopic compositions ($\Delta^{17}O = -0.76‰$), distinct from those of PMG ($\Delta^{17}O = -0.28 \pm 0.06‰$) and PES ($\Delta^{17}O = -4.68‰$) samples. However, Vermillion and Y-8451 differ from each other in modal mineral abundances and trace siderophile element concentrations; this led Boesenberg et al. (2000) to caution against placing them in the same group until more data were gathered or additional samples were found. In 2011, the 8.47-kg Choteau pallasite (which may be the third member of the pyroxene-pallasite grouplet) was purchased at an estate sale in Montana.

Vermillion contains 86 vol% metallic Fe-Ni (kamacite, taenite, and small amounts of schreibersite, troilite, and cohenite) and 14 vol% silicates. The silicate portion consists (in vol%) of 93% olivine ($Fa_{11.1-13.0}$), 4.9% orthopyroxene (averaging $Fs_{10.7}Wo_{1.7}$), 0.1% Ca-pyroxene (averaging $Fs_{4.8}Wo_{44.2}$), 1.5% chromite (present as euhedral-to-rounded grains within metal), and 0.5% merrillite. The texture of Vermillion is unique among pallasites – silicates occur within curvilinear bands (up to 1.5 × 20 cm in size) within the metallic matrix. Pyroxene is present mainly at the edges of olivine grains, and in rare cases, as crystals enclosed within olivine. The metal structure is that of a fine octahedrite (Of); 5–20-μm-thick rims of cohenite commonly surround grains of troilite, schreibersite and some grains of olivine and pyroxene.

Y-8451 contains 43 vol% metallic Fe-Ni (taenite, plessite, and kamacite as well as small amounts of troilite and schreibersite) and 57 vol% silicates. The silicate portion consists (in vol %) of 97% olivine ($Fa_{10.2-11.2}$), 2.0% polysynthetically twinned low-Ca clinopyroxene (averaging $Fs_{9.5}Wo_{0.6}$), 0.4% untwinned orthopyroxene (averaging $Fs_{10.0}Wo_{1.9}$), 0.4% Ca-pyroxene ($Fs_{4.0}Wo_{44.1}$), 0.1% merrillite, and trace amounts of submicrometer-sized chromite grains within symplectic intergrowths with augite. The meteorite has a pallasitic texture with rounded to subangular olivine grains surrounded by metallic Fe-Ni. However, the olivine grains (1–3 mm) are appreciably smaller than the centimeter-sized grains commonly present in PMG samples. The polysynthetically twinned low-Ca clinopyroxene occurs as rounded grains up to 2 mm in diameter. There are tiny inclusions of Ca-pyroxene within some of the orthopyroxene grains. No cohenite occurs in Y-8451.

Choteau is a pyroxene-rich pallasite with an O-isotopic composition ($\Delta^{17}O = -1.1‰$) very similar (but not identical) to those of Vermillion and Y-8451. Choteau consists mainly of metallic Fe-Ni (kamacite and Ni-poor taenite) and angular to subhedral olivine grains ($Fa_{10.3}$) up to 1 cm in size. Two pyroxenes are present: augite ($Fs_{4.9}Wo_{38.7}$), which occurs within olivine, and orthopyroxene ($Fs_{10.1}Wo_{1.9}$), which occurs as isolated crystals and along the margins of augite grains. Both pyroxene phases are similar in composition to those in Vermillion and Y-8451. Accessory phases in Choteau include troilite, schreibersite, anhedral merrillite, chromite, and sodic plagioclase ($Ab_{85.6}Or_{5.4}$). Chromite occurs (a) within symplectic intergrowths with orthopyroxene, (b) as euhedral to subhedral crystals at the margins of olivine grains, and (c) as small anhedral grains within olivine and pyroxene. Plagioclase is present as irregular grains within both pyroxene phases and as veins filling fractures in orthopyroxene grains.

Ungrouped pallasites. Many of the ungrouped pallasites may be the sole representatives of the core-mantle boundaries of their parent asteroids. Most are pyroxene rich: NWA 10019, NWA 1911, Zinder, and Los Vientos 263. Milton is pyroxene-free.

NWA 10019 is a pyroxene-rich pallasite with an O-isotopic composition at the $\delta^{18}O$-rich end of the PMG line and is thus not closely related to Vermillion and Y-8451. It consists (in vol%) of 46.9% olivine ($Fa_{15.7-17.2}$); 13.2% pyroxene; 0.5% moderately calcic plagioclase ($Ab_{25.9-29.8}Or_{0.4-0.8}$); 26.7% kamacite; 6.5% taenite; 0.6% schreibersite; 1.4% chromite; 1.8%

phosphate (merrillite, stanfieldite, farringtonite); and 2.4% troilite (Boesenberg et al. 2016). Pyroxene occurs as two phases: (a) subhedral-to-anhedral low-Ca pyroxene ($Fs_{6.1-7.1}Wo_{0.2-7.2}$) grains up to 5 mm across and (b) ~100-µm-sized inclusions of Ca-pyroxene ($Fs_{14.2-15.9}Wo_{44.3-48.9}$) within the low-Ca pyroxene. The olivine and low-Ca pyroxene grains are nearly as ferroan as those in the PMG meteorites Phillips County ($Fa_{18.5}$; $Fs_{16.8}Wo_{0.09}$), Springwater ($Fa_{18.0}$; $Fs_{16.6}Wo_{0.07}$), and Zaisho ($Fa_{18.3}$; $Fs_{16.8}Wo_{0.10}$). Plagioclase grains occur as (a) individual grains up to 2 mm in size within olivine and low-Ca pyroxene, (b) 10–50-µm-sized inclusions in orthopyroxene, and (c) as fine, compositionally diverse grains ($Ab_{14.3-47.4}Or_{0.3-3.8}$) in an enclave along with Mg-rich chromite, orthopyroxene, magnesian olivine, and merrillite. The metal in NWA 10019 is appreciably fractionated; it is depleted in refractory siderophiles (Re, Os, Ir, Pt) and enriched in volatile siderophiles (Au, As) relative to metal in other pallasites.

NWA 1911 is a pyroxene-rich pallasite containing (in vol%): 24.3% metallic Fe-Ni (kamacite and taenite), 40.2% olivine ($Fa_{10.8}$), 34.5% orthopyroxene ($Fs_{\sim 12}Wo_{2.1}$), and 1% troilite and chromite. Its O-isotopic composition ($\Delta^{17}O = -0.20‰$) is similar to that of PMG pallasites.

Zinder is a pyroxene-rich pallasite that contains (in vol%): 43% kamacite, 27% olivine, 28% pyroxene, and accessory or trace amounts of taenite, Ca-pyroxene, chromite, merrillite, troilite, and schreibersite. Olivine and orthopyroxene are present as (a) 1–2-mm-sized rounded grains ($Fa_{11.9}$; $Fs_{11.4}Wo_{2.0}$), (b) grain fragments, and (c) as rare inclusions within other minerals. Ca-pyroxene ($Fs_{5.5}Wo_{39.3}$) occurs as rare rims on olivine grains and as individual 20–50-µm-sized rounded grains ($Fs_{4.0-6.3}Wo_{45.6-47.8}$). Chromite is present in large grains, small inclusions within olivine and orthopyroxene, and in symplectites. The O-isotopic compositions of two silicate grains ($\Delta^{17}O = +0.01‰$ and $+0.09‰$) place Zinder a little above the terrestrial fractionation line and far above the PMG line. Boesenberg et al. (2017) reported that Zinder bulk metal is similar in composition to that of IIIF irons and suggested that Zinder could be from the IIIF asteroid's core-mantle boundary.

Los Vientos 263 is a unique pallasite composed of subequal amounts of metallic Fe-Ni, olivine ($Fa_{13.1}$), and orthopyroxene ($Fs_{11.4}Wo_{1.2}$). Some of the silicate grains contain small inclusions of chromite and troilite. The O-isotopic composition ($\Delta^{17}O = -0.187‰$) shows Los Vientos 263 to be enriched in ^{18}O relative to PMG pallasites.

Milton is a unique pallasite containing (a) ~73 vol% angular 1.6 mm olivine grains ($Fa_{17.2}$; molar Fe/Mn = 123) with ~50-µm-sized inclusions of kamacite, taenite, sulfide, phosphate, and chromite and (b) ~27 vol% metallic Fe-Ni (taenite, kamacite, tetrataenite) and ferric-oxide terrestrial weathering products. Olivine is more ferroan than in most PMG pallasites (except Phillips County, Springwater, and Zaisho). Pyroxene is absent. Milton metal is enriched in Ir and Ni relative to PMG and in Ir relative to PES pallasites. The O-isotopic composition of Milton ($\Delta^{17}O = -3.14‰$) lies close to the CCAM line and is roughly halfway between PES and PMG pallasites. McCoy et al. (2019) pointed out that Milton is similar to three related ungrouped iron meteorites, dubbed the "South Byron Trio" (South Byron, Babb's Mill (Troost's Iron), and ILD 83500), in metallography, metal composition, and O-, Mo-, and Ru-isotopic composition. They suggested that Milton and the South Byron Trio could all have been derived from the same (or a similar) differentiated asteroid.

8.6.2 *Mesosiderites*

Mesosiderites are an enigmatic group of stony-iron breccias averaging roughly 50 wt% metallic Fe-Ni, troilite, and schreibersite (with an actual range of 18 to 90 wt% metal, 0.6 to 14 wt% troilite, and 0 to 0.1 wt% schreibersite) and roughly 50 wt% silicate (with a range of 28 to 80 wt %). Silicate phases include major orthopyroxene and plagioclase and minor Ca-pyroxene, olivine, and tridymite. Also present are minor to trace amounts of merrillite, chromite, and ilmenite. Silicate occurs as (a) fine grains admixed with metal and troilite and (b) clasts and pebbles ranging in size from ~1 mm to ~12 cm. Most clasts are basaltic, gabbroic, orthopyroxenitic, or impact melts derived from other clast types; also present in some mesosiderites are rare dunitic and anorthositic clasts.

There are three *compositional classes* of mesosiderites, characterized by their silicate-rich fractions.

Class A: Samples are basaltic and relatively rich in plagioclase and Ca-pyroxene; they contain (in vol%): 55±5% orthopyroxene, 29±8% plagioclase, and 6±2% tridymite.

Class B: Samples are more ultramafic and relatively rich in orthopyroxene; they contain (in vol%): 76±5% orthopyroxene, 17±4% plagioclase, and 2±2% tridymite.

Class C: This class is represented by RKPA79015 and has very abundant orthopyroxene. One silicate-rich region was found to contain 67.2 vol% orthopyroxene, only 0.2 vol% silica, and no plagioclase or olivine. Additional phases in this region include 8.5 vol% metallic Fe-Ni, 23.2 vol% troilite, 0.1 vol% schreibersite, 0.6 vol% merrillite, and 0.2 vol% chromite.

There are five *textural subgroups* of mesosiderites; some individual meteorites have regions with different textural characteristics.

Type 0 samples are currently represented by a single meteorite, NWA 1878 (Kimura et al. 2020). The subgroup is characterized by aggregates of small spheroidal metallic Fe-Ni grains associated with phosphate and silica, brecciated igneous-textured silicate clasts, unrecrystallized low-Ca pyroxene grains (many with normal compositional zoning, some with reverse zoning, and very few with thin augite exsolution lamellae), compositionally variable chromite, generally with low TiO_2 (<0.9 wt%) and plagioclase, and olivine grains lacking corona textures (described later in this section). In addition, there is no large scale compositional equilibration among mafic silicate grains.

Type 1 samples are fine-grained, unrecrystallized rocks. They have a cataclastic texture and contain clearly distinguishable grains a few micrometers in size. Angular clasts and grain fragments are clearly visible. Type I mesosiderites are more recrystallized than NWA 1878.

Type 2 samples are even more recrystallized, although they still have a recognizable cataclastic texture. The matrix is coarser grained (mostly >10 μm); only the larger clasts are angular.

Type 3 samples are very recrystallized and have plagioclase poikilitic textures: large plagioclase grains (several millimeters across) enclose ~100-μm-sized rounded to euhedral grains of orthopyroxene.

Type 4 samples were partly melted and have been described as "melt-matrix breccias." They have intergranular textures consisting of orthopyroxene grains surrounded by plagioclase laths.

The large clasts in mesosiderites can be divided into eight types (modified from Rubin and Mittlefehldt 1992):

(1) *Polygenic and monogenic cumulates* (39%): coarse-grained gabbros highly depleted in incompatible elements. One Vaca Muerta pebble has an H-chondrite-normalized Eu/Sm ratio of 251, the largest such fractionation known among Solar System rocks.
(2) *Polygenic basalts* (30%): finer-grained rocks with positive Eu anomalies, La/Lu ratios <1, and lower REE abundances than basaltic eucrites.
(3) *Quench-textured rocks resembling impact-melted basaltic eucrites* (5%).
(4) *Quench-textured rocks resembling impact-melted cumulate eucrites* (2%).
(5) *Monogenic basalts* (11%) resembling basaltic eucrites.
(6) *Ultramafic cumulate rocks consisting of large crystals of orthopyroxene* (9%) resembling diogenites.
(7) *Ultramafic cumulate rocks consisting of large crystals (up to 10 cm) of olivine* (4%). These olivine grains vary widely in major-element composition; some may have been derived from the mantle of a differentiated asteroid.
(8) *Anorthositic clasts* ($<1\%$) several millimeters in size.

The pyroxene grains ($Fs_{18-54}Wo_{1-17}$) in many clasts are inverted pigeonites consisting of orthopyroxene with augite exsolution lamellae parallel to (001) and (100). Many plagioclase grains ($Ab_{4-9}Or_{0.2-0.7}$) exhibit polysynthetic and Carlsbad twinning. The silica grains contain small amounts of Al_2O_3 (up to 0.32 wt%), CaO (up to 0.18 wt%), and K_2O (up to 0.14 wt%). Olivine has a relatively flat compositional distribution (Fa_{8-37}); most olivine grains have FeO/MnO ratios ranging from 30 to 42, but some reach 45.

Many olivine grains are surrounded by a series of distinct mineralogical zones, forming a "corona" texture. The innermost zone consists mainly of orthopyroxene and chromite; the middle zone consists of major orthopyroxene and plagioclase, minor to major merrillite, and minor chromite. The outermost zone consists largely of pyroxene, plagioclase, merrillite, and chromite (with no metal or tridymite). Ca-pyroxene and ilmenite occur in accessory amounts in some coronas.

Many mesosiderite clasts have pyroxene grains in which Fe/Mn varies with Fe/Mg, following a trend of FeO reduction during crystallization from a melt. Reduction of FeO in pyroxene produced metallic Fe and additional free silica. It seems plausible that the main reducing agent was P brought into contact with the silicates as an alloying component in the metallic Fe-Ni. Oxidation of P produced substantial merrillite.

There are subangular to well-rounded metal nodules up to ~4 cm in size; some of the nodules in Class A mesosiderites have Widmanstätten patterns. The nodules typically contain millimeter-sized lamellar or equiaxed kamacite grains, irregular patches or bands of cloudy taenite, and ~10-μm-wide rims of tetrataenite bordering the taenite and abutting kamacite. Swathing kamacite and minor schreibersite surround many of the metal nodules. Some schreibersite grains in Emery are tabular in shape and range up to ~140 μm in size; they contain up to ~52 wt% Ni.

The bulk composition of mesosiderite metal resembles that of IIIAB irons. Metal in Class A mesosiderites generally has higher concentrations of Au, As, Ni, and Cu, and lower concentrations of W, Ir, Ga, and Ge than metal in Class B mesosiderites.

Eltanin is a unique metal-poor mesosiderite. It occurs as about two dozen small basaltic breccia fragments within Ir-rich sediments from deep-sea sediment piston cores from the Bellingshausen Sea in the southeastern Pacific Ocean. The pristine meteorite fragments among these samples are associated with vesicular impact-melt particles. Although only ~1.7 g of unmelted mesosiderite was recovered, these specimens were derived from a kilometer-sized asteroid that crashed into the Pacific in the Late Pliocene, about 2.5 Ma ago. The specimens contain major plagioclase (in some cases, maskelynite) ($Ab_{4.4-10.3}Or_{0.0-0.18}$) and low-Ca pyroxene ($Fs_{25-48}Wo_{1.5-7.7}$) with augite exsolution lamellae, a few single grains of augite, and trace amounts of troilite, metallic Fe-Ni, chromite, ilmenite, silica, and olivine ($Fa_{12.7-34.3}$). [The compositional ranges given here are based on representative analyses (Kyte 2002) and may not show the full ranges.] The sample includes an igneous rock fragment with a highly fractionated REE pattern similar to those of coarse-grained gabbroic clasts in mesosiderites. Based on the abundances of Fe, Ni, and Ir, Kyte (2002) estimated the Eltanin projectile had a total metal abundance of only 4±1 wt% (about an order of magnitude lower than in average mesosiderites).

9
Cosmomineralogy

9.1 Mercury

Mercury is the smallest and innermost major planet (Figure 9.1); from the perspective of Earth-bound observers, it never appears more than 28° from the Sun. It is locked in a state of spin-orbit coupling: its rotation period (58.646 Earth days) is exactly two-thirds of its year (87.969 Earth days). Mercury is named after the Greco-Roman messenger to the gods, perhaps because of its rapid apparent day-to-day movement with respect to the Sun. Compared to the other major planets, Mercury has the most eccentric orbit (e = 0.2056), the highest inclination relative to the ecliptic (7.007°), and the smallest axial tilt (0.01°). It has a weak magnetic field (~1.1 percent as strong as Earth's) and no natural satellites.

Messenger data have provided new insights into Mercury's internal composition (e.g., Solomon et al. 2019). The planet has a solid silicate crust (~35 km thick), a solid silicate mantle (~400 km thick), a relatively thin solid iron sulfide layer at the core–mantle boundary, a liquid iron-rich outer core, and a solid iron-rich inner core. The core may contain substantial Si in solid solution with Fe. The entire core constitutes 65 percent of the planet's mass. Consequently, Mercury's uncompressed bulk density (5.3 g cm^{-3}) is the highest of any planet in the Solar System.

The surface of Mercury is covered with impact craters, ranging from small simple bowl-shaped craters to the Caloris multi-ring basin (d = 1,550 km). Between the craters are hilly plains. Flat smooth plains, resembling in structure the lunar maria flood basalts, constitute ~6 percent of the planet's surface area; the smooth plains embay other landforms, display flow features, and partially (or completely) bury older craters. X-ray spectrometry indicates a composition intermediate between that of basalts and komatiites (ultramafic volcanic rocks rich in Mg and depleted in Si, K, and Al). Unlike the lunar maria (which are much darker than the surrounding highlands), the smooth plains on Mercury are similar in albedo to the rest of the crust. *Messenger* also identified numerous pyroclastic deposits on Mercury, some apparently associated with shield volcanoes.

About nine different geochemical terranes have been identified, ranging from komatiitic (>18 wt% MgO) to boninitic (rich in alkalis, >52 wt% SiO_2). The Mercurian surface appears to contain 38–58 wt% albite, up to 37 wt% enstatite, up to 22 wt% diopside, up to 33 wt% forsterite, and up to 8 wt% quartz along with some graphite and minor sulfide (plausibly oldhamite, niningerite, keilite, troilite, and wassonite). Fe-silicides may also occur.

Figure 9.1 Mercury. Photojournal: PIA16853. Images from the MESSENGER mission. Source: NASA/ Johns Hopkins University Applied Physics Laboratory/Carnegie Institution of Washington. (A black-and-white version of this figure will appear in some formats. For the colour version, please refer to the plate section.)

Telescopic Earth-based observations have long shown the absence of the ~1 μm absorption band in reflectance spectra from Mercury, indicating a paucity of FeO in surface materials. Recent refinements suggest that the upper limit on FeO may be ~0.5 wt%, showing that olivine and pyroxene on Mercury must be very magnesian. Most of the iron at the surface (~1.5 wt%; Weider et al. 2014) probably occurs either as metallic Fe^0 or as Fe^{2+} contained in Mg-Ca-Fe sulfides. These inferences are consistent with the low oxygen fugacity at the surface: IW – 7.3 to IW – 2.6.

Peplowski et al. (2016) noted that the surface of Mercury is darker than the surface of the Moon; *Messenger* data suggest that the darkening agent is graphite. The surface is also relatively rich in S, Na, and Cl and poor in Fe. The daytime surface temperature at perihelion is ~450°C. Nevertheless, as on the Moon, permanently shadowed regions at the poles have high radar reflectivity and are presumed to contain water ice. Ice has been inferred to occur in relatively young, laterally contiguous deposits. The water molecules on Mercury probably originate from internal volcanism as well as cometary and carbonaceous-chondrite projectiles.

Because Mercury has only a thin exosphere (composed of atoms knocked off the surface by meteoroid impacts and the solar wind), collisions would occur at cosmic velocity. About 40 percent of the craters on Mercury were produced by near-parabolic comets with impact velocities exceeding 80 km s^{-1} (Schultz 1988). Because of Mercury's high orbital velocity, micrometeoroids in retrograde orbits striking Mercury at perihelion could have impact velocities as high as ~110 km s^{-1} (Pokorný et al. 2017). It is possible that substantially larger (1–10 m in size) bodies, derived from certain groups of sungrazer comets, could also be in retrograde orbits and strike Mercury at similarly high velocities. Such enormously energetic impacts would cause extensive fracturing, brecciation, and melting of surficial rocks. There should be extensive formation of agglutinates (Cintala 1981); because agglutinate formation can cause homogenization of albedo (Pieters 1978), this could account for the lack of albedo differences on Mercury between plains material (presumed lava flows) and the remainder of the crust.

It is plausible that mercurian rocks contain high-pressure phases formed from common minerals: e.g., ringwoodite from olivine, majorite from low-Ca pyroxene, coesite, stishovite (and perhaps other high-density polymorphs) from silica, and lingunite ($NaAlSi_3O_8$) from plagioclase.

The *BepiColombo* spacecraft was launched in 2018 and is scheduled to make six close approaches to Mercury (between October 2021 and January 2025) before assuming orbit in December 2025. It will deploy two satellites: one is the *Mercury Planet Orbiter*, which will study the interior structure and surface of Mercury; the other is the *Mercury Magnetospheric Orbiter*, which will study the magnetic field. These data will place additional constraints on Mercury's geology and mineralogy.

9.2 Venus

Venus is the brightest permanent resident of the sky after the Sun and Moon (Figure 9.2). From the perspective of terrestrial observers, it always appears within 45° of the Sun. Venus is both the "Morning Star" and the "Evening Star," spending about 41 weeks as each manifestation. Perhaps because the human gestation period is about 40 weeks, Venus was long associated with the goddesses of love, beauty and fertility, e.g., Ishtar from Mesopotamia and Aphrodite from Greece.

Venus is our next-door neighbor; the mean distance between the orbits of Venus and Earth (0.2767 AU) is less than that of any other two planets in the Solar System.[1] Venus is often called

Figure 9.2 Venus. Photojournal: PIA00159. This hemispheric view of Venus is based on more than a decade of radar investigations up through the 1990–1994 *Magellan* mission. Source: NASA/JPL/USGS. (A black-and-white version of this figure will appear in some formats. For the colour version, please refer to the plate section.)

[1] Although no planet comes closer to Earth than Venus, they are often on opposite sides of the Sun. On average, Mercury is closer to Earth than is Venus.

Earth's sister planet; its diameter is 95 percent that of Earth, its mass is 82 percent that of Earth, its escape velocity is 92 percent that of Earth, and its bulk density is 95 percent that of Earth. These two planets likely accreted from broadly similar populations of planetesimals and probably have similar bulk chemical compositions. Like Earth, Venus is inferred to have an iron core and a silicate mantle. Radar data from the *Magellan* spacecraft show shield volcanoes and plateaus (presumably lava flows) on the Venusian surface. There are many more volcanoes on Venus than on Earth, and it is likely that active Venusian volcanism continues. On the order of 1,000 impact craters were discovered on Venus, many with irregular ejecta blankets.

Despite important similarities, Venus differs from Earth in extraordinary ways. (1) Venus rotates slowly retrograde (1 rotation = −243 Earth days); its day is longer than its year (225 Earth days). (2) Venus does not have a magnetic field, probably due at least in part to its slow rotation. (3) Venus receives nearly twice as much solar energy as the Earth, but the Venusian clouds reflect ~75 percent of the incident radiation back into space. (4) The Venusian atmosphere is 96.5 wt% CO_2, resulting in a runaway greenhouse effect and a surface temperature of ~470°C. The global mean atmospheric surface pressure is 92 bars. The atmosphere is extremely dry with only ~30 ppm water vapor (by volume) at the surface. There are permanent cloud layers enshrouding the planet; the clouds contain droplets of sulfuric acid. Also present in the atmosphere are SO_2, HCl, HF, and (rather questionable) phosphine (PH_3). (5) Venus does not have significant plate tectonic activity. (6) Venus has no natural satellite.

Inferences about the mineralogy of the Venusian surface (reviewed by Gilmore et al. 2017) are based on thermodynamic modeling, laboratory experiments, comparisons to terrestrial rocks, analytical data from landers, and remote observations from orbiting spacecraft. There are significant uncertainties in this approach: the *Venera* and *Vega* lander analyses are imprecise, and data on some crucial elements were not obtained. Some laboratory experiments have been flawed; others were deemed irrelevant because they did not realistically account for Venusian surface conditions (which are not well constrained). Emissivity data were obtained during nightside measurements by *Venus Express (VEX)* (which assumed a polar orbit around Venus in 2006) and *Galileo* (which had a Venus flyby in 1990). The *VEX* data were from three accessible surface windows at 1.02, 1.10, and 1.18 μm, while the *Galileo* data were in the spectral range of 0.7–5.2 μm. Uncertainties in the data arise from imprecise corrections for cloud opacity, cloud reflectivity, limb darkening, and stray sunlight. The emissivity data can be correlated with the synthetic aperture radar images previously gathered by *Magellan* (which mapped 98 percent of the surface at a resolution of ~100 m).

X-ray fluorescence (XRF) analyses of the Venusian surface by the landers *Venera 13* (Phoebe Regio), *Venera 14* (Phoebe Regio), and *Vega 2* (Rusalka Planitia) yielded bulk chemical compositions comparable to those of terrestrial basaltic rocks: alkaline basalt at *Venera 13* and tholeiitic basalt at *Venera 14* and *Vega 2.* These spacecraft appear to have landed on plains units that cover ~80 percent of the surface. It follows that 80 percent of Venus is covered by basaltic lava flows; we can infer that major minerals in the crust include plagioclase and Ca-pyroxene.

Some regions of Venus are assumed to be more felsic (based on their relatively low ~1 μm emissivity values) and might contain silica minerals. These regions are the *tessera* – tectonically deformed, topographically high terrain with high radar backscatter.

The presence of SO_2 in the atmosphere implies that anhydrite ($CaSO_4$) is stable at the surface. Pyrite (FeS_2) may also occur. Using chemical equilibria, Fegley et al. (1992) found that moderately ferroan olivine and low-Ca pyroxene are also stable. Weathering of phases containing ferrous iron could produce magnetite, hematite, and/or maghemite. All three phases may be present at the surface, perhaps with overlapping ranges in elevation; nevertheless, hematite is probably the dominant (if not exclusive) iron oxide on Venus. The presence of hematite is consistent with the visible and near-infrared spectra obtained by the landers *Venera 9* (northeast end of Beta Regio) and *Venera 10* (southeast of Beta Regio).

Some workers suggested the abundances of HCl and HF in the atmosphere may be buffered by different solid-gas reactions involving such halogen-bearing minerals as fluorphlogopite ($KMg_3(AlSi_3)O_{10}F_2$) and sodalite ($Na_4Al_3Si_3O_{12}Cl$), but it is not clear that the proposed reactions would occur on timescales sufficiently fast to buffer the atmosphere. Although no liquid water exists on Venus, chemical equilibria suggest that hydroxyl-bearing micas (e.g., eastonite or F- and OH-bearing phlogopite) might be stable at high elevations.

9.3 Earth

Earth is the most massive and geologically diverse terrestrial planet (Figure 9.3), the only one with a large moon. There are numerous excellent geology textbooks that describe Earth's physical features (e.g., Christiansen and Hamblin 2014; Plummer et al. 2019; Tarbuck and Lutgens 2019). Our planet has volcanoes, sedimentary basins, glaciers, hot and cold deserts, sand dunes and yardangs, mountain chains and island arcs, karst landforms, and playas. There are coral reefs, rift valleys, impact craters, turbidites, diapirs, moraines, mesas, and badlands

Figure 9.3 The Earth on April 17, 2019. The Earth Polychromatic Imaging Camera, a NASA camera aboard NOAA's *DSCOVR* spacecraft, returns daily images of Earth from a distance of about 1.5 million kilometers. Source: NOAA/NASA EPIC Team. (A black-and-white version of this figure will appear in some formats. For the colour version, please refer to the plate section.)

(dry terrain where soft sedimentary rocks were extensively eroded). Earth has an N_2- and O_2-rich atmosphere, a globe-encircling ocean, and a biosphere. Geomorphological processes include those caused mainly by the action of wind, water, glaciers, volcanism, tectonism, gravity, and microorganisms. There is an enormous variety of igneous, sedimentary and metamorphic rocks, each with its characteristic mineral assemblage.

Magmatic activity has produced igneous rocks as distinctive as carbonatites (>50 vol% carbonates); tholeiitic basalts (major olivine, clinopyroxene and plagioclase); quartzolites (>90 vol% quartz); foidolites (>60 vol% feldspathoids); syenites (>65 vol% feldspar, mainly orthoclase and minor plagioclase); lamproites (major Mg-rich olivine, Fe-rich leucite, Ti-rich phlogopite, K- and Ti-rich richterite, diopside, and Fe-rich sanidine); and granites (major quartz, alkali feldspar, and plagioclase, commonly with minor mica).

Sedimentary processes have produced sandstones (most are rich in quartz, feldspar, and rock fragments; the island of Hawai'i has a green-sand beach made largely of olivine); mudstones (containing abundant clay minerals, mainly illite, montmorillonite, and kaolinite, as well as major quartz and minor feldspar); carbonates (typically limestones made mostly of calcite with minor aragonite and minor to accessory quartz, clays, and chert; or dolostones composed predominantly of dolomite with calcite and aragonite, commonly containing minor to accessory gypsum, anhydrite, chert, and halite); evaporites (many composed of halite or of gypsum and anhydrite); cherts (made mainly of microcrystalline quartz); banded iron formations (made mostly of chert and hematite, but also commonly containing magnetite, siderite, and greenalite); and phosphorites (e.g., rocks rich in fluorapatite).

Metamorphism of preexisting rocks at different levels of heat and pressure have produced a myriad of mineral assemblages known as *facies*: for example, the low-pressure, low-temperature zeolite facies is characterized by the appearance of laumontite, albite, and in some cases wairakite or prehnite; the low-pressure, moderate-temperature hornfels facies (normally developed during contact metamorphism) is typically characterized by cordierite, biotite, garnet, andalusite, sillimanite and wollastonite; the rare high-pressure, high-temperature eclogite facies is usually found in rocks of basaltic composition and is characterized by omphacite and pyrope; the low-pressure, high-temperature sanidinite facies (usually a result of contact metamorphism) is typified by sanidine, mullite, pigeonite, and tridymite.

The petrographic, mineralogical and geochemical properties of a rock are functions of (1) the presence or absence of water and organic material, (2) the types and rates of applied forces, (3) the ambient pressure and temperature, (4) the chemical potentials of available elements, and (5) the variability of all these parameters during petrogenesis.

Such complex terrestrial processes have given rise to unusual phases unlikely to occur elsewhere in the Solar System. A good illustration of terrestrial mineral complexity is the set of 300-plus mineral species found among the zinc-iron-manganese deposits at Franklin and Sterling Hill, New Jersey. There are several models of the formation of these deposits (e.g., Squiller and Sclar 1980; Johnson and Skinner 2003) involving such processes as hydrothermal venting of sulfide-laden superheated water, precipitation of Fe- and Zn-rich minerals along the edge of a marine basin, transportation of sulfate-stable brines, precipitation of zincian phases (Zn carbonates, Fe-Mn-Zn oxides and Fe-Mn-Zn silicates), formation of iron-rich deposits at the boundaries between anoxic and oxygenated water, and high-grade regional metamorphism associated with the Grenville orogeny. Additional minerals were formed by late-stage hydrothermal alteration and weathering.

The phases produced through this complex series of events include common, uncommon, and unique minerals. Many are fluorescent. Franklin and Sterling Hill are the type localities for more than 60 mineral species. Representative phases include:

Native elements: arsenic, copper, gold, graphite, lead, silver, sulfur

Sulfosalts: baumhauerite ($Ag_{0.6}Pb_{11.6}As_{15.7}S_{36}$), berthierite ($FeSb_2S_4$), tennantite ($(Cu,Fe)_{12}As_4S_{13}$), zinkenite ($Pb_9Sb_{22}S_{42}$)

Halides: atacamite ($Cu_2Cl(OH)_3$), fluorite (CaF_2)

Oxides: corundum (Al_2O_3), feitnechtite (β-$MnO(OH)$), franklinite ($(Zn,Mn,Fe)(Fe,Mn)_2O_4$), hetaerolite ($ZnMn_2O_4$), magnetite (Fe_3O_4), manganite (γ-$MnO(OH)$), pyrophanite ($MnTiO_3$), rutile (TiO_2), uraninite (UO_2), woodruffite ($(Zn,Mn)Mn_3O_7 \cdot 1\text{-}2H_2O$), zincite ($(Zn,Mn)O$)

Carbonates: aragonite ($CaCO_3$), cerussite ($PbCO_3$), loseyite ($(Mn,Zn)_7(CO_3)_2(OH)_{10}$), siderite ($FeCO_3$), smithsonite ($ZnCO_3$), strontianite ($SrCO_3$)

Borates: cahnite ($Ca_2[B(OH)_4](AsO_4)$), roweite ($Ca_2Mn_2B_4O_7(OH)_6$)

Sulfates: anglesite ($PbSO_4$), charlesite ($Ca_6Al_2(SO_4)_2B(OH)_4(OH,O)_{12} \cdot 26H_2O$), torreyite ($(Mg,Mn)_9Zn_4(SO_4)_2(OH)_{22} \cdot 8H_2O$),

Molybdates: powellite ($CaMoO_4$)

Tungstates: scheelite ($CaWO_4$)

Phosphates: newberyite ($Mg(PO_3OH) \cdot 3H_2O$), niahite ($(NH_4)(Mn,Mg,Ca)PO_4 \cdot H_2O$), turneaureite ($(Ca_5(AsO_4,PO_4)_3Cl)$)

Arsenates: chlorophoenicite ($(Mn,Mg,Zn)_3Zn_2AsO_4(OH,O)_6$), jarosewichite ($Mn_4AsO_4(OH)_6$), kraisslite ($Zn_3FeMn_{24}As_5O_{18}(SiO_4)_6(OH)_{18}$), ogdensburgite ($(Ca,Zn,Mn)_4Fe_6(AsO_4)_5(OH)_{11} \cdot 5H_2O$), sterlinghillite ($Mn_3(AsO_4)_2 \cdot 4H_2O$)

Vanadates: descloizite ($PbZnVO_4(OH)$)

Silicates: albite ($NaAlSi_3O_8$), anorthite ($CaAl_2Si_2O_8$), bementite ($Mn_5Si_4O_{10}(OH)_6$), clinohedrite ($CaZnSiO_4 \cdot H_2O$), diopside ($CaMgSi_2O_6$), franklinfurnaceite ($Ca_2Mn_4FeZn_2Si_2O_{10}(OH)_8$), franklinphilite ($(K,Na)_4(Mn^{2+},Mg,Zn)_{48}(Si,Al)_{72}(O,OH)_{216} \cdot 6(H_2O)$), hemimorphite ($Zn_4Si_2O_7(OH)_2 \cdot H_2O$), orthoclase ($KAlSi_3O_8$), petedunnite ($CaZnSi_2O_6$), samfowlerite ($Ca_{28}Mn_6Zn_4(Be,Zn)_4Be_{12}[(OH)_3(SiO_4)_3(Si_2O_7)_2]_4$), willemite ($Zn_2SiO_4$), yeatmanite ($Mn_7Zn_8Sb_2^{+5}Si_4O_{28}$)

The primary zinc ore minerals are franklinite, willemite, and zincite. A total of 6.5 million metric tons of Zn were mined in this region from the mid-nineteenth century to the mid- to late twentieth century.

9.4 Moon

The Moon, characterized in Genesis 1:16 as “the lesser light to rule the night,” is the only large natural satellite among the terrestrial planets. It is the fifth largest moon (mean radius = 1,737.5 km) in the Solar System and the largest satellite relative to its planet ((mean radius of Moon)/(mean radius of Earth) = 0.27). It is tidally locked to the Earth; the Moon’s axial rotation has the same period as its orbital revolution, so that one side (the near side) always faces the Earth.

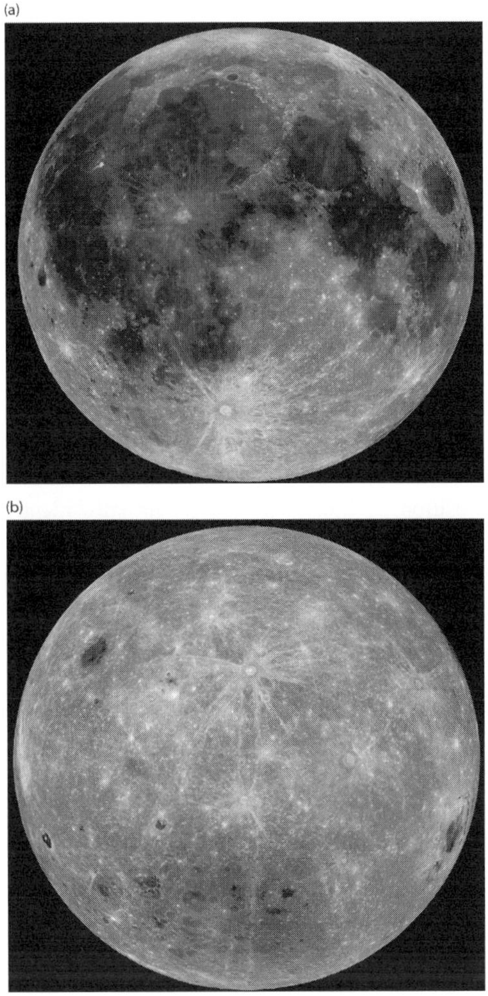

Figure 9.4 (a) The near side of the Moon, (b) the far side of the Moon. Based on data from cameras aboard NASA's robotic *Lunar Reconnaissance Orbiter* spacecraft. Source: NASA/Goddard Space Flight Center/ Arizona State University.

Photographs of the Moon's face show dark-colored maria, gray highlands, and light-colored crater rays and ejecta blankets (Figure 9.4). The maria are flood basalts, partially filling ancient impact basins; the highlands are mainly anorthosite (Heiken et al. 1991). The light-colored rays and ejecta are crushed and powdered samples of underlying rock. Numerous orbiter missions since the 1990s (e.g., *Clementine*, *Lunar Prospector*, *Chang'e*, *Lunar Reconnaissance Orbiter*, *SELENE*, *Chandrayaan*) have shown that the Moon has had a complex igneous history. This complexity is attributable to the Moon's relatively large size, which is far greater than that of 4 Vesta, the largest differentiated igneous asteroid (r = 263 km).

Ground-penetrating radar aboard the *Yutu* lunar rover (part of China's *Chang'e 3* mission, launched December 2013) detected at least nine distinct rock layers in the upper ~400 m of the crust beneath northern Mare Imbrium. The layers consist of basaltic lava flows, regolith produced by impact gardening of lava flows, and pyroclastic deposits. These results illustrate the Moon's geological complexity.

In 2019, China's *Chang'e* 4 spacecraft sent a lander and rover to the 186-km-diameter Von Kármán crater within the South Pole-Aitken Basin on the lunar far side. Spectral observations by the *Yutu-2* rover indicate that the surface materials include orthopyroxene and olivine, plausibly excavated from the lunar mantle by the impact that formed the 72-km-diameter Finsen crater near the landing site. The South Pole-Aitken Basin is an ancient structure, 2,500 km wide and 13 km deep; it is the largest basin on the Moon and among the largest in the Solar System.

Our knowledge of lunar chemistry, mineralogy, and petrology is based on five principal sources: Apollo samples (382 kg), Luna samples (0.32 kg), Chang'e-5 samples (~2 kg), lunar meteorites (>440 kg and counting), and remote sensing.

The O-isotopic composition of lunar samples is virtually identical to that of terrestrial materials (e.g., Wiechert et al. 2001; Spicuzza et al. 2007; Hallis et al. 2010; Young et al. 2016), indicating that both bodies formed from the same (or an isotopically similar) reservoir. However, Cano et al. (2020) reported distinct differences in $\Delta^{17}O$ in different lunar lithologies; samples derived from lower layers in the lunar crust have lower contributions of an isotopically light component. These deep-derived materials were suggested to be less contaminated by the hypothesized Mars-sized projectile involved in the giant impact into the Earth that formed the Moon (e.g., Hartmann and Davis 1975; Cameron and Ward 1976; Stevenson 1987; Canup 2012; Lock et al. 2018).

There are four basic types of lunar rocks (e.g., Taylor et al. 1991; Taylor 2009): (1) basaltic volcanic rocks including lava flows and pyroclastic rocks, (2) pristine nonmare rocks, (3) polymict breccias, and (4) lunar soil (Figure 9.5).

Basaltic volcanic rocks

Lava flows are basaltic rocks that erupted from fissures as low-viscosity lavas. These include high-Ti mare basalts (mainly from the Apollo 11 and Apollo 17 sites), low-Ti mare basalts (Apollo 12, Apollo 15, Luna 16), aluminous low-Ti mare basalts (Apollo 14, Luna 16) and very-low-Ti mare basalts (Apollo 17, Luna 24). All these basalts are composed principally of pyroxene, plagioclase, olivine, and oxide phases (ilmenite, spinel, and armalcolite). Minor phases include those that crystallized late: silica (tridymite or cristobalite), K-feldspar, zircon, and phosphate (apatite and merrillite). Metallic Fe and troilite are also present but may be meteorite-projectile contaminants. Some rocks contain patches of glassy mesostasis. The bulk Ti contents of the basaltic rocks correlate with the modal abundance of ilmenite. The aluminous low-Ti mare basalts have high plagioclase/pyroxene ratios. Common textures exhibited by these rocks include *intersertal* (glassy mesostasis and plagioclase occupying the interstices between pyroxene and oxide grains), *ophitic* to *subophitic* (pyroxene partially to completely enclosing plagioclase), *vitrophyric* (primarily glassy), and *gabbroic* (coarse-grained and crystalline).

Figure 9.5 Four varieties of lunar rocks. (a) Mare basalt, (b) anorthosite, (c) breccia, (d) regolith (containing orange glass beads). Source: NASA. (A black-and-white version of this figure will appear in some formats. For the colour version, please refer to the plate section.)

Pyroclastic deposits are formed from small glass-rich beads that erupted explosively along with gases from vents. The beads tend to be individually and collectively chemically homogeneous, enriched in volatile elements (e.g., F, Cl, Zn, Pb), and have high bulk Mg/Al ratios relative to mare regolith samples. Analysis of the orange beads suggests that lunar volcanic gas deposited metallic Zn and metallic Na atop these particles during fire-fountain eruptions (Figure 9.6(a), (b); Ma and Liu 2019a). Many of the beads are not entirely (or even primarily) glassy; they contain tiny crystals of Ti-rich pyroxene, olivine, chromite, ilmenite, and metallic Fe (Figure 9.6(c); Ma and Liu 2019b).

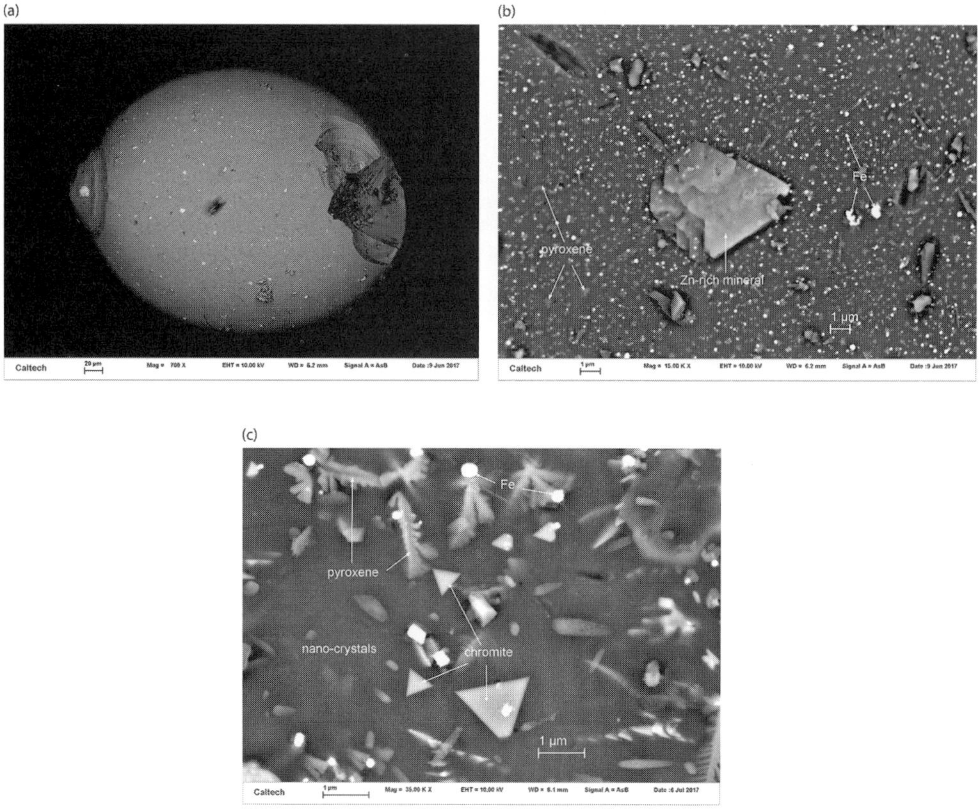

Figure 9.6 (a) An orange bead from the lunar regolith from the Apollo 17 mission. (b) A Zn-rich phase on the surface of the bead. (c) Tiny crystals of Ti-rich pyroxene, chromite and metallic Fe within the bead (Ma and Liu, 2019a, 2019b). BSE images.

Pristine nonmare rocks

Pristine nonmare rocks are mainly coarse-grained intrusive cumulate rocks that are the remnants of the original lunar crust; most are significantly brecciated. They comprise three major groups: the ferroan anorthosite suite (light-colored plagioclase-rich rocks); the magnesian suite (gabbros, norites, troctolites, and dunites); and the alkali suite. Some members of the magnesian and alkali suites are enriched in a chemical component called KREEP, derived from a trace-element-rich residual magma, part of the Procellarum KREEP Terrane. [The KREEP acronym stands for potassium (K), rare-earth elements (REE), and phosphorus (P).] The relatively high abundance of KREEP in the Apollo samples reflects the fact that all the lunar landings were confined to a small region of the Moon near this chemically unusual terrane. The pristine nonmare rocks contain low abundances of siderophile elements (e.g., Ir, Au, Os), signifying little or no meteoritic contamination. Many of the samples are brecciated.

The ferroan anorthosite suite contains the most common pristine highland rocks. They consist of Ca-plagioclase and ferroan pyroxene (mainly hypersthene or inverted pigeonite with

Figure 1.1 A Chinese early Chou Dynasty bronze weapon with meteoritic iron blade. Photo from Gettens et al. (1971); used with permission from the Smithsonian Institution.

Figure 2.1 Cut fusion-crusted individual of the Allende CV3 chondrite. The stone contains numerous CAIs of various sizes and shapes (white and light gray) and less-discernable millimeter-sized chondrules (medium-dark gray) surrounded by fine-grained silicate-rich matrix material (dark gray). Image courtesy of Darryl Pitt/Macovich Collection.

Figure 2.4 The eight planets of the Solar System. (Credit: NASA)

Figure 2.5 A meteor. Photograph ISS028-E-24847 taken on August 13, 2011, from the International Space Station. (Credits: NASA/ISS)

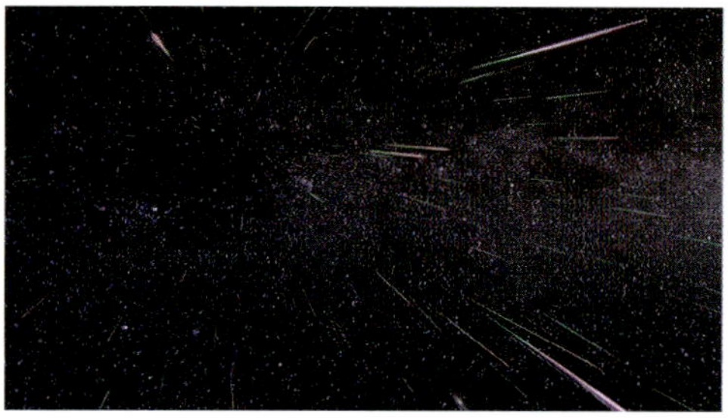

Figure 2.6 Time-lapse image of the Perseid meteor shower of August 2009. (Credits: NASA/JPL)

Figure 2.7 An individual specimen with fusion crust of the Chelyabinsk meteorite fall.

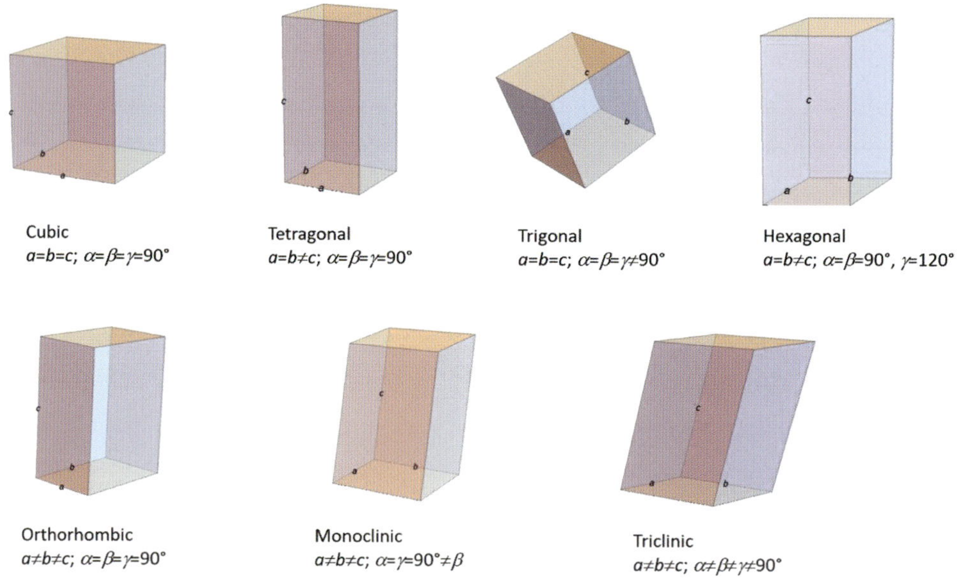

Figure 3.1 The seven crystal systems. (Credit: WOLFRAM Demonstrations Project)

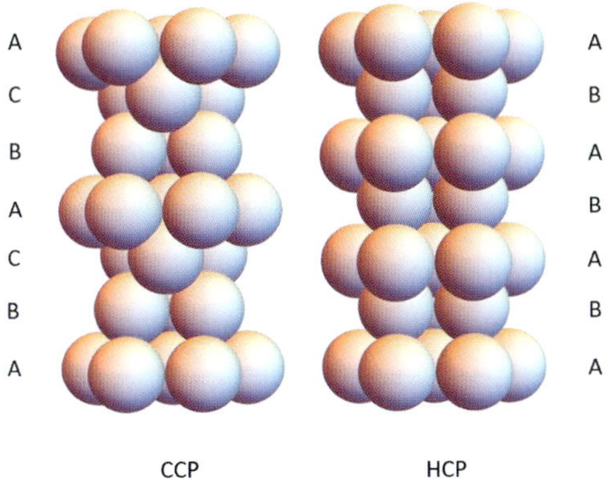

Figure 3.2 The cubic closest packing (CCP) and hexagonal closest packing (HCP) structures. (Credit: WOLFRAM Demonstrations Project)

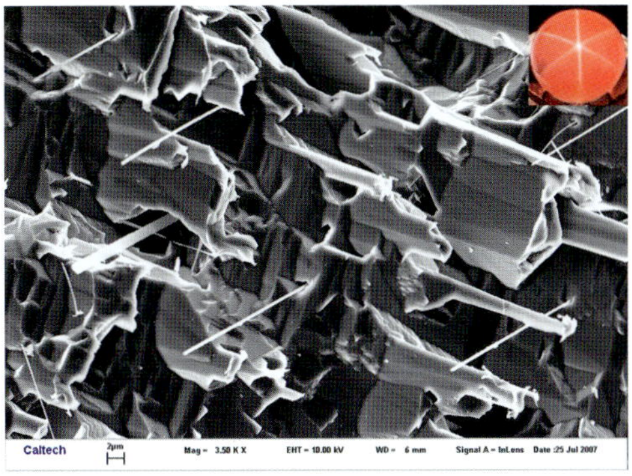

Figure 4.1 Rose quartz. Color caused by inclusions of pink dumortierite-related fibers. After Ma et al. (2002). BSE image. Inserted optical photo courtesy of George Rossman, Caltech.

Figure 4.2 Color effect of rainbow hematite due to thin-film interference by coating of Al-phosphate nanorods on the hematite surface. After Ma and Rossman (2006). Upper row: optical photos under reflected light; lower row: BSE images. Optical photo courtesy of George Rossman, Caltech.

Figure 5.1 Olivine in chondrules and melilite in a CAI from the Efremovka CV3 chondrite. Reflected light.

Figure 5.2 Phenomenon of "silicate darkening" a.k.a. "shock darkening" observed in the Tenham L6 ordinary chondrite. Reflected light.

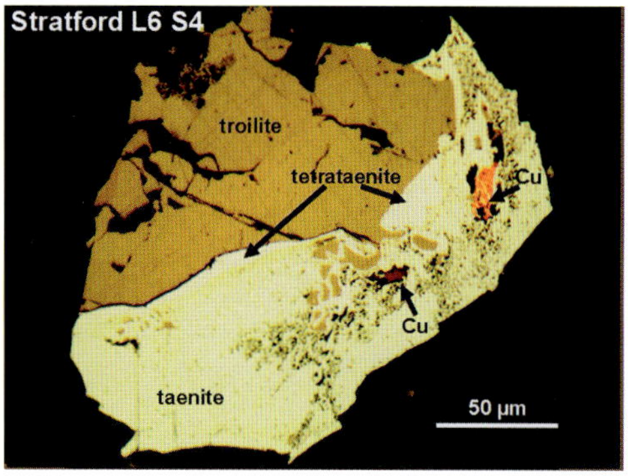

Figure 5.3 Metallic Cu in the shock stage S4 Stratford L6 chondrite. Also evident is a thin band of tetrataenite at the interface between troilite and taenite. Reflected light.

Figure 5.4 Pink spinel in a chondrule from the Allende CV3 chondrite. Reflected light.

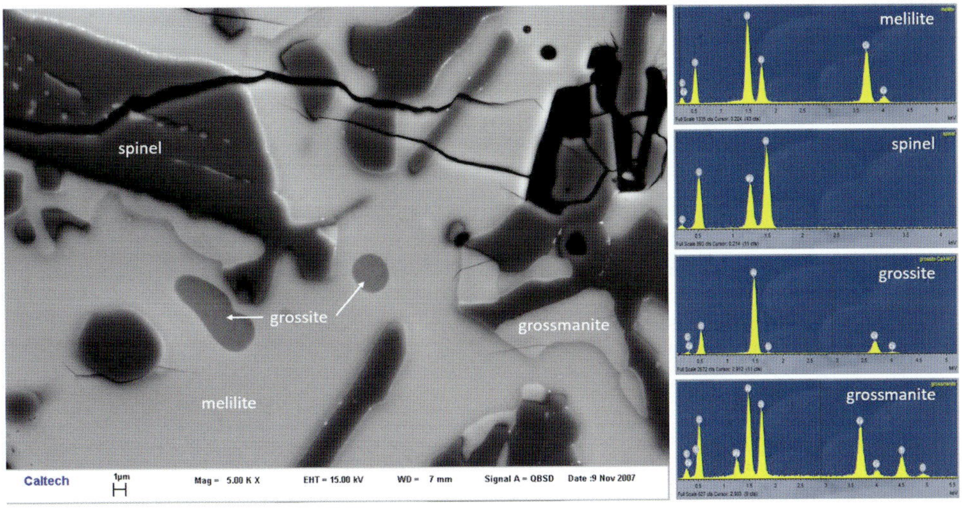

Figure 5.6 BSE image showing melilite, spinel, grossite, and grossmanite in a CAI from Allende. EDS analysis was used for phase identification.

Figure 6.2 Comet 2I/Borisov passing through the solar system. These two images, taken by NASA's Hubble Space Telescope, show the comet appearing near a background galaxy (left) and shortly after perihelion (right). Credits: NASA, ESA, and D. Jewitt (UCLA).

Figure 7.27 Thin section of a small individual Chelyabinsk stone showing an extensive network of dark shock veins. Image courtesy of Rhian Jones. Reflected light.

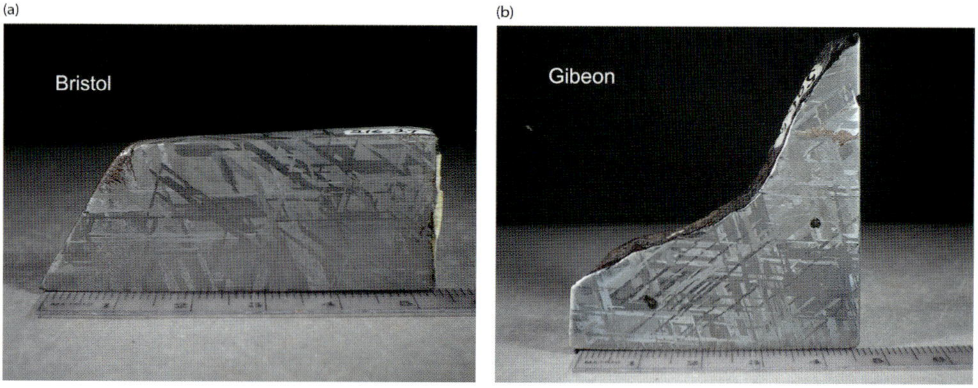

Figure 8.4 The Widmanstätten patterns of the Bristol and Gibeon IVA iron meteorites. Reflected light.

Figure 8.7 Seymchan pallasite with metal regions displaying Widmanstätten patterns. Reflected light.

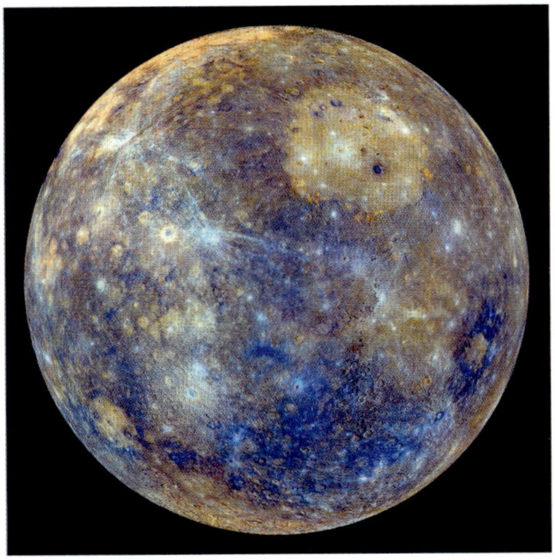

Figure 9.1 Mercury. Photojournal: PIA16853. Images from the MESSENGER mission. Source: NASA/Johns Hopkins University Applied Physics Laboratory/Carnegie Institution of Washington.

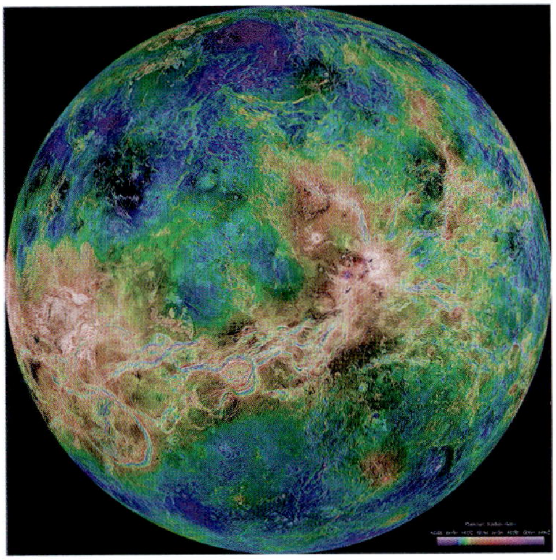

Figure 9.2 Venus. Photojournal: PIA00159. This hemispheric view of Venus is based on more than a decade of radar investigations up through the 1990-1994 *Magellan* mission. Source: NASA/JPL/USGS.

Figure 9.3 The Earth on April 17, 2019. The Earth Polychromatic Imaging Camera, a NASA camera aboard NOAA's *DSCOVR* spacecraft, returns daily images of Earth from a distance of about 1.5 million kilometers. Source: NOAA/NASA EPIC Team.

Figure 9.5 Four varieties of lunar rocks. (a) Mare basalt, (b) anorthosite, (c) breccia, (d) regolith (containing orange glass beads). Source: NASA.

Figure 9.8 The small hematite-rich spherules on the Martian surface, unofficially dubbed "blueberries," discovered by the *Opportunity* rover. Source: NASA/JPL-Caltech.

Figure 9.9 Mars. Photojournal: PIA04304. This mosaic of Mars is a compilation of images from *Viking Orbiter 1*. Source: NASA/USGS.

Figure 9.11 False-color image of Ceres, highlighting differences in surface materials. Photojournal: PIA20182. Credit: NASA/JPL-Caltech/UCLA/MPS/DLR/IDA.

Figure 9.13 Six infrared images of Titan created from data acquired by the Visual and Infrared Mapping Spectrometer instrument on board NASA's *Cassini* spacecraft. Source: NASA/JPL-Caltech/Stéphane Le Mouélic, University of Nantes; Virginia Pasek, University of Arizona.

(a)

Figure 9.14 (a) A Chinese 5-yuan coin commemorating the discovery of comets more than 4,000 years ago. (b) Silver denarius of Augustus struck 19 BCE–18 BCE in Lusitania. It shows the head of Augustus, oak-wreathed, facing left (obverse) and a stylized eight-rayed comet (reverse) representing the ascent of Julius Caesar's soul into heaven. The inscription on the reverse means "Divine Julius." According to the Roman author Pliny the Elder, a comet appeared during athletic games organized shortly after Julius Caesar's assassination in the Theater of Pompey on March 15, 44 BCE. Image 1944.100.39033 courtesy of the American Numismatic Society.

(b)

Figure 9.14 (*cont.*)

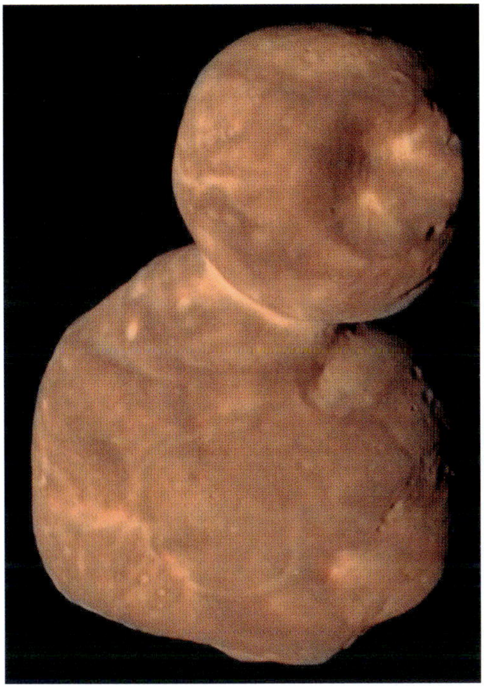

Figure 9.16 Kuiper Belt Object 486958 Arrokoth. This composite image was compiled from data obtained by NASA's *New Horizons* spacecraft in a flyby on January 1, 2019. Credits: NASA/Johns Hopkins University Applied Physics Laboratory/ Southwest Research Institute/Roman Tkachenko.

Figure 9.17 High-resolution enhanced color view of Pluto taken on July 14, 2015, by the Ralph/ Multispectral Visual Imaging Camera (MVIC) on board the *New Horizons* spacecraft. Credit: NASA/ JHUAPL/SwRI.

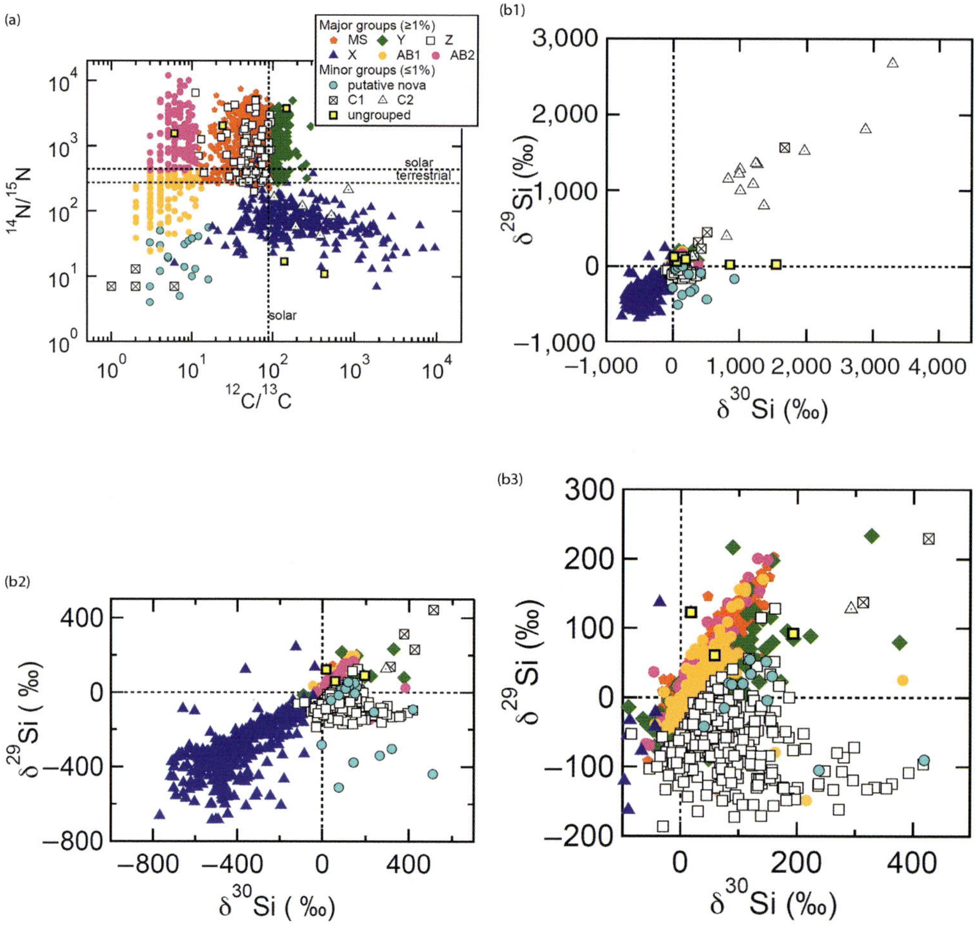

Figure 10.2 (a) Nitrogen and carbon isotopic ratios of individual presolar SiC grains (Zinner 2014 and references therein). See reviews by Zinner (2014) and Nittler and Ciesla (2016) for details. Description of new nomenclatures of AB and C grains can be found in Liu et al. (2016, 2017). (b) Silicon isotopic ratios of different groups of presolar SiC grains plotted in delta-notation. Symbols are the same as those in (a). (c) Sulfur isotopic ratios of different groups of presolar SiC grains plotted in delta-notation. Symbols are the same as those in (a). Some of the AB grains have no available N isotope data and cannot be further classified. Data sources: Gyngard et al. (2006); Hoppe et al. (2012, 2015); Fujiya et al. (2013); Xu et al. (2015); Liu et al. (2016, 2017). Courtesy of Nan Liu, WUSTL.

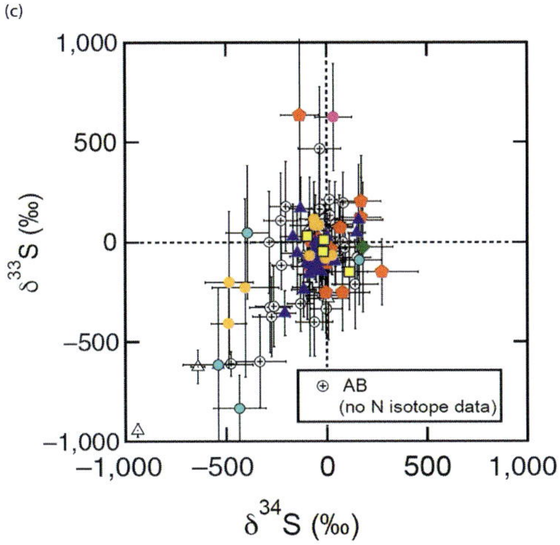

Figure 10.2 (*cont.*)

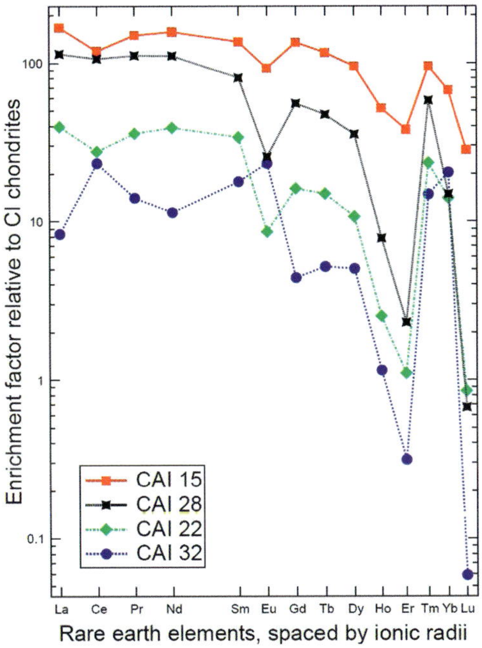

Figure 10.4 Group II REE pattern of CAIs, derived from Davis et al. (2018). Courtesy of Andrew Davis, University of Chicago.

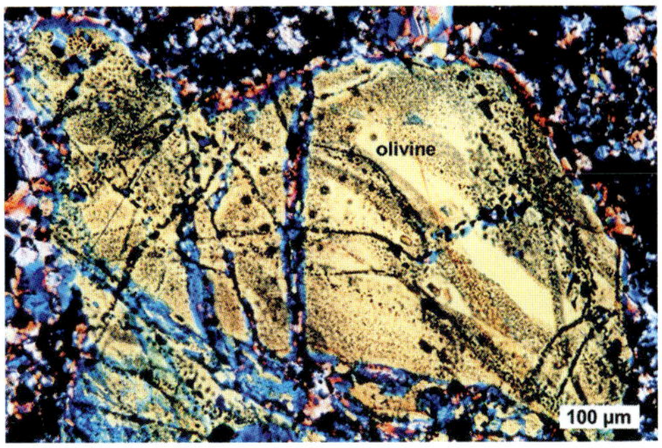

Figure 11.8 Olivine grain transected by numerous, intersecting, curvilinear trails of small blebs of metallic Fe-Ni and troilite in the MIL 99301 LL6 chondrite. The olivine exhibits sharp optical extinction because it underwent post-shock annealing. After Rubin (2002a). Transmitted light, crossed polarizers.

Figure 11.12 Portion of the Abee EH-chondrite impact-melt breccia showing a euhedral graphite grain (gr) surrounded by enstatite (en) and adjacent to an opaque assemblage consisting of kamacite (kam), troilite (troi) and keilite (keil). Reflected light.

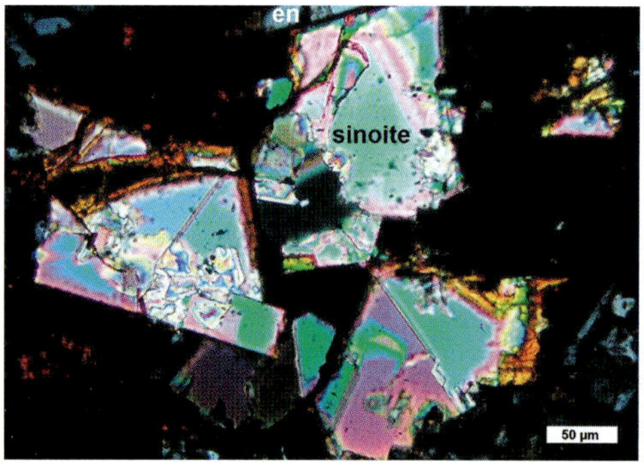

Figure 11.13 Cluster of intergrown euhedral, birefringent, optically zoned sinoite grains within an impact-melted portion of EL4 QUE 94368. An elongated enstatite (en) grain is adjacent to a large sinoite grain at top.

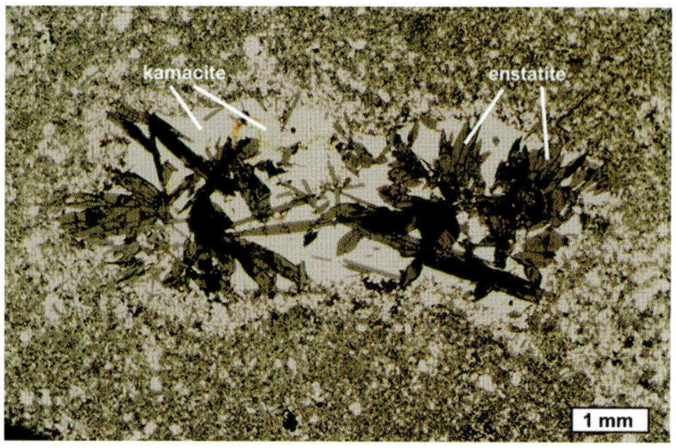

Figure 11.14 A 3 × 7 mm nodule rich in low-Ni metallic Fe in the Abee EH4 impact-melt breccia containing clusters and isolated grains of coarse euhedral enstatite. The clast has a rim, readily visible at the top center and bottom left of the clast, which consists mainly of low-Ni metallic Fe (kamacite) and fine-grained euhedral enstatite. After Rubin and Scott (1997). Reflected light.

Figure 11.23 Slab of CB_a chondrite Gujba showing large metal nodules and chondrule-like silicate nodules. There is a preferred orientation of the particles (NNE-SSW). Image courtesy of the Maine Mineral and Gem Museum. Reflected light.

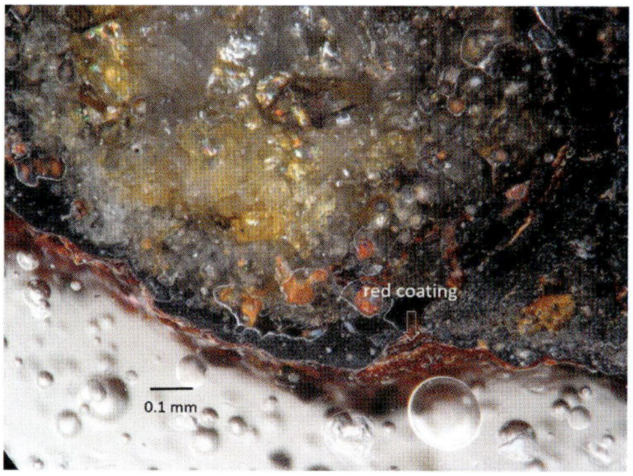

Figure 12.2 Thin red coating (possibly goethite) due to oxidation of metallic Fe on a Murchison (CM2) grain from the Caltech collection. Reflected light.

Figure 12.3 Conglomerate consisting of a cobble of the Al Haggounia 001 EL chondrite and a terrestrial rhyolite pebble cemented together by sand-sized detritus and iron hydroxides. Cube is 1 cm on a side. Photo courtesy of Greg Hupe.

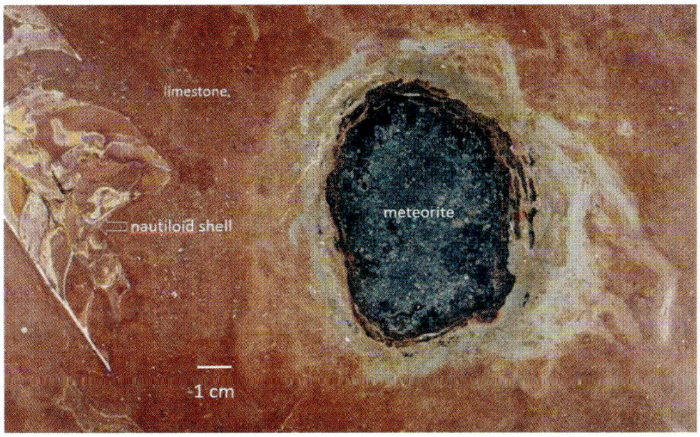

Figure 12.4 Fossil meteorite and nautiloid shell in Ordovician limestone. Photo courtesy of Birger Schmitz, Lund University.

Figure 12.6 A broken fragment of a fossil meteorite enshrouded with terrestrial clay; the sample was found in sediments retrieved from the Cretaceous–Paleogene (K–Pg) boundary in the North Pacific Ocean. This site is 9,000 km west of the Chicxulub impact structure. Image courtesy of Frank Kyte, UCLA.

traces of augite) with accessory or trace amounts in some specimens of ilmenite, Cr-spinel, metallic Fe-Ni, troilite, silica, and apatite. There are four subgroups: ferroan anorthosites (plagioclase $An_{>95}$), sodic anorthosites (An_{92-95}), mafic ferroan rocks (with >80 vol% plagioclase, $An_{>95}$), and mafic magnesian rocks (with >75 vol% plagioclase, $An_{>95}$).

The magnesian suite forms a diverse group of rocks. The most abundant samples are troctolites (consisting mainly of plagioclase and olivine) and norites (plagioclase and low-Ca pyroxene). There are also some samples of gabbronorite (plagioclase and Ca-pyroxene ± pigeonite), dunite (consisting mainly of olivine), and sodic ferrogabbro (gabbros with substantial amounts of K-feldspar). Accessory to trace phases include oxides (spinel, Cr,Fe-spinel, armalcolite, ilmenite, rutile, pyrochlore, baddeleyite, zirconolite), metallic Fe-Ni, troilite, silica, zircon, and phosphate (apatite, merrillite, and other phases).

The alkali suite includes KREEP basalts, alkali anorthosites, alkali gabbronorites, quartz monzodiorites, and rare granites (a.k.a. felsites) (Taylor 2009). KREEP basalts contain ~50 vol % plagioclase, major pigeonite and Ca-pyroxene (augite), and minor to accessory amounts of K-feldspar, phosphate, ilmenite, zircon, silica (quartz or cristobalite), metallic Fe-Ni, and troilite; olivine is rare. The granites consist of intergrown K-feldspar and quartz and trace amounts of fayalitic olivine, Ca-pyroxene, ilmenite, and metallic Fe-Ni.

Polymict breccias

Polymict breccias include impact mixtures of preexisting rock types and impact melts. There are numerous varieties.

Fragmental breccias are clastic rocks consisting mainly of rock fragments and preexisting breccias.

Glassy melt breccias contain relict rock fragments and breccias surrounded by impact-melted glass.

Crystalline melt breccias are impact-melt breccias that are slowly cooled versions of glassy melt breccias, wherein mineral grains crystallized from the interclastic melt. Grains within the melt include major feldspar, Ca-pyroxene, low-Ca pyroxene, and olivine, and minor to accessory metallic Fe-Ni, troilite, and ilmenite. Also present in some samples are mesostasis and glass.

Clast-poor impact melts are essentially impact-melt rocks that contain phenocrysts that crystallized from the melt and few (if any) relict clasts; they can be distinguished from pristine lunar rocks by their higher concentrations of siderophile elements. Major minerals include feldspar, Ca-pyroxene, low-Ca pyroxene, and olivine; minor to accessory minerals include ilmenite, ulvöspinel, spinel, metallic Fe-Ni, troilite, phosphate, schreibersite, armalcolite, tranquillityite, akaganéite, and K-Ba feldspar. Mesostasis is present in many samples, and glass is abundant in a few rocks.

Granulitic breccias are metamorphosed and recrystallized rocks formed from preexisting breccias.

Dimict breccias consist of two mutually intrusive components: (1) dark-colored impact-melt or fine-grained crystalline melt breccia (with plagioclase, olivine, and pyroxene) and (2) crushed impact debris typically consisting mainly of anorthosite (mostly plagioclase with minor low-Ca pyroxene ± olivine ± Ca-pyroxene).

Figure 9.7 A one-dollar commemorative coin showing an astronaut bootprint on the lunar regolith. Image courtesy of the U.S. Mint.

Regolith breccias formed at the lunar surface; they contain impact-produced, chondrule-like glass-rich spherules, agglutinates (vesicular glass-bonded, flow-banded, debris-laden masses), mineral grains exhibiting solar-flare particle tracks, and a minor bulk chemical component of meteoritic origin.

A few volume-percent of unconsolidated lunar regolith and lunar regolith breccias consist of submillimeter igneous spherules that consist of skeletal-to-euhedral phenocrysts of olivine, low-Ca pyroxene or plagioclase surrounded by glassy, fibrous, or microcrystalline mesostases. These objects, sometimes dubbed "lunar chondrules," have bulk chemical and O-isotopic compositions similar to common lunar materials. These chondrule-like spherules seem to have formed by a process involving impact melting of lunar rocks and soil, splattering of the impact melt, molding of free-floating melt droplets into spheroidal shapes due to surface tension, supercooling and rapid crystallization.

Lunar soil

Lunar soil is the small fraction (<1 cm) of fragmental unconsolidated material in the lunar regolith (Figure 9.7). Monomineralic particles include plagioclase, pyroxene, olivine, oxides, silica, and opaque phases. Also present are small clasts and breccias of the major lunar rock types, agglutinates and glass particles. Approximately 1–2 percent of lunar maria soils consist of a meteorite contaminant most similar to CM chondrites (Wasson et al. 1975).

Lunar meteorites

The number of recognized lunar meteorites continues to grow since the first one was found in Antarctica in 1979; hundreds are now listed in the Meteoritical Bulletin Database. Many of these specimens are actually different pieces of the same meteorite that broke apart in the atmosphere or after landing on Earth; these specimens are "fall paired." The number of separate, unpaired lunar meteorites is probably 40–50 percent of the number of individual reported specimens.

Whereas the Apollo and Luna samples were gathered from a restricted region on the lunar nearside, lunar meteorites are assumed to be representative of the entire lunar surface. This

Table 9.1 *Lunar minerals*

Native elements and metals: metallic Fe-Ni (kamacite and taenite), metallic Cu, metallic Si
Carbides: cohenite
Phosphides: schreibersite
Silicides: hapkeite, fersilicite, ferdisilicite
Sulfides: troilite, mackinawite, chalcopyrite, cubanite, bornite, pentlandite, sphalerite
Oxides: baddeleyite, rutile, ilmenite, chromite, Ti-chromite, ulvöspinel, Cr-ulvöspinel, spinel, pleonaste, Cr-pleonaste, Ti-magnetite, hercynite, zirconolite, armalcolite, loveringite, oxycalciobetafite, water-ice (detected remotely)
Phosphates: Cl- and F-rich apatite, merrillite, monazite
Nesosilicates: forsterite, olivine, fayalite, almandine
Cyclosilicates: cordierite
Inosilicates: Ca-pyroxene (diopside, hedenbergite, augite, subcalcic augite, salite), clinoenstatite, pigeonite, ferropigeonite, orthopyroxene, pyroxferroite, tranquillityite, zircon, titanite, thorite, magnesio-arfvedsonite, hornblende
Tectosilicates: cristobalite, quartz, tridymite, plagioclase, K-feldspar (orthoclase), K-Ba feldspar; yoshiokaite

assumption can be tested. Every introductory astronomy textbook divides the surface of the Moon into two distinct regions: the dark-shaded maria and the higher-albedo highlands. These provinces are not evenly distributed; the maria are confined mostly to the near side. The maria constitute 17 percent of the entire lunar surface (Hörz et al. 1991); the highlands constitute the remaining 83 percent. The global maria/highlands ratio is 0.20. If lunar meteorites are representative of the lunar surface, the ratio of mare-derived samples (basalts, gabbros, and related mafic rocks) to highland-derived samples (anorthosites, troctolites, and related feldspathic rocks) should be close to this value.

Of the several hundred lunar meteorites listed in the Meteoritical Bulletin Database, ~80 percent can be considered broadly anorthositic or feldspathic, ~15 percent are broadly basaltic or gabbroic, and ~5 percent are mixed, mingled, or polymict breccias. The basalt/anorthosite ratio is 0.19, essentially identical to the maria/highlands ratio of 0.20. This similarity supports the commonly accepted notion that lunar meteorites are random samples of the uppermost layer of the lunar crust. Their examination provides a broader and more-representative perspective of lunar igneous activity than studies restricted to Apollo, Luna and Chang'e-5 samples.

Lunar minerals

There are far fewer lunar than terrestrial minerals. The Moon has no biosphere, no tectonic plates, and is nearly devoid of water. Lunar minerals were reviewed by Smith (1974), Frondel (1975), Smith and Steele (1976), and Papike et al. (1991, 1998). A few additional phases were subsequently reported in Apollo samples and lunar meteorites. Tentatively identified phases, doubtful minerals, glasses, and terrestrial weathering products are not included in Table 9.1. Minerals formed by terrestrial alteration are present in some lunar meteorites, but not in Apollo or Luna samples. These secondary phases include akaganéite, barite, calcite, celestine, gypsum, and smectite. Also excluded here are minerals restricted to the rare meteorite samples found on the Moon: e.g., niningerite and daubréelite in the Hadley Rille EH chondrite, Apollo 15 (Haggerty 1972; Rubin 1997c); pyrrhotite, magnetite, calcite, dolomite, and saponite in the Bench Crater carbonaceous chondrite, Apollo 12 (Wood et al. 1971; McSween 1976; Zolensky 1997); and Ti-Al chromite in an unnamed possible LL chondrite among Luna 16 samples (Demidova et al. 2020).

Due to the anhydrous nature of lunar basaltic magmas, plagioclase tends to be somewhat nonstoichiometric; it contains an excess of the SiO_2 component (associated with a vacancy in the feldspar lattice) compared to terrestrial plagioclase. This coupled substitution can be expressed as $[\ ]Si_4O_8$. Plagioclase from mare basalts is commonly zoned, with 0 mole percent $[\ ]Si_4O_8$ in grain cores and 5–7 mole percent $[\ ]Si_4O_8$ in grain rims. Experiments have shown that at 1,200°C–1,500°C and 1 atm pressure, up to 10 wt% SiO_2 can be incorporated into plagioclase in the absence of water. In water-bearing terrestrial magmas, plagioclase has a lower liquidus temperature and generally cannot accommodate excess SiO_2.

With increasing levels of shock, the microtextures of apatite and merrillite in lunar-highlands samples become more deformed, starting with grain fragmentation and continuing with subgrain formation and granularization. Merrillite exhibits these features at lower shock stages than apatite.

A major difference between lunar meteorites and Apollo samples is in the abundance of maskelynite in basalts. Maskelynite forms by the shock-induced transformation of crystalline plagioclase. Because maskelynite is present in ~30 percent of basaltic lunar meteorites and only ~1 percent of mare basalts and basaltic fragments in Apollo 11, 12, 15, and 17 soils (which were never ejected from the Moon), it is clear that almost all of the maskelynite in lunar meteorites was produced during launch (Rubin 2015b). [This is consistent with the near-ubiquity of maskelynite in martian shergottites because it takes more than four times as much energy to launch a similarly sized sample off Mars than off the Moon (even neglecting the extra energy needed to push ejecta through the martian atmosphere).]

All the phases found in the Apollo and Luna samples have either already been found or are likely to occur in lunar meteorites. One possible exception is thorite ($ThSiO_4$), a single grain of which was identified in soil split 14259,97. This split is from a mature soil sample collected from the top centimeter of lunar soil at an undisturbed location ~100 m from the Apollo 14 lunar module. Haines et al. (1972) calculated that thorite contains 21.6 percent of the U in the soil sample.

The lunar atmosphere is extremely thin (about 14 orders of magnitude less dense than Earth's); its constituent atoms and molecules seldom collide. Each of the six Apollo mission landings released about the same amount of gas as was present in the ambient lunar atmosphere (Vondrak 1974). Species definitively identified include helium, argon, sodium, and potassium. Sources for the gases include (1) evaporation of surface material, (2) material released by the impacts of meteoroids, (3) interaction of surface materials with high-energy photons and solar-wind particles, and (4) outgassing from the lunar interior.

Because of the small tilt (6.68°) of the Moon's rotation axis relative to the ecliptic, the floors of some impact craters near the poles are never impinged by sunlight. These permanently shadowed regions have been postulated to harbor water ice delivered by impacting comets and water-bearing carbonaceous chondrites. (Mechanisms that can remove some of the water ice from the lunar regolith include micrometeorite impacts and the impact of solar-wind particles on water molecules.) Nevertheless, *Lunar Prospector* thermal and epithermal neutron data revealed enhanced concentrations of hydrogen in permanently shadowed craters near the lunar poles (e.g., Feldman et al. 2001); these data were interpreted to indicate the presence of water ice. It is found in ancient craters as well as relatively young craters. The presence of ice was confirmed by the near-infrared (NIR) absorption features of ice (near 1.3, 1.5, and 2.0 μm) measured by the Moon Mineralogy Mapper instrument aboard the *Chandrayaan-1* spacecraft (Li et al. 2018). The Shackleton Crater at the south pole may contain up to 22 wt% ice mixed with dry regolith. These data are consistent with the high reflectance values obtained by the Lunar Orbiter Laser Altimeter (LOLA) and the ratios of reflected UV radiation detected by the Lyman Alpha Mapping Project (LAMP) on the *Lunar Reconnaissance Orbiter*. Water ice is a potentially important resource for possible future human exploration and colonization of the Moon (e.g., Crotts 2014; Spudis 2016). Its presence on the Moon supports the notion that at least some of the water on Earth is a late veneer delivered by water-bearing asteroidal and cometary projectiles. The Sun shines constantly on the central peaks of some impact craters at the lunar poles; such "peaks of eternal light" could potentially provide energy to solar panels and enable the extraction of water ice mixed in the permanently shadowed soil below. Breathable oxygen could also be extracted from the lunar regolith by a devise using a form of molten-salt electrolysis (Lomax et al. 2020).

In his 2006 book, *Return to the Moon,* Apollo 17 astronaut Harrison Schmitt noted that undisturbed Ti-rich lunar soils probably contain 20–30 ng/g of ^{3}He, derived from the solar wind. This helium isotope could be used for the clean generation of fusion electrical power. When fused with deuterium, 100 kg of ^{3}He could power a 1,000-megawatt electric plant for a year, providing enough energy to meet the needs of a city the size of Dallas, Texas.

9.5 Mars

The first clue to the mineralogy of the Martian surface can be gleaned by observing the planet in the night sky. Because Mars appears red, it has long been linked to war and bloodshed. The ancient Babylonians called the planet "Nergal," a fiery god associated with war and judgment of the dead; the Greeks named the planet "Ares," and the Romans called it "Mars" after their god of war. The modern symbol for the planet Mars (which dates from the sixteenth century) is a shield and a spear: ♂; it was modified from the simple spear symbol used in ancient Rome to represent both the god and the planet (e.g., Bianchini's planisphere, second century CE). Modern astrologers associate Mars with fire and red gemstones: ruby (Al_2O_3), red garnet (pyrope ($Mg_3Al_2Si_3O_{12}$)), red jasper (SiO_2), and, especially, red coral ($CaCO_3$).

Human blood derives its red color from hemoglobin, an iron-bearing metalloprotein in red blood cells that transports oxygen from the lungs to the rest of the body. The most abundant red mineral at the Earth's surface also contains iron and oxygen: hematite (Fe_2O_3), a substance used since Upper Paleolithic times as a pigment for cave painting. The most abundant ferric-iron-bearing mineral in terrestrial soil is the hydroxide goethite (FeO(OH)). The common terrestrial iron ore with the highest iron content (72.4 wt% Fe) is magnetite (Fe_3O_4). Small, dust-size grains of hematite, goethite, and Ti-bearing magnetite are also present on Mars; global dust

Figure 9.8 The small hematite-rich spherules on the Martian surface, unofficially dubbed "blueberries," discovered by the *Opportunity* rover. Source: NASA/JPL-Caltech. (A black-and-white version of this figure will appear in some formats. For the colour version, please refer to the plate section.)

storms, usually starting shortly after perihelion, can spread this dust (consisting of 3–5-μm-sized iron-rich particles) all over the planet, giving it a red color. The sky on Mars during daytime has a reddish or pinkish hue due to the suspension of fine dust in the atmosphere.

Most of the hematite on Mars probably formed in the presence of water. [However, it is possible that some hematite formed by oxidation of FeO-bearing olivine or pyroxene in basaltic rocks exposed to ultraviolet (UV) radiation from the Sun. Unlike Earth, Mars has no ozone layer to absorb medium-frequency UV rays (wavelength range = 200–315 nm).] Goethite requires water for its formation. Spherules of hematite (sometimes called "blueberries" because of their appearance on false-color NASA images) were discovered by the *Opportunity* rover (Figure 9.8). They occur as loose bodies scattered on the surface as well as deeply embedded inclusions in the martian soil. The hematite blueberries appear to be concretions formed by ground water. Many of these objects weathered out of their host rock and accumulated on the surface; finer-grained material was carried away by the wind.

Two other minerals on Mars (water ice and dry ice[2]) were observed via remote sensing via telescopes and spacecraft. Telescopic observations of Mars by Christian Huygens in 1659 and Giacomo Maraldi in 1719 revealed the presence of white markings near the South Pole; these were interpreted as snow and ice by William Herschel in 1784. Spacecraft flybys and orbiters, from 1965 to the present day, have shown that the north pole of Mars has a permanent cap of water ice and a thin layer of dry ice deposited during northern winter. The south pole has a permanent cap of water ice and dry ice; this is augmented by additional dry ice deposited during southern winter. Water ice at high latitudes was confirmed in 2008 by the *Phoenix* Mars lander. But water ice is not confined to the polar regions: the neutron spectrometer aboard the *Mars*

[2] Dry ice (solid, crystalline carbon dioxide) occurs naturally on Mars as CO_2 (Phase I), the polymorph stable at low temperatures and pressures. This CO_2 phase is cubic of space group *Pa*-3 (Mangan et al. 2017). Dry ice is also present on Uranus's moons Ariel, Umbriel, and Titania, and probably on Neptune's moon Triton. However, dry ice does not occur naturally on Earth. The lowest temperatures recorded on Earth (near the East Antarctic ice divide) are about -98°C (Scambos et al. 2018), but because the partial pressure of CO_2 is so low, dry ice would not precipitate from the atmosphere at temperatures above about −150°C. There is no evidence of any CO_2-H_2O admixture in the snow on the Antarctic ice sheet (T. A. Scambos, personal communication, 2019). Thus, dry ice is a naturally occurring mineral phase on Mars but not on Earth.

Figure 9.9 Mars. Photojournal: PIA04304. This mosaic of Mars is a compilation of images from *Viking Orbiter 1*. Source: NASA/USGS. (A black-and-white version of this figure will appear in some formats. For the colour version, please refer to the plate section.)

Odyssey orbiter detected significant amounts of water ice (actually water equivalent hydrogen inferred to be water ice) under the surface at the Medusae Fossae Formation near the equator. [As is true for the Moon, water ice is a potentially important resource for possible future human exploration and colonization of Mars (e.g., Zubrin 2011; Petranek 2015).]

Mars (Figure 9.9) has been the subject of numerous review papers. Some of the best recent ones were penned by Hap McSween and coworkers (e.g., McSween et al. 2009; McSween and McLennan 2014; McSween 2015).

There seem to be globally distributed perchlorate salts on Mars, probably in the form of chlorinated hydrocarbons. The *Phoenix* lander detected perchlorates in 2008 and the *Curiosity* rover found them within Gale Crater in 2013. These results are consistent with the detection of elemental Cl by the rovers *Sojourner* (landed 1997), *Spirit* (2004), and *Opportunity* (2004). Cosmic rays and solar radiation could break perchlorates down into hypochlorite (ClO^-); this compound could be responsible for the apparent positive result for the Labeled Release biological experiment aboard the *Viking* lander in 1976.

In 2019, *Curiosity* found large enrichments in calcium sulfate within a 150-m-thick stack of sedimentary rocks within the crater; magnesium sulfate occurs as nuggets in a thinner sequence of layers. These rocks are 3.3–3.7 Ga years old, suggesting that Gale Crater hosted a briny lake that evaporated at that time.

The geologic diversity of Mars was first revealed in spectacular fashion by the *Mariner 9* orbiter (1971–1972). *Mariner 9* and its many successors (e.g., *Viking 1*, *Viking 2*, *Mars Global Surveyor*, *Mars Odyssey*, *Mars Express*, *Mars Reconnaissance Orbiter*, *Mangalyaan* a.k.a. *Mars Orbiter Mission*) sent back pictures of enormous shield volcanoes (exemplified by Olympus Mons), volcanic plains (e.g., Tharsis and Elysium), eroded mountains and debris aprons (e.g., Promethei Terra), ancient cratered terrain (much of the southern hemisphere), large multi-ring impact basins (Argyre, Hellas, and Isidis), a huge rift valley (Valles Marineris), graben (e.g., Labyrinthus Noctis), sand dunes (as in the Hellespontus region), yardangs (wind-sculpted rock ridges, e.g., near Gordii Dorsum), dried river valleys and hillside gullies (e.g., Nirgal Vallis), flood channels and teardrop-shaped islands (e.g., Marte Vallis), hummocky terrain (e.g., near Utopia Planitia), and sedimentary deposits near the south pole. Each of these regions has a unique geological history and can be expected to have a characteristic assemblage of primary and secondary minerals. The Mars InSight mission recently found the planet to be seismically active; the degree of activity resembles that of many regions on Earth distant from plate boundaries and tectonically deformed terrains.

The *Spirit*, *Opportunity*, and *Curiosity* rovers encountered many igneous rocks, mostly basalt but also some more-siliceous rocks, probably andesite, basaltic andesite, or dacite; they also came across sedimentary rocks (mudstone, aeolian sandstone, and carbonate-rich rock) and soils. The most abundant minerals in the igneous rocks are plagioclase, olivine, orthopyroxene, Ca-pyroxene, and magnetite. The sedimentary rocks and soils contain gypsum, anhydrite, smectite, Fe-sulfide, magnetite, goethite, hematite, jarosite, opal, Mg-rich carbonate (probably dolomite), siderite, and epsomite. Additional minerals that some workers have inferred to be present in the sedimentary rocks and soils include various sulfates (e.g., bassanite, kieserite, hexahydrite) and salts (e.g., halite, bischofite, antarcticite, blödite, vanthoffite, glauberite).

The Chemistry and Mineralogy (CheMin) instrument aboard *Curiosity* includes a transmission X-ray diffractometer (XRD) and X-ray fluorescence (XRF) spectrometer. It analyzed samples from the sedimentary sequence of mudstones and fine-grained sandstones exposed in the ~155-km-diameter Gale Crater; the results were reported in great detail by Rampe et al. (2020). The lower stratigraphic levels of the sequence (formed in a relatively warm and wet environment) contain smectite; the upper levels (formed when the climate was colder and dryer) contain sulfate. Phases identified by CheMin include plagioclase, felsic igneous phases (sanidine, tridymite, cristobalite, and trace amounts of quartz), mafic igneous phases (pyroxene and olivine), Fe-oxide minerals (hematite, magnetite, and minor akaganeite), phyllosilicates (smectite and, possibly, some Fe-pyrophyllite), sulfate minerals (anhydrite, bassanite, and gypsum, particularly in the upper sections, and minor to trace jarosite in the lower sections), minor fluorapatite, trace amounts of halite, and X-ray amorphous materials (probably including amorphous sulfates, nanophase Fe-oxides, and silicates).

Some modern aeolian sediments were also analyzed by CheMin. These are dominated by basaltic igneous minerals (plagioclase (An_{48-63}), pyroxene, and olivine (Fa_{40-46})), and X-ray amorphous materials. Also present are minor Fe-oxide minerals (magnetite and hematite with minor ilmenite in one sample), anhydrite, and quartz.

Ancient aeolian samples were acquired by drilling into dune-derived sandstone and associated light-colored alteration halos that lie parallel to fractures in the rock. The sandstone samples contain igneous minerals [plagioclase and pyroxene (modeled as pigeonite and orthopyroxene)], magnetite, and X-ray amorphous materials. No olivine was identified. The parent sandstone is also inferred to contain nanophase Fe-oxides, amorphous sulfates, and magnetite, along with minor sanidine, hematite, quartz, fluorapatite, and Ca-sulfate. The alteration halos have a lower abundance of igneous minerals (particularly pyroxene) and higher abundances of sulfate and SiO_2-enriched X-ray amorphous materials (probably containing opaline silica).

Until the recent acquisition of data from the *Curiosity* rover, our knowledge of martian mineralogy came mainly from the study of martian meteorites.

Nakhlites are unbrecciated olivine clinopyroxenites consisting of ~75 vol% coarse millimeter-sized augite and subcalcic augite prisms (many grains are twinned on {100}) and ~11 vol% olivine phenocrysts (Fa_{70-83}) up to 5 mm in size, surrounded by ~14 vol% finer-grained mesostasis. Minerals in nakhlites include silicates (augite, subcalcic augite, pigeonite, orthopyroxene, olivine, plagioclase, K-feldspar, silica), oxides (titanomagnetite, ulvöspinel, rutile, magnetite, hercynite), phosphates (chlorapatite, merrillite), and sulfides (pyrrhotite, pyrite with marcasite lamellae, chalcopyrite). Minor amounts of fine-grained material referred to as "iddingsite" is present within veins inside olivine crystals; this substance consists mainly of smectite, serpentine, opal, amorphous gel, stilpnomelane, Fe-oxides, and Fe-oxyhydroxide; in many cases, the iddingsite is associated with siderite, gypsum, and halite (e.g., Lee et al. 2018). These phases are formed by aqueous alteration on Mars; they are not terrestrial weathering products. Nakhlites are somewhat variable in their minor phases, so not all members contain all these minerals.

NWA 2646 is a small plagioclase-olivine clinopyroxenite containing major silicate and minor oxides, phosphate and sulfide: 41 vol% pigeonite ($Fs_{24.4}Wo_{5.7}$ zoned to $Fs_{34.4}Wo_{12.1}$), 24 vol% augite ($Fs_{19-23}Wo_{26-36}$), 22 vol% olivine (Fa_{38-44}), 11 vol% maskelynite ($An_{58.4-60.7}Or_{0.9}$), and 2 vol% minor phases (chromite, ilmenite, merrillite, and pyrrhotite). Secondary phases formed during terrestrial weathering include calcite, a hydrous Al-rich phase (possibly kaolinite), and calcium chloride.

ALH 84001 is a 1.93-kg orthopyroxenite with a cataclastic texture containing 97 vol% coarse anhedral orthopyroxene crystals of uniform composition ($Fs_{27}Wo_{3.3}$) (up to 3.5 mm), 2 vol% anhedral to euhedral Fe_2O_3-bearing chromite (up to 500 μm), and 1 vol% irregular 100- to 300-μm-sized patches of maskelynite (averaging $Ab_{63}Or_6$) and accessory augite ($Fs_{12}Wo_{42}$), apatite, and pyrite. Also present are secondary 100- to 200-μm-sized brown globules of carbonate consisting of orange ankerite cores with minor calcite±silica surrounded by rims of compositionally zoned magnesite and breunnerite. Some orthopyroxene grains contain small embayed olivine grains (Fa_{65}) that are too ferroan to be in equilibrium with the host. The rock also contains tiny (10- to 100-nm-sized) magnetite grains and trace amounts of Fe sulfate. The meteorite became famous as a possible host of martian microfossils (McKay et al. 1996), but numerous subsequent studies dispelled that idea (e.g., Bradley et al. 1997).

The most abundant variety of martian meteorite is the group of shergottites and related rocks including those classified as basalts, olivine basalts, augite basalts, olivine-phyric basalts, basaltic breccias, gabbros, microgabbros, diabases, and lherzolites. Included in this group is NWA 7034 and its numerous paired specimens, dubbed "Black Beauty" by meteorite collectors. This meteorite is a regolith breccia that contains more bulk water (~0.6 wt%) than any other martian meteorite.

Although minerals and mineral chemistry vary somewhat with rock type in the shergottites and their fellow travelers, in general, the most abundant phases are compositionally zoned pyroxene (augite, subcalcic augite, pigeonite and/or orthopyroxene), olivine (Fa_{24-40}), and maskelynite and/or plagioclase (Ab_{30-50}). NWA 7034 contains no maskelynite, only crystalline plagioclase. Magnesian olivine occurs in one lithology of EETA79001. Minor phases (not present in all these rocks) include various oxides (magnetite, titanomagnetite, ulvöspinel, ilmenite, chromite, hercynite, baddeleyite); phosphates (merrillite, chlorapatite); and sulfide (pyrrhotite, pentlandite). Late-stage minerals, occurring in the mesostasis and within melt

Table 9.2 *Martian minerals*

Primary Igneous Phases
Pyroxene: augite, subcalcic augite, pigeonite, hedenbergite, orthopyroxene
Pyroxferroite
Olivine: ferroan olivine, fayalite
Feldspar: plagioclase, K-feldspar, sanidine
Silica: quartz, cristobalite, *tridymite*
Oxide: titanomagnetite, magnetite, hercynite, ulvöspinel, chromite, ilmenite, rutile, baddeleyite
Phosphate: apatite (including chlorapatite and fluor-chlor-apatite), merrillite, monazite, xenotime, *fluorapatite*
Sulfide: pyrrhotite, pyrite, marcasite, chalcopyrite, pentlandite, troilite
Zircon
Amphibole: kaersutite, ferro-kaersutite
Phyllosilicate: biotite, phlogopite
Alteration Phases
Smectite
Possible Fe-pyrophyllite
Serpentine
Opal
Amorphous materials
Stilpnomelane
Fe-oxides: hematite, magnetite
Goethite
Akaganeite
Carbonate: calcite, siderite, magnesite, ankerite, breunnerite, dolomite
Sulfate: gypsum, bassanite, jarosite, epsomite, *anhydrite*
Halite
High-Pressure Phases
Impact glass: maskelynite, pyroxene-composition glass
Silica: stishovite, seifertite, monoclinic phase with β-ZrO_2-type structure
Akimotoite
Ahrensite
Lingunite
Majorite
Ringwoodite
Tuite
Bridgmanite
Wüstite-feo
Tissintite
Xieite
Chenmingite
Zagamiite

Table 9.2 (*cont.*)

High-Pressure Phases
Liebermannite
Stöfflerite
Tschaunerite
Feiite
Liuite

Most phases were identified within martian meteorites. Additional phases identified *in situ* on Mars, but not within martian meteorites, are in italics. Minerals formed only by terrestrial weathering are excluded. List modified from one provided by H. Y. McSween (personal communication, 2019).

inclusions, include silica, pyroxferroite, fayalite, hercynite, and ferro-kaersutite. The silica polymorphs include quartz, possibly cristobalite, and three high-pressure phases: stishovite (tetragonal), seifertite (orthorhombic), and an as-yet-unnamed monoclinic phase with the β-ZrO_2-type structure. These phases occur in shock-melt pockets, melt pools, shock-melt veins and in association with maskelynite. Ringwoodite (the high-pressure polymorph of olivine) is present in some samples. Alteration phases include illite-like clay, calcite, gypsum, Mg-phosphate, and silica.

Chassignites are dunites – cumulate rocks that contain major olivine ($\geq$90 vol%) as well as minor to accessory orthopyroxene ($Fs_{19}Wo_3$), pigeonite, augite, plagioclase, sanidine, and chromite with interstitial chlorapatite, troilite, ilmenite, rutile, baddeleyite, kaersutitic amphibole, biotite, phlogopite, pentlandite, and troilite. Some olivine grains in Chassigny contain trapped melt inclusions with augite, low-Ca pyroxene, kaersutite, chlorapatite, pyrrhotite, chromite, pentlandite, biotite, and siliceous glass. Terrestrial alteration in some samples produced marcasite, calcite, Mg-sulfate (either gypsum or bassanite), magnesite, and possibly hydromagnesite.

Martian meteorites also contain numerous high-pressure minerals produced during launch off Mars (discussed in Section 11.5). Some, possibly all, of these phases may also occur in martian impact craters.

Additional rock types and mineral assemblages will be identified on Mars by future missions. Table 9.2 provides a list of currently known martian minerals. There is a chance that Mars once hosted living organisms, so fossils might one day be discovered. If life once flourished on Mars and survives today in some underground niches, local biomineralization is likely to occur. On Earth, more than 60 different "minerals" (technically, compounds because of their organic genesis) are produced biologically; these phases include silicates, phosphates, carbonates, sulfates, and oxides.

9.6 Asteroids

The asteroid belt contains more than 10^5 bodies with diameters exceeding 2 km; their total mass is ~5×10^{-4} Earth masses (Weidenschilling 2019). Their chemical and mineralogical characteristics were initially discerned by reflectance spectroscopy, i.e., by the analysis of sunlight reflected off asteroid surfaces back to telescopes on Earth. The topic was reviewed by Burbine (2017). Many individual minerals have characteristic absorption bands where photons of particular wavelengths are absorbed (due to electronic transitions and vibrational bands) and not reflected. Prominent absorption features include those at 0.506, 0.9, and 1.9 μm (Fe^{2+} in pyroxene); 0.7 μm (Fe^{3+} in phyllosilicates); 1.0 μm (Fe^{2+} in magnetite); 1.1 μm[3] (Fe^{2+} in olivine); 2.2 μm (Fe^{2+} in spinel); 2.7 μm (-OH in phyllosilicates); 3.0 μm (H_2O); and 3.4 μm (C-H in organics). Some minerals (e.g., kamacite and taenite) do not exhibit absorption bands and tend to make the absorption bands of the silicates shallower.

The reflectance spectra of powders of different varieties of meteorites are functions of the proportions of the different constituent phases. Meteorite spectra determined in the lab can be compared directly with those of asteroids measured remotely. Several asteroid researchers have developed formulas for estimating the fayalite content of olivine, the ferrosilite content of orthopyroxene, and the bulk olivine/(olivine+pyroxene) ratio of asteroid surfaces based on their spectra. Several of these complex mineral formulae are listed in Burbine (2017).

Asteroids have been placed into different taxonomic categories based on their spectral characteristics and visual geometric albedo. There are several different generally non-mnemonic classification schemes. One of the most widely used was devised by David Tholen; it includes three major asteroid groups (C, S, X) comprising 14 individual types. The types differ spectrally in slope and absorption features; each type has a different number of members and extends across different ranges of solar distance and inclination. C-group asteroids are dark and carbonaceous and include types B (e.g., 2 Pallas), C (e.g., 324 Bamberga), D (e.g., 944 Hidalgo), F (e.g., 302 Clarissa), G (e.g., 1 Ceres), and T (e.g., 114 Kassandra). S-group asteroids are silicaceous and include types A (e.g., 1951 Lick), Q (e.g., 1862 Apollo), R (e.g., 349 Dembowska), S (e.g., 3 Juno), and V (e.g., 4 Vesta). X-group asteroids comprise a grab bag of diverse bodies including types E (e.g., 434 Hungaria), M (e.g., 16 Psyche), and P (e.g., 153 Hilda).

Other schemes have updated and extended the Tholen system and introduced new asteroid types and subtypes. The Bus–DeMeo scheme defines 25 taxonomic categories based on visual and near-infrared characteristics: S-complex (S, Sa, Sq, Sr, Sv); C-complex[4] (B, C, Cb, Cg, Cgh, Ch); X-complex (X, Xc, Xe, Xk, Xn); and types A, D, K, L, O, Q, R, T, and V. The spectral properties of these types are listed in table 4.2 of Burbine (2017).

There are close matches between meteorite varieties and asteroid types: e.g., OC (Q type); HED (V type), iron meteorites and mesosiderites (M type), carbonaceous chondrites (mainly C and D types), and aubrites (E type). These correspondences imply that these particular meteorite groups were probably derived from asteroids of these (or related) spectral types. Most A-type asteroids appear to be olivine-rich mantle samples of differentiated bodies (DeMeo et al. 2019). The numerous links between asteroids and meteorites were discussed by Cloutis et al. (2014).

[3] Olivine actually has three absorption bands caused by transitions of 3*d* electrons: those at 0.9 and 1.25 μm correspond to the M1 site; the one at 1.1 μm corresponds to the M2 site. Olivine also shows a strong absorption feature at ≤0.7 μm caused by a charge-transfer absorption between O and Fe^{2+} (Burbine 2017).

[4] Spectral observations show that about 60 percent of C-complex asteroids have hydrated silicates at the surface.

Spacecraft have flown by, orbited, landed on, or sampled several asteroids, obtaining mineralogic and compositional information.

9.6.1 Vesta

4 Vesta (V-type; mean diameter 525 km) can reach an apparent magnitude of ~5.3 at opposition and can be seen by the naked eye. It was visited by the *Dawn* spacecraft in 2011–2012 (Figure 9.10). Vesta has long been linked to HED meteorites: (1) Spectral reflectance measurements show that many regions of the surface are basaltic (rich in pyroxene and plagioclase and

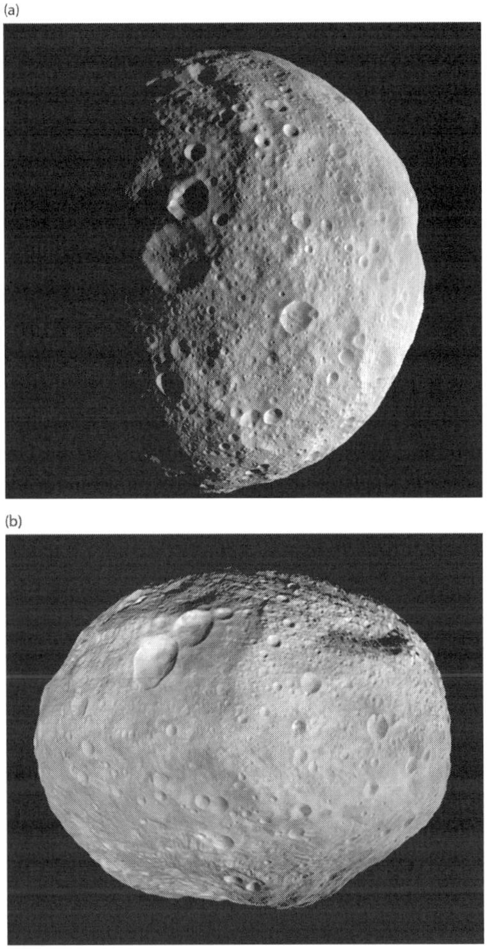

Figure 9.10 Vesta. (a) View looking down at Vesta's north pole taken by NASA's *Dawn* spacecraft as it was departing in 2012. (b) Mosaic of some of the best views of Vesta taken by the *Dawn* spacecraft. The mountain at the south pole (bottom of image) is more than twice the height of Mount Everest. Credit: NASA/JPL-Caltech/UCLA/MPS/DLR/IDA.

similar to eucrites); other regions, particularly those in the southern hemisphere, appear diogenitic (rich in orthopyroxene).[5] (2) Although Vesta itself is distant from a meteorite-supplying resonance, there are small (<10-km-diameter) V-type asteroids (dubbed vestoids) with semi-major axes between that of Vesta (2.36 AU) and 2.5 AU (the location of the 3:1 jovian resonance that can rapidly transfer small bodies toward the inner Solar System). The vestoids are probably fragments of Vesta's crust, excavated during the formation of the large impact basins Veneneia (400 km) and Rheasilvia (450 km) near the south pole. (3) The diurnal Yarkovsky effect can move small prograde-rotating bodies over distances of several tenths of an AU, e.g., from 2.36 to 2.5 AU. (4) There are V-type near-Earth asteroids (NEAs) with perihelia between 0.983 and 1.3 AU; these bodies are plausibly derived from the vestoid population and could be the immediate progenitors of at least some HED meteorites. (5) The orbit of the Sariçiçek howardite fall (Turkey, September 2, 2015) shows it to be from an inner main-belt asteroid with a semi-major axis of ~1.44 AU. The orbit of the Bunburra Rockhole eucrite fall (Australia, July 21, 2007) shows it to be from an NEA with a semi-major axis of 0.85 AU.

The magmatic history of Vesta has been inferred from HED meteorites and *Dawn* spacecraft data (McSween et al. 2019). Gamma-ray measurements have shown that Vesta has global Fe/O ratios (0.30±0.04) and Si/O ratios (0.56±0.06) in the range of HED meteorites (Fe/O: 0.17–0.40; Si/O: 0.47–0.58). Fast-neutron data from the Gamma-Ray and Neutron Detector (GRaND) instrument aboard *Dawn* gathered during low-altitude mapping showed compositional variations like those of HED meteorites. One conundrum is that although the diogenite/eucrite ratio is 0.36, there are relatively few regions of the surface of Vesta that appear rich in olivine. It is possible that such regions exist but are covered with basaltic impact ejecta.

There are low-albedo surface areas commonly associated with impact craters; they exhibit an absorption feature at ~0.7 μm that spectrally resembles CM carbonaceous chondrites. This is consistent with the occurrence of CM clasts in howardites such as Kapoeta. It is also consistent with surface water contents up to 400 μg/g determined by *Dawn* as well as a 2.8 μm absorption feature produced by OH.

Spectral data show no regions of the surface have reddish continuum slopes, low visual albedos, and relatively weak absorption bands characteristic of lunar-style space weathering. However, there appear to be signs of unusual space weathering in polymict eucrite NWA 1109. This HED meteorite contains (1) agglutinates with nanophase grains of metallic Fe-Ni and FeS, and (2) a fine-grained sulfide replacement texture (possibly formed by reaction between S-rich vapor and pyroxferroite) that consists of troilite, hedenbergite, and silica.

9.6.2 Ceres

1 Ceres (G-type; mean diameter 945 km) was visited by the *Dawn* spacecraft in 2015 (Figure 9.11). Spectral data reveal a surface apparently rich in ammoniated phyllosilicates. [Because ammonia is stable in the outer Solar System (e.g., on Enceladus), this may mean that

[5] Although most diogenites are orthopyroxenites, some have dunitic (olivine-rich), harzburgitic (olivine- and orthopyroxene-rich), or noritic (Ca-plagioclase-, orthopyroxene-, and olivine-rich) lithologies. Some noritic diogenites also contain minor pigeonite and augite.

Figure 9.11 False-color image of Ceres, highlighting differences in surface materials. Photojournal: PIA20182. Credit: NASA/JPL-Caltech/UCLA/MPS/DLR/IDA. (A black-and-white version of this figure will appear in some formats. For the colour version, please refer to the plate section.)

Ceres formed there (far from its present location) or that material from that region accreted to Ceres (e.g., Grazier et al. 2018).] An absorption band at 2.72 μm indicates the likely presence of Mg-rich serpentine (such as antigorite, chrysotile, and lizardite), a band at ~3.4 μm suggests organic material, and one at ~4 μm is probably due to carbonates. Bright areas within the Occator crater are rich in sodium carbonate (probably due to evaporation of brines). Relatively few neutrons were detected near the north pole by the GRaND instrument; this is consistent with high H abundances within the uppermost meter of the regolith, suggesting the presence of subsurface water ice. This inference is consistent with the relatively low subsolar surface temperatures expected in polar regions. Among meteorites, Ceres most closely resembles CI and CM chondrites but differs from them in containing ammoniated clays, less organic matter, and more carbonate.

Within two H-chondrite breccias, Zag and Monahans (1998), there are 4.5-Ga-old halite crystals containing fluid inclusions, organic material, and mineral grains postulated to have been derived from Ceres. Zolensky et al. (2015a, 2016a) reported that solids trapped within the halite include macromolecular carbon, aliphatic carbon compounds, Fa_{1-41} olivine, Ca-pyroxene, low-Ca pyroxene, feldspars, phyllosilicates, magnetite, sulfides, metallic Fe-Ni, lepidocrocite, carbonates, diamond, apatite, and zeolites.

9.6.3 Eros

433 Eros (S-type; 34.4 × 11.2 × 11.2 km) was the primary target of the *NEAR-Shoemaker* mission; the spacecraft flew by the C-type asteroid 253 Mathilde in 1997 and went into orbit around Eros in 2000; it touched down on Eros 12 months later. X-ray measurements showed that the surface of Eros has a chemical composition fairly close to those of ordinary chondrites for most elemental ratios: Mg/Si (0.753 vs. 0.80–0.82); Al/Si (0.069 vs. 0.062–0.066); Ca/Si (0.060 vs. 0.071–0.072); Fe/Si (1.678 vs. 1.034–1.604); Cr/Fe (0.022 vs. 0.013–0.019); Mn/Fe

(0.005 vs. 0.009–0.014); Ni/Fe (0.11 vs. 0.05–0.06). The principal exception is Eros's S/Si ratio (0.005±0.008), which is far lower than that of ordinary chondrites (0.11–0.12). [Eros data are from Lim and Nittler (2009) and Foley et al. (2006); ordinary-chondrite data are for mean H, L, and LL falls from Jarosewich (1990).] Although Eros was deemed to be of OC composition, its composition is also near the range of primitive achondrites.

The low S/Si ratio at the surface of Eros was attributed to space weathering of an ordinary chondrite via the impact of solar-wind ions and micrometeorites (e.g., McCoy et al. 2001). This was consistent with laboratory experiments that simulated space weathering of troilite involving 4 keV ions and nanosecond laser pulses (Loeffler et al. 2008). Rubin (2002b) suggested that S may have been lost from the surface of Eros from impact-induced S volatilization, a process that apparently occurred in some ordinary-chondrite impact-melt breccias. But the absence of evidence for depletion of H and K on Eros seems inconsistent with extensive volatilization and instead favors space weathering.

The inferred mineralogy of Eros is that of ordinary chondrites – major olivine, orthopyroxene, and plagioclase, and minor Ca-pyroxene, metallic Fe-Ni, and Fe-sulfide (McCoy et al. 2001). The bulk density of Eros (2.64 g cm^{-3}) is appreciably less than that of ordinary-chondrite falls (~3.4), indicating that, if Eros is indeed an OC parent body, it must have appreciable porosity.

9.6.4 Itokawa

25143 Itokawa (S-type; 0.535 × 0.294 × 0.209 km) is a small NEA visited by the *Hayabusa* spacecraft in 2005. The object is shaped like a Valencia peanut or a dozing sea otter and appears to be a contact binary.[6] The low bulk density of Itokawa (1.9 g cm^{-3}) suggests that the asteroid is a rubble pile (with a total porosity of ~44 vol%), consistent with its boulder-strewn surface.

Hayabusa captured dusty debris dislodged from the surface of Itokawa, and it brought to Earth a few thousand particles (typically 3–50 μm in size) with a total mass of <1 mg. As reported by Nakamura et al. (2011), phases include olivine ($Fa_{28.6}$), orthopyroxene ($Fs_{23.1}Wo_{1.8}$), rare lamellar low-Ca clinopyroxene ($Fs_{11.3}Wo_{0.5}$; $Fs_{20.4}Wo_{0.6}$), Ca-pyroxene ($Fs_{8.9}Wo_{43.5}$), plagioclase ($Ab_{83.9}Or_{5.5}$), albitic glass (resembling chondrule mesostasis) with embedded diopside and troilite, K-feldspar, troilite, chromite, Ca-phosphate (chlorapatite and merrillite), kamacite (with 3.8–4.2 wt% Ni and 9.4–9.9 wt% Co), and taenite (with 42–52 wt% Ni and 2.0–2.5 wt% Co). Harries and Langenhorst (2018) reported carbide-metallic-Fe-Ni assemblages similar to those described by Krot et al. (1997b) from type-3 ordinary chondrites. Different sets of collected Itokawa particles differ slightly in composition. The mineralogy matches that of ordinary chondrites and the mineral chemistry is typical of equilibrated LL chondrites (e.g., $Fa_{26–32}$ olivine; $Fs_{22–26}$ orthopyroxene; kamacite with ~5 wt% Ni and 1.42–37.0 wt% Co) with a minor component of unequilibrated LL3 material. The O-isotopic compositions of Itokawa particles ($\Delta^{17}O$ = 1.46‰ for olivine, 1.57‰ for orthopyroxene, 1.15‰ for plagioclase; Yurimoto et al. 2011) are also consistent with those of equilibrated

[6] Some binary NEAs are not in contact but orbit around their mutual center of mass (i.e., the barycenter). One such pair is 2017 YE_5, which consists of two ~900-m-wide bodies circling the barycenter every 20–24 hours.

LL chondrites ($\Delta^{17}O$ = 1.26‰). These compositional correspondences are consistent with the LL-chondrite-like near-infrared reflectance spectrum of Itokawa. Nevertheless, several particles contain organic matter similar to that in CI, CM, and CR carbonaceous chondrites.

Two-pyroxene geothermometry of Itokawa particles yields peak metamorphic temperatures of 783–837°C, somewhat lower than those estimated for LL6 chondrites (900–960°C), appreciably higher than those of type 3.7–3.8 OC (525–600°C) and in the ballpark for type-5 OC. U-Pb dating of phosphate particles yields an age of 4.64±0.18 Ga (presumably dating thermal metamorphism on Itokawa and matching the age of typical LL5 and LL6 chondrites); another shock event appears to have occurred 1.51±0.85 Ga ago (Terada et al. 2018). Many of the rocks (presumably LL chondrites) evident in high-resolution images of Itokawa's surface are the same size as LL-chondrite specimens (e.g., LL5 Paragould; 1.0 × 0.61 × 0.41 m) preserved in meteorite collections.

There is evidence of aqueous alteration. Burgess and Stroud (2020) reported a grain (RB-CV-0038) containing pyrrhotite and olivine; a ~2 μm grain of cubanite ($CuFe_2S_3$) occurs at their interface. (Cubanite has not been observed previously in ordinary-chondrite materials.) A fracture in the cubanite grain is lined with magnetite; also present are metallic Cu nanoparticles.

The Itokawa grains are enriched in solar-wind–implanted noble gases, indicating derivation from a regolith. Nanophase iron particles and Fe-sulfide particles within a thin layer at the surfaces of some grains were produced by space weathering, a process that accounts for the redder spectrum of Itokawa compared to LL chondrites. Space weathering is also responsible for particles with amorphous rims and vesicular rims.

Itokawa is not the original parent body of LL chondrites; it is a fragment of a substantially larger asteroid. Radiometric dating of shocked LL chondrites indicates that their parent asteroid suffered several energetic collisions between ~2 and ~1 Ga ago.

9.6.5 Ryugu

162173 Ryugu (Cb-type; equatorial diameter ~1.0 km; polar diameter ~0.88 km) is an NEA (shaped like a diamond or spinning top) that was visited by the *Hayabusa 2* spacecraft in 2018–2019. The asteroid has a rotation period of 7.63 hours and takes 1.30 years to orbit the Sun. Material acquired from the asteroid (including samples from a 10-m-diameter crater it created) was brought to Earth in December 2020 (Tsuda et al. 2019).

Ryugu is a rubble pile – rocks, pebbles, and boulders are strewn across the rough surface; the largest boulder (dubbed "Otohime"), located near the south pole, is ~160 m in size. Thermal imaging suggests the boulders are relatively porous and that the asteroid surface is littered with porous rock fragments >10 cm in size. There appear to be more impact craters near the equator than the poles and more craters on the eastern side of the asteroid than the western. No fine-grained regolith was observed. The bulk density is 1.19 g cm^{-3}; the total porosity is between ~50 and ~60 vol%. With an albedo of about 0.045, Ryugu is one the darkest objects in the Solar System. Although the asteroid displays little variation in color over the surface, the equator tends to be brighter and bluer than other regions.

Ryugu has a flat, nearly featureless spectrum with a slight reddish (positive) slope. A shallow absorption feature at 2.72 μm is essentially uniform across the surface; it appears to be caused by Mg phyllosilicates but is shallower than that observed in CM chondrites. Ryugu's spectral features resemble those of shocked and/or heated CI and CM chondrites; however, they are not identical to that of any known meteorite. The MASCOT lander showed some rocks to consist of millimeter-sized light-colored inclusions set in a dark matrix; the texture resembles that of CAI-bearing CM chondrites.

9.6.6 Bennu

101955 Bennu (B-type; mean diameter 0.490 km) is a top-shaped[7] NEA reached by the *OSIRIS-REx* spacecraft in December 2018; Bennu samples are scheduled to be brought to Earth in 2023. With a density of 1.19 g cm^{-3} and a total porosity of 50–60 vol%, Bennu is a geomorphologically diverse rubble pile (Figure 9.12). It has a rough surface with rocks ranging in size from pebbles to boulders up to ~100 m long; the largest boulders are partly buried. Also present are diffuse regions of dark material that are likely to be dust deposits (particulate regolith).

Figure 9.12 Four views of asteroid Bennu along with a corresponding global mosaic taken on December 2, 2018, by the *OSIRIS-REx* spacecraft's PolyCam camera. Source: NASA/Goddard/University of Arizona.

[7] Other asteroids shaped like a spinning top include 162173 Ryugu, (341843) 2008 EV5, 66391 Moshup, and 65803 Didymos. Such shapes appear to be a natural consequence of catastrophic disruptions of larger precursor bodies.

Bennu's spectrum shows an absorption feature at 2.74 μm but is otherwise featureless with a blue (negative) slope in the visible and near-infrared. Spectral features are rather uniform across the body's surface. The ~2.7 μm absorption feature is similar to that in CI and CM chondrites and is attributable to structural hydroxyl ions in phyllosilicates of the kaolinite-serpentine group (Hamilton et al. 2019). [CI and CM chondrites contain, respectively, ~11 and ~9 wt% H_2O^+ (indigenous water) bound in phyllosilicates.] There are also carbonate spectral features with absorptions at ~3.31–3.35 μm and 3.45–3.5 μm, suggesting compositions ranging from calcite to dolomite to magnesite. Based on a shallow absorption feature at ~0.55 μm, Bennu appears to contain at least a few percent magnetite. (Although magnetite constitutes ~10 vol% of CI chondrites, it is uncommon in CM chondrites, present mainly as 2–4-μm-thick rinds around rare metallic Fe-Ni grains in samples that experienced high degrees of aqueous alteration.) The global average albedo of Bennu is 4.4% at 0.55 μm, consistent with the reflectances of CI and CM chondrites. Despite these similarities to CI and CM chondrites, some of the boulders on Bennu are bright and probably unrelated to these carbonaceous-chondrite groups.

OSIRIS-REx observed 11 separate episodes of dust-plume discharge over a 6-week period in early 2019. Particles left the surface at speeds ranging from a few centimeters per second to ~3 m s^{-1}. Some particles escaped into space; others were trapped in orbit around Bennu. Dust discharges could have been triggered by the eruption of subsurface volatiles heated by the Sun. Other possible triggering mechanisms include thermal stress, rotational instability, and impacts. Other active asteroids with dust tails include 3200 Phaethon (the parent body of the Geminid meteors) and 6478 Gault.

9.6.7 Psyche

16 Psyche (M-type; mean diameter 225 km) is the target of the *Psyche* orbiter spacecraft scheduled for launch in 2022 and arrival at Psyche four years later. Radar observations indicate a metallic Fe-Ni composition, consistent with the asteroid's relatively high bulk density (3.7 g cm^{-3})[8] and very high radar albedo (0.37). Remote sensing at near-infrared wavelengths has revealed the presence of ~6 percent low-FeO orthopyroxene. A 3-μm absorption feature on Psyche suggests the presence of hydroxyl. Many researchers have modeled Psyche as an exposed metallic Fe-Ni core (or core fragment) derived from a large differentiated asteroid that was catastrophically disrupted. The orthopyroxene may represent minor adhering material from the lower mantle of the precursor body; the hydroxyl may indicate the presence of carbonaceous-chondrite debris derived from late-stage projectiles that impacted the surface of the exposed core.

[8] Although a bulk density of 3.7 g cm^{-3} is very high relative to asteroids in general and is similar to that of other M-type asteroids, it is far less than that of metallic Fe-Ni (~8 g cm^{-3}). This probably indicates that Psyche has appreciable macroporosity, perhaps a consequence of the collisional disruption of its parent body. Psyche may also contain minor amounts of lower-density phases such as troilite (4.7 g cm^{-3}) and graphite (2.2 g cm^{-3}).

9.7 Titan

Titan is Saturn's largest moon (*d* = 5,149 km) (Figure 9.13) and the second largest moon in the Solar System (after Jupiter's Ganymede, *d* = 5,268). It is larger than Mercury (*d* = 4,880 km), but only 41 percent as massive. Titan is unique among moons in having a dense, smoggy atmosphere (19 percent more massive than Earth's atmosphere and twice as thick). It is composed principally of nitrogen (97 percent) with minor amounts of methane (2.7 percent), hydrogen (0.1–0.2 percent), and trace amounts of hydrocarbons (e.g., ethane, acetylene, propane, and acetonitrile) and other gases (e.g., hydrogen cyanide, cyanogen, carbon dioxide, carbon monoxide, argon, and helium) (e.g., Lorenz and Mitton 2008). Polycyclic aromatic hydrocarbons (PAHs) reside in the upper atmosphere. A new molecular mineral phase – a co-crystal between acetylene (C_2H_2) and acetonitrile (CH_3CN) – may be moderately abundant on Titan (Cable et al. 2020).

Titan has a relatively young surface (0.1–1 Ga old). Surface features include methane lakes and seas, river channels, deltas, valleys, mountains, steep canyons, longitudinal dunes, and a few impact craters. Regions of bright and dark terrain extend thousands of kilometers. Some mountains have been interpreted as cryovolcanoes.

In 2005, the *Huygens* probe landed on Titan atop a dark plain. Scattered about were rocks and subrounded pebbles consisting of water ice coated with hydrocarbons. This ice phase (ice I_h) is the same hexagonal crystal form of ice found on Earth, but at the surface temperature of Titan (−179.3°C), it is slightly denser than ice at 0°C (0.934 vs. 0.917 g cm^{-3}). Data from the

Figure 9.13 Six infrared images of Titan created from data acquired by the Visual and Infrared Mapping Spectrometer instrument on board NASA's *Cassini* spacecraft. Source: NASA/JPL-Caltech/Stéphane Le Mouélic, University of Nantes; Virginia Pasek, University of Arizona. (A black-and-white version of this figure will appear in some formats. For the colour version, please refer to the plate section.)

Cassini spacecraft (mother ship of the *Huygens* lander) revealed a 6,300-km-long belt of water ice traversing Titan's equator; this feature could be related to massive tectonism. JPL scientists used the *Cassini* data to produce a global map of Titan: dunes are concentrated near the equator, lakes near the poles, and plains at mid-latitudes.

Calculations suggest that there is a liquid ammonia-rich water layer beneath a shell of ice I_h close to the surface. High-pressure polymorphs of water ice (ice V – monoclinic and ice VI – tetragonal) probably occur deep within the mantle. At the center of Titan is a large silicate core.

9.8 Comets and Kuiper Belt Objects

9.8.1 Comet Spectroscopy

To ancient observers, comets were unique celestial bodies. Unlike the Sun, Moon, planets, and stars, comets appeared out of nowhere in all parts of the sky, not just within the Zodiac (i.e., close to the ecliptic) (Figure 9.14). Many changed shape every few days and all faded away after

(a)

Figure 9.14 (a) A Chinese 5-yuan coin commemorating the discovery of comets more than 4,000 years ago. (b) Silver denarius of Augustus struck 19 BCE–18 BCE in Lusitania. It shows the head of Augustus, oak-wreathed, facing left (obverse) and a stylized eight-rayed comet (reverse) representing the ascent of Julius Caesar's soul into heaven. The inscription on the reverse means "Divine Julius." According to the Roman author Pliny the Elder, a comet appeared during athletic games organized shortly after Julius Caesar's assassination in the Theater of Pompey on March 15, 44 BCE. Image 1944.100.39033 courtesy of the American Numismatic Society. (A black-and-white version of this figure will appear in some formats. For the colour version, please refer to the plate section.)

a few weeks or months. [Supernovae also appeared (rarely) in all regions of the sky and vanished after a few weeks; they changed in brightness but not in shape. Ancient Chinese astronomers thought supernovae were related to comets and classified them all as different types of "guest stars."] Aristotle (384–322 BCE) regarded comets as atmospheric phenomena, originating from dry hot exhalations from fissures in the Earth, bursting into flame in the sublunary sphere. His contemporary, Apollonius of Myndus, disagreed, believing comets to be independent bodies orbiting the Sun. In one of his writings, Seneca (4 BCE–65 CE) ranked comets among "Nature's permanent creations." The Greek Stoic philosopher, Posidonius of Apameia (c. 135–51 BCE), observed a comet close to the Sun during a total solar eclipse and realized that there were more comets in the sky than were actually recorded because some were lost in the Sun's glare.

Many comets have two distinct tails: a plasma or ion tail (type I) and a dust tail (type II). Plasma tails tend to be straight with fine structural features and are capable of changing shape over the course of a few nights. Dust tails are commonly curved and relatively featureless; they change shape only slowly. By the 1530s, it was recognized that comet tails always point away from the Sun. In the nineteenth and early twentieth centuries, observers attributed the changing form of the tail (e.g., from cones to rays to fans) to fine particles reacting to radiation pressure, electrical repulsion, or bombardment by particles violently ejected by the Sun. German astronomer Ludwig Biermann suggested in 1951 that solar corpuscular radiation pushed the comets' tails in the antisolar direction. This radiation, now known as the solar wind, consists of protons, electrons, and alpha particles released from the Sun's corona due to the coronal magnetic field. The solar wind also carries the interplanetary magnetic field. When first released, solar-wind particles move slowly away from the Sun, but they quickly pick up speed. By the time they are 2–3 million kilometers from the Sun, the particles are moving at supersonic speeds (250–750 km s^{-1}).

Beginning in the 1860s, the spectroscopy of comets showed the nucleus to have a continuous spectrum with superimposed emission lines of C_2, CN, C_3, CH, and Na. In 1950, American astronomer Fred Whipple introduced the idea that comets are icy conglomerates. They were later dubbed "dirty snowballs." Whipple concluded that the observed compounds in cometary comas (CH, CH^+, CH_2, CO, CO^+, C_2, CN, NH, NH_2, OH, OH^+) formed via photodissociation of ices (H_2O, NH_3, CH_4, CO_2 or CO, C_2N_2) in cometary nuclei. He inferred that the ices coexisted with nonvolatile insulating particles that contain Na, Fe, Ni, and Cr. Modern observers found that, in many cases, ices are less abundant than nonvolatile material and have recharacterized comets as "icy dirtballs."

Spectroscopic observations of comets provided the first method of determining their mineralogy. A list of observed species in comet spectra (Wyckoff 1982) include neutral atoms, ionized atoms, neutral molecules and ionized molecules: H, O, C, C^+, S, Na, K, Ca^+, Cr, Mn, Fe, Ni, Cu, C_2, $^{12}C^{13}C$, C_3, CH, CH_3CN, CN, CO, CS, HCN, NH, NH_2, OH, CH^+, CN^+, CO^+, $CO^+{}_2$, H_2O^+, $N_2{}^+$, OH^+, and silicate (10 μm band). These species were formed by expulsion and photoionization of a variety of ices, silicates, and possibly, sulfides and metals from comet nuclei.

Encke's Comet has an orbital period of 3.3 years and has been observed many times. By 1838, it had become clear that the comet's period was *decreasing*, typically by ~0.1 day per orbit. Other comets were later found to have *increasing* periods. These orbital changes were ascribed to unknown nongravitational forces. The icy conglomerate model could account for such forces: as volatile material is expelled when the nucleus is heated, the comet is pushed in the opposite direction (thanks to Newton's third law of motion). This extra shove can alter the comet's orbital period. Such a "rocket effect" was initially proposed by German astronomer Friedrich Bessel in 1835 after observing Halley's Comet develop a sunward spike of material.

By the 1860s and 1870s, Italian astronomer Giovanni Schiaparelli (who, in 1877, famously reported observing *canali* on Mars) had connected comets to meteor showers after noticing that the Perseid and Leonid showers occurred when the Earth intersected the orbits of two particular comets – 109P/Swift-Tuttle and 55P/Tempel-Tuttle. Debris ejected at low relative velocities from a comet tend to spread out along the comet's orbital path. When the Earth crosses this path, the cometary dust particles enter the atmosphere and create a meteor shower.

Spectroscopy of meteors thus provides another way to investigate the mineralogy of comets. Observed spectral species include neutral atoms (Na, Mg, Fe, Cr, Ca, Mn, Al, Ni, Ti, H, O, N, Sr, Ba, Si); ionized atoms (Ca^+, Mg^+, Si^+, Fe^+, N^+, O^+, Sr^+, Ba^+); and molecules (N_2, FeO, MgO, CaO, CN, C_2), consistent with a source of ices, silicates, sulfides, and metals. Trigo-Rodriguez et al. (2003) analyzed the spectral characteristics of 13 fireballs observed by fixed cameras equipped with prisms or diffraction gratings. They found that the Si-normalized abundances of Mg, Fe, Ni, Cr, Mn, and Co in the fireball spectra resembled (but did exactly match) those of CI and CM chondrites as well as interplanetary dust particles.

In recent years there has been a series of close flyby missions of comet nuclei. Particles from 1P/Halley were found to consist of silicates and organic components rich in C, H, O, and N. Comet 19P/Borrelly showed no evidence of water ice and appears to be depleted in carbon compounds relative to Halley, although their abundances of NH_2 are similar. Comet 103P/Hartley contains water ice and CO_2 ice; unlike some other comets, its D/H ratio is similar to that of Earth.

The *Deep Impact* mission dropped a 370-kg impactor atop Comet 9P/Tempel at 10.2 km s^{-1} and observed the collisionally jettisoned debris. The debris had a high dust/gas ratio. Ejected dust contained amorphous olivine, crystalline olivine, amorphous pyroxene, amorphous carbon, carbonates, phyllosilicates, water ice, sulfide, and PAHs.

Comet 67P/Churyumov-Gerasimenko was visited by the *Rosetta* spacecraft in 2014 (Figure 9.15); *Rosetta* orbited the comet (probably a contact binary), took numerous photographs and eventually deployed a lander. However, the Philae lander settled in a shadowed region of the nucleus under a cliff and could not recharge its solar-powered batteries (which ran out of power after two days). The very low albedo of the comet (0.06) indicates that there is very little ice on the surface. The non-icy fraction of the dust contains ~55 wt% minerals and ~45 wt % organic compounds. An absorption feature at 3.2 μm appears to be from nitrogen-containing ammonium salts mixed with the dark surface material (Poch et al. 2020). The bulk chemical composition of the dust particles is roughly chondritic except for having a much higher abundance (more than five times higher) of C (attributable to high-molecular weight organic material).

Rosetta found that the comet changed color as it crossed inside the snow line on its approach to the Sun (Filacchione et al., 2020). Ice at the surface began to sublimate; this gaseous activity carried submicrometer-sized reddish dust grains composed of carbonaceous organic material away from the nucleus and into the coma. More-pristine, bluer ice layers were exposed on the

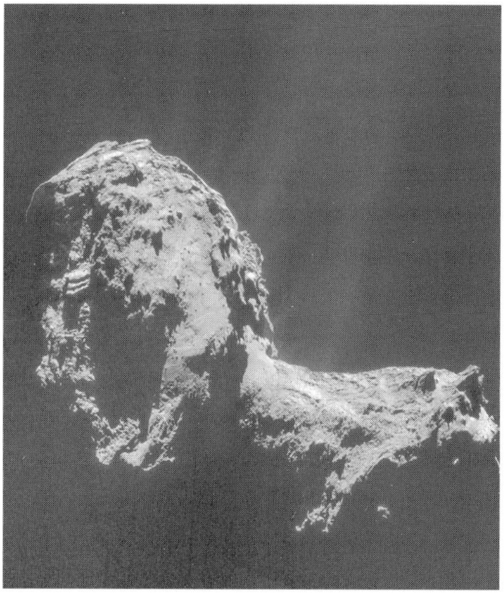

Figure 9.15 Comet 67P/Churyumov-Gerasimenko. Mosaic of four images taken by *Rosetta's* navigation camera on November 20, 2014. Photojournal: PIA18899. Source: ESA/Rosetta/NAVCAM.

nucleus. When the comet left the inner Solar System, reddish dust was redeposited on the nucleus while the coma became less red.

9.8.2 Stardust

Most of the information we have on comet mineralogy comes from the samples returned from Comet 81P/Wild 2 by the *Stardust* mission. This is an active 4.5-km-diameter comet; at least 20 kilometer-long jets were observed during the 2004 flyby. Collected particles range from <1 μm to >100 μm; larger particles were collected in silica aerogel, smaller ones lined submicrometer craters on the aluminum surfaces of the spacecraft. The particles include fragments of low-FeO and high-FeO chondrules (containing olivine, low-Ca pyroxene, pigeonite, Cr-spinel, kamacite blebs, and glassy mesostasis) and CAIs (containing fassaite, spinel, anorthite, V-rich osbornite, anorthite, gehlenite, and diopside). Also present among *Stardust* particles are magnetite, glass, metallic Fe-Ni, fayalite-tridymite assemblages, Mn-rich LIME forsterite, ^{16}O-rich forsterite, Pt-group nuggets, kosmochloric Ca-pyroxene, schreibersite, merrillite, whitlockite, and a few presolar grains (e.g., Brownlee 2014). Sulfides include cubanite, pyrrhotite-pentlandite assemblages, and pyrrhotite-sphalerite assemblages that resemble those in CI chondrites. The Wild-2 minerals are not in equilibrium. About half of the components (e.g., the CAI fragments) were probably derived from the inner Solar System and were carried by winds to the location where Comet 81P/Wild formed. The presence of magnetite attests that some aqueous alteration took place on the comet (Hicks et al. 2017), but the occurrence of anhydrous silicates, the rarity of carbonates, and the apparent absence of phyllosilicates indicate that this was only a minor process.

9.8.3 Interplanetary Dust Particles

Some of the debris shed by comets and ejected from asteroids reaches Earth. Particles 10 μm in size decelerate high in the atmosphere (at ~100 km) where the ram pressure is too low to cause particle disruption. (Ram pressure causes a drag force to be exerted on a body moving through a fluid medium.) Particles captured by high-altitude aircraft and balloons or collected from melted polar ice are known as interplanetary dust particles (IDPs). There are different kinds of IDPs. Some of the so-called chondritic porous (CP) IDPs have high atmospheric entry speeds consistent with comet-like orbits. These IDPs are heterogeneous, carbon-rich (averaging 13 wt% C), highly porous (many with porosities >30%), and composed of submicrometer-sized anhydrous minerals; hydrous silicates are absent. On the other hand, some porous IDPs are hydrous and are likely derived from asteroids.

The CP IDPs were described by Bradley (2005). Silicates include olivine and low-Ca pyroxene (commonly forsterite and enstatite, although some grains contain up to 30 mol.% Fe). The main sulfide phase is pyrrhotite (with up to ~20 atom% Ni); rare grains of troilite, pentlandite, sphalerite, and millerite also occur. There are grains of metallic Fe-Ni and Fe- and Ni-rich carbides, some of which have rims containing graphite. Also present in the CP IDPs are silicate glasses and the 0.1–0.5-μm-sized, mineralogically unequilibrated, spheroidal polycrystalline aggregates known as GEMS (Glass with Embedded Metal and Sulfides). The other major component in the CP IDPs is carbonaceous material containing PAHs; many examples of this material are vesiculated. Some varieties of CP IDPs contain nanodiamonds; at least some of these nanodiamonds are presolar (as indicated by their Xe and Te isotopes).

Nittler et al. (2019) described an unusual 100-μm-sized clast, possibly a "cometary building block," in the mildly altered LAP 02342 CR2 chondrite. The clast contains >50 wt% organic carbon, GEMS, abundant presolar silicate grains, and anomalously ^{16}O-poor Na sulfate. Its chemical and isotopic characteristics resemble those in CP IDPs and UCAMMs (ultracarbonaceous Antarctic micrometeorites) that have been inferred to be from comets. The clast may have formed in the Kuiper Belt region of the nebula and drifted into the zone where CR chondrites accreted.

9.8.4 Kuiper Belt Objects

Short-period comets (those with periods ≤200 years) are scattered icy planetesimals from the Kuiper Belt that enter the inner Solar System. There are two types of short-period comets: (1) Jupiter family comets ($P \leq 20$ years), which have prograde orbits and low inclinations relative to the ecliptic (averaging 11.5°) and (2) Halley family comets ($20 < P \leq 200$ years), whose members have either prograde or retrograde orbits and generally higher inclinations (but still <70°). Long-period comets ($P > 200$ years) have near-parabolic orbits and are perturbed toward the Sun from the Oort Cloud.

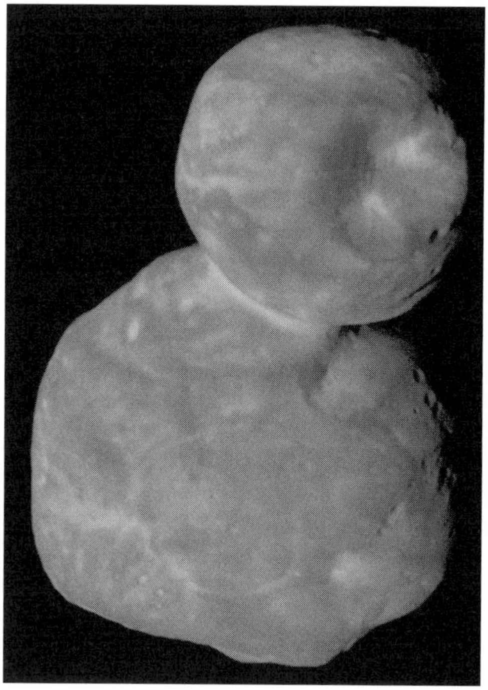

Figure 9.16 Kuiper Belt Object 486958 Arrokoth. This composite image was compiled from data obtained by NASA's *New Horizons* spacecraft in a flyby on January 1, 2019. Credits: NASA/Johns Hopkins University Applied Physics Laboratory/Southwest Research Institute/Roman Tkachenko. (A black-and-white version of this figure will appear in some formats. For the colour version, please refer to the plate section.)

The Oort Cloud is modeled as consisting of two regions: (1) a disk-shaped inner Oort Cloud (a.k.a. Hills cloud) with an inner edge at 250–5,000 AU and an outer edge at 20,000–30,000 AU and (2) a spherical outer Oort Cloud perhaps extending to 200,000 AU. Two objects (sednoids) have been identified in the inner Oort Cloud: Sedna (~995 km in diameter) with a semi-major axis of 506.8 AU, and 2012 VP_{113} (450–700 km in diameter) with a semi-major axis of 257.67 AU. Although no bodies in the outer Oort Cloud have yet been observed, there may be 10^{12}–10^{13} objects residing within it. Some astrophysicists have proposed that Oort Clouds around other stars might be detected by studying maps of the cosmic microwave background.

In contrast to the Oort Cloud, several thousand Kuiper Belt objects (KBOs) have been observed telescopically, starting with the discovery of Pluto by Clyde Tombaugh in 1930, its largest moon Charon by James Christy in 1978, and 15760 Albion (provisionally designated 1992 QB_1) (which is not part of the Pluto system) by Jewitt and Luu (1993). Recently, a 2.6-km-diameter body was found in the Kuiper Belt from a stellar occultation event (Arimatsu et al., 2019); this small object is currently impossible to observe telescopically. The KBOs are volatile rich and would develop comas like cometary nuclei if they entered the inner Solar System.

The *New Horizons* spacecraft flew by the Pluto system in 2015 and (486958) 2014 MU_{69} (nicknamed Ultima Thule and formally designated Arrokoth) in 2019 (Figure 9.16). Pluto's surface is covered mostly with molecular nitrogen ice along with small amounts of methane ice

Figure 9.17 High-resolution enhanced color view of Pluto taken on July 14, 2015, by the Ralph/Multispectral Visual Imaging Camera (MVIC) on board the *New Horizons* spacecraft. Credit: NASA/JHUAPL/SwRI (A black-and-white version of this figure will appear in some formats. For the colour version, please refer to the plate section.)

and carbon monoxide ice; water-ice bedrock is visible in many regions (Figure 9.17). Spectral data from the spacecraft showed absorption bands at 1.65 and 2.2 μm, characteristic of ammonia. Because ammoniated ices are destroyed by irradiation over geologically short time periods, it is likely that the ammonia was deposited recently, perhaps by cryovolcanic activity. The internal structure of Pluto likely consists of a rocky core and a water-ice mantle; it is possible there is a subsurface global liquid water ocean on Pluto. The surface of Charon (Pluto's largest moon) appears to be mostly crystalline water ice with small regions of ammonia hydrates; it also probably has a rocky core and water-ice mantle.

Arrokoth was discovered by the Hubble Space Telescope in 2014; it is a 36-km-long, lightly cratered, contact binary made of two lenticular lobes of ice. At visual wavelengths, the body has a substantially uniform reddish color with a bright neck between the lobes. The brighter color of the neck probably relates to textural differences resulting from the low-speed merger of the precursor planetesimals. Near-infrared spectroscopy shows the surface of Arrokoth to consist largely of methanol ice (CH_3OH) and complex organic compounds with little or no water ice. Nevertheless, amorphous water ice (containing trapped volatile species) may be abundant in the interior.

Other comets that seem to be contact binaries include 1P/Halley, 19P/Borrelly, 103P/Hartley, 67P/Churyumov-Gerasimenko, 8P/Tuttle, 45P/Honda-Mrkos-Pajdušáková, and 209P/LINEAR. These all may have formed as independent objects that merged gently with co-orbiting bodies. The relatively low abundance of small impact craters on Pluto and Charon show that violent collisions were rare in the Kuiper Belt.

The elongated interstellar interloper Oumuamua might also have been a contact binary expelled from another star's Kuiper Belt or Oort Cloud billions of years ago.

Comets were long seen as portents of disaster, signaling imminent famine, floods, war, pestilence, peasant uprisings, palace intrigue, financial hardship, or the death of a great leader.

The most famous example of a comet being associated with a leader's death is represented by the silver denarius of Augustus (63 BCE–14 CE) (Figure 9.14b). The denarius was the standard silver coin of Rome, and the reverse of this issue features a stylized eight-rayed comet with its tail pointing upwards, flanked by the words DIVVS IVLIVS (Divine Julius). It was struck to commemorate the appearance of a comet during athletic games organized by Octavian (later to become Caesar Augustus) shortly after Julius Caesar's assassination in 44 BCE. The comet was interpreted by many Romans as a sign of Caesar's soul ascending to Heaven. (Julius Caesar had maintained that his family descended from the Trojan hero Aeneas, son of Anchises and the goddess Venus.) Caesar's posthumous deification implied that Octavian (Caesar's great-nephew and adopted son and heir) could bolster his own claim of divine descent.

In his 1599 play, *Julius Caesar*, William Shakespeare wrote:

When beggars die, there are no comets seen,
The Heavens themselves blaze forth the death of Princes.

Samuel Johnson (1709–1784), the English writer and harmless-drudge lexicographer, had his own irreverent take:

If at your coming princes disappear,
Comets! Come every day and stay a year.

Some ancient Chinese records interpreted the shape of a comet's tail or a comet's position among the stars as indicators of different kinds of calamitous events. A four-tailed comet signified disease; a three-tailed comet was an omen of political turmoil. A comet's appearance in Virgo portended severe floods and famine; a comet in Orion was a sign of impending usurpation, uprisings, and war. If a comet popped up in the bowl of the Big Dipper, gold and gems would become worthless.

But now, comets are widely viewed as harbingers of a different kind. Because they contain primitive material dating back 4.5 billion years, the study of comets portends a deeper understanding of the early history of the Solar System. Much can be learned from them about the dynamics of accretion and the chemical and mineralogic nature of ancient accreted materials.

10

Formation of Meteoritic Minerals in Gas- and Dust-Rich Environments

Bob Hazen and colleagues have pointed out that as the universe evolved and as stars, planets, and life developed, the number of diverse mineral-forming environments increased (Hazen et al. 2008; Hazen and Morrison 2020; Morrison and Hazen 2020). This resulted in more complex minerals and an ever-increasing set of mineral species. The focus of the following chapters is narrower – we largely restrict the discussion to mechanisms that formed minerals within meteorites.

The basic processes listed in this and the following two chapters are not all mutually distinct: (1) condensation in gaseous envelopes around evolved stars (Section 10.1); (2) condensation in the solar nebula (Section 10.2); (3) crystallization in CAI melts (Section 10.3); (4) crystallization in AOI melts (Section 10.4); (5) crystallization in chondrule melts (Section 10.5); (6) exsolution in opaque assemblages during the cooling of CAIs (Section 10.6); (7) exsolution during the cooling of chondrules (Section 10.7); (8) exsolution during the cooling of opaque assemblages outside CAIs and chondrules (Section 10.8); (9) annealing of amorphous material in the solar nebula (Section 10.9); (10) devitrification of amorphous material on parent bodies (Section 11.1); (11) thermal metamorphism and exsolution (Section 11.2); (12) aqueous alteration, hydrothermal alteration, and metasomatism (Section 11.3); (13) precipitation from asteroidal brines (Section 11.4); (14) shock metamorphism (Section 11.5); (15) space weathering (Section 11.6); (16) condensation within impact plumes (Section 11.7); (17) crystallization from melts in differentiated or partially differentiated bodies (Section 11.8); (18) condensation from late-stage vapors in differentiated bodies (Section 11.9); (19) exsolution, inversion, and subsolidus redox effects within cooling igneous materials (Section 11.10); (20) solar heating near perihelion (Section 11.11); (21) atmospheric passage (Section 12.1); and (22) terrestrial weathering (Section 12.2).

Some meteoritic minerals form by only a single mechanism, e.g., ringwoodite and ahrensite by high-pressure shock metamorphism of olivine; other minerals form by several mechanisms, e.g., olivine by (a) condensation around red giant and asymptotic giant branch (AGB) stars, (b) condensation in the solar nebula, (c) crystallization in CAI and AOI melts, (d) crystallization in chondrule melts, (e) thermal metamorphism, (f) crystallization from impact melts, (g) condensation within impact plumes, (h) crystallization in magmatic bodies on differentiated asteroids, (i) annealing of amorphous material, (j) aqueous alteration, and (k) crystallization in meteoroid fusion crusts during atmospheric passage.

No attempt is made here to describe every mode of formation of every meteoritic mineral. That monumental task would require a multivolume treatment. Nevertheless, this chapter and the following ones (updated from Rubin and Ma 2017) provide an overview of essentially all the ways minerals in meteorites form, although some broad mechanisms (e.g., shock metamorphism, terrestrial weathering) encompass numerous different processes.

Any discussion of mineral formation must be grounded in the proper context, whether petrologic, geologic, or astrophysical. That is the guiding principle here.

10.1 Condensation in Gaseous Envelopes Around Evolved Stars

Cool, dense molecular clouds such as portions of the Eagle Nebula in the constellation Serpens contain nanoscale-sized ice particles formed by condensation from the surrounding gas. Seven species have been confirmed by infrared absorption measurements, infrared spectroscopy, and radio emission spectroscopy (Morrison and Hazen 2020): crystalline cubic water ice, amorphous H_2O, amorphous CO, crystalline CO, amorphous CO_2, amorphous methanol (CH_3OH), and amorphous methane (CH_4). These phases could not long survive the warm ambient temperatures of the inner Solar System; they do not occur in meteorites.

About 1 wt% of the molecular clouds is made up of mineral dust (Herbst 1995); some such dust was swept into the nascent solar nebula. A small fraction was incorporated into agglomerating meteorites as "presolar grains," the oldest meteoritic minerals. These grains formed in the outflows of old, evolved stars and as ejecta from classic novae and supernovae. Presolar grains may have drifted in and out of several interstellar clouds before being incorporated into the particular molecular cloud that spawned the Solar System. Cosmic ray–exposure ages of some of the largest grains show they formed ~7 Ga ago (Heck et al. 2020). These ancient grains must have avoided destruction by supernova shockwaves in the interstellar medium; they may have been parts of even larger grains or embedded within aggregates mantled by ices or organic compounds.

Although diamond is the most abundant presolar mineral species, its origin is enigmatic; this is due in large measure to the small average grain size (~2.6 nm) that has earned the phase the sobriquet "nanodiamond." Aggregates of presolar diamonds contain the noble-gas component Xe-HL, which is enriched in both heavy (H) and light (L) Xe isotopes. However, as pointed out by Zinner (2005), the concentration of Xe-HL is so low and the grains so small that only 0.0001

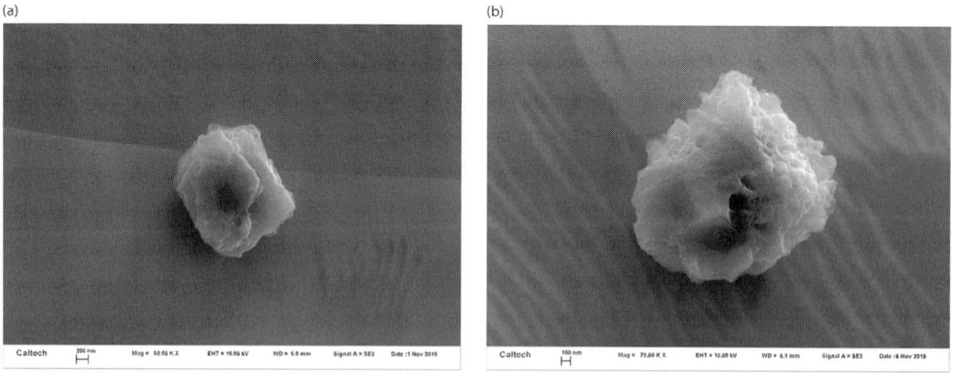

Figure 10.1 Secondary electron images showing presolar moissanite (SiC) grains from the Qingzhen EH3 chondrite.

percent of diamond grains are likely to contain a single Xe atom. The individual diamonds that do contain the Xe-HL component (and, presumably, many of their Xe-free congeners) may have condensed from the expanding gas expelled by supernovae after temperatures dipped below ~1,000 K. However, it is possible that a significant fraction of the nanodiamonds in chondrites are not presolar at all, but originated within the Solar System.

Presolar silicon carbide (moissanite (SiC)) grains range from 0.1 to 20 μm in size (Figure 10.1). Most of the grains probably condensed within the expanding envelopes around

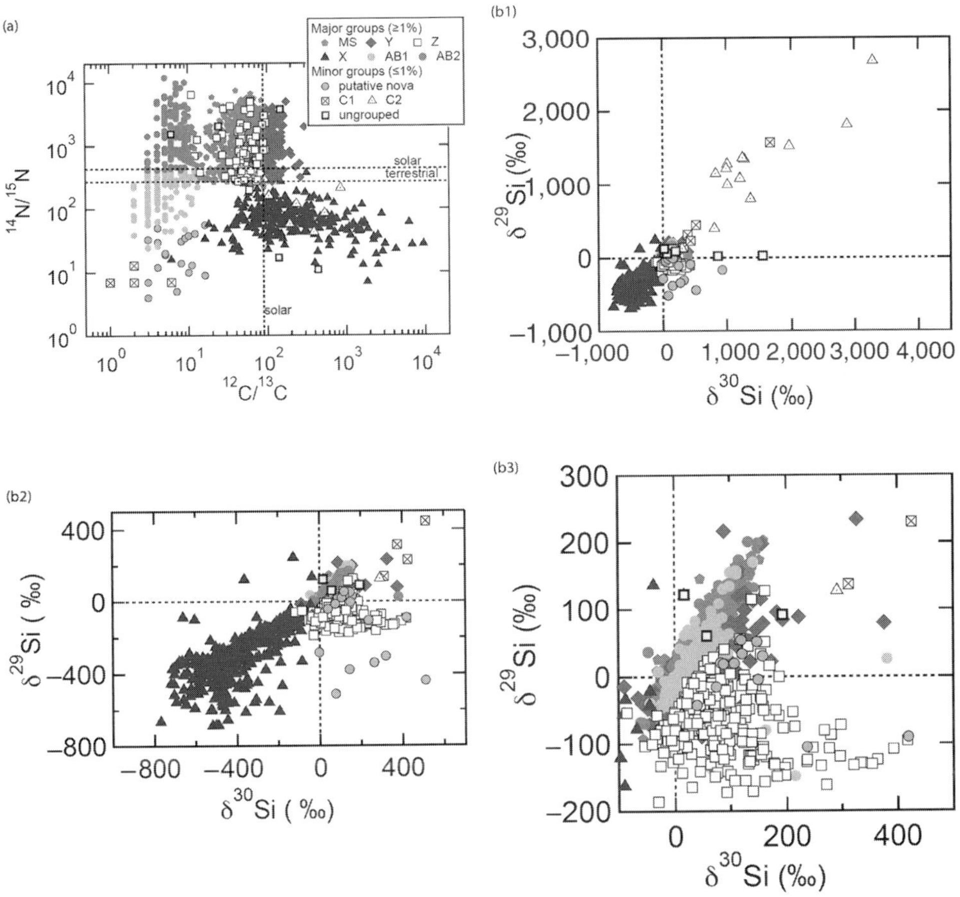

Figure 10.2 (a) Nitrogen and carbon isotopic ratios of individual presolar SiC grains (Zinner 2014 and references therein). See reviews by Zinner (2014) and Nittler and Ciesla (2016) for details. Description of new nomenclatures of AB and C grains can be found in Liu et al. (2016, 2017). (b) Silicon isotopic ratios of different groups of presolar SiC grains plotted in delta-notation. Symbols are the same as those in (a). (c) Sulfur isotopic ratios of different groups of presolar SiC grains plotted in delta-notation. Symbols are the same as those in (a). Some of the AB grains have no available N isotope data and cannot be further classified. Data sources: Gyngard et al. (2006); Hoppe et al. (2012, 2015); Fujiya et al. (2013); Xu et al. (2015); Liu et al. (2016, 2017). Courtesy of Nan Liu, WUSTL. (A black-and-white version of this figure will appear in some formats. For the colour version, please refer to the plate section.)

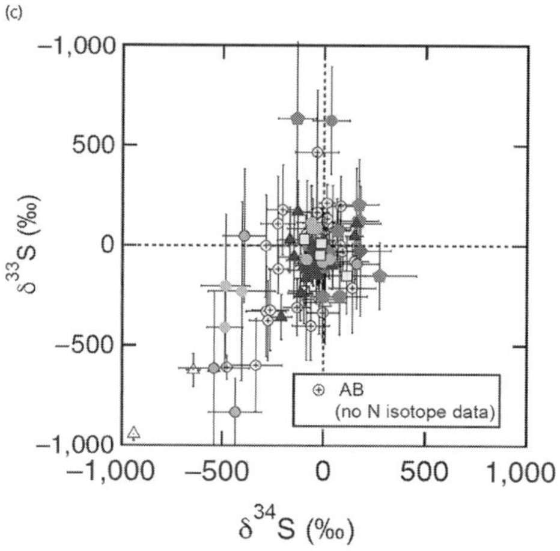

Figure 10.2 (*cont.*)

thermally pulsing, asymptotic giant branch (AGB) stars. Many of these late-type stars may have formed ~7 Ga ago during a period of enhanced star formation. They are now highly luminous red giants that have exhausted the helium fuel in their cores. They have migrated on Hertzsprung-Russell (H-R) diagrams[1] along a track subparallel to the typical red-giant track that they traversed at an earlier stage when they had exhausted their supply of hydrogen. Many SiC grains have non-solar N-, C-, Si-, and S-isotopic compositions (Figure 10.2), consistent with those measured and modeled in AGB stars.

Presolar graphite grains commonly contain inclusions of carbide and metallic Fe that must have condensed before the graphite. One variety of graphite grain that is of low density ($\leq$2.05 g cm^{-3}) has numerous isotopic anomalies consistent with a supernova origin. Typically, such grains are enriched in ^{15}N, ^{26}Al, ^{13}C, ^{18}O, ^{28}Si, and, in some cases, ^{29}Si and ^{30}Si; a few grains are enriched in ^{44}Ti, ^{49}Ti, ^{50}Ti, ^{41}K, and ^{41}Ca. Isotopic anomalies in Mo and Zr have also been reported. High-density graphite grains appear to have been derived from AGB stars and supernovae. Additional graphite grains may have originated in novae.

Isotopic anomalies in O, Al, and Ti in presolar oxide grains (corundum, spinel, hibonite, TiO_2) suggest derivation from red giant and AGB stars.

Rare presolar silicon nitride (nierite (Si_3N_4)) grains are enriched in ^{15}N, ^{28}Si, and ^{26}Al and appear to have originated in supernovae. Silicon nitride grains have also been identified in the circumstellar shells of a few extreme carbon stars. These are AGB stars that achieved a high mass-loss rate (i.e., a superwind) and in which convection dredged up C produced in the He-burning shell.

Presolar silicate grains were probably derived from AGB stars and possibly red giants.

[1] An H-R diagram is a scatter plot of stars wherein the y-axis is absolute magnitude or luminosity and the x-axis is the stellar classification (O, B, A, F, G, K, M) or effective temperature. Main-sequence stars, undergoing nuclear fusion, form a band across the diagram; other regions show concentrations of white dwarf stars, red giants, and supergiants.

10.2 Condensation in the Solar Nebula

It is probable that at least some regions of the solar nebula were hot enough to evaporate preexisting dust and that a significant fraction of nebular materials spent time in such regions. As the entire nebula cooled or as the gas from hot regions passed into cooler zones, minerals began to condense as submicrometer-sized grains. Phases predicted to condense can be modeled by assuming closed-system thermodynamic equilibrium, a fixed pressure (typically 10^{-2}–10^{-6} atm), and an initial starting gas composition (usually solar) (e.g., Ireland and Fegley 2000). Variants of the standard condensation models include those invoking significant dust enrichments (e.g., Ebel and Grossman 2000) and those involving the isolation of some fraction of the early condensates from the residual nebular gas (e.g., Petaev and Wood 1998).

Phases that appear in standard condensation calculations at temperatures between ~1,770 K and 1,100 K include corundum, hibonite, perovskite ($CaTiO_3$), spinel, gehlenite ($Ca_2Al(SiAl)O_7$), åkermanite ($Ca_2MgSi_2O_7$), Ca pyroxene (diopside), forsterite, enstatite, anorthite, albite, and metallic Fe-Ni. Most of these minerals are those most commonly found in Ca-Al-rich inclusions as primary phases; corundum and albite, however, are rare in pristine CAIs. The presence of other refractory minerals such as grossite ($CaAl_4O_7$), krotite ($CaAl_2O_4$), and rhönite ($Ca_2(Mg,Al,Ti)_6(Si,Al)_6O_{20}$)) in some CAIs suggests that, at least locally, there were minor variations in nebular conditions that allowed these phases to become stable. Refractory inclusions with clear condensation signatures include those illustrated in figure 5 of Scott and Krot (2005): e.g., in one CAI from the Adelaide ungrouped carbonaceous chondrite, corundum is replaced by hibonite which is itself corroded by grossite. Common primary phases in CAIs also include submicrometer-sized Pt-group metal nuggets that formed alloys via condensation of refractory siderophile elements at temperatures between ~1,800 and 1,300 K.

The newly identified refractory minerals tistarite (Ti_2O_3 with the corundum-type structure) and kaitianite ($Ti^{3+}{}_2Ti^{4+}O_5$ with the *C2/c* V_3O_5-type structure) in Allende along with corundum, khamrabaevite (TiC), rutile, and mullite grains (Figure 10.3), are probably condensates (formed under highly reducing conditions), sampling different environments and/or epochs of the solar nebula (Ma and Rossman 2009a; Ma 2019; Ma and Beckett 2020).

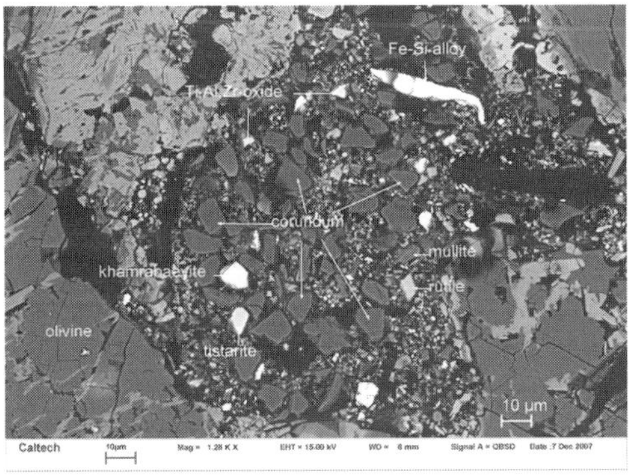

Figure 10.3 Corundum, tistarite, rutile, khamrabaevite, mullite, Fe-Si alloy, and a new Ti,Al,Zr-oxide in a cluster of refractory grains from the Allende CV3 chondrite (Ma and Rossman 2009a; Ma and Beckett 2020). BSE image.

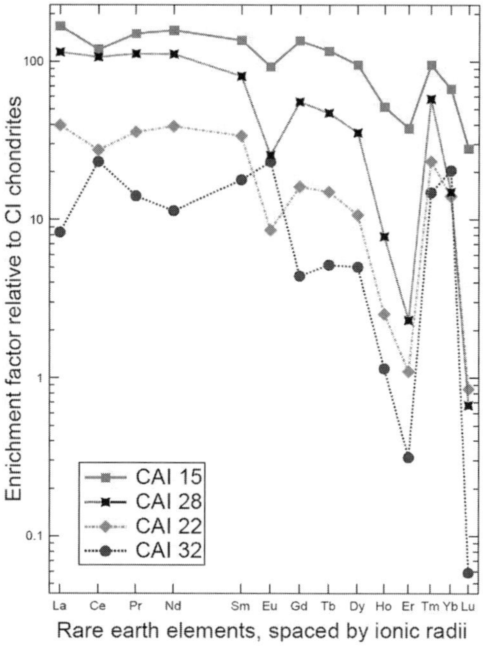

Figure 10.4 Group II REE pattern of CAIs, derived from Davis et al. (2018). Courtesy of Andrew Davis, University of Chicago. (A black-and-white version of this figure will appear in some formats. For the colour version, please refer to the plate section.)

Brearley (1993b) reported rare 0.05–0.3-μm-sized euhedral grains of Ti oxides (Ti_3O_5 with the *C2/m* Ti_3O_5-type structure and the nonstoichiometric Magnéli phases Ti_5O_9 and Ti_8O_{15}; e.g., Andersson and Magnéli 1956) in the matrices of Bells (CM2) and a CI-like carbonaceous-chondrite clast in the Nilpena polymict ureilite. Such phases were previously reported in some IDPs. It seems likely that these Ti oxides were formed by vapor-phase condensation, probably in the solar nebula. Nevertheless, a presolar environment cannot be ruled out.

The rare-earth elements (REE) provide additional evidence for condensation. CAIs can be categorized into different groups based on their bulk REE patterns. The so-called Group II pattern (Figure 10.4) is highly fractionated: light REEs (La through Sm) display a flat pattern, there is a negative Eu anomaly, the intermediate REEs (Gd through Er) are progressively depleted, there is a strong positive Tm anomaly, and Yb has an abundance similar to that of Eu.

Boynton (1975) demonstrated that this fractionated pattern is a function of the volatility of the individual REE; the pattern can be accounted for if the REE-carrier phases within the CAIs (probably initially hibonite grains) are nebular condensates formed by equilibrium gas-solid fractionation. Fine-grained CAIs with Group II patterns are the best candidates for having formed by nebular condensation processes.

As the nebular gas cooled, metallic Fe began to undergo oxidation, producing the fayalite component (Fa; Fe_2SiO_4) of olivine and the ferrosilite component (Fs; $FeSiO_3$) of low-Ca pyroxene. The very high olivine-Fa values in some FeO-rich chondrules in primitive CO3.0 chondrites (up to Fa_{77}) suggest gas temperatures <450 K. Although solar-composition gas is too reducing to produce a substantial Fa component in condensate olivine grains, evaporation of nebular fines during repeated episodes of chondrule formation at low nebular temperatures could have led to the incorporation of significant amounts of H_2O into the subsequently condensed fine-grained materials. This in turn could have resulted in the local oxidation of metallic Fe and the subsequent condensation or crystallization of FeO-rich mafic silicates.

Rubin et al. (1999) inferred that some of the individual troilite grains within chondrules in LL3.00 Semarkona were most likely primary phases (i.e., present among the chondrule precursors) if they fulfilled the following criteria: (1) they were completely embedded in a mafic silicate phenocryst, (2) they were located near the chondrule center, and (3) they were were part of an opaque assemblage with an igneous texture. About 13 percent of chondrules in Semarkona contain such troilite grains. If these grains are indeed primary, their submicrometer precursors must have condensed in the nebula after temperatures dipped below ~650 K (the 50 percent condensation temperature of S).

Tiny grains of noble Pt-group metals, metallic alloys and Ge-rich phases found within R3 lithologies and CAIs in CV chondrites have also been inferred to be condensates.

In the enstatite-chondrite-formation region of the nebula, the nebular gas may have had comparatively high C/O, S/O, and/or high pH_2/pH_2O ratios (e.g., Jacquet et al. 2018). El Goresy et al. (2011) and Lin et al. (2011) described an EL3 clast in the Almahata Sitta breccia that contains metal nodules with fine-grained condensates including sinoite (Si_2N_2O), oldhamite (CaS), graphite, enstatite, diopside, sphalerite, and troilite (Figure 10.5).

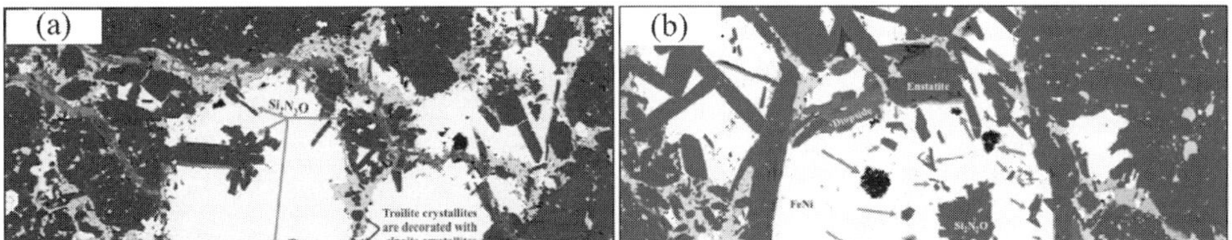

Figure 10.5 BSE images showing EL3 clast in the Almahata Sitta breccia that contains metal nodules with fine-grained condensates including sinoite (Si_2N_2O), oldhamite (CaS), graphite, enstatite, diopside, sphalerite, and troilite (El Goresy et al. 2017). Courtesy of Yangting Lin, CAS-IGG.

It is possible that the most reactive halogens in the enstatite-chondrite region of the nebula condensed as halides that were incorporated into enstatite chondrites. Lawrencite ((Fe^{+2},Ni)Cl_2), a halide identified in enstatite chondrites, occurs as thin rims around silica grains, as inclusions within kamacite and troilite, and as isolated grains associated with enstatite in the matrix of EH4 Indarch.

Although Buchwald and Clarke (1989) called lawrencite a "mineralogical chimera" and found that the Cl-bearing phase in iron meteorites and OC is akaganéite (β-FeO(OH,Cl) – a terrestrial alteration product), the presence of Fe- and Cl-bearing rims around the silica grains in Indarch (an unweathered observed fall) are unlikely to have formed by alteration. Furthermore, Lin et al. (2011) measured excess ^{36}S in small grains of lawrencite within metal nodules in an EL3 clast from Almahata Sitta; the excess ^{36}S formed from live ^{36}Cl ($t_{½}$ = 301,300 years) by electron capture in the solar nebula.

10.3 Crystallization in Ca-Al-Rich Inclusion Melts

Many of the tiny oxide, silicate, metal, and sulfide grains that formed during condensation in the nebula clumped together to form highly porous aggregates (dustballs), probably of millimeter-to-centimeter size. Early dustballs were the progenitors of many CAIs and may have been concentrated in different regions of the nebula by aerodynamic forces.

Compact type A CAIs (CTAs) (Figure 10.6) have igneous textures and spheroidal shapes, presumably due to moderate melting. Although many CTAs are essentially monomineralic melilite, some contain fassaite and rare rhönite. These grains crystallized from the melt, plausibly nucleating on relict condensates or relict phenocryst fragments from a previous generation of CTAs.

The Wark–Lovering rims (Figure 10.7) around CAIs appear to have formed by a multistage process involving flash heating of the CAI surface. This is consistent with the Wark–Lovering rims having ^{16}O enrichments like those of the underlying CAI.

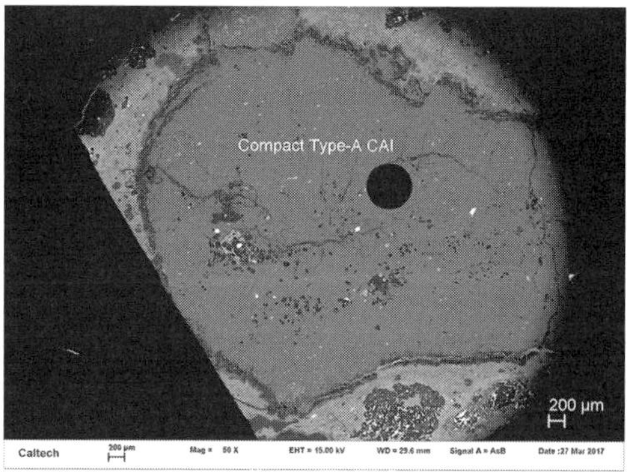

Figure 10.6 Compact Type A CAI (CTA) from the Efremovka CV3 chondrite, showing igneous texture and spheroidal shape. BSE image.

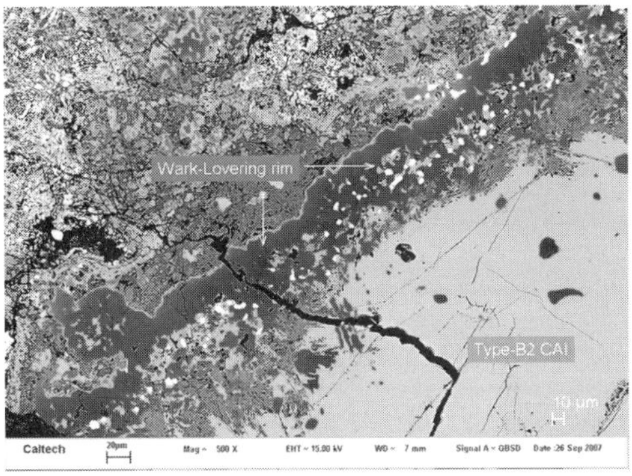

Figure 10.7 The Wark-Lovering rim on a Type B2 CAI from Allende. BSE image.

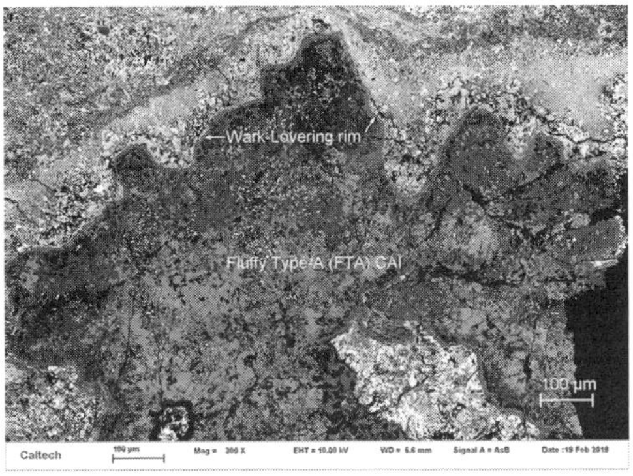

Figure 10.8 Part of a Fluffy Type A (FTA) CAI from Allende. BSE image.

Most of the centimeter-sized Fluffy Type A CAIs (FTAs) (Figure 10.8) are aggregates of numerous melilite-rich nodules that resemble individual CTAs with Wark–Lovering rims. Rubin (2012a) suggested that the individual nodules were independent CTAs (some perhaps associated with a few other CAI types that contained hibonite) that collided with each other at low relative velocity and stuck together. The aggregated CTAs experienced minor remelting. The occurrence of ^{16}O-rich and ^{16}O-poor melilite in one FTA from CV3 Vigarano suggests that this particular inclusion experienced multiple melting events.

Type B CAIs may have formed via clumping of numerous CTAs (the same basic process suggested to have formed the centimeter-sized FTAs), the incorporation of significant amounts of ^{16}O-rich mafic dust (mainly forsterite) into the assemblage, and extensive melting. Those centimeter-sized aggregates of CTAs that happened to have incorporated greater amounts of forsterite prior to melting ultimately became centimeter-sized B3 inclusions.

Because B2 inclusions (Figure 10.9) are mineralogically similar to B1 cores, it seems plausible that the B2s started out as normal B1 inclusions that acquired a melilite-rich mantle

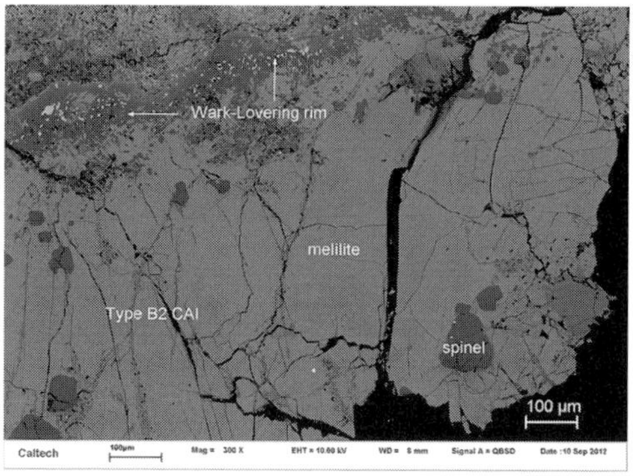

Figure 10.9 A Type B2 CAI with a melilite-rich mantle from Allende. BSE image.

through inelastic collisions with CTAs (which are very rich in melilite). These compound objects were then extensively melted. The additional melilite would have made the B1 inclusions richer in bulk Ca and Al. Relatively slow cooling allowed the minerals to grow coarser in the B1 inclusions; nonequilibrium crystallization from the melt produced compositional zoning. Differences in cooling rate and the degree of melting among different B1 inclusions caused inclusion-to-inclusion variations in mineral composition.

Type C CAIs (which contain spinel, calcic pyroxene, and anorthite, and have igneous textures) may have formed by flash melting finer-grained precursors composed of nebular condensates.

The shapes of the hibonite-silicate spherules attest to their origin as molten droplets formed from finer-grained precursors that may have contained condensates and/or fragments of collisionally disrupted preexisting CAIs.

10.4 Crystallization in Melted Portions of Amoeboid Olivine Inclusions

Weisberg et al. (2004) reported three AOIs attached to the forsterite-bearing Wark–Lovering rim around a CTA inclusion in CR2 EET 92105. This association of AOIs and CAIs is consistent with the Al-diopside, anorthite, spinel, and melilite in the interiors of AOIs being components of fine-grained CAIs. It thus seems plausible that AOIs formed from nebular condensates and fine-grained CAIs that acquired thick forsterite rims and were heated in the nebula by a flash-melting process analogous to the one thought responsible for producing igneous rims and secondary shells around chondrules. In some cases, coarse grains of forsterite and of plagioclase rimmed by calcic pyroxene crystallized near the rim, forming a hypidiomorphic-granular-like texture. Although grains in the interior of many AOIs were not extensively melted, the AOI cores were compressed during high-temperature processing. In a few AOIs, the surficial melt invaded the core, causing olivine crystallization adjacent to relict grains of plagioclase and calcic pyroxene. The AOI olivine grains in primitive chondrites have variable

CaO contents; this may have resulted from variations in the number and intensity of reheating episodes experienced by different AOIs.

10.5 Crystallization in Chondrule Melts

The spheroidal to ellipsoidal shapes of intact ferromagnesian and Al-rich chondrules were molded by surface tension; it is clear these objects formed from partially to completely molten silicate-rich droplets (e.g., King 1983; Hewins et al. 1996; Sears 2004; Russell et al. 2018).

The igneous textures of chondrules also attest to their formation from melts. For example, the most common textural variety is the set of porphyritic chondrules (84 percent of OC chondrules). Almost all of them consist of olivine and/or low-Ca pyroxene phenocrysts embedded in mesostasis. The phenocrysts in these chondrules nucleated on relict crystals and grew from the residual melt. This increased the Ca content of the residual melt sufficiently for Ca pyroxene (Figure 10.10), rather than low-Ca pyroxene, to become the liquidus phase. Granular olivine-pyroxene chondrules (3 percent of OC chondrules) are essentially finer-grained porphyritic chondrules that probably cooled at higher rates from droplets with numerous relict nuclei.

Many low-Ca-pyroxene-bearing FeO-rich (Type II) porphyritic pyroxene (PP) chondrules in type-3 ordinary chondrites contain pyroxene phenocrysts with a series of prominent overgrowth layers that may represent multiple melting/crystallization events (Figure 10.11). Although the phenocrysts have been characterized as exhibiting oscillatory zoning (a phenomenon present in terrestrial, lunar, and martian rocks), the cooling rates and length scales in these magmatic systems in planetary and subplanetary bodies are many orders of magnitude greater than those in chondrules. Detailed analytical studies led Baecker et al. (2017) to conclude that it is more likely that the compositional zoning in the pyroxene phenocrysts from chondrules resulted from multiple overgrowths during a series of melting/crystallization events. This interpretation remains controversial; many chondrule researchers view most chondrules as having formed during single events involving melting and cooling.

Figure 10.10 Portion of a porphyritic olivine chondrule in Semarkona (LL3.00). The mesostasis contains numerous acicular and skeletal Ca-pyroxene crystallites. BSE image.

Figure 10.11 Portion of a porphyritic olivine-pyroxene chondrule showing a low-Ca pyroxene phenocryst with apparent oscillatory zoning. Small Ca-pyroxene grains surround the phenocryst. From Semarkona. BSE image.

There is abundant evidence that most chondrules experienced multiple melting events. This evidence includes relict grains within chondrule phenocrysts (identified by differences in FeO/(FeO+MgO) ratio, O-isotopic composition, or cathodoluminescence), enveloping (i.e., nested) compound chondrules, igneous rims around chondrules, the set of microchondrules formed by remelting low-Ca pyroxene grains at the surfaces of normal-sized chondrules, and the presence of nonspherical, rounded, multilobate chondrules, which, if fully molten, would have collapsed into spheres long before their phenocrysts could have grown to their present sizes.

Barred olivine (BO) chondrules occur in all chondrite groups save CI (which is chondrule free). BO chondrules are rare in enstatite chondrites, but constitute 6 percent of OC chondrules. Melting experiments indicate that these chondrules formed typically by complete or near-complete melting of olivine-rich precursors. The olivine bars within these chondrules and the olivine rims around many of them tended to crystallize as optically continuous single crystals (or as adjacent sets of a few separate optically continuous crystals). Pyroxene is present in most BO chondrules in type-3 OC, averaging ~15–30 vol % and, in rare cases, ranging up to 75 vol%; feldspathic glass averages ~20–25 vol% and ranges up to 55 vol%. During crystallization of olivine, the concentrations of incompatible elements increase in the mesostasis, leading in some cases to the crystallization of opaque phases (metallic Fe-Ni and/or sulfide) between the olivine bars, e.g., in metal-rich chondrites such as NWA 5492.

Radial pyroxene (RP) chondrules consist of fan-like arrays of low-Ca pyroxene laths that, in most cases, appear to radiate from a point on the chondrule surface. This may be the site of a collision between the molten droplet and a dust grain. Cryptocrystalline (C) chondrules are very-fine-grained ($\lesssim$2 µm), normative-pyroxene-rich objects consisting of numerous domains, each of which appears to have crystallized from a quenched melt separate from its neighbors. Both RP and C chondrules have been reported in all chondrite groups except CI and CK; together these two varieties constitute 85 percent of CH3 chondrules, 18 percent of EH3 chondrules, 7 percent of H-L-LL3 chondrules, 3 percent of CO3 chondrules, 1.6 percent of

CM2 chondrules, 1.6 percent of R3 chondrules, 0.7 percent of CR chondrules, and 0.3 percent of CV chondrules.

The sulfide phases in enstatite-chondrite chondrules crystallized during chondrule formation: these include common phases such as troilite, niningerite, oldhamite, and caswellsilverite, and rare phases such as wassonite (TiS) (Figure 7.28). Kimura et al. (2005) suggested that tridymite and cristobalite in non-impact-melted EH3 and EL3 chondrites were the last phases to crystallize within chondrules during cooling.

10.6 Exsolution in Opaque Assemblages during the Cooling of Ca-Al-Rich Inclusions

Experimental studies and petrographic analyses led Blum et al. (1988, 1989) to propose that multiphase opaque assemblages within CAIs in oxidized CV3 chondrites formed by crystallization of metallic droplets within molten CAIs. The compositionally homogeneous metal grains later underwent exsolution, oxidation, and sulfidation.

Although the oxidation and sulfidation were probably caused by aqueous alteration on the parent body, the work of Simon and Grossman (1992) suggests that the exsolution probably occurred during subsolidus cooling of the CAIs. They examined two opaque assemblages in Leoville ($CV3_{Red}$) and reported that one contains sets of regularly spaced, parallel, $\leq$1-μm-thick lamellae of εRu-Fe with sharp boundaries within a larger grain of taenite (Figure 10.12). They concluded that these lamellae were crystallographically controlled and formed by exsolution. Simon and Grossman also determined that partitioning of Ir and Pt between the lamellae and host in the other Leoville opaque assemblage was consistent with equilibration at ~600°C. Because the Leoville whole rock appears to have experienced a much lower equilibration temperature (i.e., ~250°C), it seems probable that the exsolution lamellae in these opaque assemblages formed during subsolidus cooling of the CAIs, prior to agglomeration.

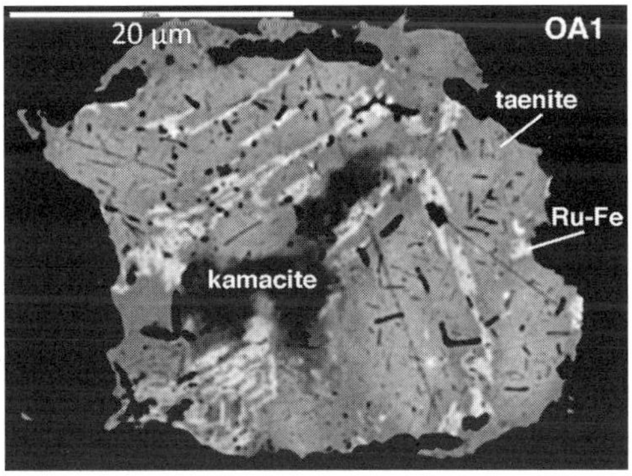

Figure 10.12 BSE image of one opaque assemblage in the Leoville $CV3_{Red}$ chondrite (Simon and Grossman 1992). Image courtesy of Steven Simon, University of New Mexico.

10.7 Exsolution during the Cooling of Chondrules

At high temperatures, minerals that exhibit solid solution have more extensive compositional fields than they have at lower temperatures. Exsolution is a solid-state process that typically occurs during cooling when a parental mineral's crystal structure can no longer accommodate substantial amounts of a particular cation (in most cases due to that cation's size); lamellae of a new mineral phase (enriched in that particular cation) nucleate and grow from the parental phase.

Minerals that formed by exsolution in chondrules during cooling include (1) pigeonite and calcic pyroxene (augite) in clinoenstatite in several type-3 OC and (2) pentlandite in pyrrhotite in LL3.00 Semarkona. Sulfide compositions are most consistent with an equilibration temperature of ~230°C; this is above the estimated peak metamorphic temperature of the Semarkona whole rock (<220°C), suggesting that pentlandite exsolution occurred during an earlier higher-temperature event such as chondrule cooling.

10.8 Exsolution during the Cooling of Opaque Assemblages Outside Ca-Al-Rich Inclusions and Chondrules

Opaque assemblages, rich in metallic Fe-Ni in type-3 ordinary and enstatite chondrites, were probably also melted during the epoch of chondrule formation. Such assemblages include so-called metallic chondrules in unequilibrated OC as well as subrounded metal- and sulfide-rich nodules in type-3 enstatite chondrites.

During the cooling of opaque assemblages in EH and EL chondrites from ≥700°C, it is likely that schreibersite and graphite exsolved from taenite due to the limited solubility of P and C in kamacite. Daubréelite exsolved from troilite, and troilite exsolved from niningerite (Figure 10.13) in some enstatite chondrites; pentlandite exsolved from pyrrhotite in some carbonaceous chondrites.

Figure 10.13 Troilite exsolved from niningerite in a sulfide-rich assemblage from the Indarch EH4 enstatite chondrite. BSE image.

10.9 Annealing of Amorphous Material in the Solar Nebula

Amorphous meteoritic materials that formed before asteroids accreted include the glassy mesostases of chondrules in unequilibrated chondrites, clumps within the matrices of unequilibrated chondrites, and portions of IDPs.

Because presolar dust grains typically spent 10^8–10^9 years in the interstellar medium, long exposure to stellar ionizing radiation is expected to have transformed most of the grains into amorphous materials. Grains with such an origin include some GEMS (glass with embedded metal and sulfides) (see Section 9.8.3). The metal and sulfide grains in the interiors of GEMS are relict crystals; other relicts include embayed forsterite and enstatite grains. Many GEMS in IDPs are surrounded by amorphous carbonaceous material that probably contains both organic and inorganic C. A presolar origin for a minority of GEMS (perhaps 1–6 percent) is indicated by their anomalous O-isotopic compositions. However, because most GEMS have heterogeneous elemental abundances and normal Solar System O-isotopic compositions, they may instead be late-stage nebular condensates.

Many GEMS appear to have undergone moderate annealing. These include those containing compositionally distinct subdomains and those with chemically distinct subparallel layers. Some of the nanophase inclusions in GEMS (e.g., kamacite, taenite, sulfide, and possibly chromite) may have formed from the amorphous material by annealing and sulfidation in the solar nebula. The heat source responsible for the annealing could be associated with chondrule formation. Annealing of amorphous material could also have occurred in circumstellar outflows to produce some of the magnesian silicate grains found within GEMS.

The matrices of primitive chondrites contain compacted nanometer- to micrometer-sized patches of amorphous material with tiny grains of magnesian olivine and low-Ca pyroxene, including some that are enriched in Mn. Heating associated with chondrule-formation events or with passage through hot regions of the solar nebula could have annealed some clumps of amorphous material and formed the Mg-Fe silicate grains that were later incorporated into chondrite matrices. Alternatively, much of the amorphous material in chondrite matrices may have formed during chondrule formation, as volatiles condensed on nearby dust grains; the associated crystalline grains may have formed later, at the earliest stages of thermal metamorphism on chondrite parent asteroids.

11

Formation of Meteoritic Minerals on Parent Bodies

11.1 Devitrification of Amorphous Material on Parent Bodies

During the earliest stages of thermal metamorphism on parent asteroids, primary feldspathic chondrule glass in primitive chondrites began to devitrify to form fine-grained clinopyroxene and crystalline feldspar (Figure 11.1). This resulted in a significant increase in the thermoluminescence (TL) sensitivity of the whole rocks. In some cases, the reaction leading to the crystallization of feldspar may have been catalyzed by water, consistent with the evidence for aqueous alteration in the most primitive ordinary and carbonaceous chondrites.

11.2 Thermal Metamorphism and Exsolution

11.2.1 Heat Sources

There are two heat sources plausibly responsible for the bulk of meteorite metamorphism and melting: (1) the decay of short-lived radionuclides such as ^{26}Al ($t_{½}$ = ~717,000 years) and (2) collisional heating. It is plausible that both mechanisms were important heat sources, e.g., collisions heated asteroids that were already warm from ^{26}Al decay. Other proposed asteroidal heating mechanisms (e.g., exothermic chemical reactions, adiabatic compression of nebular gas, an early superluminous Sun, FU Orionis-type events, explosive reconnection events in the nebula, electromagnetic induction in a fierce protosolar wind) have few current adherents.

11.2.2 Ordinary Chondrites

Every ordinary chondrite (OC) of petrologic type >3.00 (i.e., 99.99 percent of OCs) has experienced some degree of parent-body thermal metamorphism. In many cases, the early stages of thermal metamorphism include reactions with an aqueous (in some instances, also CO_2-rich) fluid, so the common pedagogical separation between thermal metamorphism, aqueous alteration, and hydrothermal alteration is somewhat arbitrary.

Textural effects of thermal metamorphism include devitrification of chondrule glass, textural integration, mineral recrystallization, homogenization of mineral compositions, loss of some primary mineral phases, destruction of presolar grains, growth of secondary minerals, transformation of low-Ca Fe-Mg pyroxene to orthopyroxene, and loss of bulk C, H_2O, noble gases,

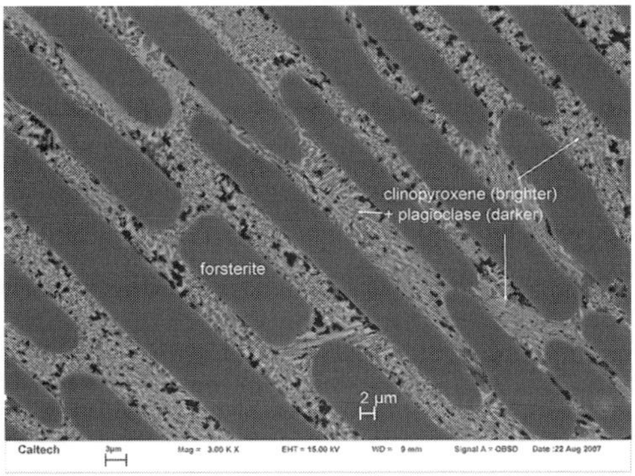

Figure 11.1 Fine-grained clinopyroxene and crystalline feldspar formed from primary feldspathic chondrule glass in a barred olivine chondrule from Allende. BSE image.

and volatile siderophile elements. Grain coarsening occurs as large grains of a phase grow at the expense of small grains; this leads to a reduction in internal energy as the total area of grain boundaries of that phase are reduced. For example, the mean sizes of opaque grains in H chondrites increase with increasing metamorphic grade (troilite: ~30 μm in H3, 80 μm in H6; metallic Fe-Ni: ~65 μm in H3, 120 μm in H6).

Whole-rock physical effects include changes in thermoluminescence (TL) properties due to the order/disorder transition of feldspar between 500°C and 600°C and changes in cathodoluminescence (CL) due to the destruction of chondrule glass in low-FeO chondrules, the growth of albite, and the FeO enrichment of mafic phases during equilibration. The myriad changes experienced by OCs during thermal metamorphism are discussed in Chapter 8.

Olivine grains in FeO-rich chondrules in type-3.00 chondrites tend to be normally zoned in Cr, with higher Cr concentrations near the grain edge, consistent with fractional crystallization. At the earliest stages of thermal metamorphism, in type 3.1–3.2 OCs, thin streaks of a Cr-rich phase (probably chromite) form by exsolution in the olivine along crystallographic axes (Grossman and Brearley 2005). By type 3.3, the olivine grains in these chondrules have developed uniformly low Cr concentrations except for scattered equant internal chromite grains.

Metallic Fe-Ni grains inside chondrules in LL3.00 Semarkona tend to consist of plessitic intergrowths with submicrometer-sized patches of kamacite surrounded by Ni-rich regions. Appreciable numbers of metal grains within low-FeO chondrules contain P, Si, and Cr alloyed with metallic Fe in solid solution; in some cases, the metal grains contain small inclusions of phosphide and silicide. At slightly higher metamorphic grades (type 3.15), the reduced P, Si, and Cr species have reacted with O, Ca, Fe, Na, and Cl from other regions of the meteorite (most likely chondrule mesostases and small matrix grains) to produce Na, Fe, and Ca phosphate, silica, and chromite. Many of these newly formed grains occur in opaque assemblages consisting of (1) troilite and merrillite with or without associated metallic Fe-Ni, (2) troilite and chlorapatite, and (3) metallic Fe-Ni and chlorapatite. Chlorapatite appears to replace some of the merrillite during metasomatic reactions.

With increasing metamorphism to type 4–6 levels, chromite and phosphate grains coarsen. Metallic Fe-Ni and sulfide grains also coarsen and become more circular; small metal and sulfide grains tend to disappear. Plessitic regions outside chondrules become surrounded by thin rims of tetrataenite; the tetrataenite is thicker in regions where it abuts troilite.

At temperatures $\geq$1,440°C, troilite can accommodate up to ~5 mg/g Ni in systems with ~100 mg/g Ni (Kullerud 1963; L. Ma et al. 1998); with decreasing temperatures, Fe diffuses into troilite whereas Ni diffuses out and into the adjacent taenite. Reuter et al. (1988) showed that taenite (γ) transforms into FeNi (γ') and kamacite just below 936°C; at 866°C, γ' undergoes an ordering reaction to form tetrataenite. Growth of tetrataenite during thermal metamorphism can account for the common occurrence of large patches of tetrataenite at the kamacite-troilite interface in metal-troilite spherules in chondrules from H/L3.6 Tieschitz (Rubin et al. 1999), a rock appreciably more metamorphosed than LL3.00 Semarkona. Patches of tetrataenite in chondrites of intermediate petrologic type (e.g., LL3.3 St. Mary's County) are generally of intermediate thickness.

Reisener and Goldstein (2003) studied 17 equilibrated ordinary chondrites and distinguished two types of metallic Fe-Ni particles: (1) zoned taenite + kamacite and (2) zoneless plessite (metal assemblages composed of tetrataenite within kamacite that are not systematically zoned in Ni). Both types of metal particles formed during whole-rock cooling from peak metamorphic temperatures. Zoned taenite + kamacite formed from polycrystalline taenite by the nucleation of kamacite at taenite/taenite grain boundaries. Zoneless plessite formed from monocrystalline taenite via the formation of martensite and its subsequent decomposition into tetrataenite and kamacite. No zoneless plessite was observed by Reisener and Goldstein in type-4 OCs because these rocks were not heated enough to produce monocrystalline taenite particles. In contrast, increasingly greater numbers of zoneless plessite grains occur in types-5 and -6 OCs because these more-metamorphosed rocks produced more precursor monocrystalline taenite particles.

Coarse grains of plagioclase are absent in type-3 OCs (outside rare Al-rich chondrules), but the compositional components of plagioclase reside in the feldspathic mesostases of ferromagnesian chondrules. In low-FeO chondrules, albite crystallizes within the mesostases by type ~3.4; in high-FeO chondrules, the abundance of albite increases progressively from type 3.00 to higher petrologic types. The sizes of albite (or oligoclase) grains increase from <1 μm in type-3 to ~2 μm in type-4 to 2–10 μm in type-5 to $\geq$50 μm in type-6 OCs.

Many chondrule feldspathic mesostases also have a normative pyroxene component. The concentration of this component increases in metamorphosed OCs after plagioclase crystallizes, leading in a few BO chondrules to the growth of optically continuous coarse pyroxene grains between the olivine bars.

Chromite in primitive type-3 chondrites (subtype $\lesssim$3.5) is present only as small grains within chondrules (in the mesostasis or within olivine phenocrysts). Some type-3.8 OCs have moderate-sized chromite grains in the matrix. Coarse chromite grains are present in type 4–6 OCs in the matrix as 10–200-μm-sized isolated blocky, equant, subhedral, anhedral, and rounded grains and in association with metallic Fe-Ni and/or troilite. With increasing petrologic type, chromite tends to become richer in FeO, MnO, and Al_2O_3. In shocked OCs, chromite also occurs in chromite-plagioclase assemblages and as small blebs or as continuous strands within veinlets in olivine grains.

The principal pyroxene phase in type-3 and some type-4 OCs is low-Ca clinopyroxene exhibiting polysynthetic twinning and inclined extinction; it occurs as phenocrysts in pyroxene-rich chondrules and chondrule fragments. This phase transforms into orthopyroxene at ~630°C;

this is above the maximum metamorphic temperature of some type-4 OCs (~600–700°C), but below those of type-5 and type-6 OCs (~700–750°C and ~820–930°C, respectively). That accounts for the absence of low-Ca clinopyroxene in unshocked type-5 and -6 chondrites. (Nevertheless, low-Ca clinopyroxene exhibiting polysynthetic twinning can form by shock heating and quenching of pyroxene grains in any rock, irrespective of metamorphic grade, at shock stage $\geq$S3.)

Type-3 OCs have heterogeneous olivine and low-Ca pyroxene compositional distributions. During thermal metamorphism, diffusion causes the grains to approach equilibrium, diminishing grain-to-grain variations in mineral chemistry. The mean olivine/low-Ca-pyroxene modal ratio covaries with oxidation state among the three OC groups: 1.4 in H, 2.0 in L, and 3.6 in LL. As more metallic Fe is transformed into FeO during oxidation, the divalent-cation/Si ratio increases, facilitating the formation of $(Mg,Fe)_2SiO_4$ olivine (which has a 2:1 ratio) at the expense of $(Mg,Fe)SiO_3$ low-Ca pyroxene (which has a 1:1 ratio).

Water with relatively high $\Delta^{17}O$ (probably present initially as ice or phyllosilicates) likely acted as an oxidizing agent during whole-rock thermal metamorphism. Phyllosilicates that formed in the matrix at low temperatures dehydrated. The liberated water oxidized metallic Fe-Ni, causing correlated changes with increasing petrologic type in all three OC groups; these include decreasing modal abundances of metallic Fe-Ni, increasing FeO/(FeO+MgO) in olivine, low-Ca pyroxene, chromite, and possibly ilmenite, increasing whole-rock olivine/low-Ca-pyroxene ratios, increasing Co and Ni concentrations in the remaining kamacite, and decreasing bulk (whole-rock) values of $\delta^{18}O$ and $\delta^{17}O$.

Minerals in OCs that formed by exsolution during metamorphism include chromite in olivine, chromite and rutile in ilmenite, calcic pyroxene in orthopyroxene, alkali feldspar in albite, and mackinawite (FeS_{1-x}) in pentlandite ($(Fe,Ni)_9S_8$).

Wasson et al. (1993) found that three type-3 OCs (Willaroy, Suwahib (Buwah), and Moorabie) and one type-4 OC (Cerro los Calvos) experienced reduction during thermal metamorphism. Although their bulk siderophile element abundances and mean chondrule sizes indicate that they are members of the H (Willaroy and Cerro los Calvos) and L (Suwahib (Buwah) and Moorabie) groups, respectively, their olivine Fa values are appreciably below those of equilibrated members of these groups.

11.2.3 Carbonaceous Chondrites

About half of the CI chondrites have undergone some thermal metamorphism; a significant fraction (perhaps ~25 percent) of CM chondrites were metamorphosed as well. In some cases, the secondary heating of CM chondrites is probably attributable to impacts. The other major carbonaceous chondrite groups that experienced significant thermal metamorphism are CO and CV-CK. In addition, the Coolidge-Loongana 001 subgroup is partly equilibrated.

CI chondrites. A typical metamorphosed CI chondrite is Y-86029. Phyllosilicates are absent; the crystalline material that is present in the rock consists of olivine, magnetite, and troilite. Mild metamorphism caused Mg-Fe-rich carbonates to break down into Ca-Mg-Fe-Mn oxides, and sulfide to break down into elongated magnetite grains with sulfide inclusions (e.g., Zolensky et al. 2005).

CM chondrites. One of the most-metamorphosed CM chondrites is Belgica 7904 (e.g., Tomeoka 1990). It contains low bulk H_2O and C. There are abundant desiccated phyllosilicates (with high analytical totals) consisting of Fe-rich serpentine, normal serpentine, and clinochlore ($(Mg,Fe^{+2})_5Al(Si_3Al)O_{10}(OH)_8$). Secondary phases in the matrix include olivine, low-Ca pyroxene, troilite, taenite, and kamacite. Tochilinite ($2[(Fe,Mg,Cu,Ni[\ ])S\cdot 1.57\text{-}1.85[(Mg,Fe,Ni,Al,Ca)(OH)_2]$) is absent. Similar metamorphic effects are evident in Y-86720 and Y-86789. Some CM chondrites that were aqueously altered and subsequently metamorphosed contained highly strained fayalite grains with voids. Metamorphism causes irreversible carbonization of insoluble organic matter (Quirico et al. 2018). Tonui et al. (2014) listed CM-like chondrites in order of extent of thermal alteration: 400°C $\leq$ Y-793321 $<$ WIS 91600 = EET 90043 = Asuka 881655 $<$ PCA 91008 $<$ Belgica 7904 $<$ Y-86720 = Y-86789 $\leq$ 700°C. The dehydrated CM chondrites do not easily fit into the petrologic types established for CM chondrites that experienced progressive aqueous alteration without significant heating, e.g., CM1 and CM2 (e.g., Zolensky et al, 1997) or CM2.0 to CM2.9 (Rubin et al. 2007). Nakamura (2005) proposed classifying them into Stages I to IV, reflecting increasing degrees of heating. King et al. (2019) proposed calling these meteorites CY chondrites and suggested they were derived from a distinct parent body.

CR chondrites. The edges of metallic Fe-Ni grains located at the margins of some chondrules in CR chondrites have low concentrations of Co and Ni, reflecting the addition of Fe produced during the reduction of FeO from mafic silicates during mild thermal metamorphism (Lee et al., 1992). The matrices of GRA 06100 and GRO 03116 have undergone extensive dehydration; the organic matter in these samples differs from that in other CR (as well as CM) chondrites, and a significant proportion of kamacite grains contains Ni-rich plessite regions. These features may be attributable to thermal metamorphism.

Coarse magnetite grains are absent in the least-altered, least-heated CR chondrites. More-altered samples contain magnetite (e.g., Renazzo, Al Rais, and NWA 2196) or iron oxides with somewhat higher concentrations of Al_2O_3 and a larger hematite compositional component (e.g., NWA 5028).

The petrographic descriptions of three chondrites listed as CR6 (NWA 7317, NWA 11561, and NWA 6921 and paired specimens) and one listed as CR7 (NWA 7531) in the Meteoritical Bulletin Database resemble those of some OC impact-melt rocks (e.g., PAT 91501), suggesting the possibility that the CR6 and CR7 chondrites are also impact melts. These rocks contain chromite, but no magnetite. They may be related to the ungrouped achondrites NWA 6693 and NWA 6704 (rapidly cooled rocks with O and Ti isotopes resembling those in carbonaceous chondrites, particularly the CR group) (Hibaya et al. 2019; Sanborn et al. 2019).

CO chondrites. Metamorphic effects in CO3 chondrites, first noted by McSween (1977), are inextricably linked to aqueous alteration. The CO3 chondrites can be divided into petrologic subtypes ranging from 3.0 to 3.8; corresponding temperatures (modeled from diffusion zoning in olivine) range from 380°C to 417°C for CO3.0 chondrites to 576°C to 587°C for CO3.8 Isna (Schwinger et al. 2016). With increasing metamorphism, CO chondrites exhibit (1) increases in the degree of textural recrystallization, mafic mineral homogeneity, the FeO contents of mafic phases, and the development and thickening of ferroan rims on olivine grains in chondrules and amoeboid olivine inclusions (AOIs), (2) increasingly positive valence states for Ti and Cr in olivine at the early stages of metamorphism, (3) decreases in the abundances of chondrule glass, metallic Fe-Ni grains, presolar diamonds, presolar SiC, and bulk volatile elements (including noble gases), (4) decreases in the abundance of presolar silicates in the matrix, (5) increases in the abundances of straight-chain amino acids (which are produced during thermal metamorphism, (6) increasingly heavy bulk O isotopes, (7) decreases in bulk C, D/H ratio and $^{15}N/^{14}N$ ratio as insoluble organic matter (IOM) becomes increasingly graphitized, (8) alteration and conversion of CAI minerals (i.e., FeO enrichment of spinel, conversion of perovskite to ilmenite, and depletion of melilite), (9) increasing TL sensitivity, (10) increasing replacement of fine-grained plessite by coarser Ni-rich metal (taenite and tetrataenite) and associated kamacite, (11) changes in the shape of chromite grains from euhedral to subhedral or anhedral, and (12) changes in the Cr_2O_3 content of ferroan olivine grains. This final change involves steady decreases in Cr_2O_3 while the standard deviation of the olivine Cr_2O_3 content increases at the earliest stages of metamorphism (subtypes 3.00–3.05) and then decreases gradually through subtype 3.2. This process is accompanied by the development of narrow ($<$0.3-μm-thick) chromite veins within the olivine grains.

Although magnetite formed in primitive CO3 chondrites from metallic Fe by parent-body aqueous alteration, the paucity or absence of magnetite in CO chondrites of subtype $\geq$3.2 suggests that magnetite is destroyed during thermal metamorphism. Thermochemical calculations suggest that magnetite is reduced by H_2 and reacts with SiO_2 to form fayalite and secondary kamacite (Zolotov et al. 2006).

CV and CK chondrites. There is controversy about whether CV3 and CK3.8–CK6 chondrites constitute a single metamorphic sequence. Magnetite-bearing carbonaceous chondrites with large chondrules and minor to negligible amounts of metallic Fe-Ni are routinely assigned either to the CV3 group (if they are relatively unequilibrated and have low mean olivine Fa contents) or to the CK group (if they are more equilibrated with olivine compositions near Fa_{30} and possess more-recrystallized textures). CK chondrites appear to differ from CV chondrites in their lower proportions of CAIs, AOIs, and chondrules with igneous rims, but these differences are probably due largely to metamorphic recrystallization of the CK chondrites.

Dunn et al. (2016) found that magnetite grains in CK3, CK4, and CV3 chondrites differ in their concentrations of Cr_2O_3, NiO, and TiO_2. The differences between CK3 and CK4 magnetite is mainly in TiO_2 and is apparently attributable to thermal metamorphism. However, CV3 magnetite is closer in composition to CK4 than to CK3 magnetite, suggesting two separate metamorphic sequences. This is consistent with the isotopic data of Yin et al. (2017) who reported (1) resolvable differences in weighted averages in $\varepsilon^{54}Cr$ between CV (+0.88±0.06) and CK (+0.66±0.06) chondrites (both with 95 percent confidence-level errors) and (2) minor differences in $\Delta^{17}O$ between the groups.

The petrologic characteristics of two meteorites add to the confusion: Camel Donga 040 is a breccia that appears to contain separate CV and CK lithologies; NWA 10588 appears to be intermediate between typical CV3 and CK4 chondrites. Additional studies are needed to sort this mess out.

Geiger and Bischoff (1995) proposed that thermal metamorphism caused (1) metallic Fe-Ni in CK chondrites to transform into magnetite at oxygen fugacities near that of the Ni-NiO buffer, (2) spinel and ilmenite to exsolve from the magnetite during cooling, and (3) noble metal–rich sulfides, arsenides, and tellurides to form [either from the oxidizing metallic Fe-Ni or from monosulfide solid solution (MSS)].

CV3 chondrites are divided into three subgroups: (1) an Allende-like oxidized subgroup ($CV3_{OxA}$), (2) a Bali-like oxidized subgroup ($CV3_{OxB}$), and (3) a reduced subgroup ($CV3_{Red}$). These subgroups have experienced different degrees of alteration and thermal metamorphism. Metamorphic effects in CV3 chondrites include (1) a diminishment in grain-to-grain compositional variation among olivines in the two oxidized subgroups, (2) progressive destruction of presolar grains in $CV3_{OxA}$ and $CV3_{Red}$, and (3) systematic changes in the Raman spectroscopic characteristics of organic material. Efremovka ($CV3_{Red}$) contains dark inclusions with dehydrated phyllosilicates and strained fayalite grains with voids, indicative of aqueous alteration followed by mild thermal metamorphism.

All the ferroan olivine in ordinary and carbonaceous chondrites had elevated levels of Cr at the time of accretion, probably a consequence of rapid cooling during chondrule formation. At the onset of metamorphism, Cr rapidly exsolved from olivine, forming fine-grained precipitates probably composed of chromite. Higher degrees of heating caused chromite to coarsen into equant grains inside olivine and form continuous rims around every grain. The concentration of Cr in olivine is an excellent classification tool for chondrites of type 3.0–3.2 and is not susceptible to modification by aqueous alteration or terrestrial weathering. Differences among type-3.0 and -3.1 chondrites can be quite large and a further subdivision of petrologic types based on Cr in olivine has been proposed to emphasize these differences.

Although CV chondrites are also not normally considered part of an extensive metamorphic sequence, the Meteoritical Bulletin Database lists one CV7 chondrite (NWA 2653). This meteorite contains chromite but not magnetite.

Coolidge-Loongana 001. These two carbonaceous chondrites constitute a grouplet and have been metamorphosed to subtype 3.8–4 levels. Both meteorites are moderately recrystallized and have fairly narrow olivine compositional distributions.

11.2.4 R Chondrites

Most R chondrites are moderately metamorphosed rocks (mainly type 3.6–4); a minority are more-extensively metamorphosed (type 5–6). About half are regolith breccias, rich in solar-wind–implanted rare gases; many of these rocks contain clasts with a range of petrologic types (including some highly unequilibrated type-3 clasts). Equilibrated R chondrites are highly oxidized rocks with mean olivine Fa contents of $Fa_{39\pm2}$. Partially equilibrated samples (e.g., PCA 91002) possess Type I chondrules containing olivine grains with forsteritic cores and ferroan rims caused by diffusion during moderate thermal metamorphism.

The modal abundance of low-Ca pyroxene is highly variable: it is abundant in unequilibrated R3 clasts, R3.6 chondrites (e.g., ALH 85151), and some R3.8–3.9 chondrites. However, it is rare to absent in other R3.8 chondrites and in many R5 and R6 lithologies. This suggests that oxidation associated with progressive thermal metamorphism could be responsible for destroying low-Ca pyroxene.

Most occurrences of Pt-rich arsenides and tellurides are in equilibrated R-chondrite materials (Bischoff et al. 2011), suggesting that these phases formed during parent-body metamorphism. The occurrence of troilite in several R chondrites suggests that troilite was the principal sulfide phase prior to oxidation. As troilite or MSS was transformed into pyrrhotite ($Fe_{1-x}S$) during the oxidation that accompanied metamorphism, it seems likely that some As and Te were removed from the sulfide and formed the noble-metal–bearing arsenides and tellurides.

Mineral phases formed during exsolution in R chondrites include flamelike lamellae of pentlandite in pyrrhotite and thin lamellae of ilmenite in pentlandite.

A few R6 chondrites contain amphibole (hornblende) and mica (phlogopite and titan-phlogopite) produced by hydrothermal alteration (McCanta et al. 2008; Rubin 2014). These rocks probably were heated to maximum temperatures of ~700–900°C at moderately high PH_2O (perhaps 250–500 bars). Sanders et al. (2017) calculated that these pressures could have been reached at depths of 50–120 km in a 400-km-diameter body. If this is correct, a giant impact or a series of smaller impacts would have been required to excavate the hornblende- and mica-bearing rocks and bring them closer to the surface.

11.2.5 Enstatite Chondrites

The enstatite chondrites comprise two major groups, EH and EL, that differ in degree of oxidation, abundance of siderophiles and halogens, sulfide mineralogy, mineral chemistry, and mean chondrule size. Each group comprises a complete metamorphic sequence from type 3 to type 6. There is a distinct possibility that all EH6 chondrites and some EL6 chondrites are actually impact-melt breccias (Rubin and Wasson 2011). Some enstatite chondrites of lower petrologic type are also impact-melt breccias (e.g., Rubin and Scott 1997; Rubin 2008, 2015a).

There are several mineralogical changes that occur in enstatite chondrites during parent-body thermal metamorphism:

(1) Additional albite forms at the expense of the feldspathic component in chondrule glass. For example, although type-3 enstatite chondrites contain little crystalline albite (e.g., <1 wt% in EH3 ALHA77156), type-6 enstatite chondrites contain appreciably more (e.g., 9 wt% in EL6 Atlanta). Additional enstatite crystallizes from the mafic-silicate component of the glass.

(2) Forsterite disappears via reaction with silica (which constitutes ~1–2 vol% of E3 chondrites) to produce additional enstatite:

$$Mg_2SiO_4 + SiO_2 = 2\ MgSiO_3$$
$$\text{forsterite} + \text{silica} = 2\ \text{enstatite}$$

The modal abundance of forsterite in EH3 and EL3 chondrites is 4.4 and 2.4 vol%, respectively; it is rare to absent in EH4 and EL4 chondrites and is absent in types-5 and -6 EH and EL chondrites.

(3) Diopside is present in type-3 enstatite chondrites (0.1–0.3 vol%), but is extremely rare in more-equilibrated enstatite chondrites. Diopside may have been destroyed through sulfidation by a reaction such as

$$CaMgSi_2O_6 + H_2S = MgSiO_3 + CaS + SiO_2 + H_2O$$
$$\text{diopside} + \text{hydrogen sulfide} = \text{enstatite} + \text{oldhamite} + \text{silica} + \text{water}$$

(4) The unusual occurrence of diopside in EL6 EET 90102 (Fogel 1997) may be due to the fact that the bulk composition of the rock lies within the stability fields of either

enstatite-diopside-oldhamite-alabandite or diopside-alabandite-enstatite. In contrast, other equilibrated enstatite chondrites have bulk compositions within the enstatite-oldhamite-alabandite stability field.

(5) Clinoenstatite in EH3 and EL3 chondrite chondrules is transformed into orthoenstatite at a temperature of ~630°C, accounting for its absence in type-5 and -6 enstatite chondrites (which, by analogy to OCs, were heated to $\geq$700°C).

(6) The CL properties of pyroxene change with increasing degrees of metamorphism. For example, EL3 pyroxene grains emit mainly violet-blue CL, whereas EL5 and EL6 pyroxene grains emit magenta CL. This may be related to the incorporation of minor-element red activators such as Cr^{+3} and Mn^{+2} into pyroxene during metamorphism.

(7) The presence of quartz in some EH5 chondrites was attributed by Kimura et al. (2005) to transformation from tridymite or cristobalite (primary phases formed in chondrules) during long-duration, low-temperature metamorphism.

(8) Perryite ($(Ni,Fe)_8(Si,P)_3$) constitutes 0.8 and 0.1 vol% of EH3 and EL3 chondrites, respectively, but is absent in more-equilibrated enstatite chondrites. It probably breaks down into schreibersite and Si-bearing kamacite. Perryite is a Ni-rich mineral, containing an average of 78.8 wt% Ni in EH3 chondrites; its breakdown during thermal metamorphism is reflected in the increase in the mean Ni content of kamacite (3.0 wt% in EH3 Qingzhen, ALHA77156, and Kota-Kota vs. 7.1 wt% in EH4 Indarch).

(9) Some grains of troilite in the matrix of type-3 enstatite chondrites contain daubréelite exsolution lamellae, probably formed during cooling after minor metamorphic heating.

(10) Sulfide undergoes changes in the EH group. (a) Djerfisherite in EH3 chondrites breaks down to form troilite, covellite (CuS), idaite (Cu_5FeS_6), bornite (Cu_5FeS_4), and other phases; (b) Fe diffuses from troilite to niningerite, causing the latter phase to become reversely zoned; and (c) sphalerite forms in EH3 Y-691 due to mobilization of Zn.

(11) Presolar silicate and carbonaceous grains (identified in EH3 chondrites) were destroyed during metamorphism.

(12) Within type-3 enstatite chondrites, low levels of thermal metamorphism also caused an increase in the structural order of organic matter and partial loss of noble gases.

11.2.6 Eucrites

The Meteoritical Bulletin Database lists more than a thousand eucrites, probably representing 500–600 separate falls; they are widely conjectured to have been derived from Vesta. Non-cumulate eucrites are basalts that crystallized with ophitic or subophitic textures. The petrologic studies of Takeda and Graham (1991) and Mayne et al. (2009) serve to divide the eucrites into six types (1–6) based on the degree of metamorphic recrystallization. Type 1 eucrites are the most pristine; they retain their primary textures, contain moderately heterogeneous pyroxenes (including some grains with ferroan compositions), have compositionally zoned pyroxene grains, do not possess microscopically identifiable exsolution lamellae in the pyroxene, and contain primary glassy mesostases. With increasing metamorphic grade, the following processes occur: (1) pyroxenes become compositionally more homogeneous, (2) pigeonite exsolves in type-5 and -6 eucrites, (3) augite exsolution lamellae coarsen, (4) the low-Ca pyroxene host flanking coarse augite exsolution lamellae in type-6 eucrites partly inverts to orthopyroxene, (5) pyroxene becomes "cloudy" due to the precipitation of small chromite grains (type 4–6), and (6) mesostasis glass disappears or recrystallizes.

Metamorphic temperatures for eucrites appear mostly to have been in the range of 800°C–900°C. The mechanisms that caused thermal metamorphism have been suggested to be impact heating, the decay of ^{26}Al, or burial by subsequent lava flows (which presumably were created after solid rock was melted either by impacts or ^{26}Al decay).

11.3 Aqueous Alteration, Hydrothermal Alteration, and Metasomatism

The proposed identification of hydrous phyllosilicates on the surfaces of asteroids of the C, D, G, F, and B classes indicates that aqueous alteration processes are likely to have affected the meteorites (i.e., carbonaceous chondrites) thought to have been derived from those bodies. Although some workers have postulated that aqueous alteration took place in the nebula (perhaps after phyllosilicates formed in the protoplanetary disk), most researchers have concluded that alteration occurred principally on asteroids. Inferred alteration temperatures for CI (0–150°C) and CM (20–80°C) chondrites are lower than those for type-3 OCs (<260°C). Aqueous alteration processes have also affected other classes of meteorites to lesser extents than the carbonaceous chondrites.

An unsolved problem is the ultimate source of the water within parent asteroids of altered chondrites. One possibility is that volatiles derived from fine nebular dust grains that had evaporated during episodes of chondrule formation reacted with water in surrounding nebular gas to produce small phyllosilicate grains. These grains later accreted with chondrules, CAIs, AOIs, metallic Fe-Ni, sulfide, and anhydrous dust to form chondrite whole rocks. Another possibility is that a few icy planetesimals accreted along with essentially anhydrous chondritic planetesimals to form hundred-kilometer-sized asteroids. If carbonaceous chondrites formed beyond the snow line, ice would have been a stable nebular phase (a situation that does not apply to OCs). Independent of the mechanism that caused incorporation of water into asteroids, the water was later mobilized by impact heating and/or the decay of ^{26}Al.

11.3.1 CI Chondrites

CI chondrites have appreciable amounts of bulk water [e.g., ~11 wt% H_2O^+ (indigenous water) in the Orgueil fall], mainly occurring as hydroxyl in Fe-Mg serpentines and as water of hydration in the interlayered saponite. These phases constitute the matrix of CI chondrites and account for ~95 vol% of these rocks. Other water-bearing minerals include sulfate [gypsum ($CaSO_4{\cdot}2H_2O$) with minor blödite ($Na_2Mg(SO_4)_2{\cdot}4H_2O$) and Ni-rich blödite ($Na_2(Ni,Mg)(SO_4)_2{\cdot}4H_2O$)] and ferrihydrite ($Fe_{4-5}(OH,O)_{12}$). Anhydrous minerals that formed in CI chondrites by aqueous alteration include magnetite (which occurs mainly in three distinct morphologies – framboids, platelets, and plaquettes), several sulfides (pyrrhotite with minor pentlandite and cubanite) and several carbonates (calcite ($CaCO_3$), dolomite ($CaMg(CO_3)_2$), breunnerite ($(Mg,Fe)CO_3$), and minor siderite ($FeCO_3$)). (By analogy to the mechanism most likely responsible for producing pyrite framboids in terrestrial sediments and low-temperature ore deposits, the magnetite framboids in CI chondrites may have formed by rapid nucleation from fluids supersaturated with oxidized iron. The fact that many magnetite framboids appear to have formed around dissolving sulfide grains, suggests that the source of the oxidized iron is the associated sulfide.)

CI chondrites contain compositionally heterogeneous iron-nickel sulfide grains (pyrrhotite and pentlandite) formed during late-stage aqueous alteration; sulfide compositions vary among the members of the group. Berger et al. (2016) concluded that Orgueil and Alais experienced different equilibration temperatures (~25°C and 100–135°C, respectively). They suggested that the aqueous alteration conditions (e.g., temperature, pH, oxygen fugacity, duration) varied on the scale of millimeters. However, because CI chondrites are breccias, it is unclear how much of this millimeter-scale variation is attributable to the incorporation of clasts that were altered to different degrees.

The relatively rare coarse isolated olivine and pyroxene grains (probably dislodged chondrule phenocrysts), refractory minerals (probably CAI fragments), and very rare altered intact CAIs (only one is known as of this writing) indicate that CI chondrites have the same set of major components as other carbonaceous chondrites, albeit in different proportions.

11.3.2 CM Chondrites

Although generally not as altered as CI chondrites, CM chondrites still contain a significant amount of bulk water (e.g., ~9 wt% H_2O^+ in Murchison). The CM group exhibits large variations in the degree of aqueous alteration. Many individual CM meteorites (e.g., Murray, Cold Bokkeveld) are breccias that contain millimeter- to centimeter-sized clasts that have been altered to different extents than the host. Several schemes have been devised to classify CM chondrites by their degree of aqueous alteration; the most widely adopted scheme is that of Rubin et al. (2007). This system considers the modal abundance of metallic Fe-Ni, the degree of alteration of mafic silicate phenocrysts in chondrules, the abundance of large clumpy intergrowths of tochilinite-cronstedtite, the bulk composition of these intergrowths, carbonate composition, and the distribution of sulfide phases.

In going from the least-altered CM chondrites (ideally subtype 3.0) to the most altered (subtype 2.0; also called CM1 in other schemes), the following changes occur in mineralogy, texture, and chemistry:

(1) glassy chondrule mesostases transform into phyllosilicate (beginning at about subtype 2.9);
(2) tiny mafic silicate grains and amorphous material in the matrix also transform into phyllosilicate (beginning at about subtype 2.9);
(3) melilite in CAIs transforms into phyllosilicates at about subtype 2.8;
(4) the modal abundance of metallic Fe-Ni systematically decreases throughout the sequence due to oxidation;
(5) thin magnetite rinds form around some of the metal grains;
(6) mafic silicate phenocrysts in chondrules transform into phyllosilicates (beginning at subtype 2.3);
(7) the modal abundance of large clumpy tochilinite-cronstedtite intergrowths decreases (beginning at about subtype 2.1);

(8) although metallic Fe^0 is oxidized to produce Fe^{3+}, bulk iron becomes increasingly reduced (Fe^{3+} is transformed into Fe^{2+}) as cronstedtite (which contains abundant Fe^{3+} in tetrahedral sites) is replaced by Mg-rich serpentine (which contains less ferric iron);

(9) the "FeO"/SiO_2 ratio of tochilinite-cronstedtite intergrowths [$2[(Fe,Mg,Cu,Ni,\square)S\cdot 1.57\text{-}1.85[(Mg,Fe,Ni,Al,Ca)(OH)_2]$ – $Fe_2^{2+}Fe^{3+}(SiFe^{3+})O_5(OH)_4$] decreases (beginning at about type 2.4) (wherein "FeO" includes FeO in mafic silicates, oxides and phyllosilicates; Fe^{3+} in cronstedtite and Fe oxide; Fe^{2+} in sulfides; and small grains of metallic Fe all combined and recalculated as FeO);

(10) the S/SiO_2 ratio of tochilinite-cronstedtite intergrowths decreases (beginning at about subtype 2.4) as S leaves the intergrowths, aiding the growth of relatively coarse Fe,Ni-sulfide grains;

(11) the modal abundance of sulfides with "intermediate" compositions (those with atomic Ni/(Fe+Ni) ratios of 0.10–0.40) increases (beginning at about subtype 2.4):

(12) the modal abundance of pyrrhotite decreases appreciably by subtype 2.2 (as primary pyrrhotite alters to secondary pentlandite, magnetite, and phyllosilicates); and

(13) Ca carbonate (major calcite, minor aragonite) forms at about subtype 2.8 or 2.9 and is partly replaced by more complex carbonates (including dolomite and breunnerite) with significant amounts of Ca, Mg, Fe, and Mn beginning at about subtype 2.1. Mn-Cr systematics show that carbonate formation occurred ~2–5 Ma after CAIs (e.g., Donohue et al. 2019).

In some cases, different aqueous-alteration products are intergrown. Maribo (CM2) contains bands of tochilinite transecting calcite (Figure 11.2) and magnetite of different morphologies (quasi-equant grains, spherulites, and framboids) occurring together with spherules of hydroxylapatite within serpentine (Figure 11.3). (Magnetite spherulites can be distinguished from spherules: spherulites have internal radial structures that spherules lack.)

Hydrothermal experiments by Vacher et al. (2019) suggest that during aqueous alteration on the CM asteroid, Fe-rich tochilinite grains formed from metallic Fe-Ni under S-bearing alkaline conditions at temperatures of 120–160°C. As temperatures declined to 50–120°C, cronstedtite formed from metallic Fe-Ni and amorphous silicates in the matrix.

One of the least-altered CM chondrite is Paris, overall subtype 2.7, but the rock is a breccia with some regions as little altered as subtype 2.9. A few chondrules contain unaltered glassy mesostases, rare CAIs contain melilite, and some amorphous material occurs in the matrix. In contrast, the most altered samples, i.e., the CM2.0 chondrites, contain chondrule pseudomorphs composed of phyllosilicates. A few CM2.0 chondrites (e.g., MET 01070) also possess tochilinite-cronstedtite lenses more than a centimeter in length. In some cases, the lenses cut across one another; they probably formed within sheer-induced fractures in the rocks.

There is a strong positive correlation between the degree of alteration of individual CM chondrites and the extent of particle alignment (i.e., the strength of the petrofabric) within them (Rubin 2012b). It is plausible that random collisions on the parent body created fractures and produced petrofabrics in some rocks. When water was later mobilized (presumably due to impact heating of accreted ice and/or water-bearing phases), the more-fractured regions had greater porosity, retained more water, and were preferentially altered.

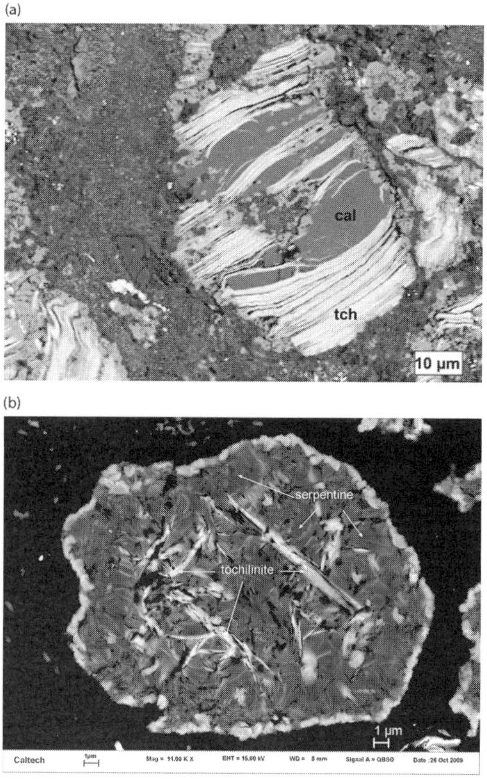

Figure 11.2 (a) Bands of tochilinite (tch) in calcite (cal) in the Maribo CM2 chondrite. BSE image courtesy of M. E. Zolensky, NASA – JSC. (b) Bands of tochilinite with serpentine in the Murchison CM2 chondrite. BSE images.

Figure 11.3 Murchison has magnetite of different morphologies (quasi-equant grains, spherulites and framboids) occurring together with spherules of hydroxylapatite. BSE image.

11.3.3 CR Chondrites

The CR chondrites vary in their degree of alteration (e.g., Weisberg and Huber 2007; Schrader et al. 2011; Harju et al. 2014). Some rocks contain chondrules with glassy mesostases, have abundant metallic Fe-Ni (up to ~17 wt%), and lack magnetite. In others, the metallic Fe-Ni has been largely transformed into oxide; nevertheless, essentially all CR2 chondrites contain iron in all three oxidation states.

In going from the least-altered CR samples (ideally subtype 3.0) to the most altered (subtype 2.0, a.k.a. CR1), the following mineralogical changes occur: (1) glassy chondrule mesostases transform into phyllosilicate (beginning at subtype 2.8); (2) tiny mafic silicate grains and amorphous material in the matrix transform into phyllosilicate, mainly serpentine intergrown with saponite (beginning at subtype 2.9); (3) magnetite grows at the expense of metallic Fe-Ni (beginning at subtype 2.7), forming rinds on coarse metal grains and framboids in the matrix; (4) mafic silicate phenocrysts in chondrules alter to phyllosilicate (beginning at subtype 2.5); (5) Ca carbonate forms (beginning at subtype 2.5); and (6) siderite and ferrous sulfate form in subtype 2.0 rocks.

Some of the least-altered samples have chondrules surrounded by phyllosilicate rims containing silica and moderately ferroan low-Ca pyroxene. The association of the silica with phyllosilicate suggests that the silica is also an alteration product, but it is possible that it predates alteration and formed as part of an igneous chondrule rim. It is not clear if the silica phase is amorphous or crystalline, but because it contains minor amounts of FeO, MgO, and Al_2O_3, it is unlikely to be quartz.

Some brecciated CR chondrites contain dark inclusions. These inclusions have sharp boundaries with their hosts and appear to be clasts of CR-chondrite matrices that have experienced more extensive aqueous alteration than their hosts. They contain microchondrules, small (<200 μm) polycrystalline fragments; sulfide laths (pentlandite and pyrrhotite); magnetite (framboids, platelets, and spherules); and accessory phases (Cr-spinel, metallic Fe-Ni, schreibersite, ilmenite, Ca-phosphate, and breunnerite) embedded in fine-grained serpentine-smectite intergrowths. Endreß et al. (1994) also reported two Os,Mo,Ir-rich particles within dark inclusions in CR2 Acfer 059 and its paired specimens. One dark inclusion in Al Rais is completely transected by a Ca-carbonate vein.

11.3.4 CO Chondrites

The CO chondrites were affected by fluid-assisted thermal metamorphism (e.g., Grossman and Brearley 2005). Unlike the case for CM and CR chondrites (which range in subtype from least-altered 2.8–3.0 samples to most-altered 2.0 samples and have experienced little thermal metamorphism), the CO chondrites range from least-altered, least-metamorphosed subtype 3.0 to most-altered, most-metamorphosed subtype 3.8 (e.g., Chizmadia et al. 2002). Perhaps the most primitive CO chondrite is CO3.00 DOM 08006 (e.g., Simon et al. 2019); this meteorite contains magnetite (which replaces metallic Fe-Ni in metal-rich assemblages) and Fe-Ni-carbide veins transecting chondrules. It has organic compounds and a high abundance of presolar grains. Hydrothermal fluids may have mobilized organic[1] compounds in primitive CO3

[1] Some of the diverse complex, high-molecular-weight, insoluble organic compounds in carbonaceous chondrites may have formed by parent-body hydrothermal alteration of hexamethylenetetramine ($C_6H_{12}N_4$). This phase can form in the interstellar

chondrites such as DOM 08006, enriching the aqueous fluid in carbon. The carbide-magnetite-bearing assemblages were modeled by Doyle et al. (2015) as having formed ~5 Ma after CAIs by oxidation during fluid-assisted metamorphism at water/rock ratios of ~0.1 to 0.2. Some CO3.1 chondrites also have fayalite veins, indicating the aqueous fluid was enriched in iron.

CO3 chondrites experienced Na-metasomatism. Chondrule mesostases in minimally altered CO3.0 chondrites contain glass and minor plagioclase. With increasing degrees of alteration, nepheline replaces glass and plagioclase. More-altered CO chondrites have larger nepheline grains and higher proportions of nepheline within chondrule mesostasis. The initial secondary phase may have been hydrous nepheline that subsequently dehydrated during parent-body heating.

Another effect of aqueous alteration in CO chondrites involves the dissolution of primitive melilite in refractory inclusions (grains with $\Delta^{17}O$ values $\leq -20‰$) and the reprecipitation of new melilite in the presence of H_2O with a high $\Delta^{17}O$ value.

The presence of Fe-Mg serpentine and minor chlorite in the matrices of some CO3.0 to CO3.4 chondrites indicates that mafic silicates and amorphous material were being destroyed at the beginning of the alteration sequence.

11.3.5 CV Chondrites

The Allende-like oxidized subgroup ($CV3_{OxA}$) contains such secondary phases as nepheline, sodalite, grossular, hedenbergite, anorthite, monticellite, wadalite, hutcheonite, and adrianite, consistent with alkali-halogen-Fe metasomatism. Phyllosilicates are relatively rare. Hutcheonite ($Ca_3Ti_2(SiAl_2)O_{12}$) is a newly discovered garnet mineral occurring in alteration areas of an Allende CAI (Figure 7.16 in Chapter 7). In one Allende CAI, larnite (Ca_2SiO_4) appears to be replacing melilite. Another phase present in many Allende chondrules is awaruite (Ni_3Fe) (Figure 11.4). The Bali-like oxidized subgroup ($CV3_{OxB}$) has undergone aqueous alteration that produced phyllosilicate, pentlandite, tetrataenite (FeNi), and awaruite grains in the matrix. Members of the reduced subgroup ($CV3_{Red}$) contain moderate amounts of metallic Fe-Ni and troilite, little magnetite, and small to negligible amounts of secondary phases; phyllosilicates are rare. The chondrules in $CV3_{Red}$ chondrites typically experienced little or no alteration. Thermodynamic calculations indicate that the hydrothermal-alteration/metasomatism experienced by the CV chondrites occurred at temperatures in the range of 200°C–300°C.

The multiphase opaque assemblages called Fremdlinge in many CAIs in oxidized CV3 chondrites were initially thought to be possible presolar grains; however, they were later interpreted as products of secondary alteration and sulfidation of metallic Fe-Ni on the CV parent body.

The presence of ^{26}Mg excesses in some primary phases (e.g., anorthite, diopside) in many non-FUN CAIs is due to the *in situ* decay of ^{26}Al. The absence of ^{26}Mg excesses in secondary phases (e.g., nepheline, grossular) in CAIs in oxidized CV chondrites implies that these minerals formed at least several million years later, after ^{26}Al had decayed away.

medium by irradiation or heating of ice-mantled dust containing formaldehyde (CH_2O), ammonia (NH_3), and formic acid (HCOOH).

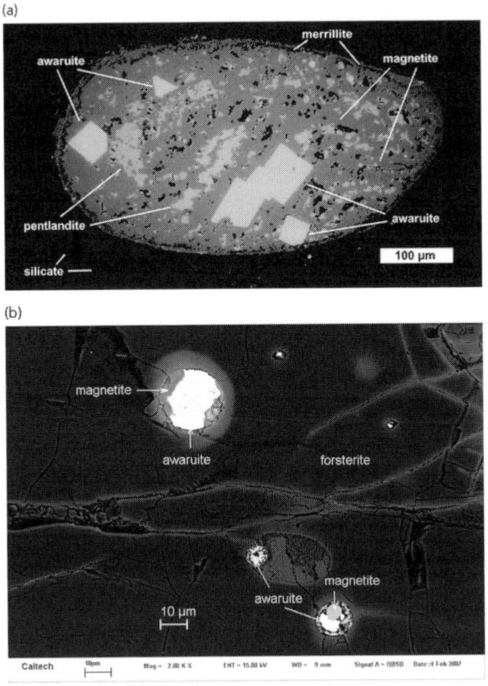

Figure 11.4 (a) Opaque nodule in one low-FeO (Type IA) porphyritic-olivine chondrule from Allende. The nodule is composed mainly of magnetite (dark gray). Embedded in the magnetite are euhedral grains of awaruite (white) and blebby pentlandite (light gray). A somewhat irregular ring of merrillite is present at the inside margin of the nodule. Black regions outside the nodule are silicate. After Rubin (1991). (b) Awaruite- and magnetite-bearing nodules within another low-FeO porphyritic chondrule from Allende. BSE images.

The reduced CV3 chondrites have experienced the least amount of alteration. Nevertheless, there are a number of indicators of alteration: (1) ferrihydrite ($Fe_{4-5}(OH,O)_{12}$) and saponite ($(Ca,Na)_{0.3}(Mg,Fe^{+2})_3(Si,Al)_4O_{10}(OH)_2{\cdot}4H_2O$) replace some ferroan olivine grains in the matrix of Vigarano; (2) the matrix contains minor magnetite and such secondary phases as hedenbergite ($CaFeSi_2O_6$), andradite ($Ca_3Fe_2(SiO_4)_3$), and kirschsteinite ($CaFe(SiO_4)$); (3) these same phases also occur at the surfaces of Wark–Lovering rims, accretionary rims, and CAIs; (4) some chondrules have fine-grained phyllosilicate-bearing rims; and (5) minor carbonate is present. Although some regions of the Vigarano matrix show correlations among elements (Ca, S, Na, K) associated with water-soluble phases (e.g., $CaSO_4$, NaCl, KCl), Vigarano is a breccia and these regions are probably oxidized-CV portions of the rock.

$CV3_{OxA}$ chondrites contain small amounts of phyllosilicates in the matrix (phlogopite ($KMg_3(Si_3Al)O_{10}(F,OH)_2$), montmorillonite ($(Na,Ca)_{0.3}(Al,Mg)_2Si_4O_{10}(OH)_2{\cdot}nH_2O$), clintonite ($Ca(Mg,Al)_3(Al,Si)_4O_{10}(OH,F)_2$), margarite ($CaAl_2(Si_2Al_2)O_{10}(OH)_2$), saponite ($(Ca,Na)_{0.3}(Mg,Fe^{+2})_3(Si,Al)_4O_{10}(OH)_2{\cdot}4H_2O$), and chlorite-group minerals). Some low-Ca-pyroxene-rich chondrules have small amounts of talc, amphibole, and disordered biopyriboles. Allende also contains the carbonate tilleyite in association with wollastonite, grossular, and monticellite. Igneous rims around many chondrules in Allende also contain secondary nepheline ($(Na,K)AlSiO_4$) and sodalite ($Na_4(Si_3Al_3)O_{12}Cl$). Many Allende low-FeO porphyritic chondrules have awaruite- and magnetite-bearing nodules which also contain pentlandite and merrillite (Figure 11.4); these nodules probably formed from metallic-Fe-Ni and troilite assemblages during parent-body alteration. As in the CO3 chondrites, the low $\Delta^{17}O$ values of melilite in Allende (-4±2‰) is consistent with the dissolution of primitive melilite and the reprecipitation of new melilite in the presence of H_2O with high $\Delta^{17}O$.

$CV3_{OxB}$ chondrites contain chondrules and CAIs in which several phyllosilicates occur: Fe-bearing saponite, Na phlogopite, Al-rich serpentine, and Na-K mica. Saponite is also abundant in the matrix of these meteorites. Secondary phases in the subgroup include magnetite, carbonates, secondary phosphates, Ni-rich sulfides, fayalite, Ca-Fe-rich pyroxene, and andradite ($Ca_3Fe_2(SiO_4)_3$).

Many CAIs (particularly fine-grained objects) in the more-altered CV and CO chondrites contain heterogeneous O-isotope compositions: some minerals (anorthite, melilite, grossite, krotite, perovskite, Zr- and Sc-rich silicates, and oxides) are depleted in ^{16}O relative to other phases (corundum, hibonite, spinel, Al-Ti diopside, forsterite, enstatite) (Krot 2019). This probably resulted from exchange with ^{16}O-poor aqueous fluids during parent-body alteration.

Oxidized CV chondrites have much higher porosities (typically 20–28 percent) than reduced CV chondrites (0.6–8 percent); MacPherson and Krot (2002; 2014) suggested the alteration of CV3 chondrites was caused by mobilized water seeping into voids, preferentially altering those particular porous rocks that became oxidized CV chondrites. Rubin (2012b) found that many $CV3_{OxB}$ chondrites are shocked and have strong petrofabrics. When water was later mobilized (presumably by impact heating), the more-fractured, more-porous rocks (i.e., $CV3_{OxB}$ chondrites) were preferentially altered. Although $CV3_{Red}$ chondrites were also shocked and also have strong petrofabrics (Rubin 2012b), they formed from low-porosity materials that inhibited subsequent aqueous alteration.

CV3 chondrites contain dark inclusions, many of which have been aqueously altered: they are enriched in heavy O isotopes; a few contain phyllosilicates; and many contain chondrule pseudomorphs, veins of Ca-rich pyroxenes, and veins with pseudomorphs after phyllosilicates. Some dark inclusions in Efremovka ($CV3_{Red}$) were thermally metamorphosed after being aqueously altered.

11.3.6 Ungrouped Carbonaceous Chondrites

Several ungrouped carbonaceous chondrites show evidence of aqueous alteration, e.g., Tagish Lake, MAC 88107, Kaidun, LEW 85332, GRA 98025, and QUE 99038. All of these samples contain phyllosilicates in their matrices; some have chondrules that have been partly altered to phyllosilicate; some contain magnetite, carbonates, fayalite, hedenbergite, and Fe,Ni sulfides. Kaidun has grains of pyrrhotite that were oxidized to form magnetite framboids (Figure 11.5); the S was apparently leached away by aqueous fluids. This meteorite also contains enstatite-chondrite clasts that have been aqueously altered; secondary phases include a poorly crystalline Si-Fe phase, serpentine, plagioclase, and calcite. (Although listed in the Meteoritical Bulletin Database as a carbonaceous chondrite, Kaidun is actually a breccia composed mainly of

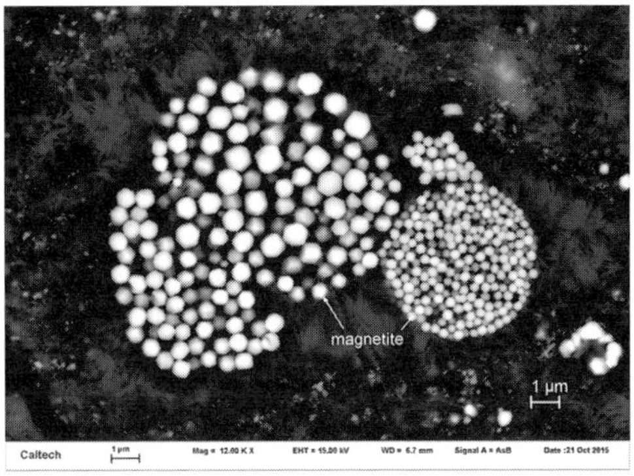

Figure 11.5 Magnetite framboids formed from grains of pyrrhotite in the Kaidun carbonaceous chondrite. BSE image.

subequal amounts of carbonaceous-chondrite and enstatite-chondrite materials and lesser amounts of ordinary-chondrite, R-chondrite, and achondrite clasts.)

These ungrouped carbonaceous chondrites are likely derived from otherwise-unsampled asteroids, attesting to the commonality of parent-body aqueous-alteration processes.

11.3.7 Ordinary Chondrites

Aqueous alteration has affected essentially all type-3 OCs (e.g., Taylor et al. 1981; Krot et al. 1997b; Grossman and Brearley 2005). This is consistent with the presence of 1.2 wt% H_2O^+ (indigenous water) in the Semarkona LL3.00 fall. The aqueous alteration of these rocks cannot be cleanly separated from the changes induced during the earliest stages of thermal metamorphism; the alteration process itself is sometimes referred to as fluid-assisted metamorphism.

The fine-grained silicate matrix of Semarkona contains Fe-rich smectite as well as opaque phases (magnetite, pentlandite, cohenite, haxonite, edscottite, Ni-rich metal) that formed from kamacite and troilite. The high-Δ^{17}O composition of magnetite grains in Semarkona and LL3.6 Ngawi is not in equilibrium with those of OCs olivine and pyroxene, indicating post-accretionary alteration of metallic Fe by Δ^{17}O-rich H_2O. Portions of the glassy mesostases of some Semarkona chondrules have been hydrated and leached; the sharp lobate boundaries between hydrated altered glass and unaltered isotropic glass is readily discerned. Smectite also occurs within some Semarkona chondrules. Nevertheless, the occurrence of pristine amorphous silicates in the Semarkona matrix (Dobrică and Brearley 2020) suggests nonuniform alteration.

Semarkona and all other type-3 OCs contain "bleached" radial pyroxene and cryptocrystalline chondrules characterized by having porous outer regions in which mesostasis was broken down, facilitating the loss of alkalis, Ca, and Al (Grossman et al. 2000). A few bleached chondrules have also been identified in type-4–-6 OCs despite the metamorphic recrystallization of these rocks. Trains of calcite crystals are present near some chondrules.

The composition of the fluid that caused alteration of OCs may have varied with time as additional phases were altered. Apatite formed from a fluid that was dry, acidic, and halogen rich (with a high Cl/F ratio). Secondary phosphates, carbides, and magnetite formed from a fluid that was P- and CO_2-bearing. Smectite (which contains water of hydration) must have formed from an H_2O-rich fluid. Albitization of metamorphic plagioclase required an alkali-rich fluid that presumably contained some K. During cooling from peak metamorphic temperatures, K-feldspar lamellae exsolved from the albite grains.

Tieschitz (H/L3.6) contains geode-like voids and veins filled with sodic-calcic amphiboles (winchite ($\square NaCa(Mg_4Al)Si_8O_{22}(OH)_2$) and barroisite ($\square NaCa(Mg_3Al_2)(Si_7Al)O_{22}(OH)_2$)) emplaced during hydrothermal alteration. The same fluid that produced the amphiboles also gave rise to the so-called white matrix (Christophe Michel-Levy 1976; Hutchison et al. 1998) of this meteorite.

11.3.8 R Chondrites

R chondrites contain abundant matrix material (an average of 42 vol%), far higher than in OCs (12 vol%). Equilibrated R chondrites have the highest mean olivine Fa values (37–42 mol%) among chondrite groups. R chondrites also have high bulk $\Delta^{17}O$ values (+2.72±0.31‰). These three properties are probably related; it seems plausible that R chondrites accreted a large amount of matrix material laden with ^{17}O-rich water (either occurring as ice or bound within phyllosilicates). During parent-body metamorphism, the water was mobilized and the rocks were oxidized; the bulk samples reflect the high $\Delta^{17}O$ values.[2] This is consistent with the presence of magnetite within sulfide-bearing intergrowths in a highly unequilibrated clast in the R-chondrite genomict breccia PCA 91241; magnetite attests to parent-body aqueous alteration of metallic Fe-Ni and sulfide. Magnetite is otherwise rare in R chondrites. Metasomatic alteration of rare CAIs produced nepheline and sodalite in R chondrites.

The strongest evidence for parent-body hydrothermal alteration of R chondrites is the occurrence of 10–15 vol% amphibole (ferri-magnesiohornblende ($Ca_2Mg_4Al_{0.75}Fe^{+3}{}_{0.25}(Si_7AlO_{22})(OH)_2$) or edenite ($NaCa_2Mg_5Si_7AlO_{22}(OH)_2$)) and accessory phlogopite in R6 MIL 11207 and R6 LAP 04840, and accessory (0.01 vol%) titan-phlogopite in R6 MIL 07440 (Figure 11.6). Another water-bearing phase present in several R chondrites is hydroxylapatite ($Ca_5(PO_4)_3OH$).

It is apparent that the amphibole- and mica-bearing R chondrites underwent hydrothermal alteration. The rocks were probably heated to maximum temperatures of 700–900°C under conditions of moderately high PH_2O (perhaps 250–500 bars). Patches within some of the hornblende grains in MIL 11207 and LAP 04840 were dehydrated during annealing, producing 2–20-μm-long subparallel pyroxene blades flanked by pore spaces. The $^{40}Ar/^{39}Ar$ plateau ages of 4,340–4,380 Ma for hornblende separates of LAP 04840 may date this annealing event. A subsequent giant impact would have been necessary to excavate the samples.

[2] It seems likely that the fluid that altered ordinary chondrites had lower $\Delta^{17}O$ values than the one that altered R chondrites.

Figure 11.6 Titan-phlogopite flanked by olivine and plagioclase in the matrix of the MIL 07440 R6 chondrite. After Rubin (2014). BSE image.

11.3.9 Enstatite Chondrites

Minor aqueous alteration has affected enstatite chondrites (e.g., Grossman et al. 1985; El Goresy et al. 1988): (1) There are hydrated NaCr sulfides in troilite-daubréelite clasts and in NaCr sulfide clasts in the Qingzhen EH3 fall. (2) Chondrule glass within Qingzhen chondrules is halogen rich, containing up to 4.4 wt% Cl. Although Qingzhen chondrules contain several primary sulfide minerals (troilite, niningerite, daubréelite, oldhamite, caswellsilverite, and two Cr-sulfides – minerals A and B of Ramdohr 1963), djerfisherite ($K_6Na_9(Fe,Cu)_{24}S_{26}Cl$) (which contains ~1.5 wt% Cl and ~8 wt% K) is virtually absent from chondrules. It seems likely that djerfisherite formed in the Qingzhen matrix after Cl and K were scavenged from chondrule mesostases, possibly by an aqueous fluid. This is consistent with the occurrence of djerfisherite inside veins throughout Qingzhen. (It is also present in millimeter-sized clasts in Qingzhen as well as in other EH3 and EL3 chondrites.)

During subsequent low-level thermal metamorphism, djerfisherite in Qingzhen and other EH3 chondrites broke down to form troilite, covellite, idaite, bornite, and additional phases.

Weisberg et al. (2020) found that EL3 NWA 8785 contains an abundant (33.7 vol%) fine-grained FeO-rich matrix that appears to have been significantly affected by aqueous alteration. The matrix contains abundant magnetite and noncrystalline material rich in SiO_2, MgO, CaO, FeO, S, and Ni. One spinel-rich inclusion is surrounded by sodalite (a common product of aqueous alteration in $CV3_{OxA}$ chondrites). [The only other enstatite-chondrite materials that show evidence of extensive hydrothermal alteration are certain clasts in the Kaidun breccia (e.g., Zolensky and Ivanov 2003).]

11.3.10 Eucrites

Eucrites contain little water and there is no evidence that these basalts were subjected to low-temperature aqueous alteration on their parent asteroid. Nevertheless, Warren et al. (2014) described higher-temperature alteration in the NWA 5738 eucrite driven by a fluid with little H_2O. The rock contains microscopic fluid-metasomatic vein deposits including microveins filled with major Ca-plagioclase and Fe^{3+}-bearing Cr-spinel (indicative of relatively high fO_2). Also present are veins of essentially Ni-free metallic Fe that may have formed from a fluid with appreciably lower fO_2. The cause of the apparent decrease in the fO_2 of the fluid is unknown. However, it is possible that foreign volatile-rich carbonaceous material buried within that region of the asteroid was displaced by an impact into a region with higher temperature and/or lower pressure where lower fluid fO_2 was caused by carbon-fueled smelting. The hypothetical occurrence of buried carbonaceous-chondrite material is consistent with the presence on Vesta (the likely eucrite parent body) of dark material that seems to have been derived from carbonaceous-chondrite projectiles.

The secondary alteration of the NWA 5738 eucrite may constitute a link between metasomatism and shock metamorphism.

11.3.11 Ureilites

The polymict ureilite EET 83309 (which contains solar-wind–implanted noble gases) has bands of opal (hydrated amorphous silica) more than 300 μm long, in some cases associated with schreibersite, olivine, or suessite. Because the O-isotopic composition of the opal is close to the terrestrial fractionation line, this could reflect the exchange of water in the Antarctic environment. If the opal is indeed preterrestrial, this indicates the presence of water during the late stages of ureilite formation. Small grains of halite (NaCl) and sylvite (KCl) are common accessory constituents of ureilites. It is possible they precipitated from a brine. Alternatively, they could have been derived from carbonaceous-chondrite projectiles, clasts of which occur in polymict ureilites.

11.3.12 Mesosiderites

Lorenz et al. (2010) suggested that an anhydrous, gaseous fluid, rich in S_2, CO, and CO_2, penetrated fractures within olivine grains in the Budulan mesosiderite. This process caused the adjacent olivine to become more reduced; it also facilitated the production of secondary mineral assemblages within fractures including aggregates of (troilite+(low-Ca pyroxene)) and ((metallic Fe-Ni)+pyroxene). A few olivine grains were replaced by aggregates of troilite, pyroxene and silica. It is not clear if the fluid was produced on the parent asteroid during differentiation or was ultimately derived from a projectile containing organic matter and sulfides.

11.3.13 Lunar Rocks

Apollo samples exhibit evidence of metasomatism. (1) Fluorapatite is present in lunar granulite breccia 79215; vapor-phase metasomatism could have added P and halogens to a typical granulite via a process akin to terrestrial rock alteration by fumaroles. (2) Primary igneous olivine in Mg-suite lithologies of subsamples of 69715 was pseudomorphically replaced by secondary silicate, sulfide, and oxide. This may have been caused by a metasomatic fluid consisting predominantly of H_2 and CH_4 with minor H_2O, CO, and H_2S; the fluid composition is consistent with late-stage degassing of a shallow intrusive magma. Although these features have not been reported in lunar meteorites as of this writing, it seems likely that additional study will reveal their presence. Although the Moon was long considered very dry (e.g., Heiken et al. 1991), recent studies have shown that the amount of water in the lunar interior may average ~745 μg/g (Saal et al. 2008), somewhat below that of glasses in terrestrial mid-ocean-ridge basalts (MORBs): 1,700–6,000 μg/g (Jambon and Zimmerman 1990).

11.3.14 Martian Meteorites

In view of the abundant evidence for the presence of liquid water on Mars in the past, it is not surprising that martian meteorites show evidence of aqueous alteration. Aqueous fluids in the Gale Crater region of Mars appear to have been enriched in Na, K, and Si, and depleted in Mg, Fe, and Al relative to modeled martian fluids.

The numerous multimillimeter-sized spheroids of hematite (dubbed "blueberries") in Meridiani Planum appear to be concretions similar to those in the Navajo Sandstone in southern Utah, USA. These are diagenetic products formed by the precipitation of aqueous fluids in pores within sediments. Although hematite blueberries have not been reported in martian meteorites, grains of hematite do occur. Some occurrences of hematite in association with jarosite may have been produced by higher-temperature magmatic-hydrothermal fluids.

The topic of aqueous alteration in martian meteorites was reviewed by Velbel (2012); the following list is derived mainly from that paper. Hydrous phases include amphibole, biotite, apatite, and opal. Nakhlites contain phyllosilicates: smectite (saponite), serpentine, and so-called iddingsite (apparently a smectite-hydroxide assemblage). Halite is also present. According to Treiman (2005), nakhlites were invaded by water ~620 Ma ago; the water altered olivine and mesostasis glass and produced "iddingsite" and halite.

Two nakhlites (MIL 03346 and NWA 5790) contain potassic-chloro-hastingsite within melt inclusions that formed after trapped chloride-rich fluids reacted with adjacent silicate. Glass within some melt inclusions in Nakhla were replaced by berthierine.

All varieties of martian meteorites contain carbonate: calcite, aragonite ($CaCO_3$), siderite-magnesite ($FeCO_3$-$(Mg,Fe)CO_3$) solid solution, and dolomite-ankerite solid solution. Some carbonates occur in veins; some fill fractures and voids. Sulfates include gypsum ($CaSO_4 \cdot 2H_2O$), anhydrite ($CaSO_4$), bassanite ($CaSO_4 \cdot \frac{1}{2}H_2O$), and jarosite ($KFe_3(SO_4)_2(OH)_6$) (in association with hematite).

The ALH 84001 orthopyroxenite experienced aqueous alteration and collisions. Aqueous alteration was responsible for forming carbonates. Subsequent decomposition of the carbonate appears to have produced magnetite and periclase.

Some REE minerals (monazite $(Ce,La,Th)PO_4$), xenotime (YPO_4), and chevkinite-perrierite ($(Ce,La,Ca,Th)_4(Fe^{+2},Mg)_2(Ti,Fe^{+3})_3Si_4O_{22}$) in the martian meteorite breccias NWA 7034 and NWA 7533 most likely formed by fluid-rock interaction in the martian crust (Liu et al. 2016).

11.4 Precipitation from Asteroidal Brines

Halite (NaCl) and sylvite (KCl) are evaporite minerals that precipitate from briny fluids after enough water has evaporated to cause the remaining liquid to become supersaturated with these salts. Dark blue crystals of halite, typically a few hundred micrometers to a few millimeters in size, occur in the Zag and Monahans (1998) H-chondrite regolith breccias. Halite grains in Monahans (1998) also contain small inclusions of sylvite. Age dating of the halite grains show them to be ~4.0–4.5 Ga old. The halite grains contain fluid inclusions (brines trapped within crystal fractures) associated with a variety of soluble and insoluble organic compounds including ketones and amino acids (e.g., Chan et al. 2018).

One possibility is that the halite grains formed on the H-chondrite asteroid. Jones et al. (2016) reported that apatite is relatively abundant in the same matrix regions of Zag where halite occurs. Both phases are Cl rich and both could have formed from a Cl-rich fluid that was heated either by impacts or the decay of ^{26}Al.

The alternative is that the halite grains are foreign materials. This is consistent with the fact that Zag and Monahans (1998) were derived from the surface of their parent asteroid. Chan et al. (2018) concluded that the halite grains formed via hydrovolcanism on a C-type asteroid, probably 1 Ceres (~945 km in diameter). According to their model, at the end of a hydrothermal cycle, the halite crystals precipitated from the fluid on the parent-body surface, trapping brines and an assortment of organic compounds. This scenario is consistent with the discovery of an extrusive cryovolcanic dome on Ceres imaged by the *Dawn* spacecraft's framing camera. This 5-km-high mountain, dubbed Ahuna Dome, is thought to have formed from a mixture of salt and mud late in the history of the Solar System.

If the halite crystals are from Ceres, they may have been expelled from the asteroid by hydrothermal activity. In that case, exposure to ionizing radiation in interplanetary space is probably responsible for causing them to become blue-purple in color (by the capture of electrons in anion vacancies). Eventually, the grains accreted to the regolith of the H-chondrite parent body and were incorporated into Zag and Monahans (1998) (Zolensky et al. 2016a).

11.5 Shock Metamorphism

There is a myriad of features produced in meteorites by shock metamorphism. High-pressure phases in meteorites, formed by shock, are listed in Table 11.1.

11.5.1 Ordinary Chondrites

Different shock stages correspond to different ranges in shock pressure: S1–S2 (<4–5 GPa), S2–S3 (5–10 GPa), S3–S4 (10–20 GPa), S4–S5 (20–35 GPa), S5–S6 (35–60 GPa), S6–S7 (>45 GPa). The higher the pre-impact temperature of the target, the lower the shock pressure necessary to produce the mineralogic and petrologic effects associated with the shock stages (Table 8.3 in Chapter 8).

Olivine grains in shocked ordinary chondrites contain numerous microstructural dislocations. The greater the degree of shock, the higher the average dislocation density (in the absence of post-shock annealing). Post-shock annealing assists in microstructural recovery of the olivine crystal lattice. Features produced by annealing include curved dislocations, small dislocation loops, so-called picket fence subgrain boundaries, destruction of dislocations, and healing of cracks (e.g., Ruzicka et al. 2015). These workers found that annealed chondrites such as H6

Table 11.1 *High-pressure phases in meteorites*

Phase	Formula
Ahrensite	Fe_2SiO_4
Akimotoite	$(Mg,Fe)SiO_3$
Albitic Jadeite	$(Na,Ca,\square_{1/4})(Al,Si)Si_2O_6$
Asimowite	Fe_2SiO_4
Bridgmanite	$(Mg,Fe)SiO_3$
Chaoite	C
Chenmingite	$FeCr_2O_4$
Coesite	SiO_2
Diamond	C
Donwilhelmsite	$CaAl_4Si_2O_{11}$
Feiite	$Fe^{2+}{}_2(Fe^{2+}Ti^{4+})O_5$
Hemleyite	$FeSiO_3$
Hiroseite	$FeSiO_3$
Icosahedrite II	$Al_{62}Cu_{31}Fe_7$
Jadeite	$NaAlSi_2O_6$
Liebermannite	$KAlSi_3O_8$
Lingunite	$NaAlSi_3O_8$
Liuite	$FeTiO_3$
Majorite	$Mg_3(MgSi)Si_3O_{12}$
Maskelynite	$(Na,Ca)(Si,Al)_4O_8$
Magnesiowüstite	$(Mg,Fe)O$
Poirierite	Mg_2SiO_4
Reidite	$ZrSiO_4$
Ringwoodite	Mg_2SiO_4
Seifertite	SiO_2
Spinelloid Silicate	$(Mg,Fe,Si)_2(Si,\square)O_4$
Spinelloid Silicate-II	$(Fe,Mg,Cr,Ti,Ca,\square)_2(Si,Al)O_4$
Stishovite	SiO_2
Stöfflerite	$CaAl_2Si_2O_8$
Tetragonal Almandine	$(Fe,Mg,Ca,Na)_3(Al,Si,Mg)_2Si_3O_{12}$
Tetragonal Majorite	$Mg_3(MgSi)Si_3O_{12}$
Tschaunerite	$Fe^{2+}(Fe^{2+}Ti^{4+})O_4$
Tuite	$Ca_3(PO_4)_2$
Vestaite	$(Ti^{4+}Fe^{2+})Ti^{4+}{}_3O_9$
Wadsleyite	$(Mg,Fe)_2SiO_4$
Wangdaodeite	$FeTiO_3$
Xieite	$FeCr_2O_4$

Kernouvé, H6 Portales Valley, and LL6 MIL 99301 have low dislocation densities, reflecting microstructural recovery caused by significant post-shock annealing.

Several minerals were produced directly or indirectly by impact processes in OCs at different shock stages.

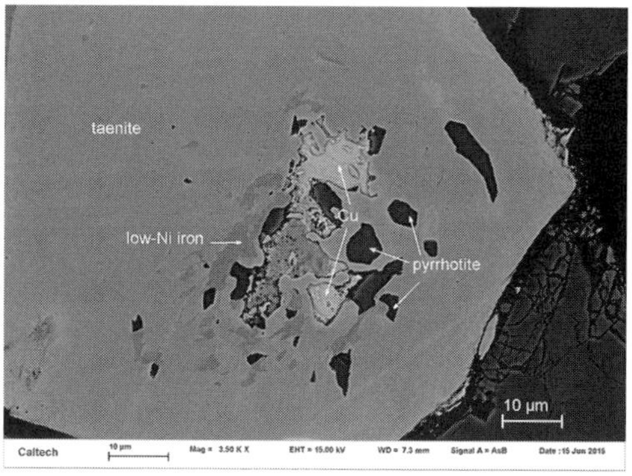

Figure 11.7 Metallic Cu within low-Ni iron and pyrrhotite in taenite in the Suizhou L6 chondrite. BSE image.

(1) Chromite and plagioclase (which occurred in proximity prior to the impact event) were shock melted, later crystallizing in chromite veinlets and chromite-plagioclase assemblages. Temperatures probably reached $\geq$1,100°C.

(2) Metallic Cu grains (Figures 5.3 and 11.7) in shocked equilibrated ordinary chondrites most commonly occur within Ni-rich metal adjacent to small, irregular grains of troilite. Because taenite can accommodate ~16 times more Cu than kamacite (Hansen 1958), metallic Cu is preferentially enriched in Ni-rich metal phases. One model for forming the metallic Cu grains (modified from Tomkins 2009) starts with troilite being heated, either directly by a shock wave or during post-shock annealing. The FeS then began to break down into metallic Fe and S_2. The liberated sulfur may have migrated into metal along concentration gradients to produce new small grains of troilite and Ni-enriched metal. As the reaction continued and additional new troilite formed at the expense of bulk metal, Cu eventually reached supersaturation in the Ni-rich metal grains and nucleated at troilite/metal grain boundaries (high-surface-energy sites).

(3) Small irregular grains of troilite crystallized within larger masses of metallic Fe-Ni after metal-sulfide assemblages were melted and quenched.

(4) Rapidly crystallized metal-sulfide melts crystallized finely intergrown globular assemblages of metal and sulfide (sometimes called "fizzed troilite").

(5) Quenched metallic Fe-Ni melts produced martensite, which, in some cases, annealed to form patches of plessite.

(6) Shock-melted metal and sulfide were injected into fractures within silicate grains to form opaque veins upon quenching.

(7) Curvilinear trails of small (metallic Fe-Ni)$\pm$troilite blebs entered narrow fractures in the silicate grains (Figure 11.8), causing the whole rocks to become darker; this phenomenon, known as "shock darkening" or "silicate darkening," can form so-called black chondrites. It results from the shock melting of metal and sulfide (which melt at a lower temperature than silicates, i.e., ~960°C vs. ~1,100°C).

(8) Molten silicate formed at $\geq$1,100°C and was also injected into fractures in silicate grains. Within some fractures, the melt quenched into dark-colored glassy (pseudotachylite-like)

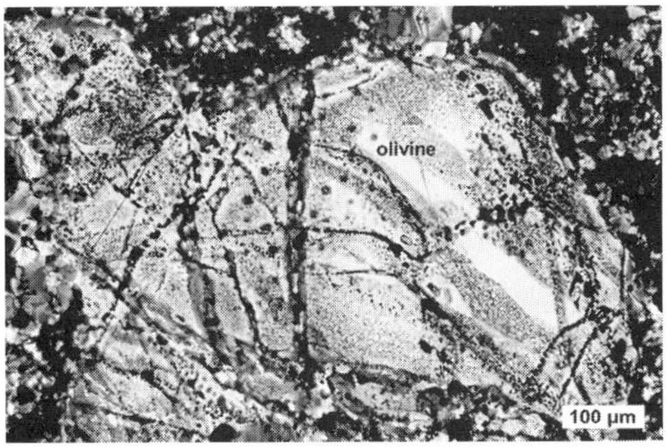

Figure 11.8 Olivine grain transected by numerous, intersecting, curvilinear trails of small blebs of metallic Fe-Ni and troilite in the MIL 99301 LL6 chondrite. The olivine exhibits sharp optical extinction because it underwent post-shock annealing. After Rubin (2002a). Transmitted light, crossed polarizers. (A black-and-white version of this figure will appear in some formats. For the colour version, please refer to the plate section.)

veins; in other fractures, fine-grained mafic silicate phases crystallized. Nanophase iron particles were identified within pyroxene glass in a shock-melt vein from an L6 chondrite.

(9) Melt pockets containing silicate glass and small grains of chromite, olivine, pyroxene, and plagioclase formed by localized shock melting at temperatures ≥1,100°C.

(10) Orthopyroxene grains were impact heated to between 950°C and 1,230°C (or higher) and entered the protoenstatite field; upon quenching, they transformed into clinoenstatite with polysynthetic twinning parallel to (100).

(11) Clinopyroxene grains in some so-called type-7 OCs (most of which are probably impact-melt breccias) contain very fine exsolution lamellae most easily detected by backscattered electron imaging. The lamellae formed during rapid cooling from high temperatures.

(12) Impact-melt rock clasts formed in many OCs by the melting of target material prior to final lithification; these clasts typically contain major olivine and low-Ca clinopyroxene and minor diopside with interstitial K-rich feldspathic glass.

(13) A few agglutinates (vesicular, glassy objects containing chunks of rock, glass, and mineral fragments) formed in some OC regolith breccias (e.g., Fayetteville).

(14) Some regions in OC breccias contain crushed silicate grains surrounded by fine-grained troilite; the troilite/(metallic Fe-Ni) modal ratios range up to ~500:1. This is much higher than the eutectic weight ratio of 7.5:1, suggesting that the troilite-rich regions formed by deposition from a shock-generated S-rich vapor (Rubin 2002b).

(15) Many shocked OCs have millimeter- to centimeter-sized vugs that contain some euhedral mineral crystals (metallic Fe-Ni, troilite, low-Ca pyroxene, albite, fayalitic olivine, merrillite, and possibly chromite) that appear to have been deposited by an impact-generated vapor (Olsen 1981). These occurrences are similar to euhedral mineral crystals (metallic Fe-Ni, troilite, ilmenite, pyroxene, plagioclase, merrillite, and apatite) inside vugs in lunar breccias that were previously interpreted to have formed through vapor deposition during shock.

(16) A euhedral tetrataenite grain within an impact-melt-rock clast in LL6 Jelica formed after impact melting and crystallization of euhedral taenite (Rubin 1994a). After cooling to 320°C, the taenite grain underwent an ordering reaction and formed tetrataenite.

(17) Pyrophanite ($MnTiO_3$) in a Mg-Al-chromite fragment in H3.8 Raguli most likely formed from an impact melt of a chromite-rich assemblage containing Mn-rich ilmenite.
(18) Plagioclase transforms into maskelynite at high shock pressures.
(19) High-pressure phases formed in the vicinity of silicate melt veins in many highly shocked OCs (summarized by Tomioka and Miyahara, 2017). These phases include ringwoodite (a polymorph of olivine with the spinel structure) (Figure 11.9(a)); wadsleyite (the β-spinel polymorph of olivine) (Figure 11.9(b)); majorite (a polymorph of orthopyroxene with the garnet structure) (Figure 11.9(c)); jadeite (monoclinic C2/c sodic pyroxene) (Figure 11.9(d)); high-pressure clinoenstatite, magnesiowüstite, ahrensite (the high-pressure polymorph of ferroan olivine) (Figure 11.9(e)); lingunite ($NaAlSi_3O_8$-hollandite) (Figure 11.9(f)); akimotoite (a polymorph of enstatite with the ilmenite structure) (Figure 11.9(g)); hemleyite (Fe analog of akimotoite), bridgmanite ($MgSiO_3$-perovskite) (Figure 11.9(h)); hiroseite (Fe analog of bridgmanite; Bindi and Xie 2019); coesite (a monoclinic polymorph of silica); stishovite (a tetragonal polymorph of silica); two post-stishovite polymorphs of silica (one with an orthorhombic α-PbO_2 structure and one with a monoclinic ZrO_2-like structure); xieite ($CaTi_2O_4$-type orthorhombic polymorph of chromite); chenmingite ($CaFe_2O_4$-type orthorhombic polymorph of chromite); tuite (a trigonal polymorph of merrillite); and a vacancy-rich, partially inverted tetragonal spinelloid silicate ($(Mg,Fe,Si)_2(Si,\square)O_4$; Ma et al. 2019b) (Figure 11.9(i)). Such phases have been reported in all three OC groups: H, L, and LL. The Château-Renard L6 chondrite also contains sodic pyroxene (with the composition and structure of omphacite) in association with ahrensite and wadsleyite (Baziotis et al. 2018).
(20) Some large metal nodules and elongated metal veins in OCs formed by a complicated process involving impact vaporization of bulk metal-sulfide assemblages, oxidation of W and Mo to form volatile oxides, partial oxidation of Re and Os, separation of the oxidized siderophiles from the refractory siderophiles, rapid fractional condensation of refractory siderophiles, transport of the residual vapor, condensation of Cu, S, and Se into vugs and fractures to form sulfide, and condensation of the remaining siderophiles to form metal nodules and veins (Widom et al. 1986; Rubin 1999). During vapor cooling and condensation, the residual W, Mo, Re, and Os oxides were reduced. One $4 \times 9 \times 12$ mm metal nodule in the Rose City impact-melt breccia is divided into three compositionally distinct zones: one end is enriched in refractory Os and Ir by 30–40 percent relative to bulk H-chondrite metal, the middle is depleted in Os and Ir by 31–35 percent, and the other end is depleted by 75 percent. This one heterogeneous metal nodule shows that the vapor transport distance was on the order of 5 mm.

Some OCs were totally (or nearly totally) impact melted; these rocks include PAT 91501, Y-74160, Ramsdorf, and the vesicular meteorite LAR 06299. The principal phases in these rocks (olivine, low-Ca pyroxene, Ca-pyroxene, Cr-rich spinels, metallic Fe-Ni, and troilite, with minor sodic plagioclase and silico-feldspathic interstitial melt) crystallized largely or entirely from the impact-generated melt.

There are also several dozen impact-melt breccias that contain shocked, but largely unmelted, chondritic clasts surrounded by impact-generated melt. In some cases, veins of impact melt penetrate the shock clasts. These samples include H chondrites (e.g., Rose City,

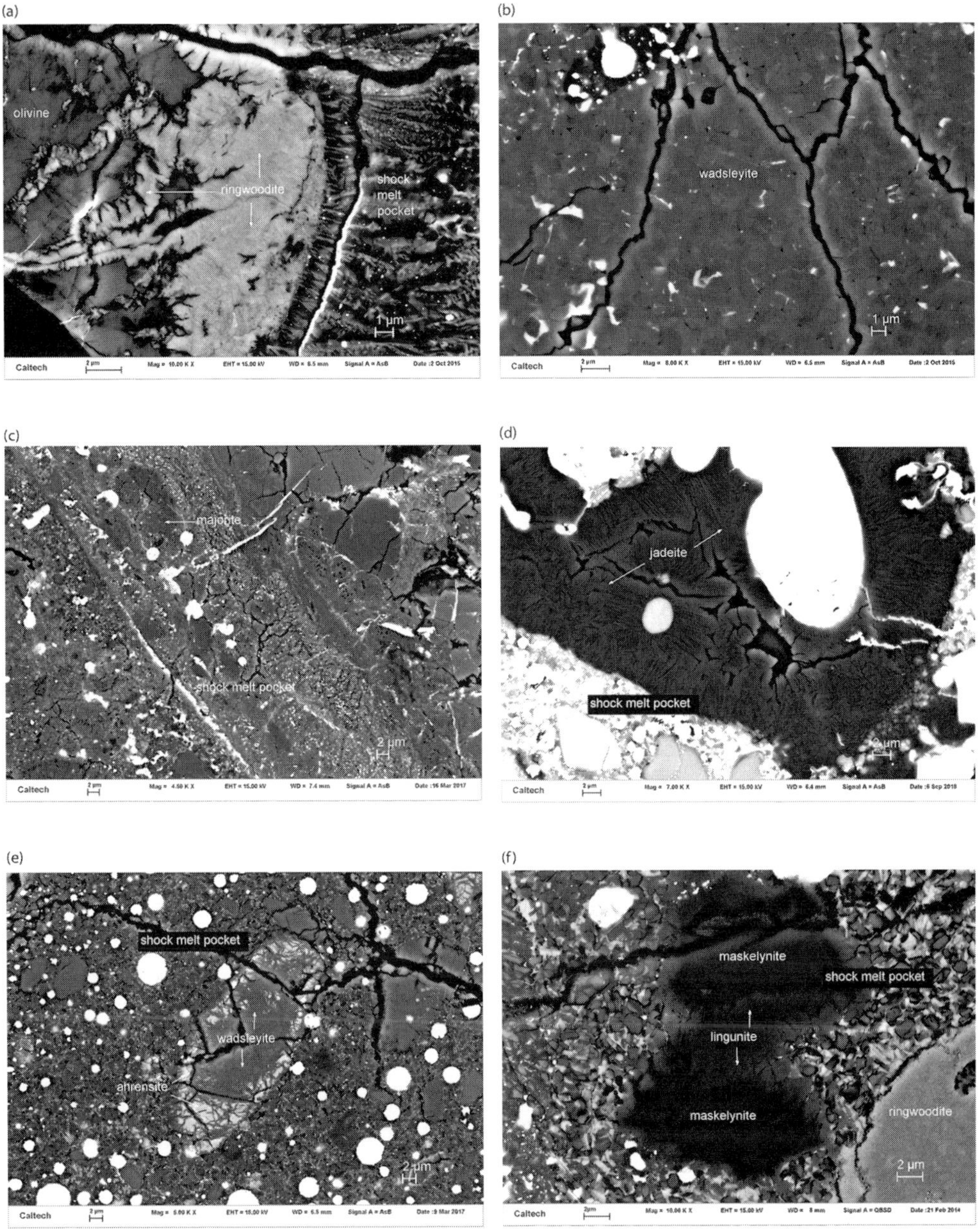

Figure 11.9 High-pressure phases identified in L6 ordinary chondrites. (a) Ringwoodite in NWA 6525, (b) wadsleyite in NWA 6525, (c) majorite in Château Renard (Baziotis et al. 2018), (d) albitic jadeite from TIL 08001 (Ma et al. 2020b), (e) ahrensite in Chateau Renard (Baziotis et al. 2018), (f) lingunite in Tenham, (g) akimotoite in Tenham, (h) bridgmanite in Tenham (Tschauner et al. 2014), (i) spinelloid silicate in Tenham (Ma et al. 2019d). BSE images.

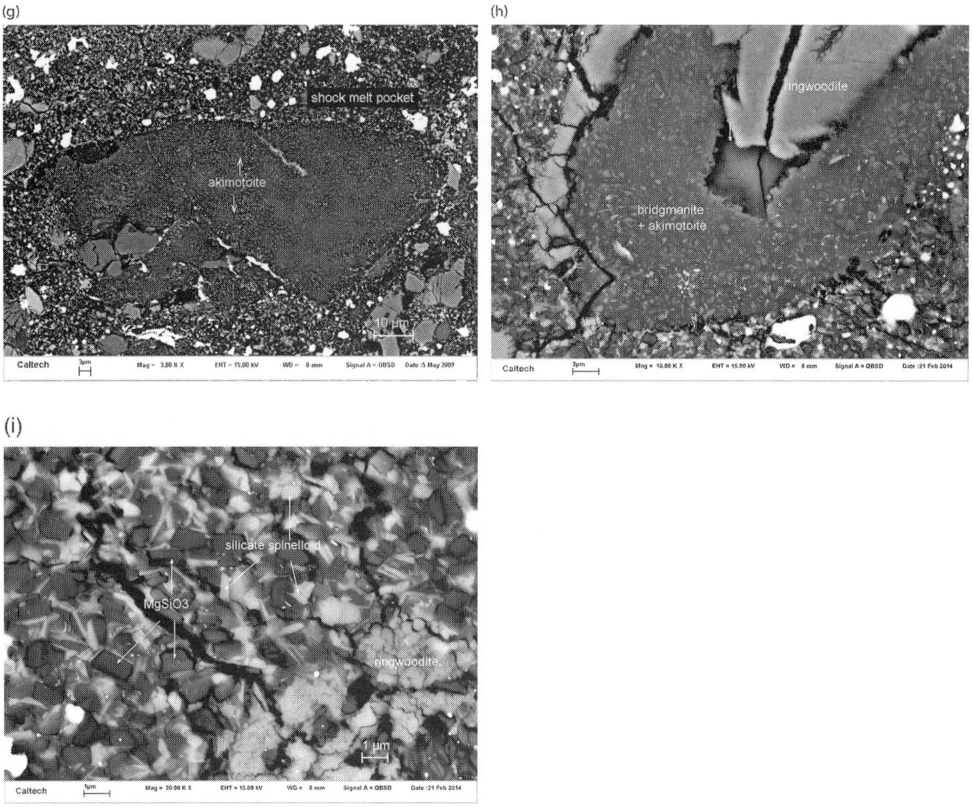

Figure 11.9 (*cont.*)

Yanzhuang), L chondrites (e.g., Cat Mountain, Chico, NWA 2085, Orvinio, Shaw), and LL chondrites (e.g., Bison, Y-790143, Y-790964).

11.5.2 R Chondrites

A few porphyritic impact-melt-rock clasts have been found in R chondrites (R3.5–6 Dar al Gani 013, R3.8–6 PCA 91241, and R3.6 LAP 031275) (e.g., Bischoff et al. 2011; Lunning et al. 2020). Most of the clasts are of R-chondrite provenance and consist of normally zoned olivine phenocrysts ($Fa_{\sim 15\text{–}50}$; Fe/Mn ≈ 65–97) within glassy mesostasis of basaltic andesite bulk chemical composition. The sole-identified clast in LAP 031275 appears to be L-chondrite-like; it contains olivine phenocrysts ($Fa_{17\text{–}27}$; Fe/Mn = 43–60), partly surrounded by smaller pyroxene grains, all within a feldspathic glassy mesostasis.

11.5.3 Carbonaceous Chondrites

Mineral phases formed by shock-metamorphic processes are relatively rare in carbonaceous chondrites. These are described in this section.

CI chondrites. The Orgueil CI chondrite contains an agglutinate with desiccated phyllosilicates and impact-melted matrix glass that has partially devitrified to olivine. Also present in this meteorite are vesicular glass beads that were interpreted as impact products (Zolensky et al. 2015b).

CM chondrites. Zolensky et al. (2015a, 2016b) reported vesicular glass beads that have been altered from glass to serpentine in CM2 Nogoya, melted sulfides in Jbilet Winselwan (CM2), and shock melt veins in CM2 lithologies in the Kaidun polymict breccia. An agglutinate grain was found in CM2-an Diepenveen.

A few CM chondrites experienced secondary heating after having been aqueously altered. Dehydration reactions in these rocks produced olivine and metallic Fe-Ni as secondary anhydrous phases. Nakamura (2005) divided the CM chondrites into five heating stages based on their abundances of phyllosilicate and secondary anhydrous phases: Unheated, I, II, III, and IV. Kimura et al. (2011) placed the CM chondrites into three categories based on the inferred degree of secondary heating: Category A (corresponding mainly to the Unheated samples of Nakamura) is characterized by kamacite and/or martensite; Category B (corresponding to Stage II and Stage III samples) is characterized by pyrrhotite containing blebs or lamellae of pentlandite; and Category C (corresponding to Stage III and Stage IV samples) is characterized by opaque assemblages consisting of kamacite, Ni-Co-rich metal, and pyrrhotite. The absence of coarse-grained assemblages of kamacite, plessite, and Ni-rich metal in CM chondrites led to the suggestion that the opaque assemblages in CM Category C formed by rapid cooling following impact heating, a view consistent with the heating experiments on Belgica 7904.

CR chondrites. The Meteoritical Bulletin Database currently lists NWA 6921 as a CR6 chondrite and suggests that it is paired with several specimens (NWA 2994, 3250, 6901, and 7317); two of these samples (NWA 3250 and 6901) are individually listed as primitive achondrites. These rocks are generally described as having poikiloblastic to porphyroblastic textures and containing some relict chondrules (particularly BO types) and chondrule fragments. These descriptions resemble those of OC impact-melt rocks such as PAT 91501. It thus seems plausible that NWA 6921 and its paired specimens are CR-related impact-melt rocks.

The Meteoritical Bulletin Database currently lists NWA 7531 as a CR7 polymict breccia. It is described as a chondrule-free rock with a poikiloblastic texture; it resembles some essentially chondrule-free OC impact-melt rocks such as Y-74160 and is likely a CR-related impact-melt rock, very similar to NWA 6921.

CO chondrites. Scott et al. (1992) reported shock melting at the interface between metal and sulfide in CO3 Lancé; at this location, there are micrometer-sized intergrowths of the two phases, indicative of melting and rapid crystallization. Although most olivine grains in this chondrite exhibit sharp optical extinction (shock stage S1), localized areas experienced impact melting where temperatures apparently exceeded the Fe-Ni-FeS eutectic.

CV-CK chondrites. The matrix in CV3 Allende contains oriented fayalitic olivine grains, forming a uniaxial crystallographic fabric (Watt et al. 2006; Bland et al. 2011). The meteorite also has a strong planar magnetic fabric, aligned with the remanent magnetization of the matrix (Muxworthy et al. 2017). (The principal magnetic mineral in Allende is pyrrhotite.) All these features were probably produced by impact compaction.

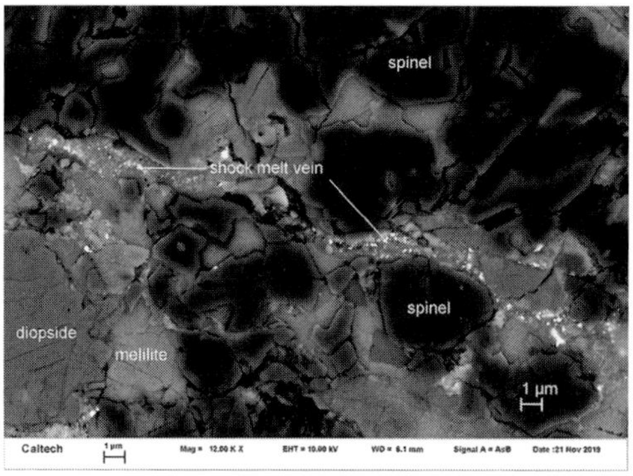

Figure 11.10 An impact-induced melt vein in a Ca-Al-rich inclusion from the Efremovka ($CV3_{Red}$) chondrite. BSE image.

Efremovka ($CV3_{Red}$) contains impact-melted cellular metal-troilite assemblages, opaque veins, and melt pockets (Figure 11.10). There are carbonaceous-chondrite impact-melt clasts in two CV3 chondrites: $CV3_{Red}$ RBT 04143 and $CV3_{oxA}$ LAR 06317. The clasts contain no evidence for water and have zoned olivine microphenocrysts embedded in vesicular glass; these features show that these objects have been melted, devolatilized, and quenched.

Many CK chondrites exhibit silicate darkening caused by curvilinear trails of abundant tiny grains of magnetite and pentlandite within silicate grains. By analogy to silicate darkening in OCs (caused by trails of small blebs of metal and sulfide), it seems plausible that darkening in CK chondrites was due to impact melting and dispersion of metal and sulfide followed by oxidation to form magnetite and pentlandite. Also present in some CK5 and CK6 chondrites are black glassy to microcrystalline shock veins several millimeters in length. These veins formed from impact-melted silicate; they contain submicrometer-sized grains of magnetite and pentlandite (that are probably postimpact oxidation products of metal and troilite).

EET 83311 (CK5; shock stage S4) has undergone extensive *in situ* melting. Some regions of this meteorite contain tiny olivine grains surrounded by a feldspathic melt; also present in the melt are relict, shocked-but-unmelted olivine grains.

CH chondrites. Dmitryivanovite is a high-pressure polymorph of $CaAl_2O_4$ found in a CAI in the NWA 470 CH3 chondrite. Krotite ($CaAl_2O_4$) was modeled as having condensed in the solar nebula under local conditions that included a high dust/gas ratio; the phase was later incorporated into the CAI. After agglomeration, an impact on the CH3 parent asteroid transformed the phase into dmitryivanovite.

CB chondrites. Patches of impact melt occur between the coarser components (chondrule-like objects and metal nodules) in several Bencubbin-like (CB) carbonaceous chondrites. Also present within the matrix and chondrule-like fragments of CB_a Gujba are high-pressure phases (majorite, possibly majorite-pyrope$_{ss}$, wadsleyite, and coesite). The occurrence of a majorite-wadsleyite assemblage suggests that pressures of $\gtrsim$19 GPa and temperatures of $\gtrsim$2,000°C were reached during the shock event (Weisberg and Kimura 2010).

11.5.4 Enstatite Chondrites

There are numerous shock features in enstatite chondrites, reviewed by Rubin (2015a).

(1) *Progressive deformation of the crystal structures of olivine, low-Ca pyroxene, and plagioclase* are reflected in the shock-stage classification scheme of Stöffler et al. (1991, 2018) and Rubin et al. (1997). Shock stages range from S1 (unshocked) to S6 (very strongly shocked). In unshocked S1 materials, olivine, low-Ca pyroxene, and plagioclase have either no fractures or irregular fractures and exhibit sharp optical extinction when viewed microscopically in transmitted light using crossed polarizers. With increasing levels of shock, olivine develops undulose extinction, planar fractures, and mosaic extinction; low-Ca pyroxene develops undulose extinction, planar fractures, clinoenstatite lamellae on (100), and mosaic extinction; plagioclase develops undulose extinction, becomes partially isotropic, transforms into maskelynite, and undergoes extensive melting.

(2) *Brecciation* is evident in several EH (e.g., Abee, Adhi Kot, Parsa) and EL (e.g., Hvittis, Atlanta, Blithfield) chondrites. These samples contain a variety of clast types including impact-melt rocks, some with high sulfide/metal ratios, some with high silica/enstatite ratios, and a few with chondritic textures. The mineral phases in all these clast types (save the chondritic clasts) crystallized from impact melts.

(3) *Petrofabrics* are evident in several EH3 (e.g., ALH 81189, Y-691) and EL3 (e.g., PCA 91020, MAC 02635, EET 90992) chondrites and one EL6 chondrite (LEW 87119), reflecting shearing during impact events. In general, no new mineral phases were produced as the rocks developed foliations.

(4) *Opaque veins* were formed by impact melting enstatite-chondrite materials. Vein types include those rich only in metallic Fe-Ni, those containing metallic Fe-Ni, sulfide, and silicate, and a 16-mm-long oldhamite vein in EL6 Jajh deh Kot Lalu.

(5). *High-pressure phases* were produced during impacts. In enstatite chondrites, only coesite (SiO_2) has been identified. It was found in a shock vein in EH3 Asuka 10164 (Kimura et al. 2017).

(6) *Silicate darkening* involves the *in situ* melting and mobilization of metallic Fe-Ni and troilite, their dispersion through fractures in neighboring silicate grains, and quenching.

(7) *Silicate-rich melt veins* (formed by impacts) occur in several enstatite chondrites: EH3 Y-791790 contains a vein similar in composition to the whole rock, reflecting total melting of the host; EH5 RKPA80259 contains veins of feldspathic glass, reflecting preferential melting of phases with low impedances to shock compression; and EH3 Asuka 10164 contains a vein enriched in fine-grained silicate including coesite, reflecting high transient shock pressures.

(8) *Silicate-rich melt pockets* in EH5 RKPA80259 are rich in feldspathic glass flanking coarse enstatite grains. The feldspathic glass formed by the impact melting of plagioclase due to its high compressibility during shock loading.

(9) *Diamonds* that formed from graphite at high shock pressures occur in the Abee EH chondrite impact melt breccia (Russell et al. 1992; Rubin 2015a).

(10) *Euhedral enstatite grains* in many enstatite chondrites, including Abee, Adhi Kot, and RKPA80259 crystallized from local pyroxene-normative impact melts.

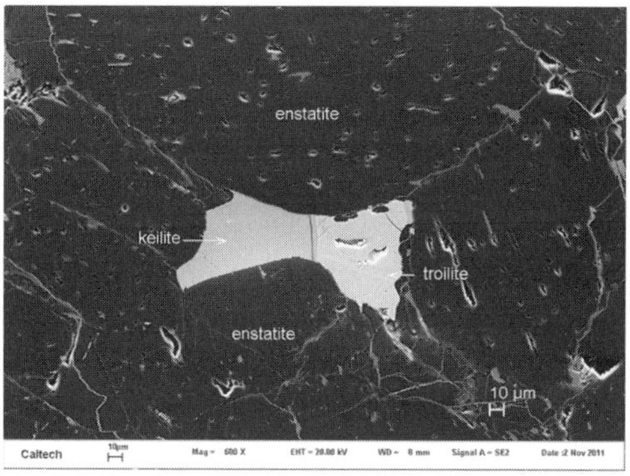

Figure 11.11 Keilite in the Zakłodzie enstatite-rich achondrite. Keilite also occurs in shocked EH chondrites. BSE image.

(11) *Nucleation of enstatite on relict chondrules* occurred in the cooling impact melt in several enstatite chondrites, most particularly Abee.

(12) *Low-MnO enstatite, high-Mn troilite, and high-Mn oldhamite* occur in Abee, Adhi Kot, and Saint-Sauveur. Rubin (2008) suggested that when MnO-bearing enstatite was impact melted under highly reducing conditions, Mn was preferentially partitioned into crystallizing sulfide phases at the expense of crystallizing enstatite. Crystallizing troilite and oldhamite acquired relatively high Mn contents due in part to enrichment of Mn in the melt from the enstatite and by the melting of preexisting Mn-bearing niningerite.

(13) *Keilite* ((Fe,Mn,Mg,Ca,Cr)S) occurs only in enstatite chondrites such as Abee, Adhi Kot, Saint-Sauveur, and RKPA80259 that appear to have been impact melted. Quenching is required for the retention of keilite (Figure 11.11); if these rocks had cooled slowly or been subsequently annealed, the keilite would have exsolved into troilite and niningerite.

(14) *Relatively abundant silica* is present in Abee; some Abee clasts contain up to 16 wt% silica. In Adhi Kot there are 3–5-mm-sized chondrule-free clasts (presumably impact products) that contain up to twice as much silica (18–28 wt%) as enstatite (12–14 wt%). (This is in contrast to most enstatite chondrites, which contain very little free silica.) As a result of impact melting, reduced Si derived from kamacite may have reacted with FeO from melted enstatite to produce silica and metallic Fe during crystallization; some silica may also have been produced (along with niningerite) by sulfidation of enstatite during crystallization from the impact melt.

(15) *Euhedral graphite* occurs as euhedral lath-like grains with pyramidal terminations in enstatite-chondrite impact-melt breccias (Figure 11.12). These grains most likely crystallized from the impact melt; their precursors probably occurred as graphite aggregates within kamacite.

(16) *Euhedral sinoite* grains are present in many EL6 chondrites as well as in impact-melted portions of the EL4 chondrites QUE 94368 (Figure 11.13) and Grein 002 (rocks that also contain euhedral grains of enstatite and graphite). It seems likely that the relatively coarse-grained sinoite in EL4 chondrites crystallized from a shock melt. (Nevertheless, some sinoite grains in primitive EL3 materials appear to be nebular condensates.)

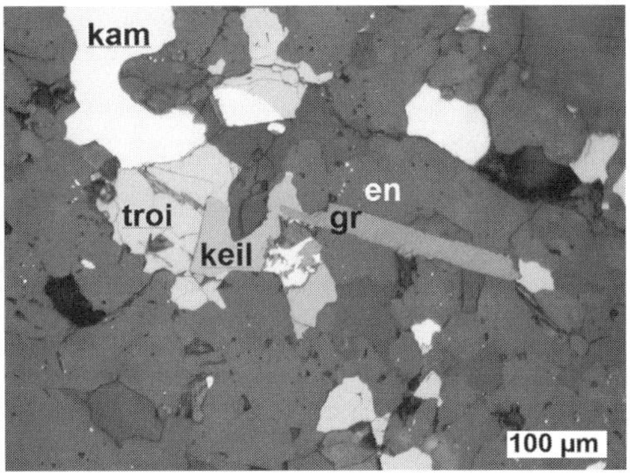

Figure 11.12 Portion of the Abee EH-chondrite impact-melt breccia showing a euhedral graphite grain (gr) surrounded by enstatite (en) and adjacent to an opaque assemblage consisting of kamacite (kam), troilite (troi) and keilite (keil). Reflected light. (A black-and-white version of this figure will appear in some formats. For the colour version, please refer to the plate section.)

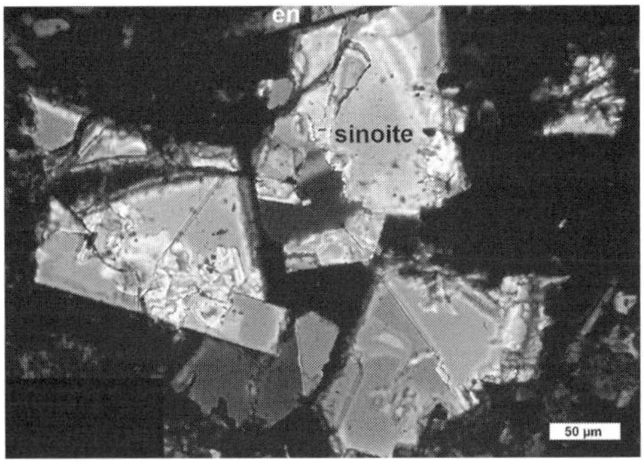

Figure 11.13 Cluster of intergrown euhedral, birefringent, optically zoned sinoite grains within an impact-melted portion of EL4 QUE 94368. An elongated enstatite (en) grain is adjacent to a large sinoite grain at top. (A black-and-white version of this figure will appear in some formats. For the colour version, please refer to the plate section.)

(17) *Fluor-richterite and fluorphlogopite* ($Na_2Ca(Mg,Fe)_5Si_8O_{22}F_2$ and $KMg_3(Si_3Al)O_{10}F_2$) occur in impact-melted enstatite chondrites: Abee contains rare 3.5-mm-long radiating acicular grains of fluor-richterite; Saint-Sauveur contains ~40 × 100-μm-sized subhedral grains of fluor-richterite; Y-82189 contains rare subhedral, 10–30-μm-sized grains of fluorphlogopite. These two F-rich phases probably crystallized from the impact melts; it is plausible that after impact melting, most of the F in Abee, Saint-Sauveur, and Y-82189 that had not been driven off during heating was scavenged by the crystallizing grains of fluor-richterite and fluorphlogopite.

(18) *Melt globules and spherules* are present in several enstatite-chondrite impact-melt breccias: Abee contains 0.1–1-mm-diameter globules of kamacite (Figure 11.14) as well as kamacite-rich nodules up to 11 mm in maximum dimension; Blithfield contains coarse, irregularly shaped nodules of kamacite (Figure 11.15); Adhi Kot contains a

Figure 11.14 A 3 × 7 mm nodule rich in low-Ni metallic Fe in the Abee EH4 impact-melt breccia containing clusters and isolated grains of coarse euhedral enstatite. The clast has a rim, readily visible at the top center and bottom left of the clast, which consists mainly of low-Ni metallic Fe (kamacite) and fine-grained euhedral enstatite. After Rubin and Scott (1997). Reflected light. (A black-and-white version of this figure will appear in some formats. For the colour version, please refer to the plate section.)

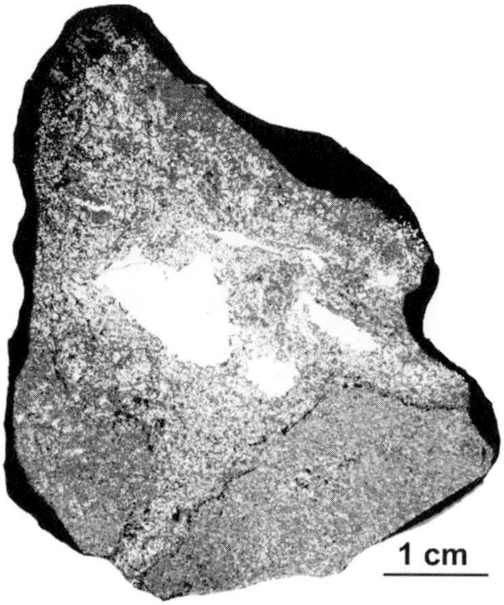

Figure 11.15 Slab of the Blithfield EL6 breccia consisting of large relatively dark-colored sulfide-rich clasts (bottom and lower left) adjacent to a metal-rich, sulfide-poor matrix containing large metal nodules (white) and a long metal vein. The sulfide-rich clasts contain 28–32 wt% sulfide and 6–8 wt% metallic Fe-Ni; the matrix contains 0.7 wt% sulfide and 64 wt% metallic Fe-Ni. Reflected light.

110-μm-diameter stoichiometric plagioclase glass spherule that is enriched in K_2O. These objects have been interpreted as having crystallized from the impact melt.

(19) *A metal nodule depleted in refractory siderophiles, but not depleted in common or volatile siderophiles* in Abee was reported by Sears et al. (1983). It seems likely that the nodule is analogous to those formed by vaporization and fractional condensation in OCs (see Widom

et al. 1986). However, the Abee nodule formed under more-reducing conditions that did not allow W to form volatile oxides and separate from the rest of the vaporized metal.

(20) *A vesicular impact-melt rock* known as Al Haggounia 001 is an incompletely melted, EL-chondrite impact melt rock (Rubin 2016) with a recovered mass of ~3 metric tons that fell ~23,000 years ago. The meteorite exhibits numerous shock effects including (1) development of undulose to weak mosaic extinction in low-Ca pyroxene; (2) dispersion of metal-sulfide blebs within silicates causing "darkening"; (3) incomplete impact melting wherein some relict chondrules survived; (4) vaporization of troilite, resulting in S_2 bubbles that infused the melt; (5) formation of immiscible silicate and metal-sulfide melts; (6) shock-induced transportation of the metal-sulfide melt to distances >10 cm; (7) partial resorption of relict chondrules and coarse silicate grains by the surrounding silicate melt; (8) crystallization of enstatite in the matrix and as overgrowths on relict silicate grains and relict chondrules; (9) crystallization of plagioclase from the melt; and (10) quenching of the vesicular silicate melt.

During cooling of enstatite chondrites that were impact melted, daubréelite exsolved from troilite (e.g., in EL Blithfield; Rubin 1984) and thin pearlitic lamellae of cohenite exsolved from kamacite (e.g., in EH Abee; Herndon and Rudee 1978).

Enstatite-rich impact-melt rocks. Rubin (2015a) listed various enstatite-rich impact-melt rocks of EH (e.g., QUE 94204, Y-82189, Y-8414); EL (e.g., Ilafegh (009), MIL 090807, NWA 6258); and ungrouped (e.g., NWA 4301, Zakłodzie) parentage. These rocks typically contain grains of plagioclase, enstatite, and sulfide that crystallized from the impact melt.

11.5.5 Howardites, Eucrites, and Diogenites

The petrologic characteristics of howardites, eucrites, and diogenites (HEDs) are discussed in Section 8.3.1. The O-isotopic compositions of meteorites in these groups are similar, indicative of a genetic relationship. Most workers believe these samples come from 4 Vesta. The majority of eucrites are monomict breccias, consisting of clasts of a single, basaltic lithology; a few eucrites are polymict breccias, containing mineralogically diverse eucritic clasts and relatively few (or no) diogenitic clasts. Many diogenites are monomict breccias, consisting of clasts with major orthopyroxene and minor to accessory plagioclase. Howardites are polymict regolith breccias containing both eucritic and diogenitic materials. Some polymict eucrites, diogenites, and howardites also contain exogenous carbonaceous-chondrite clasts, mainly with CM2, CR2, and thermally processed CI lithologies. The brecciated nature of most HEDs indicates these rocks were pummeled by projectiles. It is to be expected that shock metamorphism would have affected these meteorites.

About 5 percent of eucrites contain maskelynite, the glass formed on the parent body from crystalline plagioclase at shock pressures $\gtrsim$20 GPa. One of the most shocked eucrites is Padvarninkai; it is unusual in containing maskelynite in the unbrecciated host of the rock.

Vestaite ($(Ti^{4+}Fe^{2+})Ti^{4+}{}_3O_9$) is a newly discovered monoclinic high-pressure phase, identified in the eucrite NWA 8003, that crystallized from a shock melt at $\leq$10 GPa (Pang et al. 2018). It occurs in melt pockets and shock veins and presumably was produced after an impact on Vesta. Other high-pressure phases formed by shock in this meteorite include coesite, stishovite, tissintite ((Ca,Na,□)$AlSi_2O_6$), vacancy-rich clinopyroxene, and super-silicic garnet (Pang et al. 2016). The shocked monomict eucrite NWA 2650 contains coesite, stishovite, tissintite, vacancy-rich clinopyroxene, super-silicic garnet, and reidite ($ZrSiO_4$) (Chen et al. 2019). In addition, the meteorite contains maskelynite, anorthite glassy veins, and vesicular glass.

Glassy and crystalline impact-melt-breccia clasts constitute 30±10 percent of the lithic clasts in howardites and a few percent of those in polymict eucrites. There is a CM-chondrite impact-melted clast in the GRO 95574 howardite. Pseudotachylite-like glassy shock-melt veins occur in some eucrites. The veins in Cachari are vesicular and vary from pure glass to partly crystalline. The latter veins contain pyroxene grains that crystallized from the melt; these are surrounded by feldspathic glass. Small clasts of chromite and blebs of metallic Fe-Ni also crystallized from the shock melt. There are a few brecciated diogenites (Y-75032, Y-791199, Y-791200) with shock-melted matrices containing mineral fragments.

Also present in howardites are chondrule-like impact-melt spherules, similar to those from the Moon. Some of the spherules in the Kapoeta howardite have submicrometer-diameter micrometeorite impact craters on their surfaces (Brownlee and Rajan 1973).

11.5.6 Ureilites

Ureilites are brecciated rocks; both monomict and polymict varieties occur. The heavily shocked samples contain coarse mosaicized olivine grains. The myriad shock effects in ureilites include some in which mineral phases were formed: (1) silicate darkening (which involved melting of metallic Fe-Ni and accessory sulfide and crystallization of small blebs of these phases within curvilinear trails in coarse olivine grains); (2) the presence of the C polymorphs diamond, "lonsdaleite" (faulted and twinned cubic diamond) and chaoite (which formed from indigenous graphite at high shock pressures);[3] (3) the common occurrence of fine-grained interstitial silicates consisting of low-Ca pyroxene, augite, and Si-Al-alkali-rich glass (which appear to be plagiophile-rich shock products injected into the rocks); (4) the presence of the silicide suessite (Fe_3Si) (which probably formed by shock heating of kamacite and silicate); (5) the occurrence of Ca-, Al- and alkali-rich shock melts located at olivine grain boundaries in LEW 86216 and ALH 81101; (6) the presence of orthopyroxene in MET 78008, which probably formed from pigeonite during impact-induced annealing; and (7) the occurrence of chondrule-like crystalline spherules (probably impact-melt droplets). Some of the diverse inclusions in ureilites (e.g., enstatite-granular olivine clasts, basaltic clasts, feldspathic melt rocks) could also have been produced during impact melting.

Some ureilites contain aggregates of numerous small, unstrained olivine grains, each of which exhibits sharp optical extinction (Figure 11.16). These aggregates probably formed via post-shock annealing of coarse mosaicized olivine grains, further underscoring the important role shock has played in the petrogenetic history of ureilites.

[3] Goodrich et al. (2020) summarized evidence that ureilitic diamonds formed by shock processes; their study was a response to papers by Miyahara et al. (2015) and Nabiei et al. (2018) that had suggested the diamonds had formed at high static pressures (>20 GPa) in large parent bodies.

Figure 11.16 Numerous, moderately coarse unstrained olivine crystals with random orientations within a large olivine aggregate in the ALH 81101 ureilite. The aggregate formed from a single large mosaicized olivine grain during post-shock annealing. X-nicols.

There are many models of ureilite formation. Most assume these rocks formed in a differentiated or partially differentiated asteroid as partial melt residues, intrusive igneous cumulates or paracumulates (e.g., Goodrich 1992; McCoy et al. 1996; Keil and McCoy 2018). It is clear that many ureilites have experienced localized shock (e.g., Goodrich et al. 2004, 2020; Nakamuta and Aoki 2010; Warren and Rubin 2010) or shock followed by annealing (Rubin 2006).

In three ureilites (LAR 04315, LAP 03587, Almahata Sitta) pyroxene was reduced (and in some cases melted) by impact-induced smelting, resulting in porous grains with small blebs of metallic Fe-Ni, silica, and felsic glass.

11.5.7 Acapulcoites and Lodranites

Many workers consider acapulcoites and lodranites (which have been dubbed "primitive achondrites") to have formed by incomplete melting of chondritic material (e.g., Neumann et al. 2018). If this view is correct, then the mineral phases in these rocks may be (a) relict chondritic grains, (b) igneous grains that crystallized from a relatively large-scale melt, (c) phases that exsolved during cooling from igneous temperatures, or (d) grains formed during annealing.

Some workers have taken a different view, suggesting that acapulcoites and lodranites formed by shock-melting chondritic precursors (Rubin 2007). If this view is correct, then some of the phases in these rocks crystallized from impact melts. In acapulcoites, such phases would include metallic Fe-Ni and troilite within (a) opaque veins, (b) fine-grained rapidly cooled assemblages, and (c) assemblages consisting of small irregular troilite grains within metallic Fe-Ni (some of which also contain metallic Cu).

Irrespective of the mechanism that heated acapulcoites, it is clear that lodranites formed in a similar manner but experienced more extensive heating, loss of plagioclase, and loss of an Fe-Ni-S melt. Orthopyroxene in acapulcoites and lodranites likely formed from low-Ca clinopyroxene when the rocks were annealed to temperatures $\geq$630°C.

11.5.8 Aubrites

Approximately half of the aubrites are monomict fragmental breccias, most of which are moderately shocked (shock stage S4). The commonality of brecciation among aubrites and their moderate degree of shock indicate an extensive collisional history.

Several aubrites (e.g., LAP 03719, Bishopville, Khor Temiki, ALH 83015) contain orthopyroxene grains that exhibit more-pronounced shock effects than associated olivine grains, suggesting these rocks were shocked and buried within warm ejecta blankets or beneath fallback debris under the crater floor (Rubin 2015d). Entombed olivine crystal lattices healed (and became unstrained, reaching shock stage S1), but orthopyroxene lattices retained their S4-level shock-damaged features. (This is due to much more rapid Fe-Mg diffusion in olivine than in orthopyroxene.)

Three aubrites have experienced at least minor amounts of impact melting. Norton County has an impact-comminuted matrix and a few clasts that appear to be impact-melt breccias (Okada et al. 1988). Cumberland Falls is a polymict breccia that contains OC clasts as well as a few clasts that are OC impact-melt breccias containing shocked OC material. These clasts may all be part of a single OC projectile that impacted the aubrite parent asteroid. Mayo Belwa is an impact-melt breccia; its enstatite grains are surrounded by a melt matrix composed of 3–15-μm-sized euhedral and subhedral enstatite grains embedded in sodic plagioclase; numerous vugs are present in the melt matrix (Rubin 2010b). Fluor-richterite occurs as bundles of acicular grains in vugs. Also present in some vugs are bundles of 4–10-μm-thick laths of sodic plagioclase.

The unbrecciated aubrite Shallowater underwent a complex multistage history. One model suggests it formed by (1) impact melting of enstatite-chondrite material and concomitant partial loss of a metal-sulfide melt, (2) admixing of cold xenoliths (causing rapid cooling of the assemblage from high temperatures), (3) burial beneath warm ejecta with low thermal diffusivity (causing annealing), and (4) subsequent collisional excavation (causing rapid cooling to low temperatures).

11.5.9 Mesosiderites

Rubin and Mittlefehldt (1993) compiled available data on mesosiderite mineralogy, petrology, geochemistry, and chronology and proposed a multistage petrogenetic history, modified here: (1) accretion of the parent asteroid ~4.56 Ga ago, (2) initial melting ~4.56 Ga ago, (3) crustal remelting of the body $\geq$4.47 Ga ago, probably associated with metal-silicate mixing, (4) localized impact melting 4.5–3.9 Ga ago, and (5) impact excavation and ejection of buried breccias $\ll$3.9 Ga ago. Minerals produced in stage 2 formed by igneous processes; minerals formed during stages 3 and 4 involved shock metamorphism.

Mesosiderites contain large clasts, ~40 percent of which are coarse-grained gabbros that are highly depleted in incompatible elements relative to chondrites. These clasts have the most extreme positive Eu anomalies known among Solar System rocks (Rubin and Jerde 1987, 1988; Mittlefehldt et al. 1992). It seems likely that these abundant clasts formed during a crustal remelting event (stage 3), plausibly caused by the impact of a largely molten iron-core projectile (with some adhering silicate mantle) from a disrupted differentiated asteroid. The phases in the gabbro clasts (major pyroxene, plagioclase and silica, and minor to accessory troilite, metallic Fe-Ni, ilmenite, merrillite, and chromite) thus appear to be impact-melt products.

The most significant effect of impact melting during stage 4 is the creation of the four basic mesosiderite petrologic types, classified by the degree of recrystallization of their fine-grained silicate matrices. The Type 3/4 and Type 4 rocks are impact-melt breccias with igneous matrices (see Section 8.6.2). Impact melting at this time produced spheroidal nodules of metallic Fe-Ni and troilite in Bondoc and Estherville that range up to 10 cm in diameter. (They resemble the centimeter-sized metal-troilite nodules in the PAT 91501 L-chondrite impact-melt rock; Mittlefehldt and Lindstrom 2001.) Many mesosiderites contain pebble-sized clasts with quench textures that also formed by impact melting basaltic material during stage 4; their mineralogy is the same as that of the gabbro clasts. Annealing associated with these impact events is probably responsible for the orthopyroxene-chromite-merrillite coronas on millimeter-sized olivine grains in the Emery and Morristown mesosiderites; the coronas formed by the reaction of olivine with silica in the matrix.

Much of the metallic Fe-Ni in relatively unshocked Type I mesosiderites (e.g., Crab Orchard, Vaca Muerta) occurs as small grains, homogeneously distributed among the silicates. It seems likely that the large metallic clasts (up to 4 cm in diameter) in some type 1 mesosiderites are shock products. They may have formed by impact melting of fine-grained mesosiderite material and segregation of the molten metal.

11.5.10 Iron Meteorites

Many iron meteorites have experienced shock metamorphism (Buchwald 1975), typically involving the production of Neumann bands in kamacite, recrystallization of kamacite, mosaicism and twinning of troilite, shock melting and quenching of metallic Fe-Ni and sulfide (and in some cases daubréelite, cohenite, and schreibersite), development of martensite and/or plessite from taenite during cooling, and the development of the ε structure (a high-density hexagonal close-packed structure) from kamacite at ~13 GPa (and reversion to the α structure upon pressure release). In rare cases (e.g., Canyon Diablo and ALHA77283), high shock pressure transformed graphite into diamond. Stishovite (a high-pressure form of silica) has been reported in two irons – Tishomingo and Muonionalusta. For illustrative purposes, we discuss shock effects in the IAB Complex, Group IIE, Group IIIE, IIIAB Haig, and IVA Social Circle. Shock features in a few ungrouped irons are discussed in Section 8.5.4.

IAB Complex and winonaites. Wasson and Kallemeyn (2002) reevaluated analytical data on IAB and related iron meteorites and defined a IAB Complex in which these meteorites occupy six distinct clusters on element-Au diagrams. In contrast to magmatic iron-meteorite groups (e.g., IIAB, IIIAB, IVA) that formed by fractional crystallization in slowly cooling, well mixed magmas, the IAB Complex is designated non-magmatic; these samples likely formed from impact-generated melts of chondritic material. Their mineralogy is discussed in Section 8.5.3.

Most silicate inclusions in IAB irons have basic chondritic mineralogy and bulk chemical composition; objects that appear to be millimeter-sized, highly recrystallized POP chondrules are present in Campo del Cielo (Schrader et al. 2017). The major minerals in the IAB silicate inclusions are olivine, low-Ca pyroxene, plagioclase, Ca-pyroxene, troilite, graphite, phosphates, metallic Fe-Ni; minor phases include daubréelite and chromite. These minerals were probably inherited from chondritic precursors; in some cases, they underwent extensive metamorphic recrystallization, in others, they probably crystallized from the shock melt (along with the surrounding metallic Fe-Ni).

Some of the sulfide nodules in IAB irons contain graphite. It is likely that both the graphite and troilite crystallized from the residual S- and C-rich melt.

Winonaites are stony meteorites that were probably derived from the same parent body as the IAB irons – their silicates have similar mineralogy and near-identical bulk O-isotopic compositions. Whatever process produced the IAB silicate inclusions (presumed here to be shock related) also produced the winonaites. Most workers have concluded that both sets of objects were derived from the same parent body. In fact, winonaites might be large IAB silicate inclusions dislodged from their metallic hosts – Winona (24 kg) was recovered from the same Arizona county as IAB Canyon Diablo (30 MT); the winonaite Mount Morris (Wisconsin) (676 g) was recovered from the same Wisconsin county as IAB Pine River (3.6 kg).

Many workers consider winonaites to be "primitive achondrites"; some ascribe their heating to the decay of ^{26}Al (e.g., Ruzicka 2014). However, several features of winonaites suggest that, instead of having formed in a partially differentiated asteroid, these rocks formed by incomplete impact melting. Such features include (1) chondritic proportions of metallic Fe-Ni, sulfide, and silicate, (2) chondritic bulk compositions of many samples, (3) a component containing planetary-type rare gases, (4) several winonaites (Pontlyfni; Mount Morris (Wisconsin); NWA 1463; Y-74063) containing relict chondrules, and (5) the preservation of relict shock features including metallic Cu, chromite veinlets, chromite-plagioclase assemblages, and curvilinear metal veins. The variable Ar-Ar and Hf-W ages of winonaites seem most consistent with discrete impact-heating events. IAB silicate inclusions also have chondritic mineralogy (including relict chondrules in Campo del Cielo), chondritic bulk compositions, and contain planetary-type rare gases.

A couple of IAB irons (Caddo County and Ocotillo) have non-chondritic (basaltic and troctolitic) silicate inclusions. The origin of these chemically evolved inclusions is unclear, but it seems possible that they were produced from larger scale impact-generated magmas.

Sombrerete. The Sombrerete iron is a member of the IAB Complex, subgroup sHL (high Au, low Ni). It contains ~7 vol% silicate inclusions similar to those in the IIE irons Weekeroo Station and Miles. The Sombrerete inclusions consist primarily of sodic-plagioclase glass with acicular chlorapatite and orthopyroxene; also present are accessory kaersutite, tridymite, chromite, ilmenite, and rutile. These phases formed by impact melting and quenching of a preexisting silicate-rich assemblage.

IIE. The Elga IIE iron contains the high-pressure phase tuite (γ-$Ca_3(PO_4)_2$), formed from merrillite ($Ca_9MgNa(PO_4)_7$) by shock (Litasov and Podgornykh 2017). It also contains solidified shock melts consisting of fine-grained mixtures of silicate-phosphate and metallic Fe-Ni, and Fe-Ni-P-S melt pockets that appear to have formed at shock pressures up to ~20 GPa. There are several ungrouped irons with dendritic structures and aligned troilite nodules that appear to have cooled rapidly, most likely at the surface of the IIE parent asteroid (Schrader et al. 2010; Wasson 2017). They probably formed by impact heating and quenching.

Although Netschaëvo contains recrystallized chondrule-bearing chondritic inclusions, many IIE silicate inclusions are not chondritic in composition; they are enriched in plagiophile elements (Si, Al, Na, K). There are numerous phases that occur in one or more inclusions (orthopyroxene, clinopyroxene, olivine, albite with K-feldspar exsolution lamellae (antiperthite), sanidine ($KAlSi_3O_8$), yagiite ($(K,Na)_2(Mg,Al)_5(Si,Al)_{12}O_{30}$), silico-feldspathic glass, tridymite, phosphate); nevertheless, plagioclase and feldspathic glass are the principal phases in many inclusions.

The IIE silicate inclusions (excluding those in Netschaëvo, Techado, Garhi Yasin, and Mont Dieu) appear to have been shock melted. Many inclusions are globular, consistent with their shapes having been molded by surface tension. Quenching of the silicates in IIE Colomera is supported by the occurrence of alkali-rich glass in some inclusions and by large crystals of sanidine (the high-temperature polymorph of $KAlSi_3O_8$) and of high albite (a.k.a. analbite, a $NaAlSi_3O_8$ polymorph with a slightly larger unit cell than low albite) at the meteorite's surface. The sanidine inclusion is 11×2.5 cm in size and poikilitically encloses aggregates of Cr-rich clinopyroxene and ferroan low-Ca pyroxene with accessory silica, chromite, ilmenite, and Ca-phosphate (Wasserburg et al. 1968). Also present in the sanidine inclusion is carbonate, probably a product of terrestrial weathering. Glass-rich inclusions also occur in IIE Weekeroo Station, presumably due to remelting preexisting plagioclase and orthopyroxene.

Plagioclase has a low impedance to shock compression and is readily compressible; during shock heating, plagioclase is preferentially melted relative to mafic silicates. Because IIE inclusions have similar bulk compositions to melt-pocket glasses in ordinary chondrites that were impact melted *in situ*, it seems likely that the alkali-rich IIE silicate inclusions also formed by the preferential shock melting of preexisting broadly chondritic inclusions such as those present in some IAB irons.

IIE iron meteorites generally have metal with well-developed Widmanstätten patterns; their octahedral structures range from finest to coarse. Such IIE coarse octahedrites as Weekeroo Station and Tobychan probably cooled at ~10°C/Ma at subsolidus temperatures of ~500°C; the IIE finest octahedrites (e.g., Kodaikanal, Tarahumara) may have cooled several orders of magnitude faster at subsolidus temperatures.

Shock melting and quenching of the silicates and surrounding metallic Fe-Ni probably occurred early in the history of the Solar System; there may have been several episodes of shock during this epoch. Subsequent slow cooling at lower temperatures facilitated the formation of the Widmanstätten structures, but did not devitrify the silicate glass (e.g., as in Weekeroo Station or Colomera) or cause crystal-structure changes in the sanidine or high albite inclusions in Colomera. Late-stage shock heating is most likely responsible for the formation of metal-phosphide globules with dendritic textures in IIE Verkhne Dnieprovsk and potassic maskelynite within silicate inclusions in Kodaikanal.

IIIE. A comprehensive study of impact effects in Group IIIE irons was completed by Breen et al. (2016). Members of this C-rich magmatic group can be ordered into four categories based on their degree of shock. The weakly shocked samples in Category 1 (Aletai, Colonia Obrera, Coopertown, Porto Alegre, Rhine Villa, Staunton, Tanokami Mountain) contain haxonite within plessite, unrecrystallized kamacite grains with Neumann bands or possessing the ε structure, and sulfide inclusions consisting of polycrystalline troilite with daubréelite exsolution lamellae. (Aletai is sometimes called Armanty, Ulasitai, or Xinjiang.) Mineral transformations begin in the sole moderately shocked Category 2 sample (NWA 4704) wherein the haxonite is partially decomposed to graphite and the sulfide inclusions are partly melted. In the strongly shocked

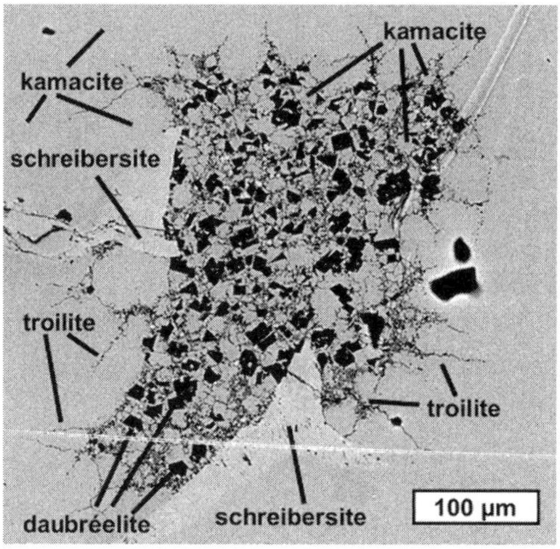

Figure 11.17 Sulfide-rich inclusion in the severely shocked Aliskerovo IIIE iron. The inclusion consists of euhedral, subhedral and fragmental grains of daubréelite (black), semi-equant grains of low-Ni metallic Fe (medium gray), rare tiny grains of tetrataenite (white specks) and spidery, filamentary troilite (dark gray), partially surrounded by schreibersite (light gray). The metallic host is composed mainly of low-Ni metallic Fe (kamacite). BSE image.

samples of Category 3 (Cachiyuyal, Kokstad, Paloduro) haxonite has fully decomposed to graphite, troilite has melted and quenched, and kamacite has been extensively recrystallized. In the severely shocked samples of Category 4 (Aliskerovo, Willow Creek), haxonite has fully decomposed to graphite. These meteorites contain sulfide assemblages that have been crushed; the troilite within these assemblages has largely vaporized, residual troilite has crystallized as spidery filaments (Figure 11.17), and low-Ni, high-Co taenite grains crystallized from the melt and transformed into kamacite at subsolvus temperatures. One or more major collisions were probably responsible for these effects.

Haig (IIIAB). Haig, a member of the magmatic IIIAB group, has a spectacular regmaglypt pattern (plate 10 of McCall 1973). There is a small troilite-daubréelite-cohenite nodule that was sheared by an impact in Haig. Finely dendritic globular masses consisting of troilite, metallic Fe-Ni, daubréelite, and cohenite associated with the nodule formed by impact melting and very rapid quenching (e.g., Bevan et al. 1981).

Social Circle (IVA). A late-stage shock event caused widespread heating of Social Circle, recrystallizing the kamacite, partially obliterating previously formed Neumann lines, partly decomposing taenite and plessite fields, and forming troilite-metal eutectic shock melts. Impact melting of the sulfide assemblages (possibly involving Mn-bearing daubréelite ($FeCr_2S_4$)) could have increased the Mn concentration in portions of the S-rich melts, leading to crystallization of the rare (and newly described) phase joegoldsteinite ($MnCr_2S_4$; Isa et al. 2016) (Figure 8.5). The shock event that affected Social Circle is unrelated to the major impact-melting event inferred to have formed the IVA magma initially from preexisting L- or LL-chondritic materials (Wasson et al. 2006).

11.5.11 Lunar Meteorites

Phases formed by shock processes in lunar meteorites include chondrule-like glassy-to-micro-crystalline spherules, agglutinate glasses, irregularly shaped glassy clumps, maskelynite formed at high shock pressure from crystalline plagioclase (mainly during launch off the Moon), and polysynthetically twinned low-Ca clinopyroxene formed from orthopyroxene. High-pressure minerals identified so far are ringwoodite, wadsleyite, coesite, stishovite, seifertite, and newly-found donwilhelmsite (Fritz et al. 2020) and reidite (Xing et al. 2020).

Lunar crystalline melt breccias among Apollo samples formed as impact melts that acquired cold clastic debris (older, igneously formed mineral grains and rocks) and cooled quickly. During cooling, fine-grained minerals (mainly plagioclase and Ca-pyroxene) crystallized to form the groundmass. Minor phases include low-Ca pyroxene, olivine, silica, ilmenite, spinel, ulvöspinel (Fe_2TiO_4), metallic Fe-Ni, troilite, phosphate, and glass.

Clast-poor melt breccias crystallized from impact-generated silicate liquids that had been heated above the liquidus (~1,200–1,300°C). The major minerals that crystallized from the shock melt in these breccias are plagioclase and Ca-pyroxene; minor phases include low-Ca pyroxene, olivine, silica, ilmenite, spinel, ulvöspinel, metallic Fe-Ni, troilite, phosphate, and glass; accessory phases include schreibersite, armalcolite ($(Mg,Fe)Ti_2O_5$), tranquillityite ($Fe^{2+}{}_8Ti_3Zr_2Si_3O_{24}$), K-Ba feldspar, and akaganéite. The occurrence of these phases in Apollo samples suggests that they are probably also present in lunar meteorites.

Phosphate grains in Apollo 17 samples display a range of shock deformation. With increasing shock, fluorapatite and merrillite remain crystalline but undergo fragmentation, subgrain formation, and granularization. Individual fragments of these minerals show increasing angles of internal crystal-plastic deformation, ranging from ~2° to ~25° for apatite and from ~2° to ~30° for merrillite.

11.5.12 Martian Meteorites

Shock effects in martian meteorites are confined largely to shergottites. The principal phase produced by shock metamorphism is maskelynite, a glass formed from crystalline plagioclase during launch off Mars. Also present in a few martian meteorites is impact-produced pyroxene-composition glass. About 93 percent of martian basaltic meteorites contain maskelynite; among the five individual shergottites in a recent compilation that lacked maskelynite (Rubin 2015b), four of them instead have vesicular plagioclase glass, produced at shock pressures even *higher* than those required to form maskelynite (i.e., >45 GPa). The fifth maskelynite-free shergottite in that compilation is NWA 7034 and its paired specimens, called "Black Beauty" by collectors. This rock is a regolith breccia (Cartwright et al. 2014) containing a variety of clasts including impact-melt rocks, plutonic rocks, basaltic rocks, contact-metamorphic rocks, and siltstones (Wittmann et al. 2015). It contains mildly shocked, but still crystalline plagioclase. Black Beauty may have experienced relatively low shock pressures (<15 GPa) during launch off Mars. That would be consistent with the high bulk water in this meteorite (~0.6 wt%), greater than that in any other martian meteorite. Rubin (2015b) suggested the sample may have been excavated as the result of a rare oblique impact (because shock pressures decrease with decreasing impact angle), but Wittmann et al. (2015) listed the round 6.9 km, ~5-Ma-old, Gratteri Crater as a possible launch site.

Shock melts in EETA79001 consist of vesicular glass containing relict grains as well as skeletal olivine and pyroxene grains that crystallized from the melt. Some lherzolitic shergottites (e.g., ALHA77005) contain melt pockets consisting of impact-produced glass with small skeletal olivine crystals and dendritic or clustered chromite grains.

There is a report of P-rich recrystallization rims on olivine grains in the Tissint shergottite that formed during impact melting and deformation of host materials.

High-pressure minerals identified in Tissint include ahrensite (the high-pressure polymorph of ferroan olivine) (Figure 11.18); ringwoodite (the high-pressure polymorph of magnesian olivine); bridgmanite ($MgSiO_3$-perovskite) (Figure 11.9(a)); wüstite (FeO); tissintite ((Ca, Na,□)$AlSi_2O_6$); xieite ($FeCr_2O_4$-CT); akimotoite (ilmenite structure of pyroxene); lingunite ($NaAlSi_3O_8$-hollandite); tuite (γ-polymorph of merrillite); and chenmingite ($FeCr_2O_4$) (Figure 11.19) (e.g., Baziotis et al. 2013; Ma et al. 2015b, 2019b; Ma 2018a). Tissintite is a newly discovered shock-produced clinopyroxene with a plagioclase composition (Figure 11.20); it is the most vacancy-rich pyroxene known. Three high-pressure polymorphs

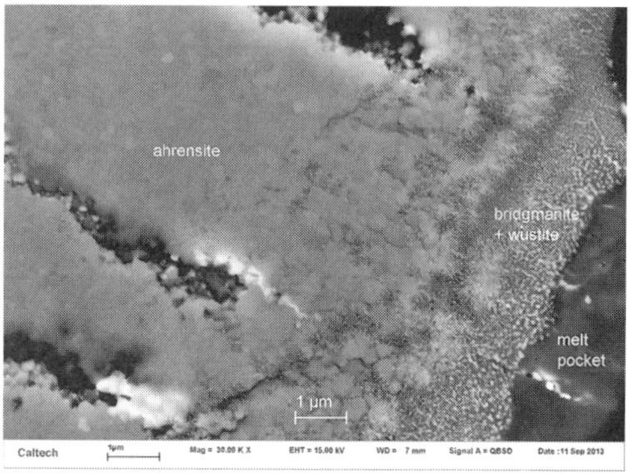

Figure 11.18 Ahrensite (γ-Fe_2SiO_4), bridgmanite ($MgSiO_3$-perovskite), and wüstite (FeO) next to a shock-melt pocket from the Tissint shergottite (Ma et al. 2016b). BSE image.

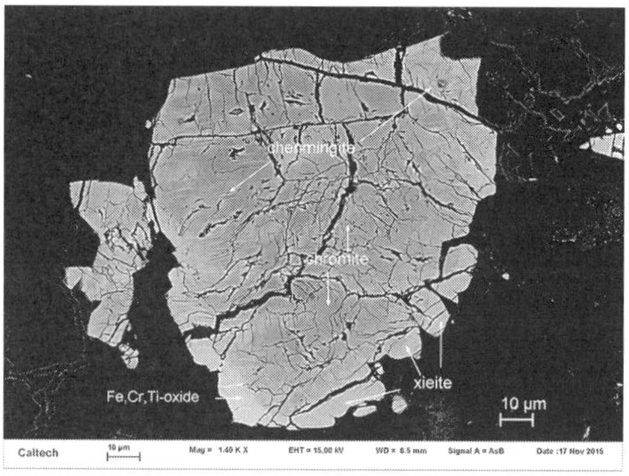

Figure 11.19 Chenmingite ($FeCr_2O_4$-CF) and xieite ($FeCr_2O_4$-CT) transformed from chromite in the Tissint shergottite (Ma et al. 2019c). BSE image.

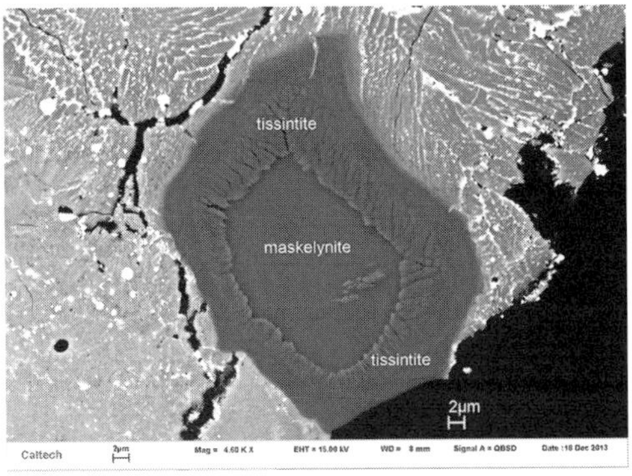

Figure 11.20 Tissintite, (Ca,Na,□)$AlSi_2O_6$, formed from maskelynite in a melt pocket. within the Tissint shergottite (Ma et al. 2015b). BSE image.

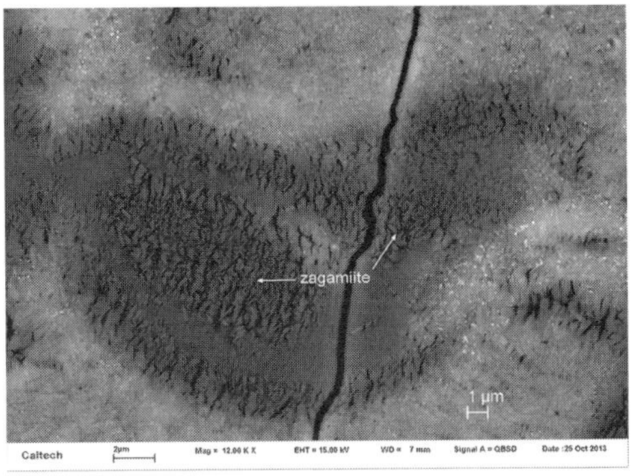

Figure 11.21 Fine-grained zagamiite ($CaAl_2Si_{3.5}O_{11}$) in a shock melt pocket from the Zagami shergottite (Ma et al. 2017c, 2019a). BSE image.

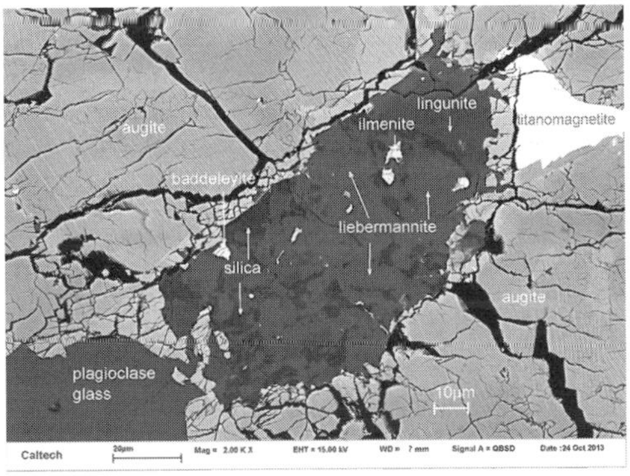

Figure 11.22 Liebermannite ($KAlSi_3O_8$) in an assemblage with lingunite ($NaAlSi_3O_8$), a silica phase, ilmenite, and baddeleyite from the Zagami shergottite (Ma et al. 2018b). BSE image.

of silica are present in many shergottites: stishovite (tetragonal), seifertite (orthorhombic), and an unnamed monoclinic phase with the β-ZrO_2-type structure. Ringwoodite occurs in Dar al Gani 670. Zagami contains black shock veins with a high-pressure assemblage consisting of tissintite, zagamiite ($CaAl_2Si_{3.5}O_{11}$) (Figure 11.21), stishovite, and liebermannite ($KAlSi_3O_8$-hollandite) (Figure 11.22) surrounded by silicate glass (e.g., Ma et al. 2017c, 2018 c, 2019a). Stöfflerite ($CaAl_2Si_2O_8$-hollandite) was discovered in NWA 856 (Tschauner and Ma 2017). Three high-pressure minerals [tschaunerite ($(Fe^{2+})(Fe^{2+}Ti^{4+})O_4$), feiite ($(Fe^{2+}{}_2)(Fe^{2+}Ti^{4+})O_5$), and liuite ($FeTiO_3$-perovskite)] were recently discovered in a Shergotty shock melt pocket (Ma and Prakapenka 2018; Ma and Tschauner 2018a, 2018b; Ma et al. 2021a, 2021b). There is a tentative identification of ringwoodite and majorite (the high-pressure polymorph of orthopyroxene) inside shock veins in EETA79001.

11.5.13 Earth

About 200 impact structures have been identified on Earth, ranging in initial diameter from ~10 m (e.g., one of the Wabar craters, Saudi Arabia) to ~300 km (Vredefort, South Africa). The highest shock pressures, reached locally during the initial contact and compression stages of a collision, are typically in the range of 100–1,000 GPa; temperatures can exceed 10^4 K (Melosh 1989). More than a dozen high-pressure phases have been identified in terrestrial impact structures; three of these phases (riesite, maohokite, and akaogiite) have not been reported in meteorites (Table 11.2). Most are high-pressure polymorphs of rock-forming minerals or accessory minerals in the target rocks.

Table 11.2 *High-pressure phases in terrestrial impact structures*

Phase	Formula
Akaogiite	TiO_2
Albitic jadeite	$(Na,Ca,\square_{1/4})(Al,Si)Si_2O_6$
Chaoite	C
Coesite	SiO_2
Diamond	C
Jadeite	$NaAlSi_2O_6$
Lingunite	$NaAlSi_3O_8$
Majorite	$Mg_3(SiMg)Si_3O_{12}$
Maohokite	$MgFe_2O_4$
Maskelynite	$(Na,Ca)(Si,Al)_4O_8$
Reidite	$ZrSiO_4$
Riesite	TiO_2
Stishovite	SiO_2
Tissintite	$(Ca,Na,\square)AlSi_2O_6$
Stöfflerite	$CaAl_2Si_2O_8$

Although moderately common in shocked meteorites and present in the Earth's upper mantle, ringwoodite (a high-pressure polymorph of forsterite with the spinel structure) has not yet been identified in terrestrial impact craters.

In addition to being transformed into high-pressure phases during impact events, common terrestrial rock-forming minerals (quartz, plagioclase, olivine, clinopyroxene) can develop planar elements and fractures. Olivine can also undergo recrystallization; clinopyroxene can produce mechanical twins.

11.6 Space Weathering

The surface features of rocky bodies that lack atmospheres can vary significantly in color and brightness. The pockmarked lunar surface has fresh impact craters (e.g., Tycho) with prominent light-colored rays as well as ancient degraded craters with no rays (e.g., Ptolemaeus). Among the samples returned from the Apollo missions, fine-grained lunar soils are darker (they have a lower albedo) and redder (they exhibit increases in reflectance at longer wavelengths) and are characterized by shallower spectral absorption bands than coarser rock fragments. These differences have been attributed to "space weathering" (e.g., Hapke 2001), a process that appears to occur on timescales of about a million years. One major space-weathering process is micrometeorite bombardment, responsible for comminuting grains in the lunar regolith, thereby making old lunar soils finer grained than freshly exposed impact debris. Other space-weathering processes include irradiation by the solar wind, solar flares, solar energetic particles, higher-energy photons (UV and X-rays), and galactic cosmic rays.

Individual grains in mature lunar soils are coated with 60–200-nm-thick amorphous rims produced by sputtering and condensation of impact-generated vapor. The spectral changes in maturing materials at the lunar surface appear to be due to the reduction of FeO in mafic silicates (olivine and pyroxene) and the production of nanophase iron particles ($npFe^0$; ~10-nm-sized grains of metallic Fe). Solar-wind bombardment of the lunar surface releases these particles (via sputtering) that then coat amorphous rims of exposed grains (Burbine 2017). Experiments showing that olivine grains subjected to pulsed laser irradiation rapidly produce $npFe^0$ particles (Sasaki et al. 2001) suggest that alteration by the solar wind is a major mechanism responsible for the formation of these particles.

Lunar agglutinates are $\lesssim$1-mm-sized aggregates of smaller particles (rocks, mineral grains, glass, and preexisting agglutinates) embedded in vesicular, flow-banded glass. Mature lunar soils contain up to ~60–70 vol% agglutinates. These objects formed by micrometeorite-induced impact melting of particles in the lunar soil; the vesicles are remnants of bubbles formed from solar-wind–implanted rare gases that had previously been adsorbed onto mineral grains and were released during impact heating. Agglutinate glass in lunar soils also contains $npFe^0$ particles.

Remnants of agglutinates occur in some lunar meteorites. In addition, the ferromagnetic resonance spectrum of lunar meteorite ALHA81005 (an anorthositic breccia) is consistent with the presence of $npFe^0$ particles.

Micrometeorite bombardment of the Moon causes impact vaporization of target materials and condensation of fine-grained phases at the surfaces of nearby mineral grains. A regolith-breccia clast within lunar meteorite Dhofar 280 (another anorthositic breccia) contains 2–35-μm-sized grains of hapkeite (Fe_2Si) and two as-yet-unnamed Fe-Si phases. All these grains were inferred to be products of space weathering, formed by reduction and vapor deposition engendered by the collision of micrometeorites with the lunar surface.

Space weathering also occurs on asteroids even though the relative impact velocity is lower and the solar wind less intense at their greater heliocentric distances compared to the Moon. Fresh craters on S-complex asteroid 243 Ida are much bluer (i.e., less red) than most of Ida's surface regions (which tend to be older). Although characterization of returned samples from the S-complex asteroid 25143 Itokawa shows that this body is made of LL-chondrite material, the asteroid exhibits shallower spectral absorption bands than typical LL chondrites. These differences are analogous to those that occur on the Moon and thus have been attributed to space weathering. [Space weathering has long been held responsible for reconciling the apparent spectral mismatch between two groups of objects rich in metallic Fe and olivine and containing different abundances of pyroxene: S asteroids (the most common taxonomic class in the inner part of the main belt) and ordinary chondrites (the most common meteorite clan among observed falls).]

Intact agglutinates occur within a few ordinary-chondrite regolith breccias (e.g., Fayetteville) and agglutinate grains have been found in CI Orgueil and a few CM2 chondrites. It is plausible that many of these agglutinates contain $npFe^0$ particles. These particles have been reported in the Kapoeta howardite, a gas-rich meteorite that formed at the surface of its parent body (probably 4 Vesta). The $npFe^0$ particles were likely produced by solar-wind–induced reduction of FeO from olivine and pyroxene grains. Among the returned samples from Itokawa, there are grains with ~5–15-nm-thick rims of amorphous material containing $npFe^0$ particles and nanophase particles of FeS and MgS.

11.7 Condensation Within Impact Plumes

The CB_a chondrites such as Bencubbin, Gujba (Figure 11.23), and Weatherford contain 40–60 vol% metal and have barred olivine (BO) and cryptocrystalline (C) chondrule-like spherules and large metallic-Fe-Ni-rich globules (and deformed globules). CB_b chondrites such as Hammadah al Hamra 237 and QUE 94411 contain ~70 vol% metal, smaller BO and C chondrules, and a few CAIs.

Large metal globules in Gujba exhibit metal-troilite quench textures and vary in their modal abundances of troilite and concentrations of volatile siderophile elements. In these globules, the less-refractory siderophile elements have different abundances than the refractory siderophiles. The metal globules may have formed as liquid droplets either via condensation in an impact-generated vapor plume on an asteroid or by evaporation of preexisting metal particles in such a plume (e.g., Campbell et al. 2002; Rubin et al. 2003; Krot et al. 2005). The large silicate chondrule-like objects probably formed in the same plume at high temperatures (e.g., Rubin et al. 2003; Hewins et al. 2018), leading to a significant loss in volatile elements. One proposal invoked a large impact involving a metal-rich body; this event generated a high-density metal-rich gas from which the CB metal nodules condensed.

The chondrule-like objects in Gujba and Hammadah al Hamra 237 all have the same relatively young age of ~4,562 Ma (Krot et al. 2005; Bollard et al. 2015) (after the Krot et al.

Figure 11.23 Slab of CB_a chondrite Gujba showing large metal nodules and chondrule-like silicate nodules. There is a preferred orientation of the particles (NNE-SSW). Image courtesy of the Maine Mineral and Gem Museum. Reflected light. (A black-and-white version of this figure will appear in some formats. For the colour version, please refer to the plate section.)

data are recalculated using an up-to-date U-isotopic composition). This young age is inconsistent with formation of these objects in the solar nebula. Krot et al. pointed out that the young ages are similar to those of the ^{182}Hf-^{182}W ages of the fractionation between metal and silicates recorded by metal condensates in CB_a and CB_b chondrites.

Several workers have suggested that the chondrule-like objects and the metallic Fe-Ni grains in CB meteorites formed in a vapor-melt plume created by a giant impact event involving planetary embryos. If these scenarios are correct, then the principal minerals in these rocks – kamacite, troilite, olivine, low-Ca pyroxene – formed by condensation in a vapor plume. Analysis of 32 silicate nodules in CB_a Gujba by Oulton et al. (2016) found that (1) the nodules vary in LREE abundances, and in U/Th, Th/Zr, Th/Sr, and Th/Sc ratios, (2) some have positive Ce anomalies and others have negative Ce anomalies, and (3) Nb/Ta correlates positively with Zr/Hf and spans the range of known lunar and martian basalts. These authors concluded that the Gujba silicates formed from a differentiated crustal target.

11.8 Crystallization from Melts in Differentiated or Partially Differentiated Bodies

The minerals in asteroidal achondrites (HED samples, angrites, aubrites), magmatic iron meteorites, most martian meteorites, and large portions of lunar meteorites crystallized from magmas inside differentiated bodies. Many workers maintain that acapulcoites (e.g., McCoy et al. 1996), lodranites (e.g., Mittlefehldt et al. 1996; McCoy et al. 1997), brachinites (e.g., Johnson et al. 1977; Warren and Kallemeyn 1989a), and ureilites (e.g., Goodrich et al. 1987, 2007; Warren and Kallemeyn 1989b) formed by igneous processes (likely in partially differentiated bodies) and should not be considered shock products.

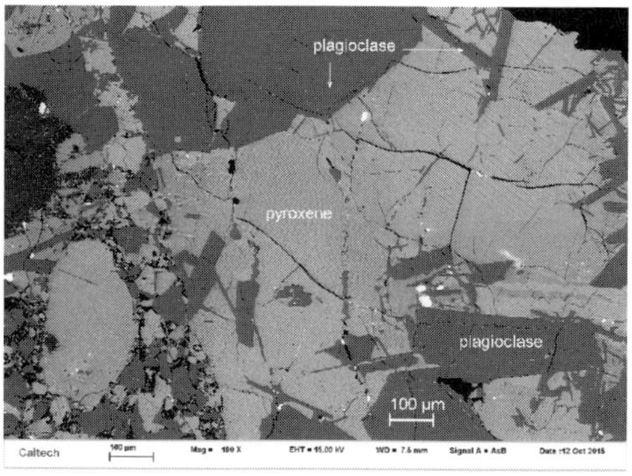

Figure 11.24 The Juvinas eucrite showing ophitic and subophitic textures with laths of plagioclase surrounded by pyroxenes. BSE image.

11.8.1 Eucrites

Eucrites typically contain major orthopyroxene (or pigeonite) and calcic plagioclase and minor to accessory silica, chromite, phosphate, ilmenite, kamacite, and troilite. Olivine is rare. Most eucrites are monomict breccias; most of the clasts with eucrites have ophitic, subophitic, variolitic, or intersertal textures[4] (Figure 11.24) consistent with igneous crystallization. Plagioclase crystals tend to be normally zoned with calcic cores and more-sodic rims. Pyroxene exhibits a variety of compositions and textures. Temperatures of the parent basaltic melt probably exceeded 1,100°C.

When eucrites are plotted on a diagram of the bulk concentration of an incompatible lithophile element such as Ti, La, Sc, or Sm versus the bulk Mg/(Mg+Fe) ratio, two compositional trends are evident: the Nuevo Laredo Trend (rocks such as Nuevo Laredo and Lakangaon that underwent fractional crystallization, most likely as partial cumulates) and the Stannern Trend (rocks such as Stannern and Bouvante that formed by partial melting) (Figure 11.25). Main-group eucrites probably belong to the Nuevo Laredo Trend. Cumulate eucrites are coarse-grained gabbros; their crystals grew slowly and settled to the bottom of a magma chamber on the eucrite parent body. They do not plot along the trends of the non-cumulate eucrites.

11.8.2 Diogenites

The diogenites are coarse-grained rocks consisting mainly of orthopyroxene; minor phases include olivine, chromite, Ca-pyroxene, plagioclase, silica, kamacite, and troilite. Phosphates are rare. Most diogenites are monomict breccias; some appear unbrecciated. These rocks

[4] These are common textures of basaltic rocks: ophitic – plagioclase laths largely or completely enclosed in pyroxene; subophitic – plagioclase laths only partly enclosed by pyroxene; variolitic – radial or sheaf-like bundles of plagioclase, in some cases intergrown with pyroxene; intersertal – angular interstices between plagioclase and pyroxene grains occupied by cryptocrystalline-to-glassy material.

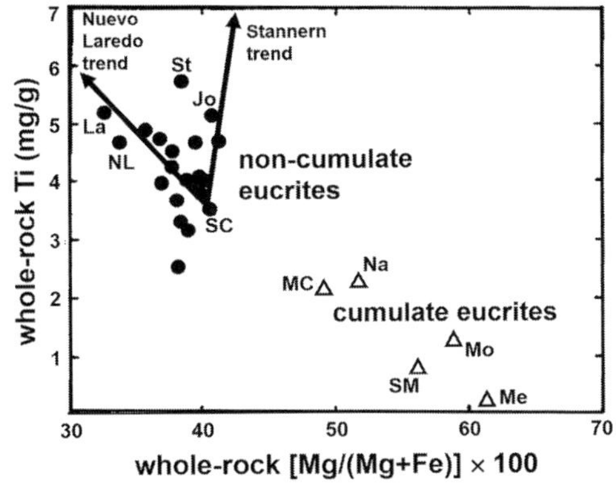

Figure 11.25 Whole-rock Ti content vs. ([Mg/(Mg+Fe)] × 100) for monomict eucrites modified from Warren and Jerde (1987). The Stannern and Nuevo Laredo compositional trends for non-cumulate eucrites likely represent origins by partial melting and fractional crystallization, respectively. Also shown is the field of cumulate eucrites. Meteorite abbreviations: St = Stannern; Jo = Jonzac; SC = Sioux County; NL = Nuevo Laredo; La = Lakangaon; MC = Moore County; Na = Nagaria; Mo = Moama; SM = Serra de Magé; Me = Medanitos.

crystallized from a Ca-poor, pyroxene-normative melt, leading to a more-feldspathic (more-basaltic) residual liquid that gave rise to the eucrites.

11.8.3 Howardites

Howardites are polymict breccias that are mechanical mixtures mainly of eucritic (cumulate and non-cumulate) and diogenitic materials. The minerals in the howardites formed in the same way as those in eucrites and diogenites. In addition to these components, some howardites contain melt rocks and glass particles formed by impact melting of eucritic and diogenitic precursors. Although some howardites contain carbonaceous-chondrite clasts, these are fragments of projectiles that impacted the surface of the HED parent body at low relative velocity.

11.8.4 Angrites

Angra dos Reis, the prototypical angrite, is a medium-to-coarse-grained igneous rock composed mainly of Al-Ti-diopside with minor olivine, spinel, and troilite, and accessory magnesian kirschsteinite, celsian, merrillite, titanomagnetite, baddeleyite, and metallic Fe-Ni. Other angrites have hypidiomorphic-granular or porphyritic textures, and, in addition to the phases in Angra dos Reis, some contain plagioclase, hercynite ($(Fe,Mg)Al_2O_4$), ulvöspinel, Ca phosphate, and silico-phosphate.

The angrites are basically basaltic in bulk composition; it is clear they formed as igneous rocks. LEW 86010 formed by partial melting of a carbonaceous-chondrite-like precursor at an oxygen fugacity of about 1 log unit above the iron-wüstite buffer. Asuka 881371 and LEW 87051 could be either contaminated partial melts or impact melts of preexisting igneous rocks.

Angra dos Reis may have formed as a cumulate with substantial trapped melt. The magmas from which D'Orbigny and Sahara 99555 crystallized could have been contaminated with batches of more-primitive melt.

11.8.5 Aubrites

The aubrites are highly reduced igneous rocks related to the enstatite chondrites. They are composed principally of enstatite with minor to accessory forsterite, diopside, albite, Si-bearing kamacite, troilite, and schreibersite, with trace amounts of sulfide (oldhamite, alabandite, niningerite, daubréelite, caswellsilverite, heideite ($(Fe,Cr)_{1+x}(Ti,Fe)_2S_4$), djerfisherite, brezinaite (Cr_3S_4)), perryite, osbornite (TiN), graphite, roedderite ($(K,Na)_2Mg_5Si_{12}O_{30}$), metallic Cu, perovskite, and geikielite ($MgTiO_3$).

Most of the clasts in Norton County are plutonic orthopyroxenites; also present are dunites, plutonic pyroxenites, and plagioclase-silica rocks. Keil (2010) modeled the aubrites as having formed from enstatite-chondrite-like material by melting and crystallization, possibly from a magma ocean. Separation of a metal-sulfide melt from the silicate melt would account for the low modal abundances of metallic Fe-Ni (averaging ~0.2 vol%) and sulfide (averaging ~0.5 vol%) in the aubrites. The negative Eu anomalies of most aubrites may reflect separation and loss of (REE-rich) oldhamite grains with positive Eu anomalies.

Two unusual unbrecciated aubrites (Shallowater and Mount Egerton) were modeled by Rubin (2015a), not as primary igneous rocks, but as impact-melt rocks derived from large cratering events on the same porous enstatite-chondrite-like asteroid.

11.8.6 Magmatic Iron Meteorites

The magmatic iron meteorite groups IC, IIAB, IIC, IID, IIF, IIG, IIIAB, IIIE, IIIF, IVA, and IVB formed by fractional crystallization in the molten metallic cores of differentiated asteroids. Each group formed within a different asteroid inside cores with different bulk chemical and isotopic compositions (with the possible exception of Group IIG, which may have formed from a P-rich magma in the IIAB core).

During initial cooling of the molten cores of these differentiated asteroids, taenite (fcc – face-centered cubic γ-iron) crystallized at high temperatures. As solid taenite crystals cooled, eventually the two-phase field was reached and kamacite (bcc – body-centered cubic α-iron) nucleated and began to grow.

The sulfide nodules within several magmatic groups probably formed from trapped S-rich liquid. Non-siderophile trace elements (e.g., O, Na, K, Ca, Mg, Mn, Pb) in the crystallizing metallic core of the parent asteroid were concentrated in the late-stage residual S-rich liquids. As the partial pressure of O increased, P and Fe from the metal host were oxidized and phosphate precipitated from the residual S-rich melt. The principal phosphate phase that crystallized directly from the residual liquid in many of the IIIAB irons was sarcopside. In IIIAB Cape York, there are five distinct phosphates including buchwaldite, maricite, and three incompletely characterized phases: phase I ($Na_4Ca_3Fe(PO_4)_4$;), phase III ($Na_4(Mn,Fe)(PO_4)_2$), and phase IV ($Na_4CaCr(PO_4)_3$) (Kracher et al. 1977).

11.8.7 Mesosiderites

Mesosiderites are fragmental breccias composed on average of ~50 vol% silicate clasts (mainly basalts, gabbros, orthopyroxenites, dunites, quench-textured impact-melt rocks, and breccias) and ~50 vol% metallic Fe-Ni. Rubin and Mittlefehldt (1992) reported that ~75 percent of the silicate clasts (gabbros, polygenic basalts, and quench-textured rocks) appear to have formed by impact-melting processes.

On the other hand, about 10 percent of the large silicate clasts are monogenic basalts that are nearly indistinguishable from eucrites in texture, mineralogy, mineral chemistry, and bulk composition. These rocks may be fragments of basalt flows on the mesosiderite parent body (MPB), their major phases (pyroxene and plagioclase) having crystallized from the cooling melt. They plot along the eucrites' Nuevo Laredo Trend and thus probably formed by fractional crystallization.

Ultramafic rocks constitute 10–15 percent of the silicate clasts. There are two varieties: (1) dunites (exemplified by 10 cm olivine crystals in Mount Padbury) are igneous cumulates formed either in the mantle of the MPB or the projectile that brought metal to the MPB surface, and (2) orthopyroxenites, which resemble diogenites and probably crystallized within the lower crust or mantle of the MPB.

The metallic Fe-Ni in the Type 1 mesosiderites (the rocks least affected by later-stage impact melting) was likely derived from the impacting metal-rich projectile. The large metallic clasts in mesosiderites have a narrow range of compositions (with near-chondritic siderophile element ratios), suggesting the impacting core was $\geq$95 percent molten (Hassanzadeh et al. 1990). This is consistent with the electrical interconnectedness of the metal (indicating that the metal flowed through preexisting fractures).

11.8.8 Pallasites

The main-group pallasites are mixtures of 65$\pm$20 vol% silicate (mostly olivine with minor to accessory low-Ca pyroxene), 35$\pm$20 vol% metallic Fe-Ni, and 0.3–1.0 vol% each of schreibersite, chromite, troilite, and phosphate (mainly farringtonite ($Mg_3(PO_4)_2$), stanfieldite ($Ca_4(Mg,Fe)_5(PO_4)_6$), and merrillite ($Ca_9MgNa(PO_4)_7$)). These phases all crystallized deep within a differentiated asteroid. Ulff-Møller et al. (1998) proposed that main-group pallasites formed by the following process: (1) Olivine grains crystallized from the molten silicate mantle at ~1,680°C and settled to the bottom, atop the molten metal core. At lower temperatures, rare pyroxene grains also crystallized in the mantle and settled. (2) The metal core crystallized from its center toward its edge; paleomagnetic studies suggest the core had an internal dynamo (Tarduno et al. 2012). (3) After temperatures had cooled to ~1,480°C, the late-stage, moderately low-viscosity, residual metallic magma invaded the silicate mantle, crystallized around large olivine-rich mantle fragments, and penetrated some individual olivine grains. Chromite and troilite probably crystallized very late from residual S-rich liquids. One or more large impact events would have been necessary to break up the asteroid and liberate core-mantle boundary material.

A few pallasites contain phosphoran olivine (with ~4–5 wt% P_2O_5); Brenham olivines range up to 7.4 wt% P_2O_5. Dynamic crystallization and isothermal experiments show that the phosphoran olivine was likely produced as a metastable phase from a P-enriched melt (e.g., Boesenberg and Hewins 2010).

Some olivine grains from pallasites are of gemstone quality and can be faceted to make jewelry; they have been labeled as peridot and chrysolite. A few display chatoyancy (cat's eye), apparently caused by light reflected at right angles to subparallel, tube-like hollow inclusions in the grains.

11.8.9 Lunar Meteorites

The major types of primary rocks returned from the Moon are mare basalts and three varieties of pristine highland rocks: ferroan anorthosites, Mg-rich rocks, and KREEP rocks. The latter samples are rich in incompatible elements including K, REE, and P. Additional rock types identified among returned lunar samples are polymict breccias (formed by impact mixing and melting of the primary rocks) and lunar soil (containing fragments of primary rocks and impact-melted materials). Lunar meteorites include unbrecciated basalts, olivine cumulates, and various types of breccias (basaltic, feldspathic, and mixed basaltic-feldspathic). One impact-melt breccia (Sayh al Uhaymir 169) contains a KREEP component. NWA 10986 contains a wide variety of highland and basalt lithologies including troctolites, gabbronorites, anorthositic norites, troctolitic anorthosites, low-Ti basalts, and very-low-Ti basalts. Also present among lunar meteorites are regolith breccias containing glass spherules (produced by impacts or by pyroclastic processes), agglutinates, and fragments of common rock types.

Lunar meteorites do not contain unmelted lunar soil samples; there are few or no clasts of ferroan anorthosites and Mg-rich rocks. It appears that KREEP samples and mafic, magnesium-suite rocks (which are abundant among Apollo samples) formed only in restricted locales on the Moon that happened to include the Apollo landing sites. Because lunar meteorites were derived from the entire lunar globe, the rarity of geographically limited samples among these meteorites is expected.

The principal minerals in lunar meteorites are anorthitic plagioclase, pyroxene, olivine, and minor ilmenite. Also abundant is impact-generated glass derived from melting the principal mineral phases. Mare basalts formed from cooling lava flows; their constituent mineral phases crystallized from the melt. The feldspathic lunar meteorite breccias have clasts of noritic anorthosites, troctolitic anorthosites, anorthitic norites, and anorthitic troctolites. (These rock-type labels depend on the normative mineral abundances.) All these rock fragments are likely derived from the lunar highlands; their constituent phases crystallized from the melt during cooling of the anorthositic crust. The high plagioclase contents of these rocks suggest the rocks are cumulates, wherein newly formed plagioclase crystals separated from the residual melt. The presence of secondary ferroan olivine veinlets (associated with accessory chromite, pyroxene, and Ca-plagioclase) in lunar highlands breccia NWA 11273 may reflect late-stage near-surface magmatic fluids penetrating preexisting fractures in the rock.

11.8.10 Martian Meteorites

Martian meteorites are all igneous rocks; the principal types include nakhlites (olivine clinopyroxenites), chassignites (dunites), the ALH 84001 orthopyroxenite, and basaltic shergottites and related rocks (including those classified as lherzolites, basaltic breccias, gabbros, microgabbros, diabases, olivine-phyric shergottites, and augite basalt).

Nakhlites are igneous augite-rich cumulate rocks formed in thick lava flows or shallow intrusions, plausibly near the large volcanoes in the Tharsis, Elysium, or Syrtis Major regions. Augite and olivine crystallized from the magma and settled to the bottom of the magma chamber. Later-stage igneous processes included crystallization of the residual magma, flow of the melt relative to the crystals, and elemental diffusion and chemical reactions among the minerals and the magma.

Chassignites are igneous olivine-chromite cumulate rocks. Major olivine and minor pyroxene crystals formed within a magma, in some cases enveloping previously crystallized chromite grains. The crystals settled to the bottom of the magma chamber and continued to grow; grains impinged on one another, squeezing out most of the intercumulus liquid. The last phases to crystallize were alkali feldspars: plagioclase (ranging from albite to andesine) and K-feldspar.

A simplified petrogenetic history of the ALH 84001 orthopyroxenite would include the rock's formation within a magma chamber in which orthopyroxene and euhedral chromite grains crystallized and settled.

The major original phases in the basaltic shergottites were pyroxene, plagioclase, and, in some cases, olivine. (Plagioclase was later transformed into maskelynite in nearly all shergottites during the collisions that launched them off Mars.) The major phase in lherzolitic shergottites is olivine, accompanied by minor Ca-poor pyroxene. Minerals in the basaltic shergottites crystallized within lava flows; the lherzolitic shergottites formed as olivine cumulates at the base of a magma chamber. In each case, minor phases crystallized from the residual melt: titanomagnetite, ilmenite, ulvöspinel, magnetite, chromite, and hercynite in the basaltic shergottites; pyroxene, plagioclase, and chromite in the lherzolitic shergottites.

11.8.11 Acapulcoites and Lodranites

Many workers consider acapulcoites and lodranites to be "primitive achondrites" and have modeled them as having formed by different degrees of partial melting of primary chondritic rocks. If these models are correct, then the phases in acapulcoites and lodranites that are not chondritic relicts crystallized from the endogenic melt; this would include some of the olivine, pyroxene, plagioclase, metal, and sulfide grains. The alternative view of the origin of these meteorites is that they are impact melt products.

11.8.12 Ureilites

Ureilites are C-rich ultramafic rocks whose origins remain poorly understood. Among the more popular models are those that have ureilites forming as (1) partial-melt residues that were infused with carbon, (2) intrusive igneous cumulates formed from C-rich magmas, or (3) paracumulates derived from disrupted partially molten asteroids. If any of these models are correct, then the principal primary phases in ureilites (olivine, pyroxene, graphite, metallic

Fe-Ni, sulfide) and, perhaps, the minor to accessory phases in related rocks (e.g., plagioclase and chlorapatite) crystallized from endogenic melts on the parent asteroid(s).

11.8.13 Brachinites

Brachinites are medium- to coarse-grained dunitic wehrlites, consisting dominantly (roughly 80–95 percent) of olivine with minor augite, chromite, sulfide, phosphate, metallic Fe-Ni, and, in some cases, plagioclase and orthopyroxene. They have been labeled "primitive achondrites" and are widely considered to be either highly metamorphosed chondritic rocks or chondrites that have undergone very low degrees of partial melting. If these scenarios are correct, then the constituent minerals of brachinites are mainly chondritic relicts. One worker modeled them as partial-melt residues from an FeO-rich asteroid; other researchers compared their textures to those of ureilites and modeled the brachinites as cumulates. If the latter model is correct, the minerals in brachinites formed igneously; some phases would have accumulated at the bottom of a magma chamber, while others crystallized from the residual liquid.

11.9 Condensation from Late-Stage Vapors in Differentiated Bodies

Tabular grains of tridymite, up to ~1 cm in maximum dimension (Figure 11.26) and containing very low concentrations of Al and Ti, occur in the IVA irons Gibeon and Bishop Canyon. Wasson et al. (2006) suggested these tridymite occurrences formed by late-stage condensation from a cooling SiO-rich vapor within fissures in solid metal in the IVA core. The SiO(g) itself may have formed by reduction of SiO_2 or $FeSiO_3$ from the asteroid mantle; the reducing agent could have been C dissolved in the metallic magma. The platelet morphology of the tridymite may have arisen as the phase minimized its surface energy.

Alternatively, SiO(g) could have formed by oxidation of minute amounts of Si dissolved in the metallic melt; the oxidizing agent could have been O dissolved in the melt or CO, CO_2, or H_2O vapors derived from the mantle. In this scenario, the tridymite grains formed from the melt and were later surrounded by solid metal.

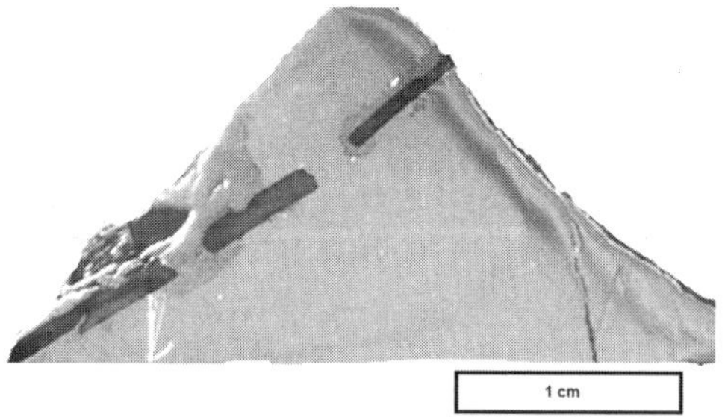

Figure 11.26 Apparently discontinuous tabular tridymite grain (dark gray) surrounded by metallic Fe-Ni (light gray) in the Gibeon IVA iron. (The tridymite grain may be continuous in three dimensions.) Modified from Wasson et al. (2006). Reflected light.

11.10 Exsolution, Inversion, and Subsolidus Redox Effects Within Cooling Igneous Materials

Minerals in all the igneous meteorites cooled from high temperatures to below the solidus and were subject to exsolution. In some cases, particularly in pyroxene, cooling causes the inversion of a high-temperature crystal structure to a low-temperature structure; this inversion is commonly attended by exsolution.

Eucrites. Basaltic eucrites originally contained pigeonite, which upon cooling below the solidus, exsolved lamellae of augite (Figure 11.27). In cumulate eucrites, primary pigeonite also underwent subsolidus exsolution of augite; in some rocks, the pigeonite inverted to orthopyroxene and subsequently exsolved augite. Some eucritic clasts in howardites show similar exsolution features.

Diogenites. Diogenites contain minor to accessory amounts of diopside (0–2 vol%) occurring mainly as ~3-μm-thick exsolution lamellae within orthopyroxene or low-Ca pigeonite. There are much thinner augite exsolution lamellae in Ibbenbüren (0.077 μm) and Johnstown (0.006 μm).

Angrites. Millimeter-sized olivine grains in the LEW 86010 angrite contain kirschsteinite exsolution lamellae up to 20 μm thick.

Aubrites. Diopside in aubrites is present as primary grains that crystallized from the melt as well as exsolution lamellae within enstatite. Schreibersite exsolved from metallic Fe-Ni during cooling when the P concentration within the metal became oversaturated.

Magmatic irons. After taenite (γ-iron, face-centered cubic) crystallized from the metallic melt in the cores of the differentiated asteroids that produced magmatic irons, the solidified cores continued to cool until the two-phase Fe-Ni field (the solvus) was reached (Figure 11.28); the temperature of this boundary (between 900°C and 500°C) depends upon the bulk Ni content of the taenite. Kamacite (α-iron, body-centered cubic) nucleated and began to crystallize along the {111} planes of the taenite. (Kamacite nucleation temperatures are dependent on the bulk Ni and P concentrations of the parent taenite grain.) Upon further cooling, both kamacite and taenite grains grew richer in Ni as the kamacite formed at the expense of taenite. The intergrowth of kamacite lamellae and Ni-rich lamellae forms the Widmanstätten pattern in most

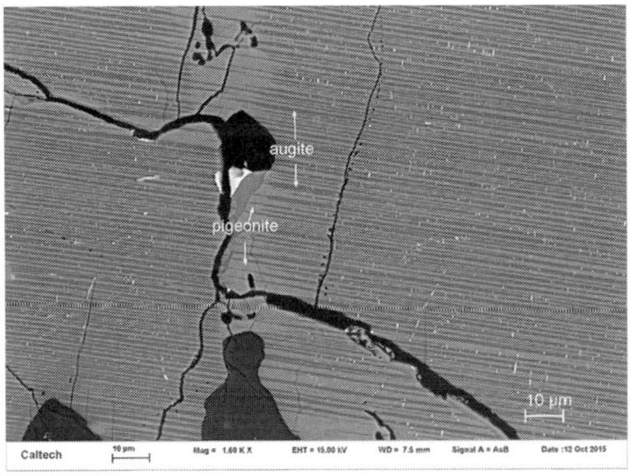

Figure 11.27 Augite lamellae exsolved from pigeonite in the Juvinas eucrite. BSE image.

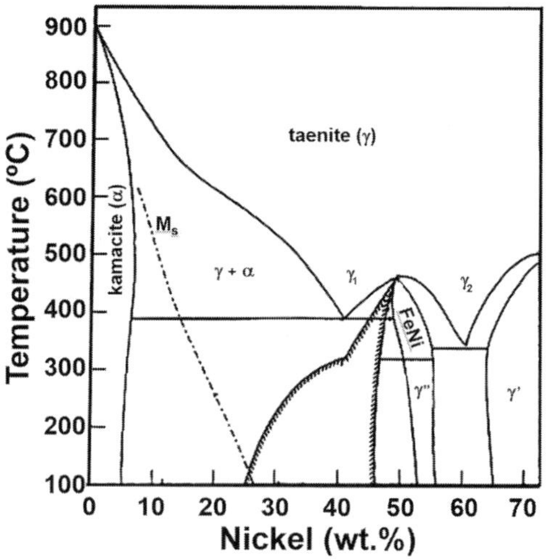

Figure 11.28 Simplified Fe-Ni phase diagram modified from Reuter et al. (1987). M_s = martensite start temperature; γ_1 = paramagnetic fcc (face-centered cubic) phase; γ_2 = ferromagnetic fcc phase; γ' = ordered Ni_3Fe; γ'' = ordered FeNi; the field labeled FeNi above γ'' is the disordered high-temperature form of γ''.

iron meteorites; because the kamacite lamellae are oriented along the planes of an octahedron, the meteorites with this pattern are known as octahedrites. (The complete thermal history may be more complicated than this; taenite may undergo decomposition at <400°C, producing tetrataenite, martensite, and awaruite.)

Solid-state diffusion of Ni is more rapid in kamacite than taenite. During cooling in the two-phase field, as diffusion becomes more sluggish, compositional traces across taenite lamellae (Figure 11.29), as determined by electron-microprobe analysis, show a buildup of Ni in taenite at the α/γ boundaries[5] and a dip in Ni concentration at the centers of the taenite grains. This classic "M-shaped" profile has been used to measure metallographic cooling rates in iron meteorites and chondrites: the Ni concentration at the centers of taenite lamellae and the distance from these centers to the nearest α/γ boundary are matched against theoretically calculated cooling rate curves (e.g., Figure 11.30). Different iron meteorites from the same chemical group can have significantly different cooling rates; e.g., IVA-iron rates vary from 100 K/Ma for high-Ni members to 6,600 K/Ma for low-Ni members (Yang et al. 2007).

Lyons et al. (2019) suggested that these differences resulted from impact events that occurred 20–70 Ma after the formation of differentiated planetesimals. The model involved stripping away much of a largely crystallized mantle, thereby bringing the hot metallic core near the surface. This was at a time when the core was at a temperature near or above its solidus. Such a scenario would lead to rapid and heterogeneous cooling rates within the core.

With the exceptions of rare very large schreibersite grains, rare multicentimeter-sized chromite occurrences, and centimeter-sized troilite and troilite-graphite nodules, nearly all the

[5] The increase in Ni concentration in taenite at the α/γ boundary is matched by a decrease in Ni in adjacent kamacite at the same boundary (Figure 11.29). This latter feature, known as the "Agrell dip," is named after Stuart Agrell, a mineralogist at Cambridge University, who discovered the phenomenon (Agrell et al. 1963).

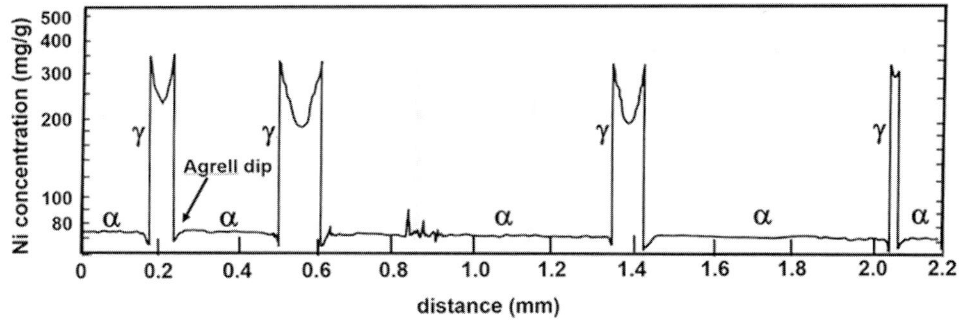

Figure 11.29 Electron microprobe trace across a slab of the Buenaventura IIIAB iron meteorite showing variations in Ni concentration. Kamacite (α) grains have low Ni contents. The Ni contents of taenite (γ) lamellae depend on the width of the grains – thin taenite grains have higher central Ni contents. During subsolidus cooling of the metal assemblage, taenite grains grow at the expense of kamacite. Because of the sluggish pace of diffusion, the concentration of Ni builds up at the edges of taenite grains and is relatively depleted at the centers of the taenite grains. This accounts for their "M-shaped" profiles. As Ni diffuses into taenite from kamacite, the Ni becomes depleted in kamacite at the boundary with taenite; this depletion is known as the "Agrell dip."

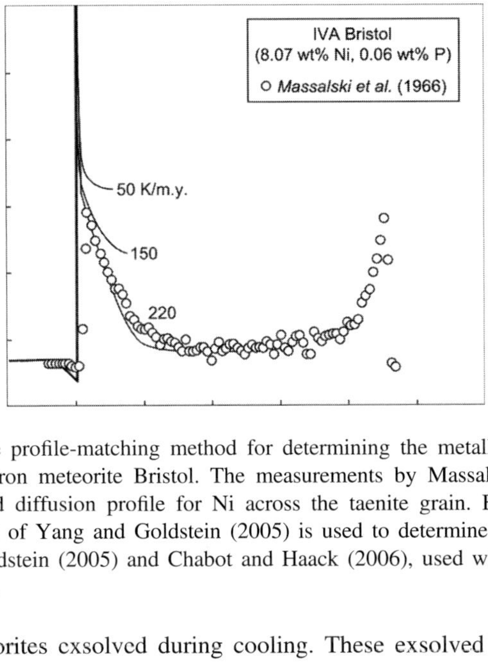

Figure 11.30 The taenite profile-matching method for determining the metallographic cooling rate, as illustrated for the IVA iron meteorite Bristol. The measurements by Massalski et al. (1966) show a characteristic "M"-shaped diffusion profile for Ni across the taenite grain. Because Bristol is low in P (0.06 wt%), the model of Yang and Goldstein (2005) is used to determine a cooling rate of 220 K/Ma. From Yang and Goldstein (2005) and Chabot and Haack (2006), used with permissions from Jijin Yang and Nancy Chabot.

minerals in iron meteorites exsolved during cooling. These exsolved phases result from the decreases in solubility of P, C, N, and S in metallic Fe-Ni with decreasing temperature (e.g., Scott 2020). In numerous irons, schreibersite (in many cases occurring in the prismatic rhabdite form) exsolved from the cooling taenite when the face-centered-cubic crystal structure could no longer accommodate appreciable amounts of P. In IIIE irons, the carbide haxonite exsolved from taenite as C became oversaturated; in some members of groups IIAB and IIIAB, cohenite formed in a similar fashion. Cliftonite is a form of graphite that displays a cubic morphology; it exsolved from the metallic Fe-Ni during cooling. In some iron groups, oriented platelets of 1–5 μm-thick chromite lamellae flanked by secondary troilite (collectively known as Reichenbach lamellae) also exsolved from taenite during cooling. Daubréelite commonly occurs as exsolution lamellae within troilite in magmatic groups IIAB, IIIAB, and IVA.

In troilite nodules in IIIAB irons, the phosphate johnsomervilleite ($Na_2Ca(Fe,Mg,Mn)_7(PO_4)_6$) occurs inside sarcopside ($(Fe,Mn)_3(PO_4)_2$) and probably exsolved from it during cooling. Some occurrences of galileiite ($NaFe_4(PO_4)_3$) also seem to have exsolved from primary graftonite ($(Fe,Mn)_3(PO_4)_2$) or sarcopside during cooling.

Mesosiderites. The ultramafic and mafic silicate clasts in mesosiderites contain pyroxene grains that have undergone exsolution and inversion, i.e., orthopyroxene inverting from pigeonite and pigeonite exsolving augite lamellae. Perryite within metallic Fe-Ni in some mesosiderites apparently exsolved from taenite during cooling from high temperatures.

Pallasites. Schreibersite is an accessory phase occurring within the metal of pallasites. It formed by exsolution from taenite at high subsolidus temperatures when the P concentration in the γ-iron exceeded the accommodation capacity of the face-centered-cubic crystal structure.

Lunar meteorites. Some pyroxene grains in Apollo samples have undergone exsolution, producing both thin pigeonite lamellae within augite and augite lamellae within pigeonite. Subsolidus reduction reactions include ulvöspinel breaking down into different products: (a) ilmenite and metallic Fe, (b) rutile and metallic Fe, or (c) chromite, rutile, and metallic Fe. In some samples, Ti-rich ulvöspinel has broken down into Ti-poor spinel, titanochromite, ilmenite, and metallic Fe. Although not all these features have so far been observed in lunar meteorites, they likely are present.

Martian meteorites. Two sets of augite lamellae have exsolved from pigeonite in the shergottites: one set parallel to (001), the other parallel to (100). Primary ulvöspinel grains in QUE 94201 underwent exsolution and oxidation to produce titanomagnetite and ilmenite. In the nakhlites, kirschsteinite lamellae within olivine grains appear to have undergone oxidation to produce augite and magnetite. Symplectic intergrowths of augite and magnetite formed by exsolution in olivine from two nakhlites (Nakhla and Governador Valadares) under oxidizing conditions at temperatures $>900°C$ during cooling. Titanomagnetite has exsolution lamellae of ilmenite and (perhaps) ulvöspinel. The chassignites contain orthopyroxene with thin augite exsolution lamellae that were inferred to have formed from primary pigeonite. Chassigny also contains symplectic intergrowths of augite and magnetite within olivine grains; this feature resembles those in nakhlites. The orthopyroxene grains in the ALH 84001 orthopyroxenite appear to be free of augite exsolution lamellae.

Experiments by Castle and Herd (2018) suggest that fine-grained Cr-spinel within olivine phenocrysts in the Tissint olivine-phyric shergottite resulted from oxidation-induced exsolution from olivine during crystallization. Other olivine-phyric shergottites also appear to have undergone a similar oxidation event during crystallization.

Acapulcoites and lodranites. The few acapulcoites that contain relict chondrules have Ca-pyroxene grains that lack low-Ca pyroxene exsolution lamellae; in contrast, many of the chondrule-free acapulcoites have Ca-pyroxene grains with thin exsolution lamellae. Lodranites contain inverted pigeonite grains transformed into orthopyroxene with coarse Ca-pyroxene exsolution lamellae.

Ureilites. Although most pyroxene grains in ureilites lack exsolution lamellae, there are a few exceptions: (a) LEW 85440 contains very thin (~0.01 μm) orthopyroxene and clinopyroxene lamellae within orthopyroxene parallel to (100), (b) ALH 82130 contains pigeonite with irregular augite lamellae, and (c) LEW 88774 contains pyroxene crystals consisting of ~50-μm-thick lamellae of orthopyroxene and augite in similar proportions.

11.11 Solar Heating Near Perihelion

Asteroids or meteoroids with elliptical orbits and small perihelion distances can be periodically heated by solar radiation. Many near-Earth objects (NEOs) undergo orbital evolution that brings them close to the Sun, leading to appreciable heating. For example, the subsolar point on asteroid 1566 Icarus (which has a perihelion distance of 0.187 AU) can experience temperatures up to ~500°C every 409 days (Ohtsuka et al., 2007). (By comparison, Mercury's perihelion is at 0.313 AU.) Near-Earth C-type asteroid 162173 Ryugu appears to be related to primitive carbonaceous chondrites in containing hydrated silicates and organic materials. The Hayabusa2 spacecraft found that the polar regions of Ryugu exhibit bluer spectral characteristics than equatorial regions, suggesting that reddening is caused by surface heating via solar radiation. This process appears to have occurred over a short time interval (after currently red craters formed and before currently blue craters formed) when Ryugu made a temporary orbital excursion near the Sun (Morota et al., 2020). Although meteorites that strike Earth spent time in NEO-like orbits, they are likely to have experienced less Sun-driven heating than the NEOs themselves. This is due to the meteorites' much smaller sizes, greater perihelion distances and, possibly, shorter average collisional lifetimes. Furthermore, it is likely that the sides of the meteoroids that were most strongly heated by the Sun (i.e., those sides close to the subsolar point of the parent NEO) are ablated and lost during atmospheric entry (Marchi et al. 2009). Nevertheless, solar heating of meteorites remains a viable, albeit minor, mechanism.

The IVA iron Social Circle is a fine octahedrite containing recrystallized kamacite grains that partially overprint lamellae with Neumann lines. Buchwald (1975) suggested that prolonged temperatures of ~500°C could have produced the recrystallization of kamacite. He also described "rhythmic growth lines" throughout the meteorite (Figure 11.31) and concluded that Social Circle was "reheated several times" to about the same temperature. He stated the meteorite could have been heat-treated by periodic close approaches to the Sun.

The IIAB Indian Valley hexahedrite contains recrystallized kamacite grains with concentric growth rings, possibly consistent with cyclic reheating. Buchwald (1975) described graphite plumes associated with rhabdite plates within certain regions of the kamacite in this meteorite. He suggested these plumes formed from cohenite that had nucleated on rhabdite and subsequently decomposed; the cause was ascribed to the same cyclic reheating events that caused the kamacite to recrystallize. If these features formed during cyclic heating events, it is plausible they were caused by close passages to the Sun.

Some CM carbonaceous chondrites may also have experienced minor thermal metamorphism caused by solar heating. For example, Sutter's Mill is a regolith breccia containing some components that have been heated sufficiently to destroy carbonates, transform phyllosilicates into fine-grained olivine, break down tochilinite to form troilite, and eliminate

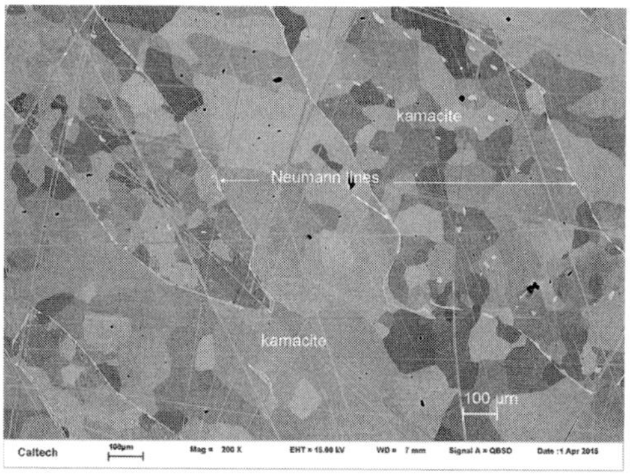

Figure 11.31 Recrystallized kamacite grains that partially overprint lamellae with Neumann lines in the IVA iron Social Circle. BSE image.

indigenous amino acids (Burton et al. 2014; Zolensky et al. 2014). The meteorite contains solar-cosmic-ray-produced nuclides acquired when the meteoroid's orbit was <1 AU from the Sun (Nishiizumi et al. 2014). If CM meteoroids ventured appreciably closer to the Sun than this, e.g., <0.1 AU, solar heating could be partly responsible for low-temperature thermal metamorphism in these rocks. But there is no evidence this is the case. Alternatively, most or all of the heating could be due to impacts, the effects of which are evident in the petrofabrics of CM chondrites (Rubin 2012b; Lindgren et al. 2015).

Chaumard et al. (2012) proposed that CK chondrites were heated by solar radiation. They used plausible NEO orbital parameters and known CK-chondrite physical properties to find that small CK meteoroids could have reached 1,050 K, consistent with estimated CK metamorphic temperatures. Nevertheless, most workers assume CK chondrites were heated by the decay of ^{26}Al and/or by collisions (e.g., Rubin 1992; Ghosh et al. 2006; Huss et al. 2006). The importance of shock metamorphism as an important cause of CK heating is illustrated by the presence of numerous shock features in CK5 and CK6 chondrites (as detailed in section 11.5).

Wittmann et al. (2011) explored the possibility that the LAP 031047 H/L ordinary chondrite resided in the near-surface environment of an Aten or Apollo asteroid that experienced transient heating to ~700–750°C during perihelion passage. Although the rock is depleted in thermal labile elements (Ga, Sb, Sn, Rb, Cs, Se, Ag, Te, Zn), contains devitrified chondrule glass, and experienced nearly complete degassing of Ar to very high temperatures, its olivine and low-Ca pyroxene grains have not equilibrated. If the rock had been shocked and annealed by burial within a warm ejecta blanket, the olivine (at least) might be expected to have equilibrated. Welten et al. (2014) concluded that the LAP 031047 meteoroid was heated ~0.5 Ma ago during one or more close solar encounters.

The dynamical evolution of the orbit of the Chelyabinsk LL5 meteoroid was calculated by Emel'yanenko et al. (2014). They found the meteoroid would likely have resided close to the Sun (<0.1 AU) between ~0.8 and ~2 Ma ago, causing significant tidal and thermal effects. Chelybinsk's orbit appears similar to that of the small Earth-crossing asteroid 2008 DJ.

12

Formation of Meteoritic Minerals in the Terrestrial Environment

12.1 Atmospheric Passage

A typical meteoroid enters the Earth's atmosphere at 18–20 km s^{-1}. During the traverse, the surface of a chondritic meteoroid reaches temperatures of 1,180–1,410°C and melts to depths of ~0.3–1.0 mm due to friction with the surrounding air. The surface of the meteoroid ablates, exposing a new surface that also melts and ablates; this process can be repeated several times.

Most meteoroids break into fragments as aerodynamic stresses increase during their descent. For objects travelling at 20 km s^{-1}, stresses rise from 10 MPa at an altitude of 30 km to 100 MPa at 15 km (Melosh 1989). This exceeds the crushing strengths of many stony meteorites (which range from ~6 to ~500 MPa and average ~200 MPa). The crushing strengths of carbonaceous chondrites are on the order of 1 MPa, consistent with the small median size of carbonaceous chondrites listed in the Meteoritical Bulletin Database (~34 g). The median size of CM chondrites (which are friable matrix-rich samples) is ~10 g. An individual stony meteoroid that does not break into fragments could lose >95 percent of its original mass by the time it reaches the Earth's surface.[1]

Deceleration of the meteoroid continues in its descent through the atmosphere until the drag on the object equals the acceleration due to gravity, at which time a constant terminal velocity of ~300 m s^{-1} is reached. By this point, frictional heating has diminished sufficiently to preclude additional melting of the meteoroid surface. The final fusion crust typically consists of three zones: (1) an outer zone consisting of vesicular black opaque glass containing submicrometer magnetite, (2) an intermediate zone containing partly melted silicate grains, and (3) an inner zone containing unmelted silicate and rapidly solidified metal-troilite intergrowths. The zones in the fusion crusts of enstatite chondrites are less developed.

Minerals that form in the fusion crusts of chondritic meteorites (e.g., Genge and Grady 1999) include eskolaite (Cr_2O_3) (formed by the reduction and breakdown of chromite), hercynite ($(Fe,Mg)Al_2O_4$) (formed possibly by the melting and oxidation of mafic silicates and plagioclase), pseudobrookite (Fe_2TiO_5) (formed by the breakdown of ilmenite), magnetite and wüstite (both formed by oxidation of metallic Fe), magnesioferrite ($Mg(Fe^{3+})_2O_4$) (formed by the combustion of metallic Fe-Ni and silicate), intergrowths of metallic Fe-Ni and troilite (formed

[1] If a meteoroid breaks up during atmospheric passage, the largest fragment could be miniscule in proportion to the original mass. For example, when it entered the atmosphere on February 15, 2013, the Chelyabinsk meteoroid (LL5) had an estimated mass of 12,000–13,000 metric tons. The largest fragment (~640 kg) that reached the surface (and plunged through the ice on Lake Chebarkul) was only ~0.005 percent of the initial meteoroid mass.

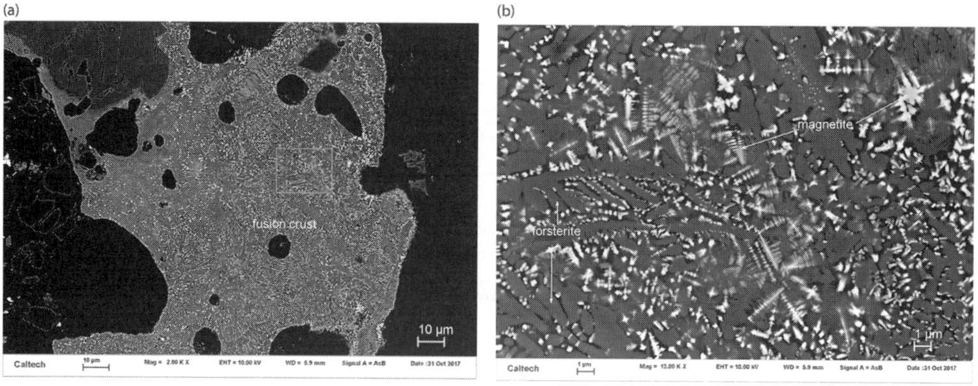

Figure 12.1 Olivine and magnetite crystals in the fusion crust of the Semarkona LL3.00 chondrite. BSE images.

by the melting and quenching of these phases) (Figure 12.1), and olivine (formed as overgrowths on relict grains and as new grains crystallized from the melt). Some of these olivine grains show reverse or oscillatory compositional zoning, indicative of complex redox effects. Most meteorite fusion crusts do not contain all these phases. Oxidation of metallic Fe-Ni during atmospheric passage has caused a decrease in the saturation magnetic moment of ordinary-chondrite fusion crusts (Maksimova et al. 2020). In some meteorites, phases produced during atmospheric passage flowed a few hundred micrometers toward the interior through fractures: e.g., wüstite in CO3 DOM 03238 (Kim et al. 2009) and metal-sulfide veins in EL6 DOM 10088 (Rubin 2015a).

Heating caused by atmospheric passage can produce back transformations of high-pressure phases to low-pressure ones. The fusion crust of the H6 chondrite Y-75267 transects a shock vein containing ringwoodite, majorite-pyrope$_{ss}$, and $NaAlSi_3O_8$ hollandite. However, portions of the vein near the fusion crust (where temperatures of ~1,400°C were reached) contain no high-pressure phases – only olivine, low-Ca pyroxene, and plagioclase glass.

Because achondrites and oxidized carbonaceous chondrites have little metallic Fe, their fusion crusts are less likely to contain rapidly cooled metal-sulfide intergrowths. Achondritic minerals melt over a relatively narrow temperature range and are likely to form a smooth glossy fusion crust.

Iron meteorite fusion crusts typically contain magnetite and wüstite. Veins of wüstite can penetrate the interior of the meteorite.

Micrometeorites (10–2,000 µm in size) and interplanetary dust particles (<10–100 µm) are also heated and partly melted during atmospheric passage. Secondary minerals formed within these small objects include magnetite, maghemite, poorly characterized Fe oxides, laihunite, and skeletal olivine grains surrounded by silicate glass.

12.2 Terrestrial Weathering

After reaching the Earth, meteorites are contaminated with fungi and bacteria as well as other organic and inorganic materials. In addition, the presence of water (atmospheric moisture,

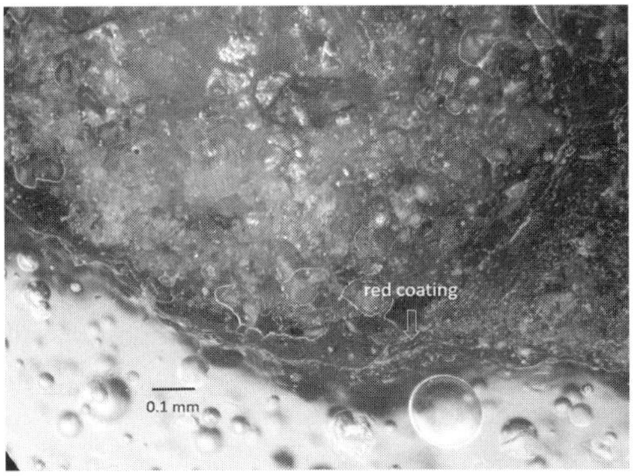

Figure 12.2 Thin red coating (possibly goethite) due to oxidation of metallic Fe on a Murchison (CM2) grain from the Caltech collection. Reflected light. (A black-and-white version of this figure will appear in some formats. For the colour version, please refer to the plate section.)

precipitation, ground water, lakes, rivers, and oceans) causes alteration of indigenous mineral phases.

The phenomenon of meteorites "rusting" in museum cabinets (at least those cabinets not filled with inert gas) is well known to curators. In many cases, slabs of fresh falls (and individual grains within fresh falls) develop thin red coatings (possibly goethite (αFeO(OH))) due to oxidation of metallic Fe from atmospheric water (Figure 12.2). Color images of weathered meteorites appear in Yanai (1981), Yanai and Kojima (1987), Killgore and Killgore (2002), and Stinchcomb (2011).

Antarctic meteorites are traditionally assigned weathering categories (A, B, C) based on the extent of surface rustiness (Table 12.1). Some samples have light-colored evaporite minerals at their surface (and, for example, are designated Ce); the evaporates include Mg carbonates (nesquehonite, hydromagnesite, amorphous Mg carbonate), Mg sulfates (epsomite, starkeyite), Ca sulfate (gypsum), and additional poorly characterized K, Fe, and Mg sulfates. Most cations are from underlying meteoritic minerals; in particular, Mg is mainly from weathered olivine. Much of the CO_2 in the carbonates probably originates in the atmosphere; in C-rich meteorites, some of the CO_2 may be derived from meteoritic carbon. In addition, the fusion crusts of the meteorites contain poorly ordered clay minerals similar in composition to smectite and mica formed by terrestrial weathering. Insolation heating of dark fusion-crusted Antarctic meteorites on calm summer days can raise the temperatures beneath the exposed surfaces to a few degrees above freezing, allowing liquid water to exist. Freezing of this water could cause evaporite minerals to precipitate.

Many meteorite finds are extensively altered. Metallic Fe-Ni and sulfide show progressive alteration with time; this is reflected by the weathering index for ordinary chondrites devised by Wlotzka (1993) and refined by Zurfluh et al. (2016) (Table 12.2). Rubin and Huber (2005) proposed a weathering index for CK and R chondrites (oxidized meteorites that contain little metallic Fe-Ni) (Table 12.3). The index is based on the percentage of brown-stained silicates, reflecting the degree of mobilization of oxidized iron produced by terrestrial weathering of Ni-bearing sulfide. This is consistent with the occurrence of weathering rinds around sulfide grains

Table 12.1 *Weathering categories for Antarctic meteorites*

Weathering Category	Description	Comments
A	Minor rustiness	Minor rust haloes on metal particles; minor rust stains along fractures
B	Moderate rustiness	Thick rust haloes on metal particles; extensive rust stains on internal fractures
C	Severe rustiness	Near-complete rust staining of metal particles
e	Evaporite minerals	Evaporite minerals visible to naked eye

After the *Antarctic Meteorite Newsletters*.

Table 12.2 *Weathering index for ordinary chondrites*

Weathering Degree	Metal Oxidation (vol%)	Troilite Oxidation (vol%)	Comments
W0.0	0	0	Fresh; some iron hydroxide staining possible
W1.0	<20	≥0	Minor oxide rims around metal and troilite; small iron oxides and iron hydroxide veins may occur
W2.0	20–60	<20	Onset of veining with iron oxides and iron hydroxides
W3.0	>60	<20	Strong oxidation of metal; troilite shows minor alteration
W3.3	>60	20–60	Strong oxidation of metal; troilite moderately altered; some troilites completely oxidized
W3.6	>60	>60	Strong oxidation of metal and troilite; most troilite grains oxidized or have reduced reflectivity
W4.0	>95	>95	Nearly complete oxidation of metal and troilite; some troilite remnants may be present
W4.5	100	100	All metal and troilite oxidized; only minor remnants of metal and troilite as inclusions in silicates; some silicate alteration (mainly olivine) possible
W5.0	100	100	Metal and troilite 100% oxidized; major alteration of silicates, mainly olivine
W6.0	100	100	Massive replacement of silicates by clay and oxides

After Zurfluh et al. (2016).

in CK- and R-chondrite finds. In these meteorites, bulk S, Se, and Ni tend to decrease with increasing degrees of weathering, mainly affecting pentlandite (the principal sulfide in CK chondrites and the second-most abundant sulfide, after pyrrhotite, in R chondrites). For example, S in pentlandite could have been converted to soluble S compounds (e.g., SO_4^{-2} and $S_2O_3^{-2}$) followed by leaching; some S could also have been converted to SO_2 or SO_3.

Table 12.3 *Weathering index for CK and R chondrites*

Weathering Degree	Description	Comments
wi-0	Unweathered	<5 vol% of silicates stained brown
wi-1	Slightly weathered	5–25 vol% of silicates stained brown
wi-2	Moderately weathered	25–50 vol% of silicates stained brown
wi-3	Significantly weathered	50–75 vol% of silicates stained brown
wi-4	Highly weathered	75–95 vol% of silicates stained brown
wi-5	Severely weathered	>95 vol% of silicates stained brown
wi-6	Extremely weathered	Nearly complete brown staining of silicates; significant replacement of mafic silicates by phyllosilicates

After Rubin and Huber (2005).

Selenium behaves somewhat differently than S during terrestrial weathering: some Se can be converted into SeO_2, a solid that can remain within the meteorites. In Antarctic CK chondrites, bulk Cl increases with weathering index, reflecting terrestrial contamination of the rocks with Cl derived from airborne sea mist.

There are about 70 minerals known to have formed in meteorites by terrestrial weathering. Those formed in iron meteorites by weathering include (1) oxides and hydroxides formed directly from metallic Fe-Ni by oxidation and the incorporation of H_2O, Cl^{-1} and CO_3^{-2} (Cl-bearing akaganéite, bunsenite, goethite, ferrihydrite, hematite, hibbingite, lepidocrocite, maghemite, magnetite, trevorite, and possibly zaratite); (2) graphite and kamacite formed by the decomposition of iron carbide; (3) phosphates formed by the oxidation of schreibersite (apatite, arupite, cassidyite, collinsite, lipscombite, vivianite); and (4) sulfates formed from troilite (honessite, jarosite). Additional phases formed in irons by terrestrial weathering include metallic Cu, opal, various sulfides (pentlandite, bornite, chalcopyrite, heazlewoodite, isocubanite), carbonates (reevesite, siderite), phyllosilicate (pecoraite), and elemental S (Rubin and Ma 2017).

In enstatite meteorites, the unusual sulfides rich in cations that are found only in silicate or oxide minerals under less-reducing conditions (e.g., Na, Mg, K, Ca, Ti, Cr, Mn) are unstable at the Earth's surface. The sulfides break down to produce an assortment of weathering products: (a) schöllhornite forms from caswellsilverite; (b) bassanite, vaterite, calcite and portlandite form from oldhamite (CaS); (c) Nakamura-Messenger et al. (2012) examined a small BO chondrule in the Y-691 EH3 chondrite and found a tiny grain of an unknown Ti-rich layer phase associated with wassonite (TiS); they also found a few small patches of an unknown S-rich layer phase associated with schöllhornite. These uncharacterized phases are probably terrestrial weathering products of wassonite or of additional as-yet-undiscovered primary titanium-sulfide phases in that chondrule.

The $\geq$3-metric-ton fossil EL chondrite impact-melt rock Al Haggounia 001 contains a plethora of terrestrial alteration phases including calcite, barite, jarosite, gypsum, halite, sylvite, melanterite, native S, and clay (Bunch et al. 2014). Al Haggounia 001 has a ^{14}C age of 23,000 ± 2,000 years (Chennaoui-Aoudjehane et al. 2009) and has been called a fossil meteorite. Some

Figure 12.3 Conglomerate consisting of a cobble of the Al Haggounia 001 EL chondrite and a terrestrial rhyolite pebble cemented together by sand-sized detritus and iron hydroxides. Cube is 1 cm on a side. Photo courtesy of Greg Hupe. (A black-and-white version of this figure will appear in some formats. For the colour version, please refer to the plate section.)

specimens are cemented together with terrestrial limestone clasts and rhyolite pebbles within conglomerates (Figure 12.3).

The Danebury H5 chondrite fell to Earth ~2,350 years ago. It was excavated from a grain pit dug into chalk by prehistoric peoples at the site of an Iron Age hillfort in Hampshire, England. Although the interior of the stone is not appreciably weathered (W1/2), there is a 100–200-μm-thick weathering rind on the sample consisting of (1) lepidocrocite and goethite that experienced minor Ni substitution and (2) fragments of chalk derived from prehistoric infilling of the pit.

Sulfate veins within CI chondrites and sulfate patches at the surfaces of these meteorites include epsomite, blödite, gypsum, and poorly characterized Fe-, Mg-, Na-, and Ca-sulfates (some containing Ni). These veins are not indigenous to the meteorites but instead formed on Earth in museum cabinets via reactions with atmospheric water. The terrestrial sulfates in these rocks formed mainly by the dissolution and remobilization of indigenous sulfate grains as well as by weathering of indigenous pyrrhotite.

The fine-grained matrices of some CR chondrites have feathery-textured regions with abundant hisingerite ($Fe_2^{3+}Si_2O_5(OH)_4{\cdot}2(H_2O)$). This Mg-free phyllosilicate appears to form from saponite by oxidation, leading to the loss of octahedral Mg ions from the lattice.

A major impact appears to have disrupted the L-chondrite parent asteroid 470±6 Ma ago and sent fragments to Earth (e.g., Heymann 1967; Korochantseva et al. 2007). Two possibly paired achondrites (NWA 11042 and NWA 4284) may be ancient impact melts whose Ar-Ar ages were reset by this breakup event (Váci et al., 2020b). The Lockne-Målingen (7.5 km, 0.7 km) doublet impact crater in Central Sweden formed 458 Ma ago and may be a product of the same breakup. This same event may also have flooded the stratosphere with L-chondrite dust, possibly triggering the mid-Ordovician ice age.

Dozens of highly altered L chondrites from this asteroid disruption have been recovered as fossil meteorites in a quarry of Ordovician limestone in Sweden (e.g., Schmitz et al. 2001). One of these fossil meteorites is Brunflo (Figure 12.4). Except for relict chromite grains, all primary minerals in this meteorite have been replaced. The secondary phases, formed by extensive terrestrial alteration, include carbonate (major calcite); sulfate (barite); a disordered illite-type phyllosilicate; rare grains of elemental Ni; sulfarsenides (cobaltite, gersdorffite); arsenides (rammelsbergite, safflorite, nickeline, maucherite, orcelite); sulfides (chalcopyrite, sphalerite, galena, pyrite, and several poorly characterized Cu sulfides); TiO_2 (probably anatase or brookite); quartz; and carbonate-fluorapatite (francolite) (Figure 12.5). Alteration phases in a

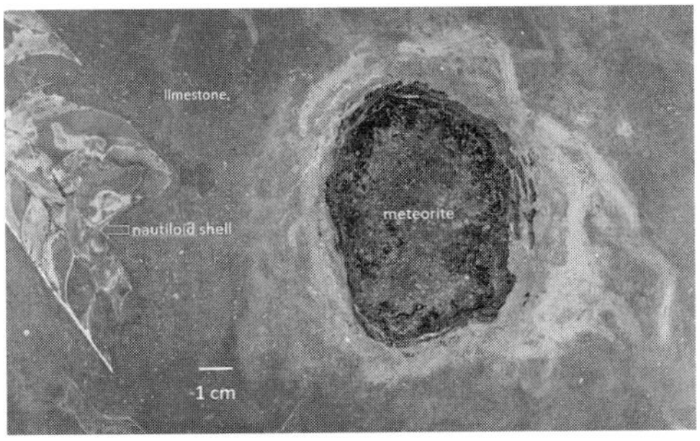

Figure 12.4 Fossil meteorite and nautiloid shell in Ordovician limestone. Photo courtesy of Birger Schmitz, Lund University. (A black-and-white version of this figure will appear in some formats. For the colour version, please refer to the plate section.)

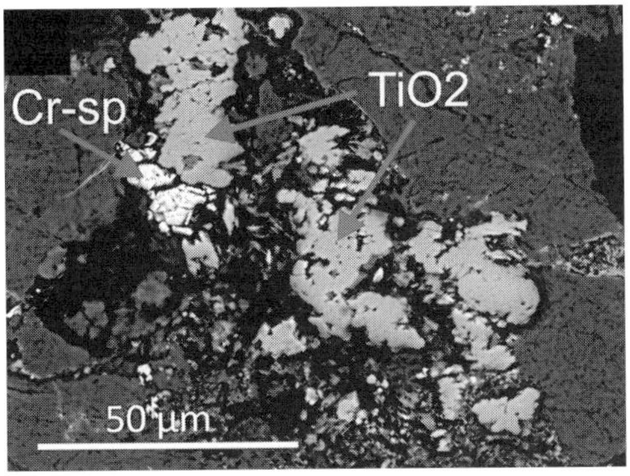

Figure 12.5 TiO_2 and chrome spinel in the fossil L-chondrite Österplana. BSE image courtesy of Birger Schmitz, Lund University.

different fossil L chondrite (Österplana) found in this limestone quarry include calcite, Cr-bearing layered silicate, barite, apatite, and TiO_2 (Nyström et al. 1988). Numerous additional highly altered fossil L chondrites and isolated relict spinel-group grains have been found in the past few decades (e.g., Schmitz 2013); the list of secondary phases will undoubtedly increase.

A relict iron meteorite (probably IIIAB) named Gove was excavated in 1979 from a bauxite mine in Northern Territory, Australia (Bevan et al. 2019). Its terrestrial age is likely to be 2.5–23 Ma. Remaining primary phases in the iron include rare grains of kamacite, taenite, and daubréelite. Secondary phases formed by terrestrial weathering include magnetite, goethite, hematite, akaganéite, pyrite, siderite, and böhmite (AlO[OH]). A new, as-yet-unnamed, sulfide ($CuCrS_2$) replaces daubréelite within former kamacite lamellae. Some of the Cu-Cr-sulfide grains are compositionally zoned; a few have daubréelite cores. (The phase has been known previously only as a synthetic compound.)

Kyte (1998) reported a ~2.5-mm-sized fossil carbonaceous chondrite from a deep-sea drill core at the Cretaceous–Paleogene (K–Pg) boundary (formerly the Cretaceous/Tertiary, KT

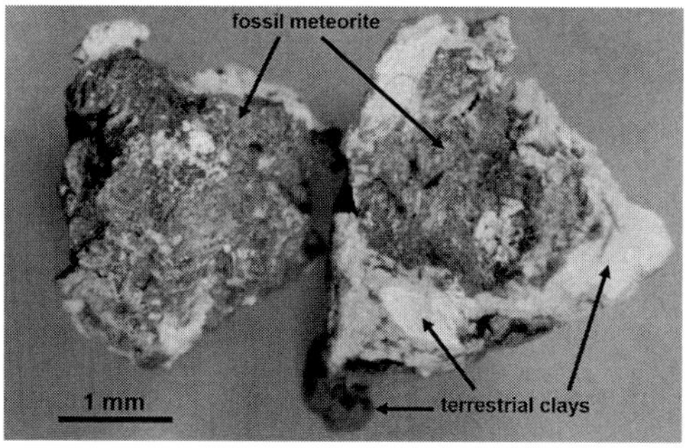

Figure 12.6 A broken fragment of a fossil meteorite enshrouded by terrestrial clay; the sample was found in sediments retrieved from the Cretaceous–Paleogene (K–Pg) boundary in the North Pacific Ocean. This site is 9,000 km west of the Chicxulub impact structure. Image courtesy of Frank Kyte, UCLA. (A black-and-white version of this figure will appear in some formats. For the colour version, please refer to the plate section.)

boundary) (Figure 12.6). The object is probably a fragment of the ~10 km asteroid that produced the Chicxulub impact structure in the Yucatán ~66 Ma ago and triggered the mass extinction that wiped out ~75 percent of all species (including, most famously, the non-avian dinosaurs). Terrestrial alteration phases in this fossil meteorite include hematite, clay (saponite and inter-layered glauconite and smectite), Ni-Fe sulfide, and a Ni-rich metal grain ($Ni_{87}Fe_{13}$). Also present are $\leq$50-µm-sized phosphate grains probably derived from fish scales and fish bones.

Fossil micrometeorites have also been extensively altered (e.g., Suttle and Genge 2017). Many of the particles contain spherical vesicles, resulting from the weathering away of metallic Fe-Ni blebs. Fossil micrometeorites found within chalk beds contain magnetite with small to moderate concentrations of Al, Si, and Mn; these particles were altered under oxidizing conditions. Other fossil micrometeorites contain suessite with small amounts of Cr and Mn; these particles were altered under reducing conditions.

13

The Strange Case of the Aluminum-Copper Alloys

As discussed in Chapter 2, quasicrystals are an unusual form of matter with orderly, but nonperiodic atomic arrangements. They can display diffraction patterns with otherwise forbidden rotational symmetry, e.g., fivefold, eightfold, and higher axes of symmetry (Levine and Steinhardt 1984; Shechtman et al. 1984). Princeton physicist Paul Steinhardt spent decades searching for natural quasicrystals, a quest documented in his book, *The Second Kind of Impossible* (Steinhardt 2019).

Steinhardt and his colleagues eventually got their hands on the Khatyrka $CV3_{OxA}$ chondrite. Published reports (e.g., Bindi et al. 2009, 2011, 2015, 2016; Steinhardt and Bindi 2012; Hollister et al. 2014; Lin et al. 2017; Ma et al. 2017b; Bindi and Steinhardt 2018; Andronicos et al. 2018) indicate this meteorite (which occurs as fragments up to a few millimeters in size) is associated with Al-rich quasicrystals (icosahedrite ($Al_{63}Cu_{24}Fe_{13}$) and $Al_{62}Cu_{31}Fe_7$[(i-phase II]; decagonite ($Al_{71}Ni_{24}Fe_5$)); metallic Al alloys (khatyrkite ($CuAl_2$), cupalite (CuAl), stolperite (AlCu), hollisterite (Al_3Fe), kryachkoite ($(Al,Cu)_6(Fe,Cu)$), steinhardite (Al,Ni,Fe), Fe-poor steinhardite, Al-bearing taenite (γ-(Fe,Ni))); Al-bearing sulfide ($Fe_{0.84}Al_{0.04}S_{1.12}$); Al-oxide (corundum ($Al_2O_3$)); and Al-bearing oxide (trevorite ($NiFe_2O_4$)). [The icosahedrite composition matched that of a quasicrystal synthesized two decades earlier (Tsai et al. 1987).] Khatyrka specimens contain primary silicates (diopside, clinoenstatite, forsterite); secondary silicates and oxide (sodalite, nepheline, magnetite); sulfide (pentlandite, Cu-bearing troilite); and high-pressure silicate phases formed by intense shock (coesite (SiO_2), stishovite (SiO_2), and ahrensite (γ-Fe_2SiO_4)). There is petrographic evidence that portions of the Khatyrka chondrite are in direct contact with the Al-Cu alloys (Hollister et al. 2014; MacPherson et al. 2016; Lin et al. 2017). There appear to be cross-cutting relationships and redox reactions between silicate and Al-Cu alloys, suggesting the alloys and the meteoritic material are co-genetic.

Although icosahedrite is a liquidus phase in the Al-Cu-Fe system in a restricted temperature range (700–850°C) at ambient pressure, experiments by Stagno et al. (2017) demonstrated this phase is stable at high temperature and pressure (along with other Al-bearing phases) in equilibrium with a residual melt. Shock recovery experiments (Asimow et al. 2016; Hu et al. 2019) showed that AlCuFe quasicrystals could form along with khatyrkite and stolperite during hypervelocity collisions. It thus seems possible that impacts produced these Al-rich phases along with coesite, stishovite, and ahrensite.

Meier et al. (2018) measured He and Ne in Khatyrka olivine grains. They determined a cosmic-ray exposure age of 2–4 Ma (assuming a meteoroid <3 m across); they also interpreted the U, Th-He ages of the olivine grains as indicating a major shock event occurred <600 Ma ago.

They suggested the large K-type asteroid 89 Julia (d = 151 km) was a plausible parent body of the Khatyrka meteorite. The asteroid is close to the 3:1 and ν_6 orbital resonances, allowing efficient delivery of meteoroids to the inner Solar System.

Suttle et al. (2019) described a unique spheroidal porphyritic micrometeorite (d = ~130 μm), dubbed KT01, that contains a rapidly cooled mixture of dendritic khatyrkite and stolperite. The principal phases in this small object are dusty-olivine phenocrysts and less-abundant equant magnetite crystallites embedded within a Ca-rich, Na-bearing silicate glass with a pyroxene-normative composition.

Two basic models were proposed for the formation of the Khatyrka assemblage: (1) An energetic impact event transformed preexisting silicates into high-pressure phases and simultaneously created an extremely reducing environment. The Cu was derived from metal and sulfide; the Al came from Al-bearing metallic Fe-Ni, and (2) there were two separate impact events: the first one formed the Al-Cu alloys; the second transformed silica and fayalitic olivine into coesite, stishovite, and ahrensite.

There are problems with these formation models. (1) Coesite and stishovite are very rare in meteorites. (Coesite has been identified only in ordinary, EH3, and CB chondrites, and in eucrites; stishovite has been found only in ordinary chondrites, eucrites, martian meteorites, and a couple of irons.) (2) Free silica (the precursor phase of coesite and stishovite) has not been reported previously in CV3 chondrites. (3) Aluminum-bearing metallic Fe-Ni is unknown as a natural phase outside of Khatyrka. (However, there are commercially available aluminized steels.) (4) The formation of metallic Al requires exceedingly reducing conditions (far more reducing than those pertaining to enstatite chondrites and aubrites). (5) Aluminum (a refractory lithophile element) and Cu (a moderately volatile siderophile/chalcophile element) have different geochemical affinities and condense at very different temperatures (1,670 K and 1,170 K, respectively, from a gas of solar composition at 10^{-4} bar). These problems have led some researchers to question the provenance of the Khatyrka assemblage.

Suggested alternatives to a natural origin for the Al-Cu alloys include (1) *Industrial contamination*: man-made Al-Cu alloys may have been inadvertently admixed with CV-chondrite material in the field during gold prospecting in eastern Siberia (e.g., Ivanova et al. 2017). (Analogous scenarios are unlikely to apply to the Al-Cu alloys in the KT01 micrometeorite.) This scenario by itself cannot account for the high-pressure phases: coesite, stishovite, ahrensite, i-phase II. (2) *Laboratory contamination* (e.g., during thin-section preparation): there are laboratory materials made of aluminum-bronze alloys (e.g., AMPCO 45 ($Al_{0.10}Cu_{0.81}Ni_{0.05}Fe_{0.025}Mn_{0.01}X_{\leq 0.005}$)) similar in composition to some of the Al-Cu alloys in Khatyrka and KT01. This scenario also cannot account for the high-pressure phases. (3) *Hoax*: MacPherson et al. (2016) acknowledged that some members of the cosmochemistry community have concluded that the association of the Khatyrka CV3 chondrite with Al-Cu alloys is fraudulent. However, if this were the case, there likely would be nanoscale inclusions of synthetic Al alloys within the Al-Cu phases in Khatyrka (Paar et al. 2019); none has been observed.

The discovery of additional, coarser-grained, Al-Cu samples that experienced lesser amounts of secondary thermal processing could support the validity of these phases as natural species. MacPherson et al. (2016) summarized the situation: “In short, the argument against a natural origin is that because these things can’t happen, they didn’t. Our contrary view, based on all the evidence, is that if things did happen, they can happen." They could have mentioned that Alfred Wegener's evidence for continental drift (e.g., Wegener 1924) was prematurely dismissed by many contemporary geologists and geophysicists (e.g., Jeffreys 1924, 1929) because no then-known force could move the continents about.

At present, the origin of the weird assemblages in Khatyrka remains controversial, and Sherlock Holmes is unavailable. Few researchers are willing to embrace Tertullian's dictum: *Prorsus credibile est, quia ineptum est* – It is wholly believable, because it is incongruous.[1]

[1] Quintus Septimius Florens Tertullianus (c. 155/160 to c. 225/240 CE) was a Christian writer and apologist. The quotation, from *De Carne Christi*, has been paraphrased as *certum est quia impossibile est* (It is certain because it is impossible) and *credo quia absurdum* (I believe because it is absurd).

Summary

The Formation of Meteoritic Minerals

Meteorites are delivered to Earth from at least 100 different asteroids as well as from the Moon and Mars. A significant fraction of micrometeorites and interplanetary dust particles are from comets. Some of the meteorite parent bodies melted and differentiated; they developed metal cores and basaltic crusts. Other bodies never melted, preserving materials formed in the solar nebula. Some unmelted asteroids were thermally metamorphosed; many were aqueously altered. Meteorites suffered shock damage on their parent bodies, ranging from fracturing and brecciation to impact melting and devolatilization. After their formation, many samples experienced thermal metamorphism, aqueous alteration, and shock metamorphism to different extents; some were altered and/or shocked more than once.

Martian meteorites were subjected to high shock pressures when they were launched off Mars; in contrast, meteorites ejected from asteroids typically experienced little shock during launch. Some meteoroids were probably in highly elliptical orbits and had relatively small perihelion distances; they may have been subjected to periodic heating from solar radiation.

All the meteorites in our collections (except for a couple of small meteorite specimens brought back from the Moon along with Apollo samples) traversed the Earth's atmosphere and developed fusion crusts. The meteorites that were not immediately collected after falling experienced some degree of terrestrial weathering. All meteoritic samples, even witnessed falls stored in museum cabinets, have reacted with atmospheric water.

About 470 minerals have been identified in meteorites, about 8.3 percent of the total number of well-characterized mineral phases. Meteorite mineral species include native elements, metals and metallic alloys, carbides, nitrides and oxynitrides, phosphides, silicides, sulfides and hydroxysulfides, tellurides, arsenides and sulfarsenides, halides, oxides, hydroxides, carbonates, sulfates, molybdates, tungstates, phosphates and silico-phosphates, oxalates, and silicates from all six structural groups.

The minerals in meteorites formed by numerous processes that are not all mutually distinct: (1) condensation in gaseous envelopes around evolved stars; (2) condensation in the solar nebula; (3) crystallization in CAI melts; (4) crystallization in melted portions of AOIs; (5) crystallization in chondrule melts; (6) exsolution in opaque assemblages during the cooling of CAIs; (7) exsolution during the cooling of chondrules; (8) exsolution during the cooling of opaque assemblages outside CAIs and chondrules; (9) annealing of amorphous material in the solar nebula; (10) devitrification of amorphous material on parent bodies; (11) thermal metamorphism and exsolution; (12) aqueous alteration, hydrothermal alteration, and metasomatism; (13) precipitation from asteroidal brines; (14) shock metamorphism; (15) space weathering; (16) condensation within impact plumes; (17) crystallization from melts in differentiated or partially differentiated bodies; (18) condensation from late-stage vapors in differentiated bodies; (19) exsolution, inversion, and subsolidus redox effects within cooling igneous materials; (20) solar heating near perihelion; (21) atmospheric passage; and (22) terrestrial weathering.

Epilogue

More than half a century ago, Buzz Aldrin surveyed the lunar landscape, noted its starkness and ash-colored hues, and declared it a site of "magnificent desolation." Extraterrestrial mineralogy is the study of minerals from our parental molecular cloud as well as from the Moon, Mars,[1] asteroids, and similarly desolate places. The information gathered from detailed analyses of these phases has provided a deeper understanding of the myriad processes that can produce minerals in the Solar System and its erstwhile stellar neighborhood.

Recent decades have seen the recovery of tens of thousands of new meteorites, the development of sophisticated instrumentation, the discovery of scores of new meteoritic minerals, and the publication of tens of thousands of meteorite research papers. Such achievements have allowed us to gain a greater understanding of meteorite petrogenesis and mineral formation. The research community has coupled these advances with experimental and theoretical studies, extraterrestrial sample returns, planetary rovers, telescopic observations, and spacecraft fly-by missions. Mineralogy now encompasses the microscopic and the telescopic: its purview ranges from ancient grains predating the Solar System by billions of years to volcanic rocks forming before our eyes, from the dust beneath our feet to nascent stars enshrouded by dusty debris disks with embedded exoplanets.

More than two centuries ago, the poet William Blake was able to see a world in a grain of sand. Today we can survey many grains from many worlds.

[1] After examining *Viking* orbiter photos, planetary scientist Steve Squyres described Mars as a "beautiful, terrible, desolate place" (Squyres 2005).

References

Abe, K., Sakamoto, N., Krot, A. N., and Yurimoto, H. (2017) Occurrences, abundances, and compositional variations of cosmic symplectites in the Acfer 094 ungrouped carbonaceous chondrite. *Geochemical Journal* 51, 3–15.

Abreu, N. M. (2016) Why is it so difficult to classify Renazzo-type (CR) carbonaceous chondrites? – Implications from TEM observations of matrices for the sequences of aqueous alteration. *Geochimica et Cosmochimica Acta* 194, 91–122.

Abreu, N. M. and Brearley, A. J. (2010) Early solar system processes recorded in the matrices of two highly pristine CR3 carbonaceous chondrites, MET 00426 and QUE 99177. *Geochimica et Cosmochimica Acta* 74, 1146–1171.

Abron, A.-M. (2019) Building other worlds. *Griffith Observer* 83(4), 2–16.

Ackermand, D. and Raase, P. (1973) Die mineralogische Zusammensetzung des Meteoriten von Kiel. *Contributions to Mineralogy and Petrology* 39, 289–300.

Acquadro, J. J., MacPherson, G. J., Corrigan, C. M., and Lunning, N. G. (2019) Evidence for impact-induced shock melting in carbonaceous chondrites. *Lunar and Planetary Science* 50, Abstract #2529.

Afiattalab, F. and Wasson, J. T. (1980) Composition of the metal phases in ordinary chondrites: Implications regarding classification and metamorphism. *Geochimica et Cosmochimica Acta* 44, 431–446.

Agrell, S. O., Long, J. V. P., and Ogilvie, R. E. (1963) Nickel content of kamacite near the interface with taenite in iron meteorites. *Nature* 198, 749–750.

Akaiwa, H. (1966) Abundances of selenium, tellurium, and indium in meteorites. *Journal of Geophysical Research* 71, 1919–1923.

Alexander, C. M. O'D., Hutchison, R. H., Graham, A. L., and Yabuki, H. (1987) Discovery of scapolite in the Bishunpur (LL3) chondritic meteorite. *Mineralogical Magazine* 51, 733–735.

Alexander, C. M. O'D., Barber, D. J., and Hutchison, R. H. (1989) The microstructure of Semarkona and Bishunpur. *Geochimica et Cosmochimica Acta*, 53, 3045–3057.

Alexander, C. M. O'D., Prombo, C. A., Swan, P. D., and Walker, R. M. (1991) SiC and Si_3N_4 in Qingzhen (EH3) (abstract). *Lunar and Planetary Science* 22, 5–6.

Alexander, C. M. O'D., Swan, P., and Prombo, C. A. (1994) Occurrence and implications of silicon nitride in enstatite chondrites. *Meteoritics* 29, 79–85.

Anand, M., Taylor, L. A., Nazarov, M. A., Shu, J., Mao, H.-K., and Hemley, R. J. (2004) Space weathering on airless planetary bodies: Clues from the lunar mineral hapkeite. *Proceedings of the National Academy of Sciences* 101, 6847–6851.

Anders, A. and Zinner, E. (1993) Interstellar grains in primitive meteorites: Diamond, silicon carbide, and graphite. *Meteoritics* 28, 490–514.

Andersen, C. A., Keil, K., and Mason, B. (1964) Silicon oxynitride: A meteoritic mineral. *Science* 146, 256–257.

Andersson, S. and Magnéli, A. (1956) Diskrete Titanoxydphasen im Zusammensetzungsbereich $TiO_{1,75}$-$TiO_{1,90}$. *Naturwissenschaften*, 43, 495–496.

Andronicos, C. L., Bindi, L., Distler, V. V., Hollister, L. S., Lin, C., MacPherson, G. J., Steinhardt, P. J., and Yuduvskaya, M. (2018) Comment on "Compositon and origin of holotype Al–Cu–Zn minerals in relation to quasicrystals in the Khatyrka meteorite" by M. Ivanova et al. (2017). *Meteoritics & Planetary Science* 53, 2430–2440.

Arimatsu, K., Tsumura, K., Usui, F., Shinnaka, Y., Ichikawa, K., Ootsubo, T., Kotani, T., Wada, T., Nagase, K., and Watanabe, J. (2019) A kilometre-sized Kuiper belt object discovered by stellar occulation using amateur telescopes. *Nature Astronomy* 3, 301–306. https://doi.org/10.1038/s41550-018-0685-8.

Armstrong, J. C., Wells, L. E., and Gonzalez, G. (2002) Rummaging through Earth's attic for remains of ancient life. *Icarus* 160, 183–196.

Armstrong, J. T., Hutcheon, I. D., and Wasserburg, G. J. (1987) Zelda and company: Petrogenesis of sulfide-rich Fremdlinge and constraints on solar nebula processes. *Geochimica et Cosmochimica Acta* 51, 3155–3173.

Asimow, P. D., Lin, C., Bindi, L., Ma, C., Tschauner, O., Hollister, L. S., Steinhardt, P. J. (2016) Shock synthesis of quasicrystals with implications for their origin in asteroid collisions. *Proceedings of the National Academy of Sciences* 113, 7077–7081.

Baecker, B., Rubin, A. E., and Wasson, J. T. (2017) Secondary melting events in Semarkona chondrules revealed by compositional zoning in low-Ca pyroxene. *Geochimica et Cosmochimica Acta* 211, 256–279.

Barber, D. J. (1981) Matrix phyllosilicates and associated minerals in C2M carbonaceous chondrites. *Geochimica et Cosmochimica Acta* 45, 945–970.

Barber, D. J., Beckett, J. R., Paque, J. M., and Stolper, E. (1994) A new titanium-bearing calcium aluminosilicate phase: II. Crystallography and crystal chemistry of grains formed in slowly cooled melts with bulk compositions of calcium-aluminum-rich inclusions. *Meteoritics* 29, 682–690.

Barnatowicz, T. J., Messenger, S., Pravdivtseva, O., Swan, P., and Walter, R. M. (2003) Pristine presolar silicon carbide. *Geochimica et Cosmochimica Acta* 67, 4679–4691.

Baziotis, I. P., Liu, Y., DeCarli, P. S., Melosh, H. J., McSween, H. Y., Bodnar, R. J. and Taylor, L. A. (2013) The Tissint martian meteorite as evidence for the largest impact excavation. *Nature Communications* 4, 1–7, Article 1404.

Baziotis, I., Asimow, P. D., Hu, J., Ferrière, L., Ma, C., Cernok, A., Anand, M., and Topa, D. (2018) High pressure minerals in the Château-Renard (L6) ordinary chondrite: Implications for collisions on its parent body. *Scientific Reports* 8, 9851.

Becker, R. H. and Pepin, R. O. (1984) Solar composition noble gases in the Washington County iron meteorite. *Earth & Planetary Science Letters* 70, 1–10.

Bellucci, J. J., Nemchin, A. A., Grange, M., Robinson, K. L., Collins, G., Whitehouse, M. J., Snape, J. F., Norman, M. D., and Kring, D. A. (2019) Terrestrial-like zircon in a clast from an Apollo 14 breccia. *Earth & Planetary Science Letters* 510, 173–185.

Benedix, G. K., McCoy, T. J., Keil, K., and Love, S. G. (2000) A petrologic study of the IAB iron meteorites: Constraints on the formation of the IAB-Winonaite parent body. *Meteoritics & Planetary Science* 35, 1127–1141.

Berger, E., Lauretta, D. S., Zega, T. J., and Keller, L. P. (2016) Heterogeneous histories of Ni-bearing pyrrhotite and pentlandite grains in the CI chondrites Orgueil and Alais. *Meteoritics & Planetary Science* 51, 1813–1829.

Berkley, J. L., Taylor, G. J., Keil, K., Harlow, G. E., and Prinz, M. (1980) The nature and origin of ureilites. *Geochimica et Cosmochimica Acta* 44, 1579–1597.

Bermingham, K. R., Worsham, E. A., and Walker, R. J. (2018) New insights into Mo and Ru isotope variation in the nebula and terrestrial planet accretionary genetics. *Earth and Planetary Science Letters* 487, 221–229.

Bernatowicz, T. J., Amari, S., Zinner, E. K., and Lewis, R. S. (1991) Interstellar grains within interstellar grains. *Astrophysical Journal* 373, L73–L76.

Bernatowicz, T. J., Cowsik, R., Gibbons, P. C., Lodders, K., Fegley, B., Amari, S., and Lewis R. S. (1996) Constraints on stellar grain formation from presolar graphite in the Murchison meteorite. *Astrophysical Journal* 472, 760–782.

Bevan, A. W. R., Bevan, J. C., and Francis, J. G. (1977) Amphibole in the Mayo Belwa meteorite: First occurrence in an enstatite achondrite. *Mineralogical Magazine* 41, 531–534.

Bevan, A. W. R., Kinder, J., and Axon, H. J. (1981) Complex shock-induced Fe–Ni–S–Cr–C melts in the Haig (IIIA) iron meteorite. *Meteoritics* 16, 261–267.

Bevan, A. W. R., Downes, P. J., Henry, D. A., Verrall, M., and Haines, P. W. (2019) The Gove relict iron meteorite from Arnhem Land, Northern Territory, Australia. *Meteoritics & Planetary Science* 54, 1710–1719.

Bhandari, N., Shah, V. B., and Wasson, J. T. (1980) The Parsa enstatite chondrite. *Meteoritics* 15, 225–233.

Bindi, L. and Steinhardt, P. J. (2018) How impossible crystals came to Earth: A short history. *Rocks and Minerals* 93, 50–57.

Bindi, L. and Xie, X. (2018): Shenzhuangite, $NiFeS_2$, the Ni-analogue of chalcopyrite from the Suizhou L6 chondrite. *European Journal of Mineralogy* 30, 165–169.

Bindi, L. and Xie, X. (2019) Hiroseite, IMA 2019–019. CNMNC Newsletter No. 50, June-July 2019, page 617; *Mineralogical Magazine* 83, 615–620 https://doi.org/ 10.1180/mgm.2019.46.

Bindi, L., Steinhardt, P. J., Yao, N., and Lu, P. J. (2009) Natural quasicrystals. *Science* 324, 1306–1309.

Bindi, L., Steinhardt, P. J., Yao, N., and Lu, P. J. (2011) Icosahedrite, $Al_{63}Cu_{24}Fe_{13}$, the first natural quasicrystal. *American Mineralogist* 96, 928–931.

Bindi, L., Yao, N., Lin, C., Hollister, L.S., MacPherson, G.J., Poirier, G.R., Andronicos, C.L., Distler, V.V., Eddy, M.P., Kostin, A., Kryachko, V., Steinhardt, W.M., Yudovskaya, M. (2014) Steinhardtite, a new body-centered-cubic allotropic form of aluminum from the Khatyrka CV3 carbonaceous chondrite. *American Mineralogist* 99, 2433–2436.

Bindi, L.,Yao, N., Lin, C., Hollister, L. S., Andronicos, C. L., Distler, V. V., Eddy, M. P., Kostin, A. Kryachko, V., MacPherson, G. J., Steinhardt, W. M., Yudovskaya, M., and Steinhardt, P. J. (2015) Decagonite, $Al_{71}Ni_{24}Fe_5$, a quasicrystal with decagonal symmetry from the Khatyrka CV3 carbonaceous chondrite. *American Mineralogist* 100, 2340–2343.

Bindi, L., Lin, C., Ma, C., and Steinhardt, P. J. (2016) Collisions in outer space produced an icosahedral phase in the Khatyrka meteorite never observed previously in the laboratory. *Scientific Reports* 6, 38117.

Bindi, L., Chen, M. and Xie, X. (2017) Discovery of the Fe-analogue of akimotoite in the shocked Suizhou L6 chondrite. *Scientific Reports* 7, 42674.

Bindi, L., Pham, J., and Steinhardt, P. J. (2018) Previously unknown quasicrystal periodic approximant found in space. *Scientific Reports* 8, 16271.

Bindi, L., Brenker, F. E., Nestola, F., Koth, T. E., Prior, D. J., Lilly, K., Krot, A. N., Bizzarro, M, and Xie, X. (2019) Discovery of asimowite, the Fe-analog of wadsleyite, in shock-melted silicate droplets of the Suizhou L6 and the Quebrada Chimborazo 001 CB3.0 chondrites. *American Mineralogist* 104, 775–778.

Binns, R. (1967) Stony meteorites bearing maskelynite. *Nature* 213, 1111–1112.

Binzel, R. P. (2001) Forging the fourth link between planetary worlds: Vesta and the HEDs. *Meteoritics & Planetary Science* 36,479480.

Binzel, R. P. and Xu, S. (1993) Chips off of asteroid 4 Vesta: Evidence for the parent body of basaltic achondrite meteorites. *Science* 260, 186–191.

Bischoff, A. and Keil, K. (1984) Al-rich objects in ordinary chondrites: Related origin of carbonaceous and ordinary chondrites and their constituents. *Geochimica et Cosmochimica Acta* 48, 693–709.

Bischoff, A. and Palme, H. (1987) Composition and mineralogy of refractory metal-rich assemblages from a Ca,Al-rich inclusion in the Allende meteorite. *Geochimica et Cosmochimica Acta* 51, 2733–2748.

Bischoff, A., Palme, H., Schultz, L., Weber, D., Weber, H. W., and Spettel, B. (1993) Acfer 182 and paired samples, an iron-rich carbonaceous chondrite: Similarities with ALH 85085 and relationship to CR chondrites. *Geochimica et Cosmochimica Acta* 57, 2631–2648.

Bischoff, A., Geiger, T., Palme, H., Spettel, B., Schultz, L., Scherer, P., Loeken, T., Bland, P., Clayton, R. N., Mayeda, T. K., Herpers, U., Meltzow, B., Michel, R., and Dittrich-Hannen, B. (1994) Acfer 217-A new member of the Rumuruti chondrite group (R). *Meteoritics* 29, 264–274.

Bischoff, A., Scott, E. R. D., Metzler, K., and Goodrich, C. A. (2006) Nature and origins of meteoritic breccias. In *Meteorites and the Early Solar System II*, Tucscon: University of Arizona Press, 679–712.

Bischoff, A., Vogel, N., and Roszjar, J. (2011) The Rumuruti chondrite group. *Chemie der Erde – Geochemistry* 71, 101–133.

Bland, P. A., Howard, L. E., Prior, D. J., Wheeler, J., Hough, R. M., and Dyl, K. A. (2011) Earliest rock fabric formed in the solar system preserved in a chondrule rim. *Nature Geoscience* 4, 244–247.

Bloss, F. D. (1971) *Crystallography and Crystal Chemistry: An Introduction*, New York: Holt, Rinehart and Winston, 545 pp.

Blum, J. D., Wasserburg, G. J., Hutcheon, I. D., Beckett, J. R., and Stolper, E. M. (1988) 'Domestic' origin of opaque assemblages in refractory inclusions in meteorites. *Nature* 331, 405–409.

Blum, J. D., Wasserburg, G. J., Hutcheon, I. D., Beckett, J. R., and Stolper, E. M. (1989) Origin of opaque assemblages in CV3 meteorites: Implications for nebular and planetary processes. *Geochimica et Cosmochimica Acta* 53, 543–556.

Boesenberg, J. S. and Hewins, R. H. (2010) An experimental investigation into the metastable formation of phosphoran olivine and pyroxene. *Geochimica et Cosmochimica Acta*, 74, 1923–1941.

Boesenberg, J. S., Prinz, M., Weisberg, M. K., Davis, A. M., Clayton, R. N., Mayeda, T. K., and Wasson, J. T. (1995) Pyroxene pallasites: A new pallasite grouplet (abstract). *Meteoritics* 30, 488–489.

Boesenberg, J. S., Davis, A. M., Prinz, M., Weisberg, M. K., Clayton, R. N., and Mayeda, T. K. (2000) The pyroxene pallasites, Vermillion and Yamato 8451: Not quite a couple. *Meteoritics & Planetary Science* 35, 757–769.

Boesenberg, J. S., Mayne, R. G., Humayun, M., Silver, A. P., Greenwood, R. C., and Franchi, I. A. (2016) Pyroxene-plagioclase pallasite Northwest Africa 10019: Where does it belong? *Lunar and Planetary Science* 47, Abstract #2297.

Boesenberg, J. S., Humayun, M., and Van Niekerk, D. (2017) Zinder: The first mantle sample from the IIIF iron parent body. *Lunar and Planetary Science*, 48, Abstract #2319.

Boesenberg, J. S., Humayun, M., Windmill, R., Greenwood, R. C., and Franchi, I. A. (2018) Sericho: A new main group pallasite with two types of chromite. *Lunar and Planetary Science* 49, Abstract#1556.

Bogard, D. D. and Johnson, P. (1983) Martian gases in an Antarctic meteorite. *Science* 221, 651–654.

Bollard, J., Connelly, J. N., and Bizzarro, M. (2015) Pb-Pb dating of individual chondrules from the CBa chondrite Gujba: Assessment of the impact plume formation model. *Meteoritics & Planetary Science* 50, 1197–1216.

Boynton, W. V. (1975) Fractionation in the solar nebula: Condensation of yttrium and the rare earth elements. *Geochimica et Cosmochimica Acta* 39, 569–584.

Bradley, J. P. (2005) Interplanetary dust particles. In *Treatise on Geochemistry, Volume 1: Meteorites, Comets and Planets*, ed. Davis, A. M. Amsterdam: Elsevier, pp. 689–711.

Bradley, J. P., Harvey, R. P., and McSween, H. Y. (1997) No 'nanofossils' in martian meteorite. *Nature* 390, 454.
Brearley, A. J. (1989) Nature and origin of matrix in the unique type 3 chondrite, Kakangari. *Geochimica et Cosmochimica Acta* 53, 2395–2411.
Brearley, A. J. (1990) Carbon-rich aggregates in type 3 ordinary chondrites: Characterization, origins, and thermal history. *Geochimica et Cosmochimica Acta* 54, 831–850.
Brearley, A. J. (1991) Mineralogical and chemical studies of matrix in the Adelaide meteorite, a unique carbonaceous chondrite with affinities to ALH A77307 (CO3) (abstract). *Lunar and Planetary Science* 22, 133–134.
Brearley, A. J. (1993a) Matrix and fine-grained rims in the unequilibrated CO3 chondrite, ALHA77307: Origins and evidence for diverse, primitive nebular dust components. *Geochimica et Cosmochimica Acta* 57, 1521–1550.
Brearley, A. J. (1993b) Occurrence and possible significance of rare Ti oxides (Magneli phases) in carbonaceous chondrite matrices. *Meteoritics* 28, 590–595.
Brearley, A. J. (1995) Aqueous alteration and brecciation in Bells, an unusual, saponite-bearing, CM chondrite. *Geochimica et Cosmochimica Acta* 59, 2291–2317.
Brearley, A. J. (1996) Disordered biopyriboles in the Allende meteorite: First extraterrestrial occurrence. *Geological Society of America Abstracts with Program* 28, A103.
Brearley, A. J. (1997) Phyllosilicates in the matrix of the unique carbonaceous chondrite, LEW 85332 and possible implications for the aqueous alteration of CI chondrites. *Meteoritics & Planetary Science* 32, 377–388.
Brearley, A. J. (2006) The action of water. In *Meteorites and the Early Solar System II*, eds. Lauretta, D. S. and McSween, H. Y. Tucson: University of Arizona Press, pp. 587–624.
Brearley, A. J. and Jones, R. H. (1998) Chondritic meteorites. In *Planetary Materials, Reviews in Mineralogy & Geochemistry, Vol. 36*, ed. Papike, J. J. Washington, DC: Mineralogical Society of America, pp. 3-1–3-398.
Breen, J. P., Rubin, A. E., and Wasson, J. T. (2016) Variations in impact effects among IIIE iron meteorites. *Meteoritics & Planetary Science* 51, 1611–1631.
Brigham, C. A., Yabuki, H., Ouyang, Z., Murrell, M. T., El Goresy, A., and Burnett, D. S. (1986) Silica-bearing chondrules and clasts in ordinary chondrites. *Geochimica et Cosmochimica Acta* 50, 1655–1666.
Britvin, S. N., Kolomensky, V. D., Boldyreva, M. M., Bogdanova, A. N., Kretser, Y. L., Boldyreva, O. N., and Rudashesky, N. S. (1999) Nickelphosphide $(Ni,Fe)_3P$, the nickel analog of schreibersite. *Zapiski Vserossijskogo Mineralogicheskogo Obshchestva* 128, 64–72.
Britvin, S. N., Guo, X. Y., Kolomensky, V. D., Boldyreva, M. M., Kretser, Y. L., and Yagovkina, M. A. (2001) Cronusite, $Ca_{0.2}(H_2O)_2CrS_2$, a new mineral from the Norton County enstatite achondrite. *Zapiski Vserossijskogo Mineralogicheskogo Obshchestva* 130, 29–36.
Britvin, S. N., Rudashevsky, N. S., Krivovichev, S. V., Burns, P. C., and Polekhovsky, Y. S. (2002) Allabogdanite,$(Fe, Ni)_2P$, a new mineral from the Onello meteorite: The occurrence and crystal structure. *American Mineralogist* 87, 1245–1249.
Britvin, S. N., Bogdanova, A. N., Boldyreva, M. M., and Aksenova, G. Y. (2008) Rudashevskyite, the Fe-dominant analogue of sphalerite, a new mineral: Description and crystal structure. *American Mineralogist* 93, 902–909.
Britvin, S. N., Krivovichev, S. V., and Armbruster, T. (2016) Ferromerrillite, $Ca_9NaFe^{2+}(PO_4)_7$, a new mineral from the Martian meteorites, and some insights into merrillite–tuite transformation in shergottites. *European Journal of Mineralogy* 28, 125–136.
Britvin, S. N., Galuskina, I. O., Vlasenko, N. S., Vereshchagin, O.S., Bocharov, V. N., Krzhizhanovskaya, M. G., Shilovskikh, V. V., Galuskin, E. V., Vapnik, Y., and Obolonskaya, E. V. (2020a) Keplerite, IMA 2019-108, CNMNC Newsletter No. 54, page 277; *European Journal of Mineralogy* 32, 275–283.

Britvin, S. N., Murashko, M. N., Vapnik, Y., Polekhovsky, Y. S., Krivovichev, S. V., Krzhizhanovskaya, M. G., Vereshchagin, O. S., Shilovskikh, V. V., and Vlasenko, N. S. (2020b) Transjordanite, Ni_2P, a new terrestrial and meteoritic phosphide, and natural solid solutions barringerite-transjordanite (hexagonal Fe_2P-Ni_2P). *American Mineralogist* 105, 428–436.

Brown, P. G., Revelle, D. O., Tagliaferri, E., and Hildebrand, A. R. (2002) An entry model for the Tagish Lake fireball using seismic, satellite and infrasound records. *Meteoritics & Planetary Science* 37, 661–675.

Brownlee, D. (2014) The Stardust mission: Analyzing samples from the edge of the solar system. *Annual Review of Earth and Planetary Science* 42, 179–205.

Brownlee, D. E. and Rajan, R. S. (1973) Micrometeorite craters discovered on chondrule-like objects from Kapoeta meteorite. *Science* 182, 1341–1344.

Buchwald, V. F. (1975) *Handbook of Iron Meteorites*. Berkeley: University of California Press, 1,418 pp.

Buchwald, V. F. (1977) The mineralogy of iron meteorites. *Philosophical Transactions of the Royal Society of London A* 286, 453–491.

Buchwald, V. F. (1989) *Mineralogi og Reaktionsmodeller ved Korrosion of Jordfundne Jergenstande (Meteoritter og Oldsager)*, Lyngby, Denmark: Technical University of Denmark.

Buchwald, V. F. and Clarke, R. S. (1988) Akaganeite, not lawrencite, corrodes Antarctic iron meteorites (abstract). *Meteoritics* 23, 261.

Buchwald, V. F. and Clarke, R. S. (1989) Corrosion of Fe-Ni alloys by Cl-containing akaganeite (beta-FeOOH): The Antarctic meteorite case. *American Mineralogist* 74, 656–667.

Buchwald, V. F. and Scott, E. R. D. (1971) First nitride (CrN) in iron meteorites. *Nature* 233, 113–114.

Buddhue, J. D. (1957) *The Oxidation and Weathering of Meteorites*. Albuquerque: University of New Mexico, 161 pp.

Budde, G., Burkhardt, C., Brennecka, G. A., Fischer-Gödde, M., Kruijer, T. S., and Kleine, T. (2016) Molybendum isotopic evidence for the origin of chondrules and a distinct genetic heritage of carbonaceous and non-carbonaceous meteorites. *Earth and Planetary Science Letters*, 454, 293–303.

Budde, G., Burkhardt, C., and Kleine, T. (2017) The distinct genetics of carbonaceous and non-carbonaceous meteorites inferred from molybdenum isotopes. 80th Annual Meeting of the Meteoritical Society, Abstract #6271.

Budde, G., Kruijer, T. S., and Kleine, T. (2018) Hf-W chronology of CR chondrites: Implications for the timescales of chondrule formation and the distribution of 26Al in the solar nebula. *Geochimica et Cosmochimica Acta*, 222, 284–30C4.

Bunch, T. E., Keil, K., and Olsen, E. (1970) Mineralogy and petrology of silicate inclusions in iron meteorites. *Contributions to Mineralogy and Petrology* 25, 297–340.

Bunch, T., Wittke, J., and Irving, A. (2014) The Al Haggounia "fossil or paleo" meteorite problem, accessed July 3, 2020, www.cefns.nau.edu/geology/naml/Meteorite/Al_Haggounia.html.

Burbine, T. H. (2014) Asteroids. In *Treatise on Geochemistry*, eds. Holland, H. and Turekian, K. San Diego: Elsevier Pergamon, pp. 365–415.

Burbine, T. H. (2017) *Asteroids: Astronomical and Geological Bodies*. Cambridge: Cambridge University Press, 367 pp.

Burbine, T. H., Buchanan, P. C., Binzel, R. P., Bus, S. J., Hiroi, T., Hinrichs, J. L., Meibom, A., and McCoy, T. J. (2001) Vesta, Vestoids, and the howardite, eucrite, diogenite group: Relationships and the origin of spectral differences. *Meteoritics & Planetary Science* 36, 761–781.

Burgess, K. D. and Stroud, R. M. (2020) Space weathering of three Itokawa grains and presence of cubanite. *Lunar and Planetary Science* 51, Abstract#1133.

Burke, J. G. (1986) *Cosmic Debris: Meteorites in History*. Berkeley: University of California Press, 445 pp.

Burton, A. S., Glavin, D. P., Elsila, J. E., Dworkin, J. P., Jenniskens, P., and Yin, Q.-Z. (2014) The amino acid composition of the Sutter's Mill CM2 carbonaceous chondrite. *Meteoritics & Planetary Science* 49, 2074–2086.

Buseck, P. R. (1968) Mackinawite, pentlandite and native copper from the Newport pallasite. *Mineralogical Magazine* 36, 717–725.

Buseck, P. R. (1977) Pallasite meteorites—mineralogy, petrology and geochemistry. *Geochimoca et Cosmochimica Acta* 41, 711–740.

Buseck, P. R. and Hua, X. (1993) Matrices of carbonaceous chondrite meteorites. *Annual Review of Earth & Planetary Sciences* 21, 255–305.

Cable, M. L., Vu, T. H., Maynard-Casely, H. E., Malaska, M., Choukroun, M., and Hodyss, R. (2020) Evidence for a new Titan molecular mineral: A co-crystal between acetylene and acetonitrile. *Lunar and Planetary Science* 51, Abstract #1769.

Caillet Komorowski, C., El Goresy, A., Miyahara, M., Boudouma, O., and Ma, C. (2012) Discovery of Hg–Cu-bearing metal-sulfide assemblages in a primitive H-3 chondrite: Towards a new insight in early solar system processes. *Earth and Planetary Science Letters* 349–350, 261–271.

Cameron, A. G. W. and Ward, W. (1976) The origin of the Moon (abstract). *Lunar Science* 7, 120–122.

Campbell, A. J., Humayun, M., and Weisberg, M. K. (2002) Siderophile element constraints on the formation of metal in the metal-rich chondrites Bencubbin, Weatherford, and Gujba. *Geochimica et Cosmochimica Acta* 66, 647–660.

Campbell, A. J. and Humayun, M. (2005) Compositions of group IVB iron meteorites and their parent melt. *Geochimica et Cosmochimica Acta* 69, 4733–4744.

Campins, H., Hargrove, K., Pinilla-Alonso, N., Howell, E. S., Kelley, M. S., Licandro, J., Mothé-Diniz, T., Fernández, Y., and Ziffer, J. (2010) Water ice and organics on the surface of the asteroid 24 Themis. *Nature* 464, 1320–1321.

Cano, E. J., Sharp, Z. D., and Shearer, C. K. (2020) Distinct oxygen isotopic compositions of the Earth and Moon. *Nature Geoscience* 13, 270–274. https://doi.org/10.1038/s41561-020-0550-0.

Canup, R. M. (2012) Forming a Moon with an Earth-like composition via a giant impact. *Science* 338, 1052–1055.

Carter, N. L., Raleigh, C. B., and DeCarli, P. S. (1968) Deformation of olivine in stony meteorites. *Journal of Geophysical Research* 73, 5439–5461.

Cartwright, J. A., Ott, U., Herrmann, S., and Agee, C. B. (2014) Modern atmospheric signatures in 4.4 Ga martian meteorite NWA 7034. *Earth and Planetary Science Letters* 400, 77–87.

Castle, N. and Herd, C. D. K. (2018) Experimental investigation into the effects of oxidation during petrogenesis of the Tissint meteorite. *Meteoritics & Planetary Science* 53, 1341–1363.

Chabot, N. L. and Haack, H. (2006) Evolution of asteroidal cores. In *Meteorites and the Early Solar System II*, eds. Lauretta, D. S. and McSween, H. Y. Tucson: University of Arizona Press, pp. 747–771.

Chan, Q. H. S., Zolensky, M. E., Kebukawa, Y., Fries, M., Ito, M., Steele, A., Rahman, Z., Nakato, A., Kilcoyne, A. L. D., Suga, H., Takahashi, Y., Takeichi, Y., and Mase, K. (2018) Organic matter in extraterrestrial water-bearing salt crystals. *Science Advances* 4, eaao3521. https://doi: 10.1126/sciadv.aao3521.

Chao, E. C. T., Fahey, J. J., Littler, J., and Milton, D. J. (1962) Stishovite, SiO_2, a very high pressure new mineral from Meteor Crater, Arizona. *Journal of Geophysical Research* 67, 419–421.

Chaumard, N., Devouard, B., Delbo, M., Provost, A., and Zanda, B. (2012) Radiative heating of carbonaceous near-Earth objects as a cause of thermal metamorphism for CK chondrites. *Icarus* 220, 65–73.

Chen, D.-L., Zhang, A.-C., Pang, R.-L., Chen, J.-N., and Li, Y. (2019) Shock-induced phase transformation of anorthitic plagioclase in the eucrite meteorite Northwest Africa 2650. *Meteoritics & Planetary Science* 54, 1548–1562.

Chen, M., Sharp, T. G., El Goresy, A., Wopenka, B., and Xie, X. (1996) The majorite-pyrope-magnesiowustite assemblage: Constraints on the history of shock veins in chondrites. *Science* 271, 1570–1573.

Chen, M., Shu, J., and Mao, H. K. (2008) Xieite, a new mineral of high-pressure $FeCr_2O_4$ polymorph. *Chinese Science Bulletin* 53, 3341–3345.

Chennaoui-Aoudjehane, H., Jambon, A., Rjimati, E., Jull, A. J. T., and Leclerc-Giscard, M. D. (2009) The Late Quaternary fall at Al Haggounia (Morocco): The ^{14}C evidence. *Meteoritics & Planetary Science* 44, A100.

Chizmadia, L., Rubin, A. E., and Wasson, J. T. (2002) Mineralogy and petrology of amoeboid olivine inclusions: Evidence for CO3 parent-body aqueous alteration. *Meteoritics & Planetary Science* 37, 1781–1796.

Choe, W. H., Huber, H., Rubin, A. E., Kallemeyn, G. W., and Wasson, J. T. (2010) Compositions and taxonomy of 15 unusual carbonaceous chondrites. *Meteoritics & Planetary Science* 45, 531–554.

Christiansen, E. H. and Hamblin, K. (2014) *Dynamic Earth: An Introduction to Physical Geology*. Burlington, MA: Jones and Bartlett, 838 pp.

Christophe Michel-Levy, M. (1976) La matrice noire et blanche de la chondrite de Tieschitz (H3). *Earth and Planetary Science Letters* 30, 143–150.

Chukanov N., Pekov I., Levitskaya L., and Zadov A. (2009) Droninoite, $Ni_3Fe^{3+}{}_2Cl(OH)_8 \cdot 2H_2O$, a new hydrotalcite-group mineral species from the weathered Dronino meteorite. *Geology of Ore Deposits* 51, 767–773.

Cintala, M. (1981) The Mercurian regolith: An evaluation of impact glass production by micrometeoroid impact. *Lunar and Planetary Science* XII, 141–143.

Clarke, R.S. and Scott, E. R. D. (1980) Tetrataenite–ordered FeNi, a new mineral in meteorites. *American Mineralogist* 65, 624–630.

Clarke, R. S., Buchwald, V. F., and Olsen E. (1990) Anomalous ataxite from Mount Howe, Antarctica. *Meteoritics* 25, 354.

Clayton, D. D. and Nittler, L. R. (2004) Astrophysics with presolar stardust. *Annual Review of Astronomy & Astrophysics* 42, 39–78.

Clayton, R. N., Mayeda, T. K., Goswami, J. N., and Olsen, E. J. (1991) Oxygen isotope studies of ordinary chondrites. *Geochimica et Cosmochimica Acta* 55, 2317–2337.

Cloutis, E. A., Binzel, R. P., and Gaffey, M. J. (2014) Establishing asteroid-meteorite links. *Elements* 10, 25–30.

Comelli, D., D'Orazio, M., Folco, L., El-Halwagy, M., Frizzi, T., Alberti, R., Capogrosso, V., Elnaggar, A., Hassan, H., Nevin, A., Porcelli, F., Rashed, M. G., and Valentini, G. (2016) The meteoritic origin of Tutankhamun's iron dagger blade. *Meteoritics & Planetary Science* 51, 1301–1309.

Connelly, J. N., Bizzarro, M., Krot, A. N., Nordlund, Å., Wielandt, D., and Ivanova, M. A. (2012) The absolute chronology and thermal processing of solids in the solar protoplanetary disk. *Science* 338, 651–655.

Connolly, H. C., Zipfel, J., Grossman, J. N., Folco, L., Smith, C., Jones, R. H., Righter, K., Zolensky, M., Russell, S. S., and Benedix, G. K. (2006) The Meteoritical Bulletin, No. 90, 2006 September. *Meteoritics & Planetary Science Archives* 41, 1383–1418.

Consolmagno, G. J. and Drake, M. J. (1977) Composition and evolution of the eucrite parent body: Evidence from rare earth elements. *Geochimica et Cosmochimica Acta* 41, 1271–1282.

Consolmagno, G. J., Macke, R. J., Rochette, P., Britt, D. T., and Gattacceca J. (2006) Density, magnetic susceptibility, and the characterization of ordinary chondrite falls and showers. *Meteoritics & Planetary Science* 41, 331–342.

Corrigan, C. M., Rumble, D., Ash, R. D., McDonough, W. F., Honesto, J., and Walker, R. J. (2005) The Tishomingo iron: Relationship to IVB irons, CR clan chondrites and angrites and implications for the origin of volatile-depleted iron meteorites. *Lunar and Planetary Science* 36, Abstract #2062.

Croat, T. K., Berg, T., Bernatowicz, T., Groopman, E., and Jadhav, M. (2013) Refractory metal nuggets within presolar graphite: First condensates from a circumstellar environment. *Meteoritics & Planetary Science* 48, 686–699.

Crotts, A. (2014) *The New Moon: Water, Exploration and Future Habitation*. Cambridge, UK: Cambridge University Press, 522 pp.

Cruikshank, D. P., Tholen, D. J., Hartmann, W. K., Bell, J. H., and Brown, R. H. (1991) Three basaltic Earth-approaching asteroids and the source of basaltic meteorites. *Icarus* 89, 1–13.

Daly, L., Bland, P. A., Dyl, K. A., Forman, L. V., Evans, K. A., Trimby, P. W., Moody, S., Yang, L., Liu, H., Ringer, S. P., Ryan, C. G., and Saunders, M. (2017) *In situ* analysis of refractory metal nuggets in carbonaceous chondrites. *Geochimica et Cosmochimica Acta* 216, 61–81.

Davis, A. M. (1991) Ultrarefractory inclusions and the nature of the group II REE fractionation (abstract). *Meteoritics* 26, 330.

Davis, A. M., Zhang, J., Greber, N. D., Hu, J., Tissot, F. L. H., and Dauphas, N. (2018) Titanium isotopes and rare earth patterns in CAIs: Evidence for thermal processing and gas-dust decoupling in the protoplanetary disk. *Geochimica et Cosmochimica Acta* 221, 275–295.

Delaney, J. S., Prinz, M., and Takeda, H. (1984) The polymict eucrites. *Proceedings Lunar and Planetary Science Conference*, 15, C251–C288. https://doi.org/10.1029/JB089iS01p0C251.

DeMeo, F. E., Polishook, D., Carry, B., Burt, B. J., Hsieh, H. H., Binzel, R. P., Moskovitz, N. A., and Burbine, T. H. (2019) Olivine-dominated A-type asteroids in the main belt: Distribution, abundance and relation to families. *Icarus* 322, 13–30.

Demidova, S. I., Merle, R., Kenny, G. G., Nemchin, A. A., Whitehouse, M. J., Brandstätter, F., and Ntaflos, Th. (2020) Possible LL chondrite projectile in Luna-16 soil samples. *Lunar and Planetary Science* 51, Abstract#1368.

Dobrică, E. and Brearley, A. J. (2014) Widespread hydrothermal alteration minerals in the fine-grained matrices of the Tieschitz unequilibrated ordinary chondrite. *Meteoritics & Planetary Science* 49, 1323–1349.

Dobrică, E. and Brearley, A. J. (2020) Amorphous silicates in the matrix of Semarkona: The first evidence for the localized preservation of pristine matrix materials in the most unequilibrated ordinary chondrites. *Meteoritics & Planetary Science* 55, 1–20. https://doi.org/ 10.1111/maps.13458.

Dodd, R. T. (1981) *Meteorites: A Petrologic-Chemical Synthesis*. New York: Cambridge, 368 pp.

Dodd, R. T. and Jarosewich, E. (1979) Incipient melting in and shock classification of L-group chondrites. *Earth and Planetary Science Letters* 44, 335–340.

Dodd R. T. and Jarosewich E. (1982) The composition of incipient shock melts in L6 chondrites. *Earth and Planetary Science Letters* 59, 355–363.

Dodd R. T., Van Schmus W. R., and Marvin U. B. (1965) Merrihueite, a new alkali-ferromagnesian silicate from the Mezö-Madaras chondrite. *Science* 149, 972–974.

Dodd R. T., Van Schmus W. R., and Marvin U. B. (1966) Significance of iron-rich silicate in the Mezö-Madaras chondrite. *American Mineralogist* 51, 1177–1191.

Dodd, R. T., Van Schmus, W. R., and Koffman, D. M. (1967) A survey of the unequilibrated ordinary chondrites. *Geochimica et Cosmochimica Acta* 31, 921–951.

Dones, H., Zahnle, K. J., and Alvarellos, J. L. (2018) Asteroids and meteorites from Venus? Only the Earth goddess knows (abstract). *American Astronomical Society, Division on Dynamical Astronomy Meeting #49*, i.d. 102.02. https://ui.adsabs.harvard.edu/#abs/2018DDA....4910202D/abstract

Donohue, P. H., Huss, G. R., and Nagashima, K. (2019) Manganese-chromium systematics of calcite in the CM chondrites QUE 93005 and MET 01070 determined using a new matrix-matched standard. *Lunar and Planetary Science* 50, Abstract #1949.

Doyle, P. M., Jogo, K., Nagashima, K., Krot, A. N., Wakita, S., Ciesla, F. J., and Hutcheon, I. D. (2015) Early aqueous activity on the ordinary and carbonaceous chondrite parent bodies recorded by fayalite. *Nature Communications* 6, 1–10. https://doi.org/ 10.1038/ ncomms8444.

Dunn, T. L. and Gross, J. (2017) Reclassification of Hart and Northwest Africa 6047: Criteria for distinguishing between CV and CK3 chondrites. *Meteoritics & Planetary Science* 52, 2412–2423.

Dunn, T. L., Gross, J., Ivanova, M. A., Runyon, S. E., and Bruck, A. M. (2016) Magnetite in the unequilibrated CK chondrites: Implications for metamorphism and new insights into the relationship between the CV and CK chondrites. *Meteoritics & Planetary Science* 51, 1701–1720.

Dunn, T. L., Battifarano, O. K., Gross, J., and O'Hara, E. J. (2018) Characterization of matrix material in Northwest Africa 5343: Weathering and thermal metamorphism of the least equilibrated CK chondrite. *Meteoritics & Planetary Science* 53, 2165–2180.

Dye, S. T., Huang, Y., Lekic, V., McDonough, W. F., and Šrámek, O. (2015) Geo-neutrinos and Earth models. *Physics Procedia* 61, 310–318.

Ebata, S., Nagashima, K., Itoh, S., Kobayashi, S., Sakamoto, N., Fagan, T. J., and Yurimoto, H. (2006) Presolar silicate grains in enstatite chondrites. *Lunar and Planetary Science* 37, Abstract#1619.

Ebata, S., Fagan, T. J., and Yurimoto, H. (2007) Identification of silicate and carbonaceous presolar grains in the type 3 enstatite chondrite ALHA81189 (abstract). *Meteoritics & Planetary Science* 42, A38.

Ebel, D. S. and Grossman, L. (2000) Condensation in dust enriched systems. *Geochimica et Cosmochimica Acta* 64, 339–366.

El Goresy, A. (1976) Opaque oxide minerals in meteorites. In *Oxide Minerals*, ed. Rumble, D. Blacksburg, Virginia: Mineralogical Society of America, Southern Printing Company, pp. EG47–EG72.

El Goresy, A. and Ottemann, J. (1966) Gentnerite, $Cu_8Fe_3Cr_{11}S_{18}$, a new mineral from the Odessa meteorite. *Zeitschrift für Naturforschung* 21, 1160–1161.

El Goresy, A., Nagel, K., Dominik, B., and Ramdohr, P. (1977) Fremdlinge: Potential presolar material in Ca-Al-rich inclusions of Allende. *Meteoritics* 12, 215–216.

El Goresy, A., Nagel, K., and Ramdohr, P. (1978) Fremdlinge and their noble relatives. *Proceedings Lunar and Planetary Science Conference* 9, 1279–1303.

El Goresy, A., Wopenka, B., Chen, M., Weinbruch, S., and Sharp, T. (1997) Evidence for two different shock induced high-pressure events and alkali-vapor metasomatisme in Peace River and Tenham (L6) chondrites *Lunar and Planetary Science* 28, Abstract#1044.

El Goresy, A., Yabuki, H., Ehlers, K., Woolum, D., and Pernicka, E. (1988) Qingzhen and Yamato-691: A tentative alphabet for the EH chondrites. *Proceedings of the NIPR Symposium on Antarctic Meteorites* 1, 65–101.

El Goresy, A., Dera, P., Sharp, T. G., Prewitt, C. T., Chen, M., Dubrovinsky, L., Wopenka, B., Boctor, N. Z., and Hemley, R. J. (2008) Seifertite, a dense orthorhombic polymorph of silica from the Martian meteorites Shergotty and Zagami. *European Journal of Mineralogy* 20, 523–528.

El Goresy, A., Boyer, M., and Miyahara, M. (2011) Almahata Sitta MS-17 EL-3 chondrite fragment: Contrasting oldhamite assemblages in chondrules and matrix and significant oldhamite REE-patterns. *Meteoritics & Planetary Science* 46, A63, Abstract#5079.

El Goresy, A., Lin, Y., Miyahara, M., Gannoun, A., Boyet, M., Ohtani, E., Gillet, P., Trieloff, M., Simionovici, A., Feng, L., and Lenelle, L. (2017) Origin of EL3 chondrites evidence for variable C/O ratios during their course of formation – A state of the art scrutiny. *Meteoritics & Planetary Science* 52, 781–806.

Emel'yanenko, V. V., Naroenkov, S. A., Jenniskens, P., and Popova, O. P. (2014) The orbit and dynamical evolution of the Chelyabinsk object. *Meteoritics & Planetary Science* 49, 2169–2174.

Endress M. and Bischoff A. (1996) Carbonates in CI chondrites: Clues to parent body evolution. *Geochimica et Cosmochimica Acta* 60 489–507.

Endreß, M., Keil, K., Bischoff, A., Spettel, B., Clayton, R. N., and Mayeda, T. K. (1994) Origin of dark clasts in the Acfer/El Djouf 001 CR2 chondrite. *Meteoritics* 29, 26–40.

Eugster, O. (2003) Cosmic-ray exposure ages of meteorites and lunar rocks and their significance. *Geochemistry* 63, 3–30.

Fagan, T. J., Scott, E. R. D., Keil, K., Cooney, T. F., and Sharma, S. K. (2000) Formation of feldspathic and metallic melts by shock in enstatite chondrite Reckling Peak A80259. *Meteoritics & Planetary Science* 35, 319–329.

Farrington, O. C. (1915) *Meteorites*. Published by the author. Chicago, 233 pp.

Faust G. T., Fahey J. J., Mason B. H., and Dwornik, E. J. (1973) The disintegration of the Wolf Creek meteorite and the formation of pecoraite, the nickel analog of clinochrysotile. *United States Geological Survey Professional Paper* 384-C, 107–135.

Fegley, B., Treiman, A. H., and Sharpton, V. L. (1992) Venus surface mineralogy: Observational and theoretical constraints. *Proceedings of Lunar and Planetary Science* 22, 3–19.

Feldman, W. C., Maurice, S., Lawrence, D. J., Little, R. C., Lawson, S. L., Gasnault, O., Wiens, R. C., Barraclough, B. L., Elphic, R. C., Prettyman, T. H., Steinberg, J. T., and Binder, A. B. (2001) Evidence for water ice near the lunar poles. *Journal of Geophysical Research* 106 (E10), 23231–23251.

Fesenkov, V. G. (1958) Progress in meteoritics (in Russian) *Meteoritika* 16, 5–10.

Filacchione, G., Capaccioni, F., Ciarniello, M., Raponi, A., Rinaldi, G., Cristina De Sanctis, M., Bockelèe-Morvan, D., Erard, S., Arnold, G., Mennella, V., Formisano, M., Longobardo, A., and Mottola, S. (2020) An orbital water-ice cycle on comet 67P from colour changes. *Nature* 578, 49–52.

Flight W. (1887) *A Chapter in the History of Meteorites*. London: Dulau and Co.

Floran, R. J., Prinz, M., Hlava, P. F., Keil, K., Nehru, C. E., and Hinthorne, J. R. (1978) The Chassigny meteorite: A cumulate dunite with hydrous amphibole-bearing melt inclusions. *Geochimica et Cosmochimica Acta* 42, 1213–1229.

Fodor, R. V. and Keil, K. (1975) Implications of poikilitic textures in LL-group chondrites. *Meteoritics* 10, 325–339.

Fodor, R. V. and Keil, K. (1976) Carbonaceous and non-carbonaceous lithic fragments in the Plainview, Texas chondrite: Origin and history. *Geochimica et Cosmochimica Acta* 40, 177–189.

Fodor, R. V. and Keil, K. (1978) *Catalog of Lithic Fragments in LL-Group Chondrites*. Special Publication of the University of New Mexico. Albuquerque: Dept. of Geology & Institute of Meteoritics, University of New Mexico, pp. 1–38.

Fogel, R. A. (1997) On the significance of diopside and oldhamite in enstatite chondrites and aubrites. *Meteoritics & Planetary Science* 32, 577–591.

Foley, C. N., Nittler, L. R., McCoy, T. J., Lim, L. F., Brown, M. R. M., Starr, R. D., and Trombka, J. I. (2006) Minor element evidence that asteroid 433 Eros is a space-weathered ordinary chondrite parent body. *Icarus* 184, 338–343.

Franza, A. and Pratesi, G. (2020) Julius Obsequen's book, *Liber Prodigiorum*, A Roman era record of meteorite falls, fireballs, and other celestial phenomena. *Meteoritics & Planetary Science* 55, 1697–1708 https://doi.org/10.1111/maps.13525.

Friedrich, J. M., Weisberg, M. K., Ebel, D. S., Biltz, A. E., Corbett, B. M., Iotzov, I. V., Khan, W. S., and Wolman, M. D. (2015) Chondrule size and related physical properties:

A compilation and evaluation of current data across all meteorite groups. *Chemie der Erde – Geochemistry* 75, 419–443.

Fritz, J., Greshake, A., Klementova, M., Wirth, R., Palatinus, L., Assis Fernandes, V., Böttger, U., and Ferrière, L. (2019) Donwilhelmsite, IMA 2018-113. CNMNC Newsletter No. 47, February 2019, page 145; *Mineralogical Magazine* 83, 143–147.

Fritz, J., Greshanke, A., Klementova, M., Wirth, R., Palatinus, L., Trønnes, R. G., Fernandes, V. A., Böttger, U., and Ferrière, L. (2020) Donwilhelmsite, [$CaAl_4Si_2O_{11}$], a new lunar high-pressure Ca-Al-silicate with relevance for subducted terrestrial sediments. *American Mineralogist* 105, 1704–1711.

Frondel, J. W. (1975) *Lunar Mineralogy*. New York: Wiley.

Fuchs, L. H. (1966a) Djerfisherite, alkali copper-iron sulfide: A new mineral from enstatite chondrites. *Science* 153, 166–167.

Fuchs, L. H. (1966b) Roedderite, a new mineral from the Indarch meteorite. *American Mineralogist* 51, 949–955.

Fuchs, L. H. (1969) *The Phosphate Mineralogy of Meteorites*. The Netherlands: Reidel, Dordrecht.

Fuchs, L. H. (1971) Occurrence of wollastonite, rhonite, and andradite in the Allende meteorite. *American Mineralogist* 56, 2053–2068.

Fuchs, L. H. and Olsen, E. (1965) The occurrence of chlorapatite in the Mount Stirling octahedrite (abstract). *Transactions of the American Geophysical Union* 46, 122.

Fuchs, L. H., Olsen, E., and Henderson E. P. (1967) On the occurrence of brianite and panethite, two new phosphate minerals from the Dayton meteorite. *Geochimica et Cosmochimica Acta* 31, 1711–1719.

Fuchs, L. H., Olsen E., and Jensen, K.J. (1973) Mineralogy, mineral-chemistry, and composition of the Murchsion (C2) meteorite. *Smithsonian Contributions to the Earth Sciences* 10, 1–39.

Fujimaki, H., Matsu-ura M., Sunagawa I., and Aoki K. (1981) Chemical compositions of chondrules and matrices in the ALH-77015 chondrite (L3). Proceedings of the Symposium on Antarctic Meteorites 6, 161–174.

Fujiya, W., Hoppe, P., Zinner, E., Pignatari, M., and Herwig, F. (2013) Evidence for radiogenic sulfur-32 in Type AB presolar silicon carbide grains. *The Astrophysical Journal Letters* 776, L29, 6 pp.

Garvie, L. A. J., Németh, P., and Buseck, P. R. (2011) Diamond, bucky-diamond, graphite-diamond, Al-silicate, and stishovite in the Gujba CB chondrite. 74th Annual Meeting of the Meteoritical Society, held August 8–12, 2011, in London, UK Abstract #5227. Published in Meteoritics and Planetary Science Supplement.

Garvie, L. A. J., Knauth, L. P., and Morris, M. A. (2017) Sedimentary laminations in the Isheyevo (CH/CBb) carbonaceous chondrite formed by gentle impact-plume sweep-up. *Icarus* 292, 36–47.

Garvie, L. A. J., Ma, C., Ray, S., Domanik, K., Wittmann, A., and Wadhwa, M. (2021a) Carletonmooreite, Ni_3Si, a new silicide from the Norton County, aubrite meteorite. *American Mineralogist* 106, in press. DOI:10.2138/am-2021-7645

Garvie, L. A. L., Ma, C., and Wittmann, A. (2021b) Location and speciation of germanium in the Butler and Northwest Africa 859 ungrouped iron meteorites. *Lunar and Planetary Science* 52, Abstract #2398.

Geiger T. and Bischoff A. (1989) Mineralogy of metamorphosed carbonaceous chondrites (abstract). *Meteoritics* 24, 269–270.

Geiger T. and Bischoff A. (1990) Exsolution of spinel and ilmenite in magnetites from type 4-5 carbonaceous chondrites–indications for metamorphic processes (abstract). *Lunar and Planetary Science* 21, 409–410.

Geiger, T. and Bischoff, A. (1995) Formation of opaque minerals in CK chondrites. *Planetary and Space Science* 43, 485–498.

Genge, M. J. and Grady, M. M. (1999) The fusion crust of stony meteorites: Implications for the atmospheric reprocessing of extraterrestrial materials. *Meteoritics & Planetary Science*, 34, 341–356.

George, J., Waraquiers, D., Di Stefano, D., Petretto, G., Rignanese, G.-M., and Hautier, G. (2020) The limited predictive power of the Pauling rules. *Angewandte Chemie* 132, 7639–7645. https://doi.org/10.1002/ange.202000829.

Gettens, R. J., Clarke, Jr., R. S. and Chase, W. T. (1971) Two early Chinese bronze weapons with meteoritic iron blades. Freer Gallery of Art, Washington, DC, Occasional Papers, 4, No. 1.

Ghosh, A., Weidenschilling, S. J., McSween, H. Y., and Rubin, A. (2006) Asteroidal heating and thermal stratification of the asteroid belt. In *Meteorites and the Early Solar System II* eds. Lauretta, D., Leshin, L. A., and McSween, H. Y. Tucson: University of Arizona Press, pp. 555–566.

Gillet, P., Chen, M., Dubrovinsky, L., and El Goresy A. (2000) Natural $NaAlSi_3O_8$-hollandite in the shocked Sixiangkou meteorite. *Science* 287, 1633–1636.

Gilmore, M., Treiman, A., Helbert, J., and Smrekar, S. (2017) Venus surface composition constrained by observation and experiment. *Space Science Reviews* 212, 1511–1540.

Gladman, B. and Coffey, J. (2009) Mercurian impact ejecta: Meteorites and mantle. *Meteoritics & Planetary Science* 44, 285–291.

Gladman, B. J., Burns, J. A., Duncan, M., Lee, P., and Levison, H. F. (1996) The exchange of impact ejecta between terrestrial planets. *Science* 271, 1387–1392.

Glass, B. P., Liu, S., and Leavens P. B. (2002) Reidite: An impact-produced high-pressure polymorph of zircon found in marine sediments. *American Mineralogist* 87, 562–565.

Goldstein, J. I. and Michael, J. R. (2006) The formation of plessite in meteoritic metal. *Meteoritics & Planetary Science* 41, 553–570.

Gomes, C. B. and Keil, K. (1980) *Brazilian Stone Meteorites*. Albuquerque: University of New Mexico Press, 162 pp.

Gooding, J. L. (1981) Mineralogical aspects of terrestrial weathering effects in chondrites from Allan Hills, Antarctica. *Proceedings of the Lunar and Planetary Science Conference* 12B, 1105–1122.

Gooding, J. L. (1992) Soil mineralogy and chemistry on Mars: Possible clues from salts and clays in SNC meteorites. *Icarus* 99, 28–41.

Gooding, J. L. and Keil, K. (1981) Relative abundances of chondrule primary textural types in ordinary chondrites and their bearing on conditions of chondrule formation. *Meteoritics* 16, 17–43.

Gooding, J. L., Wentworth, S. J., and Zolensky, M. E. (1991) Aqueous alteration of the Nakhla meteorite. *Meteoritics & Planetary Science* 26, 135–143.

Goodrich, C. A. (1992) Ureilites: A critical review. *Meteoritics* 27, 327–352.

Goodrich, C. A., Keil, K., Berkley, J. L., Laul, J. C., Smith, M. R., Wacker, J. F., Clayton, R. N., and Mayeda, T. K. (1987) Roosevelt County 027: A low-shock ureilite with interstitial silicates and high noble-gas concentration. *Meteoritics* 22, 191–218.

Goodrich, C. A., Scott, E. R. D., and Fioretti, A. M. (2004) Ureilitic breccias: Clues to the petrologic structure and impact disruption of the ureilite parent asteroid. *Chemie der Erde – Geochemistry* 64, 283–327.

Goodrich, C. A., Van Orman, J. A., and Wilson, L. (2007) Fractional melting and smelting on the ureilite parent body. *Geochimica et Cosmochimia Acta* 71, 2876–2895.

Goodrich, C. A., Hartmann, W. K., O'Brien, D. P., Weidenschilling, S. J., Wilson, L., Michel, P., and Jutzi, M. (2015) Origin and history of ureilitic material in the solar system: The view from asteroid 2008 TC_3 and the Almahata Sitta meteorite. *Meteoritics & Planetary Science* 50, 782–809.

Goodrich, C. A., Kita, N. T., Yin, Q.-Z., Sanborn, M. E., Williams, C. D., Nakashima, D., Lane, M. D., and Boyle, S. (2017) Petrogenesis and provenance of ungrouped achondrite Northwest Africa 7325 from petrology, trace elements, oxygen, chromium and titanium isotopes, and mid-IR spectroscopy. *Geochimica et Cosmochimia Acta* 203, 381–403.

Goodrich, C. A., Nestola, F., Jakubek, R., Erickson, T., Fries, M., Fioretti, A. M., Ross, D. K., and Brenker, F. E. (2020) The origin of diamonds in ureilites. *Lunar and Planetary Science* 51, Abstract#1411.

Goreva, J. S., Ma, C., and Rossman, G. R. (2001) Fibrous nanoinclusions in massive rose quartz: The source of rose coloration. *American Mineralogist* 86, 466–472.

Gradie, J. and Tedesco, E. (1982) Compositional structure of the asteroid belt. *Science* 216, 1405–1407.

Grady, M. M. (2000) *Catalogue of Meteorites*, 5th ed. Cambridge, UK: Cambridge University Press, 689 pp.

Grady, M. M., Pratesi, G., and Moggi-Cecchi, V. (2015) *Atlas of Meteorites*. Cambridge, UK: Cambridge University Press, 373 pp.

Grazier, K. R., Castillo-Rogez, J. C., and Horner, J. (2018) It's complicated: A big data approach to exploring planetesimal evolution in the presence of Jovian planets. *Astronomical Journal* 156, 232s (19 pp).

Greenland, L. (1965) The abundances of selenium, tellurium, silver, palladiumm, cadmium, and zinc in chondritic meteorites. *Geochimica et Cosmochimica Acta* 31, 849–860.

Greenwood, R. C., Burbine, T. H., Miller, M. F., and Franchi, I. A. (2017) Melting and differentiation of early-formed asteroids: The perspective from high precision oxygen isotope studies. *Chemie der Erde – Geochemistry* 77, 1–43.

Greenwood, R. C., Burbine, T. H., and Franchi, I. A. (2020) Linking asteroids and meteorites to the primordial planetesimal population. *Geochimica et Cosmochimica Acta* 277, 377–406. https://doi.org/10.1016/j.gca.2020.02.004.

Greshake, A. (1997) The primitive matrix components of the unique carbonaceous chondrite Acfer 094: A TEM study. *Geochimica et Cosmochimica Acta* 61, 437–452.

Greshake, A. and Bischoff, A. (1996) Chromium-bearing phases in Orgueil (CI): Discovery of magnesiochromite ($MgCr_2O_4$), ureyite ($NaCrSi_2O_6$), and chromium-oxide (Cr_2O_3) (abstract). *Lunar and Planetary Science* 27, 461–462.

Greshake A., Bischoff A., and Putnis A. (1996a) Pure CaO, MgO (periclase), TiO_2 (rutile), and Al_2O_3 (corundum) in Ca,Al-rich inclusions from carbonaceous chondrites (abstract). *Lunar and Planetary Science* 27, 463–464.

Greshake A., Bischoff A., Putnis A., and Palme H. (1996b) Corundum, rutile, periclase, and CaO in Ca,Al-rich inclusions from carbonaceous chondrites. *Science* 272, 1316–1318.

Grew, E. S., Yates, M. G., Beane, R. J., Floss, C., and Gerbi, C. (2010) Chopinite-sarcopside solid solution,[(Mg, Fe) 3□]$(PO_4)_2$, in GRA95209, a transitional acapulcoite: Implications for phosphate genesis in meteorites. *American Mineralogist* 95, 260–272.

Grossman, J. N. and Brearley, A. J. (2005) The onset of metamorphism in ordinary and carbonaceous chondrites. *Meteoritics & Planetary Science* 40, 87–122.

Grossman, J. N. and Rubin, A. E. (1986) The origin of chondrules and clasts bearing calcic plagioclase in ordinary chondrites (abstract). *Lunar and Planetary Science* 17, 293–294.

Grossman, J. N. and Wasson, J. T. (1985) The origin and history of the metal and sulfide components of chondrules. *Geochimica et Cosmochimica Acta* 49, 925–939.

Grossman, J. N., Alexander, C. M. O'D., and Brearley, A. J. (2000) Bleached chondrules: Evidence for widespread aqueous processes on the parent asteroids of ordinary chondrites. *Meteoritics & Planetary Science* 35, 467–486.

Grossman, J. N., Rubin, A. E., Rambaldi, E. R., Rajan, R. S., and Wasson, J. T. (1985) Chondrules in the Qingzhen type-3 enstatite chondrite: Possible precursor components and comparison to ordinary chondrite chondrules. *Geochimica et Cosmochimica Acta* 49, 1781–1795.

Gyngard, F., Amari, S., Jadhav, M., Zinner, E., and Lewis, R. S. (2006) Carbon, nitrogen, and silicon isotopic ratios in KJG presolar SiC grains from Murchison. *Lunar and Planetary Science Conference* 37, Abstract #2194.

Gyngard, F., Ávila, J. N., Ireland, T. R., and Zinner, E. (2014) More interstellar exposure ages of large presolar SiC grains from the Murchison meteorite. *Lunar and Planetary Science* 45, Abstract #2348

Gyngard, F., Jadhav, M., Nittler, L. R., Stroud, R. M., and Zinner, E. (2018) Bonanza: An extremely large dust grain from a supernova. *Geochimica et Cosmochimica Acta* 221, 60–86.

Haba, M. K., Wotzlaw, J.-F., Lai, Y.-J., Yamaguchi, A., Schönbächler, M. (2019) Mesosiderite formation on asteroid 4 Vesta by a hit-and-run collision. *Nature Geoscience* 12, 510–515.

Haggerty, S. E. (1972) An enstatite chondrite from Hadley Rille (absract). In *The Apollo 15 Lunar Samples*, eds. Chamberlain, J. W. and Watkins, C. Houston, Texas: Lunar Science Institute, pp. 85–87.

Harries, D. and Langenhorst, F. (2018) Carbide-metal assemblages in a sample returned from asteroid 25143 Itokawa: Evidence for methane-rich fluids during metamorphism. *Geochimica et Cosmochimica Acta* 222, 53–73.

Haines, E. L., Gancarz, A. J., Albee, A. L., and Wasserburg, G. J. (1972) The uranium distribution in lunar soils and rocks 12013 and 14310. *Lunar and Planetary Science Conference*, Vol. 3, Abstract#1127, p. 350.

Hallis, L. J., Anand, M., Greenwood, R. C., Miller, M. F., Franchi, I. A., and Russell, S. S. (2010) The oxygen isotope composition, petrology and geochemistry of mare basalts: Evidence for large-scale compositional variation in the lunar mantle. *Geochimica et Cosmochimica Acta* 74, 6885–6899.

Hamilton, V. E., Simon, A. A., Christensen, P. R., Reuter, D. C., Clark, B. E., Barucci, M. A., Bowles, N. E., Boynton, W. V., Brucato, J. R., Cloutis, E. A., Connolly, H. C., Donaldson Hanna, K. L., Emery, J. P., Enos, H. L., Fornasier, S., Haberle, C. W., Hanna, R. D., Howell, E. S., Kaplan, H. H., Keller, L. P., Lantz, C., Li, J.-Y., Lim, L. F., McCoy, T. J., Merlin, F., Nolan, M. C., Praet, A., Rozitis, B., Sandford, S. A., Schrader, D. L., Thomas, C. A., Zou, X,-D., Lauretta, D. S., and the OSIRIS-REx Team (2019) Evidence for widespread hydrated minerals on asteroid (101955) Bennu. *Nature Astronomy* 3, 332–340.

Hansen, M. (1958) *Constitution of Binary Alloys*. New York: McGraw-Hill, 1,305 pp.

Hapke, B. (2001) Space weathering from Mercury to the asteroid belt. *Journal of Geophysical Research: Planets* 106, 10039–10073.

Harju, E. R., Rubin, A. E., Choi, B.-G., Ahn, I., Ziegler, K., and Wasson, J. T. (2014) Progressive aqueous alteration of CR carbonaceous chondrites. *Geochimica et Cosmochimica Acta* 139, 267–292.

Hartmann, W. K. and Davis, D. R. (1975) Satellite-sized planetesimals and lunar origin. *Icarus* 24, 504–515.

Hartmann, W. K., Forte, A., and Sabyr, A. (2018) Comment on "Chelyabinsk, Zond IV, and a possible first-century fireball of historical importance." *Meteoritics & Planetary Science* 53, 2181–2192.

Hassanzadeh, J., Rubin, A. E., and Wasson, J. T. (1990) Compositions of large metal nodules in mesosiderites: Links to iron meteorite group IIIAB and the origin of mesosiderite subgroups. *Geochimica et Cosmochimica Acta* 54, 3197–3208.

Hazen, R. M. and Morrison, S. M. (2020) An evolutionary system of mineralogy, Part I: Stellar mineralogy (>13 to 4.6 Ga). *American Mineralogist* 105, 627–651. https://doi.org/10.2138/am-2020-7173.

Hazen, R. M., Papineau, D., Bleeker, W., Downs, R. T., Ferry, J. M., McCoy, T. J., Sverjensky, D. A., and Yang, H. (2008) Mineral evolution. *American Mineralogist* 93, 1693–1720.

Heck, P. R., Greer, J., Kööp, L., Trappitsch, R., Gyngard, F., Busemann, H., Maden, C., Ávila, J. N., Davis, A. M., and Wieler, R. (2020) Lifetimes of interstellar dust from cosmic ray exposure ages of presolar silicon carbide. *Proceedings of the National Academy of Sciences* 17, 1884–1889. pnas.org/cgi/doi/10.1073/pnas.1904573117.

Heiken, G., Vaniman, D., and French, B. M., eds. (1991) *Lunar Sourcebook: A User's Guide to the Moon*. Cambridge, UK: Cambridge University Press, 736 pp.

Herbst, E. (1995) Chemistry in the interstellar medium. *Annual Review of Physical Chemistry* 46, 27–54.

Herndon, J. M. and Rudee, M. L. (1978) Thermal history of the Abee enstatite chondrite. *Earth and Planetary Science Letters* 41, 101–106.

Hewins, R. H., Jones, R. H., and Scott, E. R. D., eds. (1996) *Chondrules and the Protoplanetary Disk*. Cambridge, UK: Cambridge University Press, 360 pp.

Hewins, R. H., Condie, C., Morris, M., Richardson, M. L. A., Ouellette, N., and Metcalf, M. (2018) Thermal history of CB_b chondrules and cooling rate distributions of ejecta plumes. *Astrophysical Journal Letters* 855, L17, 7 pp.

Hey, M. H. (1973) Mineral analysis and analysts. *Mineralogical Magazine* 39, 4–24.

Heymann, D. (1967) On the origin of hypersthene chondrites: Ages and shock effects of black chondrites. *Icarus* 6, 189–221.

Hibaya, Y., Archer, G. J., Tanaka, R., Sanborn, M. E., Sato, Y., Iizuka, T., Ozawa, K., Walker, R. J., Yamaguchi, A., Yin, Q.-Z., Nakamura, T., and Irving, A. J. (2019) The origin of the unique achondrite Northwest Africa 6704: Constraints from petrology, chemistry and Re–Os, O and Ti isotope systematics. *Geochimica et Cosmochimica Acta* 245, 597–627.

Hicks, L. J., MacArthur, J. L., Bridges, J. C., Price, M. C., Wickham-Eade, J. E., Burchell, M. J., Hansford, G. M., Butterworth, A. L., Gurman, S. J., and Baker, S. H. (2017) Magnetite in comet Wild 2: Evidence for parent body aqueous alteration. *Meteoritics & Planetary Science* 52, 2075–2096.

Hilton, C. D., Bermingham, K. R., Ash, R. D., Walker, R. J., and McCoy, T. J. (2018) Genetics, age, and crystallization sequence of the South Byron Trio and the potential relation to the Milton pallasite. *Lunar and Planetary Science* 49, Abstract #1186.

Hiyagon, H., Sugiura, N., Kita, N. T., Kimura, M., Morishita, Y., and Takehana, Y. (2016) Origin of the eclogitic clasts with graphite-bearing and graphite-free lithologies in the Northwest Africa 801 (CR2) chondrite: Possible origin from a Moon-sized planetary body inferred from chemistry, oxygen isotopes and REE abundances. *Geochimica et Cosmochimica Acta* 186, 32–48.

Hollister, L. S., Bindi, L., Yao, N., Poirier, G. R., Andronicos, C. L., MacPherson, G. J., Lin, C., Distler, V. V., Eddy, M. P., Kostin, A., Kryachko, V., Steinhardt, W. M., Yudovskaya, M., Eiler, J. M., Guan, Y., Clarke, J. J., and Steinhardt P. J. (2014) Impact-induced shock and the formation of natural quasicrystals in the early solar system. *Nature Communications* 5, 4040.

Hoppe, P., Fujiya, W., and Zinner, E. (2012) Sulfur molecule chemistry in supernova ejecta recorded by silicon carbide stardust. *The Astrophysical Journal Letters* 745, L26, 5 pp.

Hoppe, P., Lodders, K., and Fujiya W. (2015) Sulfur in presolar silicon carbide grains from asymptotic giant branch stars. *Meteoritics & Planetary Science* 50, 1122–1138.

Hörz, F., Grieve, R., Heiken, G., Spudis, P., and Binde, R. A. (1991) Lunar surface processes. In *Lunar Sourcebook: A User's Guide to the Moon*, eds. Heiken, G. H., Vaniman, D. T., and French, B. M. Cambridge: Cambridge University Press, pp. 61–120.

Hu, J., Asimow, P. D., and Ma, C. (2019) First shock synthesis of khatyrkite, stolperite and a newly-found natural quasicrystal: Implications for the impact origin of quasicrystals from the Khatyrka meteorite. *Lunar and Planetary Science* 50, Abstract #3126.

Hua, X., Eisenhour, D. D., and Buseck, P. R. (1995) Cobalt-rich, nickel-poor metal (wairauite) in the Ningqiang carbonaceous chondrite. *Meteoritics* 30, 106–109.

Hunt, A. C., Benedix, G. K., Hammond, S. J., Bland, P. A., Rehkämper, M., Kreissig, K., and Strekopytov (2017) A geochemical study of the winonaites: Evidence for limited partial melting and constraints on the precursor composition. *Geochimica et Cosmochimica Acta* 199, 13–30.

Hurlbut, C. S. and Klein, C. (1977) *Manual of Mineralogy*, 19th ed., New York: Wiley, 532 pp.

Huss, G. R. (1990) Ubiquitous interstellar diamond and SiC in primitive chondrites: Abundances reflect metamorphism. *Nature* 347, 159–162.

Huss, G. R., Keil, K., and Taylor, G. J. (1981) The matrices of unequilibrated ordinary chondrites: Implications for the origin and history of chondrites. *Geochimica et Cosmochimica Acta* 45, 33–51.

Huss, G. R., Rubin, A. E., and Grossman, J. N. (2006) Thermal metamorphism in chondrites. *Meteorites and the Early Solar System II*. Tucson: University of Arizona Press, 567–586.

Hutchison, R. (1982) Meteorites – Evidence for the interrelationships of materials in the solar system of 4.55 Ga ago. *Physics of the Earth and Planetary Interiors* 29, 199–208.

Hutchison, R. (2004) *Meteorites: A Petrologic, Chemical and Isotopic Synthesis*. Cambridge: Cambridge University Press, 506 pp.

Hutchison, R. and Bevan, A. W. R. (1983) Conditions and time of chondrule accretion. In *Chondrules and Their Origins*, ed. King, E. A. Houston: Lunar and Planetary Institute, pp. 162–179.

Hutchison, R., Alexander, C. M. O'D., and Bridges, J. C. (1998) Elemental redistribution in Tieschitz and the origin of white matrix. *Meteoritics & Planetary Science* 33, 1169–1179.

Hwang, S. L., Shen, P., Chu, H. T., Chui, T. F., Varela, M. E., and Iizuka, Y. (2014) Kuratite (IMA 2013-109): The "unknown" Fe-Al-Ti silicate from the angrite D'Orbigny *Lunar and Planetary Science* 45, Abstract #1818.

Hwang, S.-L., Shen, P., Chu, H.-T., Yui, T.-F., Varela, M. E., and Iizuka, Y. (2016a) Matyhite, IMA 2015-121. CNMNC Newsletter No. 31, June 2016, page 692. *Mineralogical Magazine* 80, 691–697.

Hwang S. L., Shen P., Chu H. T. Y., T. F., Varela M.E. and Iizu. (2016b) Tsangpoite: The unknown calcium silico phosphate phase in the angrite D'Orbigny. *Lunar and Planetery Science* 47, Abstract #1466.

Imae, N., Kimura, M., Yamaguchi, A., and Kojima, H. (2019) Primordial, thermal, and shock features of ordinary chondrites: Emulating bulk X-ray diffraction using in-plane rotation of polished thin sections. *Meteoritics & Planetary Science* 54, 919–937.

Ireland, T. R. and Fegley, B. (2000) The solar system's earliest chemistry: Systematics of refractory inclusions. *International Geology Review* 42, 865–894.

Ireland, T. R. and Wlotzka, F. (1992) The oldest zircons in the solar system. *Earth and Planetary Science Letters* 109, 1–10.

Irving, A. J., Kuehner, S. M., Bunch, T. E., Ziegler, K., Chen, G., Herd, C. D. K., Conrey, R. M., and Ralew, S. (2013) Ungrouped mafic achondrite Northwest Africa 7325: A reduced, iron poor cumulate olivine gabbro from a differentiated planetary body. *Lunar and Planetary Science* 44, Abstract #2164.

Isa, J., Rubin, A. E., and Wasson, J. T. (2014) R-chondrite bulk-chemical compositions and diverse oxides: Implications for parent-body processes. *Geochimica et Cosmochimica Acta* 124, 131–151.

Isa, J., Ma, C., and Rubin, A. E. (2016) Joegoldsteinite: A new sulfide mineral ($MnCr_2S_4$) from the Social Circle IVA iron meteorite. *American Mineralogist* 101, 1217–1221.

Ishii, H. A., Krot, A. N., and Bradley, J. P. (2010) Discovery, mineral paragenesis, and origin of wadalite in a meteorite. *American Mineralogist* 95, 440–448.

Itoh, S., Russell, S. S., and Yurimoto, H. (2007) Oxygen and magnesium isotopic compositions of amoeboid olivine aggregates from the Semarkona LL3.0 chondrite. *Meteoritics & Planetary Science* 42, 1241–1247.

Ivanov, A. V., Zolensky, M. E., Saito, A., Ohsumi, K., Yang, V., Kononkova, N. N., and Mikouchi, T. (2000) Florenskyite, FeTiP, a new phosphide from the Kaidun meteorite. *American Mineralogist* 85, 1082–1086.

Ivanov, A. V., Kononkova, N. N., Zolensky, M. E., Migdisova, L. F., and Stroganov, I. A. (2001) The Kaidun meteorite: A large albite crystal-fragment of an alkaline rock. *Lunar and Planetary Science* 32, Abstract #1080.

Ivanova, M. A., Kononkova, N. N., Franchi, I. A., Verchovsky, A. B., Korochantseva, E. V., Trieloff, M., and Brandstaetter, F. (2006) Isheyevo meteorite: Genetic link between CH and CB chondrites? *Lunar and Planetary Science* 37, Abstract #1100.

Ivanova, M. A., Lorenz, C. A., Ma, C., and Ivanov, A. V. (2016) The Kaidun breccia material variety: New clasts and updated hypothesis on a space trawl origin. *Meteoritics and Planetary Science* 51, Abstract #6100.

Ivanova, M. A., Lorenz, C. A., Borisovskiy, S. E., Burmistrov, A. A., Korost, D. V., Korochantsev, A. V., Logunova, M. N., Shornikov, S. I., and Petaev, M. I. (2017) Composition and origin of holotype Al-Cu-Zn minerals in relation to quasicrystals in the Khatyrka meteorite. *Meteoritics & Planetary Science* 52, 869–883.

Ivanova, M. A., Lorenz, C. A., Humayun, M., Richter, K., Corrigan, C. M., Franchi, I. A., Verchovsky, A. B., Korochantseva, E. V., Kozlov, V. V., Teplyakova, S. N., Kononkova, N. N., and Korochantsev, A. V. (2019a) Properties of a new grouplet of G metal-rich chondrites. 82nd Annual Meeting of the Meteoritical Society, held July 7–12, 2019, Sapporo, Japan. Abstract #6143.

Ivanova, M. A., Ma, C., Lorenz, C. A., Franchi, I. A., Kononkova, N. N. (2019b) A new unusual bencubbinite (CBa), Sierra Gorda 013, with unique V-rich sulfides. *Meteoritics & Planetary Science* 54, Abstract #6149.

Izawa, M. R. M., Flemming, R. L., Banerjee, N. R., and McCausland, P. J. A. (2011) Micro-X-ray diffraction assessment of shock stage in enstatite chondrites. *Meteoritics & Planetary Science* 46, 638–651.

Jabeen, I., Ali, A., Banerjee, N. R., Osinski, G. R., Ralew, S., and DeBoer, S. (2014) Oxygen isotope compositions of mineral separates from NWA 7325 suggest a planetary (Mercury?) origin. *Lunar and Planetary Science* 45, Abstract #2215.

Jacquet, E., Piani, L., and Weisberg, M. K. (2018) Chondrules in enstatite chondrites. In *Chondrules: Records of Protoplanetary Disk Processes*, eds. Russell, S. S., Connolly, H. C., and Krot, A. N. Cambridge, UK: Cambridge University Press, pp. 175–195; 450 pp.

Jambon, A. and Zimmerman, J. L. (1990) Water in oceanic basalts: Evidence for dehydration of recycled crust. *Earth and Planetary Science Letters* 101, 323–331.

Jarosewich, E. (1990) Chemical analyses of meteorites: A compilation of stony and iron meteorite analyses. *Meteoritics* 25, 323–337.

Jeffreys, H. (1924) *The Earth, Its Origin, History, and Physical Constitution*. Cambridge, UK: Cambridge University Press.

Jeffreys, H. (1929) *The Future of the Earth*. New York: Norton and Company.

Jenniskens, P., Shaddad, M. H., Numan, D., Elsir, S., Kudoda, A. M., Zolensky, M. E., Le, L., Robinson, G. A., Friedrich, J. M., Rumble, D., Steele, A., Chesley, S. R., Fitzsimmons, A., Duddy, S., Hsieh, H. H., Ramsay, G., Brown, P. G., Edwards, W. N., Tagliaferri, E., Boslough, M. B., Spalding, R. E., Dantowitz, R., Kozubal, M., Pravec, P., Borovicka, J., Charvat, Z., Vaubaillon, J., Kuiper, J., Albers, J., Bishope, J. L., Mancinelli, R. L., Sandford, S. A., Milam, S. N., Nuevo, M., and Worden, S. P. (2009) The impact and recovery of asteroid 2008 TC_3. *Nature* 458, 485–488.

Jewitt, D. and Luu, J. (1993) Discovery of the candidate Kuiper Belt object 1992 QB1. *Nature* 362, 730–732.

Jin, Z. and Bose, M. (2019) New clues to ancient water on Itokawa. *Science Advances* 5, eaav8106.

Johnson, C. A. and Skinner, B. J. (2003) Geochemistry of the Furnace Magnetite Bed, Franklin, New Jersey, and the relationship between stratiform oxide ores and stratiform zinc oxide-silicate ores in the New Jersey Highlands. *Economic Geology* 98, 837–854.

Johnson, J. E., Scrymgour, J. M., Jarosewich, E., and Mason, B. (1977) Brachina meteorite – a chassignite from South Australia. *Records of the South Australia Museum* 17, 309–319.

Johnson, M. C., Rutherford, M. J., and Hess, P.C. (1991) Chassigny petrogenesis: Melt compositions, intensive parameters, and water contents of martian(?) magmas. *Geochimica et Cosmochimica Acta* 55, 349–366.

Jones, R. H., Mccubbin, F. M., and Guan, Y. (2016) Phosphate minerals in the H-group of ordinary chondrites, and fluid activity recorded by apatite heterogeneity in the Zag H3-6 regolith breccia. *American Mineralogist* 101, 2452–2467.

Kallemeyn, G. W. and Rubin, A. E. (1995) Coolidge and Loongana 001: Members of a new carbonaceous chondrite grouplet. *Meteoritics* 30, 20–27.

Kallemeyn, G. W., Rubin, A. E., Wang, D., and Wasson, J. T. (1989) Ordinary chondrites: Bulk compositions, classification, lithophile-element fractionations, and composition-petrographic type relationships. *Geochimica et Cosmochimica Acta* 53, 2747–2767.

Kallemeyn, G. W., Rubin, A. E., and Wasson, J. T. (1991) The compositional classification of chondrites: V. The Karoonda (CK) group of carbonaceous chondrites. *Geochimica et Cosmochimica Acta* 55, 881–892.

Kallemeyn, G. W., Rubin, A. E., and Wasson, J. T. (1996) The compositional classification of chondrites: VII. The R chondrite group. *Geochimica et Cosmochimica Acta* 60, 2243–2256.

Komatsu, M., Fagan, T. J., Krot, A. N., Nagashima, K., Petaev, M. I., Kimura, M., and Yamaguchi, A. (2018) First evidence for silica condensation within the solar protoplanetary disk. *Proceedings of the National Academy of Sciences* 115, 7497–7502. https://doi .org/10.1073/pnas.1722265115

Karwowski L. and Muszyński, A. (2008) Multimineral inclusions in the Morasko coarse octahedrite. *Meteoritics & Planetary Science* 43, A71–A71.

Karwowski Ł., Kusz J., Muszyński, A., Kryza R., Sitarz M., and Galuskin E. V. (2015) Moraskoite, $Na_2Mg(PO_4)F$, a new mineral from the Morasko IAB-MG iron meteorite (Poland). *Mineralogical Magazine* 79, 387–398.

Karwowski, Ł., Kryza, R., Muszyński, A., Kusz, J., Helios, K., Drożdżewski, P., and Galuskin, E. V. (2016) Czochralskiite, $Na_4Ca_3Mg(PO_4)_4$, a second new mineral from the Morasko IAB-MG iron meteorite (Poland). *European Journal of Mineralogy* 28, 969–977.

Keil, K. (1968) Mineralogical and chemical relationships among enstatite chondrites. *Journal of Geophysical Research* 73, 6945–6976.

Keil, K. (1982) Composition and origin of chondritic breccias. In *Workshop on Lunar Breccias and Soils and Their Meteoritic Analogs*, eds. Taylor, G. J. and Wilkening, L. L. *LPI Technical Report* 82-02. Houston: Lunar and Planetary Institute, pp. 65–83.

Keil, K. (1989) Enstatite meteorites and their parent bodies. *Meteoritics & Planetary Science* 24, 195–208.

Keil, K. (2007) Occurrence and origin of keilite,(Fe$>$ 0.5, Mg$<$ 0.5) S, in enstatite chondrite impact-melt rocks and impact-melt breccias. *Chemie der Erde – Geochemistry* 67, 37–54.

Keil, K. (2010) Enstatite achondrite meteorites (aubrites) and the histories of their asteroidal parent bodies. *Chemie der Erde – Geochemistry* 70, 295–317.

Keil, K. (2012) Angrites, a small but diverse suite of ancient, silica-undersaturated volcanic-plutonic mafic meteorites, and the history of their parent asteroid. *Chemie der Erde – Geochemistry* 72, 191–218.

Keil, K. (2014) Brachinite meteorites: Partial melt residues from an FeO-rich asteroid. *Chemie der Erde – Geochemistry* 74, 311–329.

Keil, K. and Brett, R. (1974) Heideite, $(Fe,Cr)_{1+x}(Ti,Fe)_2S_4$, a new mineral in the Bustee enstatite achondrite. *American Mineralogist* 59, 465–470.

Keil, K. and McCoy, T. J. (2018) Acapulcoite-lodranite meteorites: Ultramafic asteroidal partial melt residues. *Chemie der Erde – Geochemistry* 78, 153–203.

Keil, K. and Snetsinger, K. G. (1967) Niningerite: A new meteoritic sulfide. *Science* 155, 451–453.

Keil, K., Berkley, J. L., and Fuchs L. H. (1982) Suessite, Fe_3Si: A new mineral in the North Haig ureilite. *American Mineralogist* 67, 126–131.

Keil, K., Ntaflos, Th., Taylor, G. J., Brearley, A. J., Newsom, H. E., and Romig, A. D. (1989) The Shallowater aubrite: Evidence for origin by planetesimal impacts. *Geochimica et Cosmochimica Acta* 53, 3291–3307.

Kerridge, J. F. and Matthews, M. S. (1988) *Meteorites and the Early Solar System*. Tucson: University of Arizona Press, 1,269 pp.

Kerridge, J. F., MacKay, A. L., and Boynton, W. V. (1979) Magnetite in CI carbonaceous meteorites: Origin by aqueous activity on a planetesimal surface. *Science* 205, 395–397.

Killgore, K. and Killgore, M. (2002) *Southwest Meteorite Collection: A Pictorial Catalog*. Payson, Arizona: Southwest Meteorite Press, 201 pp.

Kim, H. Y., Choi, B.-G., and Rubin, A. E. (2009) Wüstite in the DOM 03238 magnetite-rich CO3.1 chondrite: Formation during atmospheric passage. *Meteoritics & Planetary Science* 44, A109.

Kimura, M. (1996) Meteorite minerals (in Japanese). *Kobutsugaku Zassi* 25, 49–60.

Kimura, M. and El Goresy, A. (1989) Discovery of E-chondrite assemblages, SiC, and silica-bearing objects in ALH85085: Link between E- and C-chondrite (abstract). *Meteoritics* 24, 286.

Kimura, M. and Ikeda, Y. (1995) Anhydrous alteration of Allende chondrules in the solar nebula II: Alkali-Ca exchange reactions and formation of nepheline, sodalite and Ca-rich phases in chondrules. *Proceedings of the NIPR Symposium on Antarctic Meteorites* 8, 123–138.

Kimura, M., Tsuchiyama, A., Fukuoka, T., and Iimura, Y. (1992) Antarctic primitive achondrites, Yamato-74025, -75300, and -75305: Their mineralogy, thermal history, and the relevance to winonaite. *Proceedings of the NIPR Symposium on Antarctic Meteorites* 5, 165–190.

Kimura, M., Weisberg, M., Lin, Y., Suzuki, A., Ohtani, E., and Okazaki, R. (2005) Thermal history of the enstatite chondrites from silica polymorphs. *Meteoritics & Planetary Science* 40, 855–868.

Kimura, M., Mikouchi, T., Suzuki, A., Miyahara, M., Ohtani, E., and El Goresy, A. E. (2009) Kushiroite, $CaAlAlSiO_6$: A new mineral of the pyroxene group from the ALH 85085 CH chondrite, and its genetic significance in refractory inclusions. *American Mineralogist* 94, 1479–1482.

Kimura, M., Grossman, J. N., and Weisberg, M. K. (2011) Fe-Ni metal and sulfide minerals in CM chondrites: An indicator for thermal history. *Meteoritics & Planetary Science* 46, 431–442.

Kimura, M., Sugiura, N., Mikouchi, T., Hirajima, T., Hiyagon, H., and Takehana, Y. (2013) Eclogitic clasts with omphacite and pyrope-rich garnet in the NWA 801 CR2 chondrite. *American Mineralogist* 98, 387–393.

Kimura, M., Yamaguchi, A., and Miyahara, M. (2017) Shock-induced thermal history of an EH3 chondrite, Asuka 10164. *Meteoritics & Planetary Science* 52, 24–35.

Kimura, M., Sugiura, N., Yamaguchi, A., and Ichimura, K. (2020) The most primitive mesosiderite Northwest Africa 1878, subgroup 0. *Meteoritics & Planetary Science* 55. https://doi.org/10.1111/maps.13474.

King, E. A., ed. (1983) *Chondrules and their Origins*. Houston: Lunar and Planetary Institute, 377 pp.

King, A. J., Bates, H. C., Krietsch, D., Busemann, H., Clay, P. L., Schofield, P. F., and Russell, S. S. (2019) The Yamato-type (CY) carbonaceous chondrite group: Analogues for the surface of asteroid Ryugu? *Chemie der Erde – Geochemistry* 79, 125531.

Klein, C. and Dutrow, B. (2007) *Manual of Mineral Science*, 23rd ed., New York: Wiley, 716 pp.

Koefoed, P., Amelin, Y., Yin, Q.-Z., Wimpenny, J., Sanborn, M. C., Iizuka, T., and Irving, A. J. (2016) U–Pb and Al–Mg systematics of the ungrouped achondrite Northwest Africa 7325. *Geochimica et Cosmochimica Acta* 183, 3145.

Korochantseva, E. V., Trieloff, M., Lorenz, C. A., Buykin, A. I., Ivanova, M. A., Schwartz, W. H., Hopp, J., and Jessberger, E. K. (2007) L-chondrite asteroid breakup tied to Ordovician meteorite shower by multiple isochron ^{40}Ar-^{39}Ar dating. *Meteoritics & Planetary Science* 42, 113–130.

Kracher, A., Kurat, G., and Buchwald, V. F. (1977) Cape York: The extraordinary mineralogy of an ordinary iron meteorite and its implication for the genesis of all IIIAB irons. *Geochemical Journal* 11, 207–217.

Krot, A. N. (2016) Machiite, IMA 2016-067. CNMNC Newsletter No. 34, December 2016, page 1317; *Mineralogical Magazine* 80, 1315–1321.

Krot, A. N. (2019) Refractory inclusions in carbonaceous chondrites: Records of early solar system processes. *Meteoritics & Planetary Science* 54, 1647–1691.

Krot, A. N. and Keil, K. (2002) Anorthite-rich chondrules in CR and CH carbonaceous chondrites: Genetic link between calcium-aluminum-rich inclusions and ferromagnesian chondrules. *Meteoritics & Planetary Science* 37, 91–111.

Krot, A. N. and Rubin, A. E. (1993) Chromite-rich mafic silicate chondrules in ordinary chondrites: Formation by impact melting. *Lunar and Planetary Science* 24, 827–828.

Krot, A. N. and Rubin, A. E. (1994) Glass-rich chondrules in ordinary chondrites. *Meteoritics* 29, 697–706.

Krot, A. N. and Wasson, J. T. (1994) Silica-merrihueite/roedderite-bearing chondrules and clasts in ordinary chondrites: New occurrences and possible origin. *Meteoritics* 29 707–718.

Krot, A., Rubin, A. E., and Kononkova, N. N. (1993) First occurrence of pyrophanite ($MnTiO_3$) and baddeleyite (ZrO_2) in an ordinary chondrite. *Meteoritics* 28, 232–239.

Krot, A. N., Scott, E. R. D., and Zolensky, M. E. (1995) Mineralogical and chemical modification of components in CV3 chondrites: Nebular or asteroidal processing? *Meteoritics* 30, 748–775.

Krot, A. N., Rubin, A. E., Keil, K., and Wasson, J. T. (1997a) Microchondrules in ordinary chondrites: Implications for chondrule formation. *Geochimica et Cosmochimica Acta* 61, 463–473.

Krot, A. N., Zolensky, M. E., Wasson, J. T., Scott, E. R. D., Keil, K., and Ohsumi, K. (1997b) Carbide-magnetite assemblages in type-3 ordinary chondrites. *Geochimica et Cosmochimica Acta* 61, 219–237.

Krot, A., Scott, E. R. D., and Zolensky, M. (1997c) Origin of fayalitic olivine rims and lath-shaped matrix olivine in the CV3 chondrite Allende and its dark inclusions. *Meteoritics & Planetary Science* 32, 31–49.

Krot, A. N., Brearley, A. J., Ulyanov, A. A., Biryukov, V. V., Swindle, T. D., Keil, K., Mittlefehldt, D. W., Scott, E. R. D., Clayton R. N., and Mayeda, T. K. (1999) Mineralogy, petrography, bulk chemical, iodine-xenon, and oxygen-isotopic compositions of dark inclusions in the reduced CV3 chondrite Efremovka. *Meteoritics & Planetary Science* 34, 67–89.

Krot, A. N., Petaev, M. I., Meibom, A., and Keil, K. (2000a) *In situ* growth of Ca-rich rims around Allende dark inclusions. *Geochemistry International* 38, S351–S368.

Krot, A. N., Meibom, A., and Keil, K. (2000b) A clast of Bali-like oxidized CV material in the reduced CV chondrite breccia Vigarano. *Meteoritics & Planetary Science* 35, 817–825.

Krot, A. N., Amelin, Y., Cassen, P., and Meibom, A. (2005) Young chondrules in CB chondrites from a giant impact in the early Solar System. *Nature* 436, 989–992.

Krot, A. N., Ma, C., Nagashima, K., Davis, A. M., Beckett, J. R., Simon, S. B., Komatsu, M., Fagan, T. J., Brenker, F., Ivanova, M. A., and Bischoff, A. (2019) Mineralogy, petrography, and oxygen isotopic compositions of ultrarefractory inclusions from carbonaceous chondrites. *Chemie der Erde – Geochemistry* 79, https://doi.org/10.1016/j.chemer.2019.07.001.

Krot, A. N., Nagashima, K., and Rossman, G. R. (2020) Machiite, $Al_2Ti_3O_9$, a new oxide mineral from the Murchison carbonaceous chondrite: A new ultra-refractory phase from the solar nebula. *American Mineralogist* 105, 239–243.

Kruijer, T. S., Burkhardt, C., Budde, G., and Kleine, T. (2017) Age of Jupiter inferred from the distinct genetics and formation times of meteorites. *Proceedings of the National Academy of Sciences* 114, 6712–6716.

Ksanda, C. J. and Henderson, E. P. (1939) Identification of diamond in the Canyon Diablo iron. *American Mineralogist* 24, 677–680.

Kullerud, G. (1963) The Fe-Ni-S system. *Carnegie Institute of Washington Yearbook* 62, 175–189.

Kurat, G., Brandstatter, H., Palme, H., and Michel-Levy, M. C. (1981) Rusty Ornans. *Meteoritics* 16, 343–344.

Kurat, G., Varela, M. E., Zinner, E., and Brandstätter, F. (2010). The Tucson ungrouped iron meteorite and its relationship to chondrites. *Meteoritics & Planetary Science* 45, 1982–2006.

Kyte, F. T. (1998) A meteorite from the Cretaceous/Tertiary boundary. *Nature* 396, 237–239.

Kyte, F. T. (2002) Unmelted meteoritic debris collected from Eltanin ejecta in *Polarstern* cores from expedition ANT XII/4.*Deep Sea Research Part II: Topical Studies in Oceanography* 49, 1063–1071.

Le Guillou, C., Changela, H. G., and Brearley, A. J. (2015) Widespread oxidized and hydrated amorphous silicates in CR chondrites matrices: Implications for alteration conditions and H_2 degassing of asteroids. *Earth and Planetary Science Letters* 420, 162–173.

Lee, M. R., Russell, S. S., Arden, J. W., and Pillinger C. T. (1995) Nierite (Si_3N_4), a new mineral from ordinary and enstatite chondrites. *Meteoritics* 30, 387–398.

Lee, M. R., Lindgren, P., and Sofe, M. R. (2014) Aragonite, breunnerite, calcite and dolomite in the CM carbonaceous chondrites: High fidelity recorders of progressive parent body aqueous alteration. *Geochimica et Cosmochimica Acta* 144, 126–156.

Lee, M. R., Daly, L., Cohen, B. E., Hallis, L. J., Griffin, S., Trimby, P., Boyce, A., and Mark, D. F. (2018) Aqueous alteration of the Martian meteorite Northwest Africa 817: Probing fluid-rock interaction at the nakhlite launch site. *Meteoritics & Planetary Science* 53, 2395–2412.

Lee, M. S., Rubin, A. E., and Wasson, J. T. (1992) Origin of metallic Fe-Ni in Renazzo and related chondrites. *Geochimica et Cosmochimica Acta* 56, 2521–2533.

Lehner, S. W., Buseck, P. R., and McDonough, W. F. (2010) Origin of kamacite, schreibersite, and perryite in metal-sulfide nodules of the enstatite chondrite Sahara 97072 (EH3). *Meteoritics & Planetary Science* 45, 289–303.

Leitner, J., Vollmer, C., Harries, D., Kodolányi, J., Ott, U., and Hoppe, P. (2020) Investigation of nitrides in carbonaceous chondrites: A window to early solar nebula processes? *Lunar and Planetary Science* 52, Abstract#1937.

Levin, H. L. (1990) *Contemporary Physical Geology*, 3rd ed. Philadelphia: Saunders College Publishing, 623 pp.

Levine, D. and Steinhardt, P. J. (1984) Quasicrystals: A new class of ordered structures. *Physical Review Letters* 53, 923–925.

Lewis, J. A. and Jones, R. H. (2016) Phosphate and feldspar mineralogy of equilibrated L chondrites: The record of metasomatism during metamorphism in ordinary chondrite parent bodies. *Meteoritics & Planetary Science* 51, 1886–1913.

Lewis, J. A. and Jones, R. H. (2019) Primary feldspar in the Semarkona LL3.00 chondrite: Constraints on chondrule formation and secondary alteration. *Meteoritics & Planetary Science* 54, 72–89.

Li, S., Lucey, P. G., Milliken, R. E., Hayne, P. O., Fisher, E., Williams, J.-P., Hurley, D. M., and Elphic, R. C. (2018) Direct evidence of surface exposed water ice in the lunar polar regions. *Proceedings of the National Academy of Sciences* 115, 8907–8912.

Lim, L. F. and Nittler, L. R. (2009) Elemental composition of 433 Eros: New calibration of the NEAR-Shoemaker XRS data. *Icarus* 200, 129–146.

Lin, C., Hollister, L. S., MacPherson, G. J., Bindi, L., Ma, C., Andronicos, C. L., and Steinhardt, P. J. (2017) Evidence of cross-cutting and redox reaction in Khatyrka meteorite reveals metallic-Al minerals formed in outer space. *Scientific Reports* 7, 1–14, Article #1637.

Lin, Y. and Kimura, M. (1996) Discovery of complex titanium oxide associations in a plagioclase-olivine inclusion (POI) in the Ningqiang carbonaceous chondrite (abstract). *Lunar and Planetary Science* 27, 755–756.

Lin, Y. and Kimura, M. (1997) Titanium-rich oxide-bearing plagioclase-olivine inclusions in the unusual Ningqiang carbonaceous chondrite. *Antarctic Meteorite Research* 10, 227–248.

Lin, Y., El Goresy, A., Boyer, M., Feng, L., Zhang, J., and Hao, J. (2011) Earliest solid condensates consisting of the assemblage oldhamite, sinoite, graphite and excess 36S in lawrencite from Almahata Sitta MS-17 EL3 chondrite. *Workshop on Formation of the First Solids in the Solar System*, Abstract #9040.

Lindgren, P., Hanna, R. D., Dobson, K. J., Tomkinson, T., and Lee, M. R. (2015) The paradox between low shock-stage and evidence of compaction in CM carbonaceous chondrites explained by multiple low-intensity impacts. *Geochimica et Cosmochimica Acta* 148, 159–178.

Lindstrom, M. M., ed. (1990) New Meteorites. *Antarctic Meteorite Newsletter* 13, 9–24.

Litasov, K. D. and Podgornykh, N. M. (2017) Raman spectroscopy of various phosphate minerals and occurrence of tuite in the Elga IIE iron meteorite. *Journal of Raman Spectroscopy* 48, 1518–1527.

Liu, N., Nittler, L. R., Pignatari, M., Alexander, C. M. O'D., and Wang J. (2017) Stellar origin of ^{15}N-rich presolar SiC grains of Type AB: Supernovae with explosive hydrogen burning. *The Astrophysical Journal Letters* 842, L1, 8 pp.

Liu, Y., Ma, C., Beckett, J. R., Chen, Y., and Guan, Y. (2016) Rare-earth-element minerals in martian breccia meteorites NWA 7034 and 7533: Implications for fluid-rock interaction in the martian crust. *Earth and Planetary Science Letters* 451, 251–262.

Lock, S. J., Stewart, S. T., Petaev, M. I., Leinhardt, Z., Mace, M. T., Jacobsen, S. B., and Cuk, M. (2018) The origin of the Moon within a terrestrial synestria. *Journal of Geophysical Research: Planets* 123, 910–951.

Lodders, K. and Amari, S. (2005) Presolar grains from meteorites: Remnants from the early times of the solar system. *Chemie der Erde – Geochemistry* 65, 93–166.

Loeffler, M. J., Dukes, C. A., Chang, W. Y., McFadden, L. A., and Baragiola, R. A. (2008) Laboratory simulations of sulfur depletion at Eros. *Icarus* 195, 622–629.

Lomax, B. A., Conti, M., Khan, N., Bennett, N. S., Ganin, A. Y., and Symes, M. D. (2020) Proving the viability of an electrochemical process for the simultaneous extraction of oxygen and production of metal alloys from lunar regolith. *Planetary and Space Science* 180, 104748. https://doi.org/10.1016./j.pss.2019.104748.

Lorenz, C. A., Nazarov, M. A., Brandstaetter, F., and Ntaflos, T. (2010) Metasomatic alterations of olivine inclusions in the Budulan mesosiderite. *Petrology* 18, 461–470.

Lorenz, R. and Mitton, J. (2008) *Titan Unveiled: Saturn's Mysterious Moon Exlpored.* Princeton, NJ: Princeton University Press, 264 pp.

Love, S. G. and Keil, K. (1995) Recognizing mercurian meteorites. *Meteoritics* 30, 269–278.

Lovering, J. F., Wark, D. A., and Sewell, D. K. B. (1979) Refractory oxide, titanate, niobate and silicate accessory mineralogy of some type B Ca-Al inclusions in the Allende meteorite (abstract). *Lunar and Planetary Science* 10, 745–746.

Lunning, N. G., Corrigan, C. M., McSween, H. Y., Tenner, T. J., Kita, N. T., and Bodnar, R. J. (2016) CV and CM chondrite impact melts. *Geochimica et Cosmochimica Acta* 189, 338–358.

Lunning, N. G., Bischoff, A., Gross, J., Patzek, M., Corrigan, C. M., and McCoy, T. J. (2020) Insights into the formation of silica-rich achondrites from impact melts in Rumuruti-type chondrites. *Meteoritics & Planetary Science* 55, 130–148.

Lyons, R. J., Bowling, T. J., Ciesla, F. J., Davison, T. M., and Collins, G. S. (2019) The effects of impacts on the cooling rates of iron meteorites. *Meteoritics & Planetary Science* 54, 1604–1618.

Ma, C. (2010) Hibonite-(Fe),(Fe, Mg)$Al_{12}O_{19}$, a new alteration mineral from the Allende meteorite. *American Mineralogist* 95, 188–191.

Ma, C. (2011) Discovery of meteoritic lakargiite ($CaZrO_3$), a new ultra-refractory mineral from the Acer 094 carbonaceous chondrite. *Meteoritics & Planetary Science*, 46 (S1), A144.

Ma, C. (2012) Discovery of meteoritic eringaite, $Ca_3(Sc,Y,Ti)_2Si_3O_{12}$, the first solar garnet? *Meteoritics & Planetary Science*, 47 (S1), A256.

Ma, C. (2015) Nanomineralogy of meteorites by advanced electron microscopy: Discovering new minerals and new materials from the early solar system. *Microscopy and Microanalysis* 21, 2353–2354.

Ma, C. (2018a) A closer look at shock meteorites: Discovery of new high-pressure minerals. *American Mineralogist* 103, 1521–1522.

Ma, C. (2018b) Discovery of meteoritic baghdadite, $Ca_3(Zr,Ti)Si_2O_9$, in Allende: The first solar silicate with structurally essential zirconium? *Meteoritics & Planetary Science*, 53 (S1), Abstract #6358.

Ma, C. (2019) Discovery of kaitianite, $Ti^{3+}{}_2Ti^{4+}O_5$, in Allende: A new refractory mineral from the solar nebula. *Meteoritics & Planetary Science* 54 (S1), Abstract #6098.

Ma, C. (2020) Discovery of meteoritic calzirtite in Leoville: A new ultrarefractory phase from the solar nebula. *Goldschmidt,* Abstract #1674. DOI: 10.46427/gold2020.1674

Ma, C. (2021) Zolenskyite, IMA 2010-070. CNMNC Newsletter 59, European Journal oif Mineralogy 32.

Ma, C. and Beckett, J. R. (2016a) Burnettite, $CaVAlSiO_6$, and paqueite, $Ca_3TiSi_2(Al_2Ti)O_{14}$, two new minerals from Allende: Clues to the evolution of a V-rich Ca-Al-rich inclusion. *Lunar and Planetary Science* 47, Abstract#1595.

Ma, C. and Beckett, J. R. (2016b) Majindeite, $Mg_2Mo_3O_8$, a new mineral from the Allende meteorite and a witness to post-crystallization oxidation of a Ca-Al-rich refractory inclusion. *American Mineralogist* 101, 1161–1170.

Ma, C. and Beckett, J. R (2018) Nuwaite (Ni_6GeS_2) and butianite (Ni_6SnS_2), two new minerals from the Allende meteorite: Alteration products in the early solar system. *American Mineralogist* 103, 1918–1924.

Ma, C. and Beckett, J. R. (2020) Kaitianite, $Ti^{3+}{}_2Ti^{4+}O_5$, a new titanium oxide mineral from Allende. *Meteoritics & Planetary Science*, early view. DOI: https://doi.org/10.1111/maps.13576.

Ma, C. and Krot, A. N. (2014) Hutcheonite, $Ca_3Ti_2(SiAl_2)O_{12}$, a new garnet mineral from the Allende meteorite: An alteration phase in a Ca-Al-rich inclusion. *American Mineralogist* 99, 667–670.

Ma, C. and Krot, A. N. (2018) Adrianite, $Ca_{12}(Al_4Mg_3Si_7)O_{32}Cl_6$, a new Cl-rich silicate mineral from the Allende meteorite: An alteration phase in a Ca-Al-rich inclusion. *American Mineralogist* 103, 1329–1334.

Ma, C. and Liu, Y. (2019a) Discovery of a zinc-rich mineral on the surface of lunar orange pyroclastic beads. *American Mineralogist* 104, 447–452.

Ma, C. and Liu, Y. (2019b) Nanomineralogy of lunar orange beads: Discovery of a Zinc-rich mineral (probably gordaite), derived from volcanic vapor condensates on the Moon. *Lunar and Planetary Science* 50, Abstract #1463.

Ma, C. and Prakapenka, V. (2018) Tschaunerite, IMA 2017-032a. CNMNC Newsletter No. 46, December 2018, page 1378; *Mineralogical Magazine* 82, 1369–1379.

Ma, C. and Rossman, G. R. (2006) Low voltage FESEM of geological materials. *Microscopy Today* 14 (1), 20–22.

Ma, C. and Rossman, G. R. (2008) Discovery of tazheranite (cubic zirconia) in the Allende Meteorite. *Geochimica et Cosmochimica Acta* 72, A577.

Ma, C. and Rossman, G. R. (2009a) Tistarite, Ti_2O_3, a new refractory mineral from the Allende meteorite. *American Mineralogist* 94, 841–844.

Ma, C. and Rossman, G. R. (2009b) Davisite, $CaScAlSiO_6$, a new pyroxene from the Allende meteorite. *American Mineralogist* 94, 845–848.

Ma, C. and Rossman, G. R. (2009c) Grossmanite, $CaTi^{3+}AlSiO_6$, a new pyroxene from the Allende meteorite. *American Mineralogist* 94, 1491–1494.

Ma, C. and Rubin, A. E. (2019) Edscottite, Fe_5C_2, a new iron carbide mineral from the Ni-rich Wedderburn IAB iron meteorite. *American Mineralogist* 104, 1351–1355.

Ma, C. and Tschauner, O. (2016) Discovery of tetragonal almandine, $(Fe,Mg,Ca,Na)_3(Al,Si,Mg)_2Si3O_{12}$, a new high-pressure mineral in Shergotty. *Meteoritics & Planetary Science* 51, Abstract#6124.

Ma, C. and Tschauner, O. (2017) Zagamiite, IMA 2015-022a. CNMNC Newsletter No. 36, April 2017, page 409. *Mineralogical Magazine* 81, 403–409.

Ma, C. and Tschauner, O. (2018a) Feiite, IMA 2017-041a. CNMNC Newsletter No. 46, December 2018, page 1378; *Mineralogical Magazine* 82, 1369–1379.

Ma, C. and Tschauner, O. (2018b) Liuite, IMA 2017-042a. CNMNC Newsletter No. 46, December 2018, page 1378; *Mineralogical Magazine* 82, 1369–1379.

Ma, C., Goreva, J. S., and Rossman, G. R. (2002) Fibrous nanoinclusions in massive rose quartz: HRTEM and AEM investigations. *American Mineralogist* 87, 269–276.

Ma, C., Simon, S. B., Rossman, G. R., and Grossman, L. (2009) Calcium Tschermak's pyroxene, $CaAlAlSiO_6$, from the Allende and Murray meteorites: EBSD and micro-Raman characterizations. *American Mineralogist* 94, 1483–1486.

Ma, C., Beckett, J. R., and Rossman, G. R. (2011a) Murchisite, Cr_5S_6, a new mineral from the Murchison meteorite. *American Mineralogist* 96, 1905–1908.

Ma, C., Beckett, J. R., Tschauner, O., and Rossman, G. R. (2011b) Thortveitite ($Sc_2Si_2O_7$), the first solar silicate? *Meteoritics & Planetary Science* 46, A144.

Ma, C., Connolly, H. C., Beckett, J. R., Tschauner, O., Rossman, G. R., Kampf, A. R., Zega, T. J., Smith, S. A. S. and Schrader, D. L. (2011c) Brearleyite, $Ca_{12}Al_{14}O_{32}Cl_2$, a new alteration mineral from the NWA 1934 meteorite. *American Mineralogist* 96, 1199–1206.

Ma, C., Kampf, A. R., Connolly, H. C., Beckett, J. R., Rossman, G. R., Smith, S. A. S. and Schrader, D. L. (2011d) Krotite, $CaAl_2O_4$, a new refractory mineral from the NWA 1934 meteorite. *American Mineralogist* 96, 709–715.

Ma, C., Beckett, J. R., and Rossman, G. R. (2012a) Buseckite, (Fe,Zn,Mn)S, a new mineral from the Zakłodzie meteorite. *American Mineralogist* 97, 1226–1233.

Ma, C., Beckett, J. R., and Rossman, G. R. (2012b) Browneite, MnS, a new sphalerite-group mineral from the Zakłodzie meteorite. *American Mineralogist* 97, 2056–2059.

Ma, C., Tshauner, O., Beckett, J. R., Rossman, G. R., and Liu, W. (2012c) Panguite, $(Ti^4+,Sc,Al,Mg,Zr,Ca)_{1.8}O_3$, a new ultra-refractory titania mineral from the Allende meteorite: Synchrotron micro-diffraction and EBSD. *American Mineralogist* 97, 1219–1225.

Ma, C., Beckett, J. R., Connolly, H. C., and Rossman, G. R. (2013a) Discovery of meteoritic loveringite, $Ca(Ti,Fe,Cr,Mg)_{21}O_{38}$, in an Allende chondrule: Late-stage crystallization in a melt droplet. *Lunar and Planetary Science* 44, Abstract #1443.

Ma, C., Krot, A. N., and Bizzarro, M. (2013b) Discovery of dmisteinbergite (hexagonal $CaAl_2Si_2O_8$) in the Allende meteorite: A new member of refractory silicates formed in the solar nebula. *American Mineralogist* 98, 1368–1371.

Ma, C., Tschauner, O., Beckett, J. R., Rossman, G. R., and Liu, W. (2013c) Kangite, $(Sc,Ti,Al,Zr,Mg,Ca,\square)_2O_3$, a new ultra-refractory scandia mineral from the Allende meteorite: Synchrotron micro-Laue diffraction and electron backscatter diffraction. *American Mineralogist* 98, 870–878.

Ma, C., Beckett, J. R., and Rossman, G. R. (2014a) Allendeite ($S_{c4}Zr_3O_{12}$) and hexamolybdenum (Mo,Ru,Fe), two new minerals from an ultrarefractory inclusion from the Allende meteorite. *American Mineralogist* 99, 654–666.

Ma, C., Beckett, J. R., and Rossman, G. R. (2014b) Monipite, MoNiP, a new phosphide mineral in a Ca-Al-rich inclusion from the Allende meteorite. *American Mineralogist* 99, 198–205.

Ma C., Krot A. N., Beckett, J. R., Nagashima, K., and Tschauner O. (2015a) Discovery of warkite, $Ca_2Sc_6Al_6O_{20}$, a new Sc-rich ultra-refractory mineral in Muchison and Vigarano. *Meteoritics & Planetary Science* 50, Abstract #5025.

Ma, C., Tschauner, O., Beckett, J. R., Liu, Y., Rossman, G. R., Zhuravlev, K., Prakapenka, V., Dera, P., and Taylor, L. A. (2015b) Tissintite, (Ca, Na,□)$AlSi_2O_6$, a highly-defective, shock-induced, high-pressure clinopyroxene in the Tissint martian meteorite. *Earth and Planetary Science Letters* 422, 194–205.

Ma, C., Paque, J. M., and Tschauner, O. (2016a) Discovery of beckettite, $Ca_2V_6Al_6O_{20}$, a new alteration mineral in a V-Rich Ca-Al-rich inclusion from Allende. *Lunar and Planetery Science* 47, Abstract #1704.

Ma, C., Tschauner, O., Beckett, J. R., Liu, Y., Rossman, G. R., Sinogeikin, S. V., Smith, J. S., and Taylor, L. A. (2016b) Ahrensite, γ-Fe_2SiO_4, a new shock-metamorphic mineral from the Tissint meteorite: Implications for the Tissint shock event on Mars. *Geochimica et Cosmochimica Acta* 184, 240–256.

Ma, C., Krot, A. N., and Nagashima, K. (2017a) Addibischoffite, $Ca_2Al_6Al_6O_{20}$, a new calcium aluminate mineral from the Acfer 214 CH carbonaceous chondrite: A new refractory phase from the solar nebula. *American Mineralogist* 102, 1556–1560.

Ma, C., Lin, C., Bindi, L., and Steinhardt, P. J. (2017b) Hollisterite (Al_3Fe), kryachkoite $(Al, Cu)_6(Fe,Cu)$, and stolperite (AlCu): Three new minerals from the Khatyrka CV3 carbonaceous chondrite *American Mineralogist* 102, 690–693.

Ma, C., Tschauner, O., and Beckett, J. R. (2017c) A new high-pressure calcium aluminosilicate ($CaAl_2Si_{3.5}O_{11}$) in martian meteorites: Another after-life for plagioclase and connections to the CAS phase. *Lunar and Planetary Science*, 48, Abstract #1128.

Ma, C., Yoshizaki, T., Krot, A. N., Beckett, J. R., Nakamura, T., Nagashima, K., Muto, J., and Ivanova, M. A. (2017d) Discovery of rubinite, $Ca_3Ti^{3+}{}_2Si_3O_{12}$, a new garnet mineral in refractory inclusions from carbonaceous chondrites. *Meteoritics & Planetary Science* 52 (S1), Abstract #6023.

Ma, C., Garvie, L. A. J., and Wittmann, A. (2018a) Carletonmooreite, IMA 2018-068. CNMNC Newsletter No. 45, October 2018, page 1042; *European Journal of Mineralogy* 30, 1037–1043.

Ma C., Tschauner O., Beckett, J. R., Rossman G. R., Prescher, C., Prakapenke V. B., Bechtel H. A., and McDowell, A. (2018b) Liebermannite, $KAlSi_3O_8$, a new shock-metamorphic, high-pressure mineral from the Zagami martian meteorite. *Meteoritics & Planetary Science* 53, 50–61.

Ma, C., Tschauner, O., and Beckett, J. R. (2019a) A closer look at Martian meteorites: Discovery of the new mineral zagamiite, $CaAl_2Si_3.5O_{11}$, a shock-metamorphic, high-pressure, calcium aluminosilicate. Ninth International Conference on Mars, Abstract #6138.

Ma, C., Tschauner, O., and Beckett, J. R. (2019b) Discovery of a new high-pressure silicate phase, (Fe,Mg,Cr,Ti,Ca,□)2(Si,Al)O4 with a tetragonal spinelloid structure, in a shock melt pocket from the Tissint Martian meteorite. *Lunar and Planetary Science* 50, Abstract #1460.

Ma, C., Tschauner, O., Beckett, J. R., Liu, Y., Greenberg, E., and Prakapenka, V. B. (2019c) Chenmingite, $FeCr_2O_4$ in the $CaFe_2O_4$-type structure, a shock-induced, high-pressure mineral in the Tissint Martian meteorite. *American Mineralogist* 104, 1521–1525.

Ma, C., Tschauner, O., Bindi, L., Beckett, J. R., and Xie, X. (2019d) A vacancy-rich, partially inverted spinelloid silicate, $(Mg,Fe,Si)_2(Si,□)O_4$, as a major matrix phase in shock melt veins of the Tenham and Suizhou L6 chondrites. *Meteoritics & Planetary Science*, 54, 1907–1918.

Ma, C., Krot, A. N., Beckett, J. R., Nagashima, K., Tschauner, O., Rossman, G. R., Simon, S. B., and Bischoff, A. (2020a) Warkite, $Ca_2Sc_6Al_6O_{20}$, a new mineral in carbonaceous chondrites and a key-stone phase in ultrarefractory inclusions from the solar nebula. *Geochimica et Cosmochimica Acta* 277, 52–86.

Ma, C., Tschauner, O., Kong, M., Beckett, J. R., Greenberg, E., Prakapenka, V. B., and Lee, Y. (2020b) Discovery of a highly-defective, shock-induced, high-pressure albitic jadeite, (Na,

Ca,□$_{1/4}$)(Al,Si)Si_2O_6: Natural occurrence of a clinopyroxene with excess Si. *Lunar and Planetary Science* 51, Abstract #1712.

Ma, C., Beckett, J. R., and Prakapenka, V. (2021a) Discovery of new high-pressure mineral tschaunerite, $(Fe^{2+})(Fe^{2+}Ti^{4+})O_4$, a shock-induced, post-spinel phase in the Martian meteorite Shergotty. *Lunar and Planetary Science* 52, Abstract #1720.

Ma, C., Tschauner, O. Beckett, J. R., and Prakapenka, V. (2021b) Discovery of feiite $(Fe^{2+}{}_2(Fe^{2+}Ti^{4+})O_5)$ and liuite ($GdFeO_3$-type $FeTiO_3$), two new shock-induced, high-pressure minerals in the Martian meteorite Shergotty. *Lunar and Planetary Science* 52, Abstract #1681.

Ma, L., Williams, D. B., and Goldstein, J. I. (1998) Determination of the Fe-rich portion of the Fe-Ni-S phase diagram. *Journal of Phase Equilibria* 19, 299–309.

MacPherson, G. J. and Delaney, J. S. (1985) A fassaite-two olivine-pleonaste-bearing refractory inclusion from Karoonda (abstract). *Lunar and Planetary Science* 16, 515–516.

MacPherson, G. J. and Krot, A. N. (2002) Distribution of Ca-, Fe-rich silicates in CV3 chondrites: Possible controls by parent-body compaction (abstract). *Meteoritics & Planetary Science* 37, A91.

MacPherson, G. J. and Krot, A. N. (2014) The formation of Ca-, Fe-rich silicates in reduced and oxidized CV chondrites: The roles of impact-modified porosity and permeability and heterogeneous distribution of water ices. *Meteoritics & Planetary Science* 49, 1250–1270.

MacPherson, G. J., Wark, D. A., and Armstrong, J. T. (1988) Primitive material surviving in chondrites: Refractory inclusions. In *Meteorites and the Early Solar System*, eds. Kerridge J. F. and Matthews M. S. Tucson: University of Arizona, pp. 746–807.

MacPherson, G. J., Huss, G. R., and Davis, A. M. (2003) Extinct ^{10}Be in Type A calcium-aluminum-rich inclusions from CV chondrites. *Geochimica et Cosmochimica Acta*, 67, 3165–3179.

MacPherson, G. J., Lin, C., Hollister, L. S., Bindi, L., Andronicos, C. L., and Steinhardt, P. J. (2016) The Khatyrka meteorite: A summary of evidence for a natural origin of its remarkable Cu-Al alloys. *Lunar and Planetary Science* 47, Abstract #2655.

Maier, W. D., Andreoli, M. A. G., McDonald, I., Higgins, M. D., Boyce, A. J., Shukolyukov, A., Lugmair, G. W., Ashwal, L. D., Gräser, P., Ripley, E. M., and Hart, R. J. (2006) Discovery of a 25-cm asteroid clast in the giant Morokweng impact crater, South Africa. *Nature* 441, 203–206.

Mangan, T. P., Salzmann, C. G., Plane, J. M. C., and Murray, B. J. (2017) CO_2 ice structure and density under Martian atmospheric conditions. *Icarus* 294, 201–208.

Manser, C. J., Gänsicke, B. T., Eggl, S., Hollands, M., Izquierdo, P., Koester, D., Landstreet, J. D., Lyra, W., Marsh, T. R., Meru, F., Mustill, A. J., Rodríguez-Gil, P., Toloza, O., Veras, D., Wilson, D. J., Burleigh, M. R., Davies, M. B., Farihi, J., Gentile Fusillo, N., de Martino, D., Parsons, S. G., Quirrenbach, A., Raddi, R., Reffert, S., Del Santo, M., Schreiber, M. R., Silvotti, R., Toonen, S., Villaver, E., Wyatt, M., Xu, S., and Portegies Zwart, S. (2019) A planetesimal orbiting within the debris disc around a white dwarf star. *Science* 364, 66–69.

Maksimova, A. A., Petrova, E. V., Chukin, A. V., Karabanalov, M. S., Felner, I., Gritsevich, M., and Oshtrakh, M. I. (2020) Characterization of the matrix and fusion crust of the recent meteorite fall Ozerki L6. *Meteoritics & Planetary Science* 55, 231–244.

Marchi, S., Delbo, M., Morbidelli, A., Paolicchi, P., and Lazzarin, M. (2009) Heating of near-Earth objects and meteoroids due to close approaches to the Sun. *Monthly Notices of the Royal Astronomical Society* 400, 147–153.

Marvin, U. B. (1962) Cristobalite in the Carbo iron meteorite. *Nature* 196, 634–636.

Marvin, U. B. (1983) The discovery and initial characterization of Allan Hills 81005: The first lunar meteorite. *Geophysical Research Letters*, 10, 775–778.

Marvin, U. B. (2006) Meteorites in history: An overview from the Renaissance to the 20th century. In *The History of Meteoritics and Key Meteorite Collections: Fireballs, Falls and Finds*, eds. McCall, G. J. H., Bowden, A. J., and Howarth, R. J. Geological Society,

London, Special Publication 256. Bath, UK: The Geological Society Publishing House, pp. 15–71.
Marvin, U. B. and Klein, C. (1964) Meteoritic zircon. *Science* 146, 919–920.
Mason, B. (1962) *Meteorites*. New York: Wiley, 274 pp.
Mason, B. (1963) The pallasites. *American Museum Novitates* 2163, 19 pp.
Mason, B. (1982) In *Antarctic Meteorite Newsletter* 5, No. 4, November 1982.
Mason, B. and Jarosewich, E. (1967) The Winona meteorite. *Geochimica et Cosmochimica Acta* 31, 1097–1099.
Mason, B. and Wiik, H. B. (1966) The composition of the Bath, Frankfort, Kakangari, Rose City and Tadjera meteorites. *American Museum Novitates* 2272, 1–24.
Massalski, T. B., Park, F. R., and Vassamillet, L. F. (1966) Speculations about plessite. *Geochimica et Cosmochimica Acta* 30, 649–662.
Matsumoto, M., Tsuchiyama, A., Nakato, A., Matsuno, J., Miyake, A., Kataoka, A., Ito, M., Tomioka, N., Kodama, Y., Uesugi, K., Takeuchi, A., Nakano, T., and Vaccaro, E. (2019) Ultra porous lithology, a fossil asteroidal ice, in carbonaceous chondrite Acfer 094: Implications for parent body formation by icy dust agglomeration. 82nd Annual Meeting of the Meteoritical Society, Abstract #6089.
Matsumoto, M., Tsuchiyama, A., Miyake, A., Ito, M., Tomioka, N., Kodama, Y., Uesugi, K., Takeuchi, A., Nakano, T., Matsuno, J., Nakato, A., and Vaccaro, E. (2020) Three dimensional microstructure and mineralogy of a cosmic symplectite in the Acfer 094 carbonaceous chondrite. *Lunar and Planetary Science*, 51, Abstract #1035.
Mayne, R. G., McSween, H. Y., McCoy, T. J., and Gale, A. (2009) Petrology of the unbrecciated eucrites. *Geochimica et Cosmochimica Acta* 73, 794–819.
McCall, G. J. H. (1973) *Meteorites and Their Origins*. Newton Abbot, UK: David & Charles, 352 pp.
McCanta, M. C., Treiman, A. H., Dyar, M. D., Alexander, C. M. O'D., Rumble III, D. and Essene, E. J. (2008) The LaPaz Icefield 04840 meteorite: Mineralogy, metamorphism, and origin of an amphibole- and biotite-bearing R chondrite. *Geochimica et Cosmochimica Acta* 72, 5757–5780.
McCord, T. B., Adams, J. B., and Johnson, T. V. (1970) Asteroid Vesta: Spectral reflectivity and compositional implications. *Science* 168, 1445–1447.
McCoy, T. J., Steele, I. M., Keil, K., Leonard, B. F., and Endress, M. (1994) Chladniite, $Na_2CaMg_7(PO_4)_6$: A new mineral from the Carlton (IIICD) iron meteorite. *American Mineralogist* 79, 375–380.
McCoy, T. J., Keil, K., Clayton, R. N., Mayeda, T. K., Bogard, D. D., Garrison, D. H., Huss, G. R., Hutcheon, I. D., and Wieler, R. (1996) A petrologic, chemical, and isotopic study of Monument Draw and comparison with other acapulcoites: Evidence for formation by incipient partial melting. *Geochimica et Cosmochimica Acta* 60, 2681–2708.
McCoy, T. J., Keil, K., Clayton, R. N., Mayeda, T. K., Bogard, D. D., Garrison, D. H., and Wieler, R. (1997) A petrologic and isotopic study of lodranites: Evidence for early formation as partial melt residues from heterogeneous precursors. *Geochimica et Cosmochimica Acta* 61, 623–637.
McCoy, T. J., Burbine, T. H., McFadden, L. A., Starr, R. D., Gaffey, M. J., Nittler, L. R., Evans, L. G., Izenberg, N., Lucey, P. G., Trombka, J. I., Bell, J. F., Clark, B. E., Clark, P. E., Squyres, S. W., Chapman, C. R., Boynton, W. V., and Veverka, J. (2001) The composition of 433 Eros: A mineralogical-chemical synthesis. *Meteoritics & Planetary Science* 36, 1661–1672.
McCoy, T. J., Corrigan, C. M., Nagashima, K., Reynolds, V. S., Ash, R. D., McDonough, W. F., Yang, J., Goldstein, J. I., and HIlton, C. D. (2019) The Milton pallasite and South Byron Trio irons: Evidence for oxidation and core crystallization. *Geochimica et Cosmochimica Acta* 259, 358–370.
McCubbin, F. M., Tosca, N. J., Smirnov, A., Nekvasil, H., Steele A., Fries M., and Lindsley, D. H. (2009) Hydrothermal jarosite and hematite in a pyroxene-hosted melt inclusion in

martian meteorite Miller Range (MIL) 03346: Implications for magmatic-hydrothermal fluids on Mars. *Geochimica et Cosmochimica Acta* 73, 4907–4917.
McKay, D. S., Gibson, E. K., Thomas-Keprta, K. L., Vali, H., Romanek, C. S., Clemett, S. J., Chillier, X. D. F., Maechling, C. R., and Zare, R. N. (1996) Search for life on Mars: Possible relic biogenic activity in Martian meteorite ALH84001. *Science* 273, 924–930.
McKinley, S. G., Scott, E. R. D., Taylor, G. J., and Keil, K. (1981) A unique type 3 ordinary chondrite containing graphite-magnetite aggregates – Allan Hills A77011. *Proceedings of the Lunar and Planetary Science Conference* 12B, 1039–1048.
McSween, H. Y. (1976) A new type of chondritic meteorite found in lunar soil. *Earth and Planetary Science Letters* 31, 193–199.
McSween, H. Y. (1977) Carbonaceous chondrites of the Ornans type: A metamorphic sequence. *Geochimica et Cosmochimica Acta* 41, 477–491.
McSween, H. Y. (1994) What we have learned about Mars from SNC meteorites. *Meteoritics* 29, 757–779.
McSween, H. Y. (2015) Petrology on Mars. *American Mineralogist* 100, 2380–2395.
McSween, H. Y. and Huss, G. R. (2010) *Cosmochemistry*. Cambridge: Cambridge University Press, 549 pp.
McSween, H. Y. and McLennan, S. M. (2014) Mars. In *Treatise on Geochemistry, Vol. 2*, 2nd ed., eds. Holland, H. D. and Turekian, K. K. Oxford: Elsevier, pp. 251–300.
McSween, H. Y., Taylor, G. J., and Wyatt, M. B. (2009) Elemental composition of the martian crust. *Science* 324, 736–739.
McSween, H. Y., Binzel, R. P., De Sanctis, M. C., Ammanito, E., Prettyman, T. H., Beck, A. W., Reddy, V., Le Corre, L., Gaffey, M. J., McCord, T. B., Raymond, C. A., Russell, C. T., and the Dawn Science Team (2013) Dawn; the Vesta–HED connection; and the geologic context for eucrites, diogenites, and howardites. *Meteoritics & Planetary Science* 48, 2090–2104.
McSween, H. Y., Raymond, C. A., Stolper, E. M., Mittlefehldt, D. W., Baker, M. B., Lunning, N. G., Beck, A. W., and Hahn, T. M. (2019) Differentiation and magmatic history of Vesta: Constraints from HED meteorites and Dawn spacecraft data. *Chemie der Erde – Geochemistry* 79, 125526.
Meier, M. M. M., Bindi, L., Heck, P. R., Neander, A. I., Spring, N. H., Riebe, M. E. I., Maden, C., Baur, H., Steinhardt, P. J., Wieler, R., and Busemann, H. (2018) Cosmic history and a candidate parent asteroid for the quasicrystal-bearing meteorite Khatyrka. *Earth and Planetary Science Letters* 490, 122–131.
Melosh, H. J. (1989) *Impact Cratering: A Geologic Process*. Oxford, UK: Oxford University Press, 245 pp.
Mikouchi, T., Zolensky, M., Ivanova, M., Tachikawa, O., Komatsu, M., Le, L., and Gounelle, M. (2009) Dmitryivanovite: A new high-pressure calcium aluminum oxide from the Northwest Africa 470 CH3 chondrite characterized using electron backscatter diffraction analysis. *American Mineralogist* 94, 746–750.
Mikouchi, T., Hagiya, K., Sawa, N., Kimura, M., Ohsumi, K., Komatsu, M., and Zolensky, M. (2016) Synchrotron radiation XRD analysis of indialite in Yamato-82094 ungrouped carbonaceous chondrite. *Lunar and Planetery Science* 47, Abstract #1919.
Mittlefehldt, D. W. (2015) Asteroid (4) Vesta: I. The howardite-eucrite-diogenite (HED) clan of meteorites. *Chemie der Erde – Geochemistry* 75, 155–183.
Mittlefehldt, D. W. and Lindstrom, M. M. (2001) Petrology and geochemistry of Patuxent Range 91501, a clast-poor impact melt from the L-chondrite parent body and Lewis Cliff 88663, an L7 chondrite. *Meteoritics & Planetary Science* 36, 439–457.
Mittlefehldt, D. W., Rubin, A. E., and Davis, A. M. (1992) Mesosiderite clasts with the most extreme positive europium anomalies among solar system rocks. *Science* 257, 1096–1099.
Mittlefehldt, D. W., Lindstrom, M. M., Wang, M.-S., and Lipschutz, M. E. (1995) Geochemistry and origin of achondritic inclusions in Yamato-75097, -793241 and -

794046 chondrites. *Proceeding of the NIPR Symposium on Antarctic Meteorites* 8, 251–271.

Mittlefehldt, D. W., Lindstrom, M. M., Bogard, D. D., Garrison, D. H., and Field, S. W. (1996) Acapulco- and Lodran-like achondrites: Petrology, geochemistry, chronology and origin. *Geochimica et Cosmochimica Acta* 60, 867–882.

Mittlefehldt, D. W., McCoy, T. J., Goodrich, C. A., and Kracher, A. (1998) Non-chondritic meteorites from asteroidal bodies. In *Planetary Materials, Reviews in Mineralogy & Geochemistry, Vol. 36*, ed. Papike, J. J. Washington, DC: Mineralogical Society of America, pp. 4-1 – 4-195.

Mittlefehldt, D. W., Bogard, D. D., Berkley, J. L., and Garrison, D. H. (2003) Brachinites: Igneous rocks from a differentiated asteroid. *Meteoritics & Planetary Science* 38, 1601–1625.

Miyahara, M., Ohtani, E., El Goresy, A. E., Lin, Y., Feng, L., Zhang, J.-C., Gillet, P., Nagase, T., Muto, J., and Nishijima, M. (2015) Unique large diamonds in a ureilite from Almahata Sitta 2008 TC3 asteroid. *Geochimica et Cosmochimica Acta* 163, 14–26.

Miyake, G. T. and Goldstein, J. I. (1974) The Tucson meteorite. *Geochimica et Cosmochimica Acta* 38, 1201–1212.

Morota, T., Sugita, S., Cho, Y., Kanamaru, M., Tatsumi, E., Sakatani, N., Honda, R., Hirata, N., Kikuchi, H., Yamada, M., Yokota, Y., Kameda, S., Matsuoka, M., Sawada, H., Honda, C., Kouyama, T., Ogawa, K., Suzuki, H., Yoshioka, K., Hayakawa, M., Hirata, N., Hirabayashi, M., Miyamoto, H., Michikami, T., Hiroi, T., Hemmi, R., Barnouin, O. S., Ernst, C. M., Kitazato, K., Nakamura, T., Riu, L., Senshu, H., Kobayashi, H., Sasaki, S., Komatsu, G., Tanabe, N., Fujii, Y., Irie, T., Suemitsu, M., Takaki, N., Sugimoto, C., Yumoto, K., Ishida, M., Kato, H., Moroi, K., Domingue, D., Michel, P., Pilorget, C., Iwata, T., Abe, M., Ohtake, M., Nakauchi, Y., Tsumura, K., Yabuta, H., Ishihara, Y., Noguchi, R., Matsumoto, K., Miura, A., Namiki, N., Tachibana, S., Arakawa, M., Ikeda, H., Wada, K., Mizuno, T., Hirose, C., Hosoda, S., Mori, O., Shimada, T., Soldini, S., Tsukizaki, R., Yano, H., Ozaki, M., Takeuchi, H., Yamamoto, Y., Okada, T., Shimaki, Y., Shirai, K., Iijima, Y., Noda, H., Kikuchi, S., Yamaguchi, T., Ogawa, N., Ono, G., Mimasu, Y., Yoshikawa, K., Takahashi, T., Takei, Y., Fujii, A., Nakazawa, S., Terui, F., Tanaka, S., Yoshikawa, M., Saiki, T., Watanabe, S., Tsuda, Y. (2020) Sample collection from asteroid (162173) Ryugu by Hayabusa2: Implications for surface evolution. *Science* 368, 654–659.

Morris, M. A., Garvie, L. A. J., and Knauth, L. P. (2015) New insight into the solar system's transition disk phase provided by the metal-rich carbonaceous chondrite Isheyevo. *Astrophysical Journal Letters* 801, L22, 5 pp.

Morrison, S. M. and Hazen, R. M. (2020) An evolutionary system of mineralogy, Part II: Interstellar and solar nebula primary condensation mineralogy (>4.565 Ga). *American Mineralogist*, 105, 1508–1535. https://doi.org/10.2138/am-2020-7447.

Munayco P., Munayco, J., Varela, M. E., and Scorzelli, R. B. (2013) The new Peruvian meteorite Carancas: Mössbauer spectroscopy and X-ray diffraction studies. *Earth, Moon, Planets* 110, 1–9.

Muxworthy, A. R., Bland, P. A., Davison, T. M., Moore, J., Collins, G. S., and Ciesla, F. J. (2017) Evidence for an impact-induced magnetic fabric in Allende, and exogenous alternatives to the core dynamo theory for Allende magnetization. *Meteoritics & Planetary Science* 52, 2132–2146.

Nabiei, F., Badro, J., Dennenwaldt, T., Oveisi, E., Cantoni, M., Hébert, C., El Goresy, A., Barrat, J.-A., and Gillet, P. (2018) A large planetary body inferred from diamond inclusions in a ureilite meteorite. *Nature Communications* 9, 1–6. https://doi.org/ 10.1038/s41467-018-03808-6.

Nakamura, T. (2005) Post-hydration thermal metamorphism of carbonaceous chondrites. *Journal of Mineralogical and Petrological Sciences* 100, 260–272.

Nakamura, T., Noguchi, T., Tanaka, M., Zolensky, M. E., Kimura, M., Tsuchiyama, A., Nakato, A., Ogami, T., Ishida, H., Uesugi, M., Yada, T., Shirai, K., Fujimura, A., Okazaki, R., Sandford, S. A., Ishibashi, Y., Abe, M., Okada, T., Ueno, M., Mukai, T., Yoshikawa, M., and Kawaguchi, J. (2011) Itokawa dust particles: A direct link between S-Type asteroids and ordinary chondrites. *Science* 333, 1113–1116.

Nakamura, A., Miyahara, M., Suga, H., Yamaguchi, A., Daisuke, W., Yamashita, S., Takeichi, Y., and Ohtani E. (2020) The discovery of Mn-precipitates in nakhlite Yamato 000802. Eleventh Symposium on Polar Science, abstract.

Nakamura-Messenger, K., Keller, L. P., Clemett, S. J., Messenger, S., Jones, J. H., Palma, R. L., Pepin, R. O., Klöck, W., Zolensky, M. E., and Tatsuoka H. (2010) Brownleeite: A new manganese silicide mineral in an interplanetary dust particle. *American Mineralogist* 95, 221–228.

Nakamura-Messenger, K., Clemett, S. J., Rubin, A. E., Choi, B.-G., Zhang, S., Rahman, Z., Oikawa, K., and Keller, L.P. (2012) Wassonite: A new titanium monosulfide mineral in the Yamato 691 enstatite chondrite. *American Mineralogist* 97, 807–815.

Nakamuta Y. and Aoki Y. (2000) Mineralogical evidence for the origin of diamond in ureilites. *Meteoritics & Planetary Science* 35, 487–493.

Nakashima, D., Nakamura, T., and Okazaki, R. (2006) Cosmic-ray exposure age and heliocentric distance of the parent bodies of enstatite chondrites ALH 85119 and MAC 88136. *Meteoritics & Planetary Science* 41, 851–862.

Nelson, V. E. and Rubin, A. E. (2002) Size-frequency distributions of chondrules and chondrule fragments in LL3 chondrites: Implications for parent-body fragmentation of chondrules. *Meteoritics & Planetary Science* 37, 1361–1376.

Németh, P., Garvie, L. A. J., Aoki, T., Dubrovinskaia, N., Dubrovinsky, L., and Buseck, P. R. (2014) Lonsdaleite is faulted and twinned cubic diamond and does not exist as a discrete mineral. *Nature Communications* 5, 5447. https://doi.org/10.1038/ncomms6447.

Nesse, W. D. (2012) *Introduction to Mineralogy*, 4th ed. New York: Oxford University Press, 384 pp.

Neumann, W., Henke, S., Breuer, D., Gail, H.-P., Schwarz, W. H., Trieloff, M., Hopp, J., and Spohn, T. (2018) Modeling the evolution of the parent body of acapulcoites and lodranites: A case study for partially differentiated asteroids. *Icarus* 311, 146–169.

Newton, J., Bischoff, A., Arden, J. W., Franchi, I. A., Geiger, T., Greshake, A., and Pillinger, C. T. (1995) Acfer 094, a uniquely primitive carbonaceous chondrite from the Sahara. *Meteoritics* 30, 47–56.

Nickel, E. H. and Graham, J. (1987) Paraotwayite: A new nickel hydroxide mineral from Western Australia. *Canadian Meteorologist* 25, 409–411.

Nielsen, H. P. and Buchwald, V. F. (1981) Roaldite, a new nitride in iron meteorites. *Proceedings Lunar and Planetary Science Conference* 12, 1343–1348.

Nishiizumi, K., Caffee, M. W., Hamajima, Y., Reedy, R. C., and Welton, K. C. (2014) Exposure history of the Sutter's Mill carbonaceous chondrite. *Meteoritics & Planetary Science* 49, 2056–2063.

Nittler, L. R. and Ciesla, F. (2016) Astrophysics with extraterrestrial materials. *Annual Review of Astronomy and Astrophysics* 54, 53–93.

Nittler, L. R., Stroud, R. M., Trigo-Rodriguez, J. M., De Gregorio, B. T., Alexander, C. M. O'D., Davidson, J., Moyano-Cambero, C. E., and Tanbakouei, S. (2019) A cometary building block in a primitive asteroidal meteorite. *Nature Astronomy* 3, 659–666. https://doi.org/10.1038/s41550-019-0737-8.

Nittler, L. R., Alexander, C. M. O'D., Foustoukos, D., Patzer, A., and Verdier-Paoletti, M. J. (2020) Asuka 12236, the most pristine CM chondrite to date. *Lunar and Planetary Science* 51, Abstract #2276.

Norton, O. R. and Chitwood, L. A. (2008) *Field Guide to Meteors and Meteorites*. London: Springer, 287 pp.

Nyström, J. O. and Wickman, F. E. (1991) The Ordovician chondrite from Brunflo, central Sweden, II. Secondary minerals. *Lithos* 27, 167–185.

Nyström, J. O., Lindström, M., and Wickman, F. E. (1988) Discovery of a second Ordovician meteorite using chromite as a tracer. *Nature* 336, 572–574.

Ohtsuka, K., Arakida, H., Ito, T., Kasuga, T., Watanabe, J., Kinoshita, D., Sekiguchi, T., Asher, D. J., and Nakano, S. (2007) Apollo asteroids 1566 Icarus and 2007 MK_6: Icarus family members? *Astrophysical Journal* 668, L71–L74.

Okada, A. and Keil K. (1982) Caswellsilverite, $NaCrS_2$: A new mineral in the Norton County enstatite achondrite. *American Mineralogist* 67, 132–136

Okada, A., Keil, K., and Taylor, G. J. (1981) Unusual weathering products of oldhamite parentage in the Norton County enstatite achondrite. *Meteoritics* 16, 141–152.

Okada, A., Keil, K., Leonard, B. F., and Hutcheon, I. D. (1985) Schollhornite, $Na_{0.3}(H_2O)_1[CrS_2]$, a new mineral in the Norton County enstatite achondrite. *American Mineralogist* 70, 638–643.

Okada, A., Keil, K., Taylor, G. J., and Newsom, H. (1988) Igneous history of the aubrite parent asteroid: Evidence from the Norton County enstatite achondrite. *Meteoritics* 23, 59–74.

Okada, A., Kobayashi, K., Ito, T., and Sakurai, T. (1991) Structure of synthetic perryite $(Ni, Fe)_8(Si,P)_3$. *Acta Crystallographica* C47, 1358–1361.

Olsen, E. (1979) Meteor-wrongs. *Field Museum of Natural History Bulletin* 50, No. 4, 18–21.

Olsen, E. (1981) Vugs in ordinary chondrites. *Meteoritics* 16, 45–59.

Olsen, E. and Fredriksson, K. (1966) Phosphates in iron and pallasite meteorites. *Geochimica et Cosmochimica Acta* 30, 459–470.

Olsen, E. and Fuchs, L. H. (1968) Krinovite, $NaMg_2CrSi_3O_{10}$: A new meteorite mineral. *Science* 161, 786–787.

Olsen, E. and Jarosewich, E. (1971) Chondrules: First occuurrennce in an iron meteorite. *Science* 174, 583–585.

Olsen, E. J. and Steele, I. M. (1997) Galileiite: A new meteoritic phosphate mineral. *Meteoritics & Planetary Science* 32, A155–A156.

Olsen, E., Huebner, J. S., Douglas, J. A. V., and Plant, A. G. (1973) Meteoritic amphiboles. *American Mineralogist* 58, 869–872.

Olsen, E., Erlichman, J., Bunch, T. E., and Moore, P. B. (1977) Buchwaldite, a new meteoritic phosphate mineral. *American Mineralogist* 62, 362–364.

Olsen, E. J., Kracher, A., Davis, A. M., Steele, I. M., Hutcheon, I. D., and Bunch, T. E. (1999) The phosphates of IIIAB iron meteorites. *Meteoritics & Planetary Science* 34, 285–300.

Ott U. (1996) Interstellar grains: Some facts, implications, and ideas (abstract). *Meteoritics & Planetary Science* 31, A102–A103.

Oulton, J., Humayun, M., Fedkin, A., and Grossman, L. (2016) Chemical evidence for differentiation, evaportation and recondensation from silicate clasts in Gujba. *Geochimica et Cosmochimica Acta* 177, 254–274.

Paar, W. H., Ma, C., Topa, D., Culetto, F. J., Hammer V. F. M., Guan, Y., Braithwaite, R. S. W. (2019) Discovery of native aluminum on Variscan metagranitoids in Upper Carinthia, Austria: Natural or anthropogenic origin? *Rendiconti Lincei. Scienze Fisiche e Naturali* 30, 167–184.

Palme, H. and Jones, A. (2005) Solar system abundances of the elements. In *Meteorites, Comets and Planets*, ed. Davis, A. Amsterdam: Elsevier, pp. 41–61.

Pang, R.-L., Zhang, A.-C., Wang, S.-Z., Wang, R.-C., and Yurimoto, H. (2016) High-pressure minerals in eucrite suggest a small source crater on Vesta. *Scientific Reports* 6, 26063. https://doi.org/10.1038/srep26063.

Pang, R.-L., Harries, D., Pollok, K., Zhang, A.-C., and Langenhorst, F. (2018) Vestaite, $(Ti^{4+}Fe^{2+})Ti^{4+}{}_3O_9$, a new mineral in the shocked eucrite Northwest Africa 8003. *American Mineralogist* 103, 1502–1511.

Pang, R. L., Du, W., Zhang, A. C., Liu, J., and Qin, L. (2020) Unique achondrite Dhofar 778: A mantle-derived fragment from a new differentiated body? *Lunar and Planetary Science* 51, Abstract #1947.

Papike, J., Taylor, L., and Simon, S. (1991) Lunar minerals. In *Lunar Sourcebook: A User's Guide to the Moon*, eds. Heiken, G. H., Vaniman, D. T., and French, B. M. Cambridge: Cambridge University Press, pp. 121–181.

Papike, J. J., Ryder, G., and Shearer, C. K. (1998) *Lunar samples*. In *Planetary Materials*, ed. Papike J. J., Washington, DC: Mineralogical Society of America, pp. 5-1–5-234.

Paque J. M., Beckett, J. R., Barber, D. J., and Stolper, E. M. (1994) A new titanium-bearing calcium aluminosilicate phase: I. Meteoritic occurrences and formation in synthetic systems. *Meteoritics* 29, 673–682.

Patzer, A., Hill, D. H., and Boynton, W. V. (2001) Itqiy: A metal-rich enstatite meteorite with achondritic texture. *Meteoritics & Planetary Science* 36, 1495–1505.

Pederson, T. P. (1999) Schwertmannite and awaruite as alteration products in iron meteorites. *Meteoritics* 62, 5117.

Pekov, I. (1998) *Minerals First Discovered on the Territory of the Former Soviet Union*. Moscow: Ocean Pictures Ltd., 369 pp.

Pekov, I. V., Perchiazzi, N., Merlino, S., Kalachev, V. N., Merlini, M., and Zadov, A. E. (2007) Chukanovite, $Fe_2(CO_3)(OH)_2$, a new mineral from the weathered iron meteorite Dronino. *European Journal of Mineralogy* 19, 891–898.

Pepin, R. O. (1985) Evidence of martian origins. *Nature*, 317, 473–475.

Peplowski, P. N., Lawrence, D. J., Prettyman, T. H., Yamashita, N., Bazell, D., Feldman, W. C, Le Corre, L., McCoy, T. J., Reddy, V., Reedy, R. C., Russell, C. T., and Toplis, M. J. (2013) Compositional variability on the surface of 4 Vesta revealed through GRaND measurements of high-energy gamma rays. *Meteoritics & Planetary Science* 48, 2252–2270.

Peplowski, P. N., Klima, R. L., Lawrence, D. J., Ernst, C. M., Denevi, B. W., Frank, E. A., Goldsten, J. O., Murchi, S. L., Nittler, L. R. and Solomon, S. C. (2016) Remote sensing evidence for an ancient carbon-bearing crust on Mercury. *Nature Geoscience* 9, 273–276.

Petaev, M. I. and Wood, J. A. (1998) The condensation with partial isolation (CWPI) model of condensation in the solar nebula. *Meteoritics & Planetary Science* 33, 1123–1137.

Petaev, M. I., Clarke, R. S., Olsen, E. J., Jarosewich, E., Davis, A. M., Steele, I. M., Lipschutz, M. E., Wang, M.-S., Clayton, R. N., Mayeda, T. K., and Wood, J. A. (1993) Chaunskij: The most highly metamorphosed, shock-modified and metal-rich mesosiderite abstract. *Lunar and Planetary Science Conference* 24, 1131–1132.

Petranek, S. L. (2015) *How We'll Live on Mars*. New York: TEDBooks, Simon & Schuster, 96 pp.

Phillips, W. R. and Griffen, D. T. (1981) *Optical Mineralogy: The Nonopaque Minerals*. San Francisco: Freeman, 677 pp.

Pickersgill, A. E., Flemming, R. L., and Osinski, G. R. (2015) Toward quantification of strain-related mosaicity in shocked lunar and terrestrial plagioclase by in situ micro-X-ray diffraction. *Meteoritics & Planetary Science* 50, 1851–1862.

Pieters, C. M. (1978) Mare basalt types on the front side of the Moon: A summary of spectral reflectance data. *Lunar and Planetary Science Conference Proceedings* 9, 2825–2849.

Plummer, C., Carlson, D., and Hammersley, L. (2019) *Physical Geology*, 16th ed., New York: McGraw-Hill, 672 pp.

Poch, O., Istiqomah, I., Quirico, E., Beck, P., Schmitt, B., Theulé, P., Faure, A., Hily-Blant, P., Bonal, L., Raponi, A., Ciarniello, M., Rousseau, B., Potin, S., Brissaud, O., Flandinet, L., Filacchione, G., Pommerol, A., Thomas, N., Kappel, D., Mennella, V., Moroz, L.,

Vinogradoff, V., Arnold, G., Erard, S., Bockelée-Morvan, D., Leyrat, D., Capaccioni, F., De Sanctis, M. C., Longobardo, A., Mancarella, F., Palomba, E., and Tosi, F. (2020) Ammonium salts are a reservoir of nitrogen on a cometary nucleus and possibly on some asteroids. *Science* 367, eaaw7462. https://doi.org/ 10.1126/science.aaw7462.

Pokorný, P., Sarantos, M., and Janches, D. (2017) Reconciling the dawn-dusk asymmetry in Mercury's exosphere with the micrometeoroid impact directionality. *The Astrophysical Journal Letters* 842, L17.

Pratesi, G., Bindi, L., and Moggi-Cecchi V. (2006) Icosahedral coordination of phosphorus in the crystal structure of melliniite, a new phosphide mineral from the Northwest Africa 1054 acapulcoite. *American Mineralogist* 94, 451–454.

Price G. D., Putnis, A., and Agrell, S. O. (1979) Electron petrography of shock-produced veins in the Tenham chondrite. *Contributions to Mineralogy and Petrology* 71, 211–218.

Prinz, M., Nehru, C. E., Weisberg, M. K., and Delaney, J. S. (1984) Type 3 enstatite chondrites: A newly recognized group of unequilibrated enstatite chondrites (UEC's). *Lunar and Planetary Science* 15, Abstract #1331, 653–654.

Prior, G. T. (1920) The classification of meteorites. *Mineralogical Magazine* 19, 51–63.

Prior, G. T. (1953) *Catalogue of Meteorites*. British Museum, London, 432 pp.

Przylibski, T. A., Zagożdżon, P. P., Kryza, R., and Pilski, A. S. (2005) The Zakłodzie enstatite meteorite: Mineralogy, petrology, origin, and classification. *Meteoritics & Planetary Science* 40, A185–A200.

Quirico, E., Bonal, L., Alexander, C. M. O'D., Yabuta, H., Nakamura, T., Nakato, A., Flandinet, L., Montagnac, G., Schmitt-Kopplin, P., and Herd, C. D. K. (2018) Prevalence and nature of heating processes in CM and C2-ungrouped chondrites as revealed by insoluble organic matter. *Geochimica et Cosmochimica Acta* 241, 17–37.

Ramdohr, P. (1963) Opaque minerals in stony meteorites. *Journal of Geophysical Research* 68, 2011–2036.

Ramdohr, P. (1967) Die Schmelzkruste der Meteoriten. *Earth and Planetary Science Letters* 2, 197–209.

Ramdohr, P. (1969) *The Ore Minerals and Their Intergrowths*. Oxford: Pergamon Press, 1,174 pp.

Ramdohr, P. (1973) *The Opaque Minerals in Stony Meteorites*. Amsterdam: Elsevier, 245 pp.

Rampe, E. B., Blake, D. F., Bristow, T. F., Ming, D. W., Vaniman, D. T., Morris, R. V., Achilles, C. N., Chipera, S. J., Morrison, S. M., Tu, V. M., Yen, A. S., Castle, N., Downs, G. W., Downs, R. T., Grotzinger, J. P., Hazen, R. M., Treiman, A. H., Peretyazhko, T. S., Des Marais, D. J., Walroth, R. C., Craig, P. I., Crisp, J. A., Lafuente, B., Morookian, J. M., Sarrazin, P. C., Thorpe, M. T., Bridges, J. C., Edgar, L. A., Fedo, C. M., Freissinet, C., Gellert, R., Mahaffy, P. R., Newsom, H. E., Johnson, J. R., Kah, L. C., Siebach, K. L., Schieber, J., Sun, V. Z., Vasavada, A. R., Webster, C., Wellington, D., and Wiens, R. C. (2020) Mineralogy and geochemistry of sedimentary rocks and eolian sediments in Gale crater, Mars: A review after six earth years of exploration with Curiosity. *Chemie der Erde – Geochemistry* 80, 125605. https://doi.org/10.1016/j.chemer.2020.125605.

Rancourt, D. G. and Scorzelli, R. B. (1995) Low-spin γ-Fe-Ni (γLS) proposed as a new mineral in Fe-Ni-bearing meteorites: epitaxial intergrowth of γLS and tetrataenite as a possible equilibrium state at$\sim$ 20–40 at% Ni. *Journal of Magnetism and Magnetic Materials* 150, 30–36.

Reisener, R. J. and Goldstein, J. I. (2003) Ordinary chondrite metallography: Part 2. Formation of zoned and unzoned metal particles in relatively unshocked H, L, and LL chondrites. *Meteoritics & Planetary Science* 38, 1679–1696.

Reuter, K. B., Williams, D. B., and Goldstein, J. I. (1987) Determination of the Fe-Ni phase diagram below 400°C. *Metallurgical Transactions A* 20A, 719–725.

Reuter, K. B., Williams, D. B., and Goldstein, J. I. (1988) Low temperature transformations in the metallic phases of iron and stony-iron meteorites. *Geochimica et Cosmochimica Acta* 52, 617–626.

Rickwood, P. C. (1981) The largest crystals. *American Mineralogist* 66, 885–907.
Rivkin, A. S. and Emery, J. P. (2010) Detection of ice and organics on an asteroidal surface. *Nature* 464, 1322–1323.
Rubin, A. E. (1983a) Impact melt-rock clasts in the Hvittis enstatite chondrite breccia: Implications for a genetic relationship between EL chondrites and aubrites. *Proceedings of the Fourteenth Lunar and Planetary Science Conference,* B293–B300.
Rubin, A. E. (1983b) The Adhi Kot breccia and implications for the origin of chondrules and silica-rich clasts in enstatite chondrites. *Earth and Planetary Science Letters* 64, 201–212.
Rubin, A. E. (1984) The Blithfield meteorite and the origin of sulfide-rich, metal-poor clasts and inclusions in brecciated enstatite chondrites. *Earth and Planetary Science Letters* 67, 273–283.
Rubin, A. E. (1985) Impact melt products of chondritic material. *Reviews of Geophysics* 23, 277–300.
Rubin, A. E. (1990) Kamacite and olivine in ordinary chondrites: Intergroup and intragroup relationships. *Geochimica et Cosmochimica Acta* 54, 1217–1232.
Rubin, A. E. (1991) Euhedral awaruite in the Allende meteorite: Implications for the origin of awaruite- and magnetite-bearing nodules in CV3 chondrites. *American Mineralogist* 76, 1356–1362.
Rubin, A. E. (1992) A shock-metamorphic model for silicate darkening and compositionally variable plagioclase in CK and ordinary chondrites. *Geochimica et Cosmochimica Acta* 56, 1705–1714.
Rubin A. E. (1994a) Euhedral tetrataenite in the Jelica meteorite. *Mineralogical Magazine* 58, 215–221.
Rubin, A. E. (1994b) Metallic copper in ordinary chondrites. *Meteoritics* 29, 93–98.
Rubin, A. E. (1995) Fractionation of refractory siderophile elements in metal from the Rose City meteorite. *Meteoritics* 30, 412–417.
Rubin, A. E. (1997a) Mineralogy of meteorite groups. *Meteoritics & Planetary Science* 32, 231–247.
Rubin, A. E. (1997b) Mineralogy of meteorite groups: An update. *Meteoritics & Planetary Science* 32, 733–734.
Rubin, A. E. (1997c) The Hadley Rille enstatite chondrite and its agglutinate-like rim: Impact melting during accretion to the Moon. *Meteoritics & Planetary Science* 32, 135–141.
Rubin, A. E. (1999) Formation of large metal nodules in ordinary chondrites. *Journal of Geophysical Research – Planets* 104, 30799–30804.
Rubin, A. E. (2000) Petrologic, geochemical and experimental constraints on models of chondrule formation. *Earth Science Reviews* 50, 3–27.
Rubin, A. E. (2002a) Post-shock annealing of MIL99301 (LL6): Implications for impact heating of ordinary chondrites. *Geochimica et Cosmochimica Acta* 66, 3327–3337.
Rubin, A. E. (2002b) The Smyer H-chondrite impact-melt breccia and evidence for sulfur vaporization. *Geochimica et Cosmochimica Acta* 66, 683–695.
Rubin, A. E. (2003) Chromite-plagioclase assemblages as a new shock indicator; implications for the shock and thermal histories of ordinary chondrites. *Geochimica et Cosmochimica Acta* 67, 2695–2709.
Rubin, A. E. (2004) Aluminian low-Ca pyroxene in a Ca-Al-rich chondrule from the Semarkona meteorite. *American Mineralogist* 89, 867–872.
Rubin, A. E. (2006) Shock, post-shock annealing and post-annealing shock in ureilites. *Meteoritics & Planetary Science* 41, 125–133.
Rubin, A. E. (2007) Petrogenesis of acapulcoites and lodranites: A shock-melting model. *Geochimica et Cosmochimica Acta* 71, 2383–2401.
Rubin, A. E. (2008) Explicating the behavior of Mn-bearing phases during shock melting and crystallization of the Abee EH-chondrite impact-melt breccia. *Meteoritics & Planetary Science* 43, 1481–1485.

Rubin, A. E. (2010a) Physical properties of chondrules in different chondrite groups: Implications for multiple melting events in dusty environments. *Geochimica et Cosmochimica Acta* 74, 4807–4828.
Rubin, A. E. (2010b) Impact melting in the Cumberland Falls and Mayo Belwa aubrites. *Meteoritics & Planetary Science* 45, 265–275.
Rubin, A. E. (2011) Origin of the differences in refractory-lithophile-element abundances among chondrite groups. *Icarus* 213, 547–558.
Rubin, A. E. (2012a) A new model for the origin of Type-B and Fluffy Type-A CAIs: Analogies to remelted compound chondrules. *Meteoritics & Planetary Science* 47, 1062–1074.
Rubin A. E. (2012b) Collisional facilitation of aqueous alteration of CM and CV carbonaceous chondrites. *Geochimica et Cosmochimica Acta* 90, 181–194.
Rubin, A. E. (2013a) Multiple melting in a four-layered barred-olivine chondrule with compositionally heterogeneous glass from LL3.0 Semarkona. *Meteoritics & Planetary Science* 48, 445–456.
Rubin, A. E. (2013b) An amoeboid olivine inclusion (AOI) in CK3 NWA 1559, comparison to AOIs in CV3 Allende, and the origin of AOIs in CK and CV chondrites. *Meteoritics & Planetary Science* 48, 432–444.
Rubin, A. E. (2014) Shock and annealing in the amphibole- and mica-bearing R chondrites. *Meteoritics & Planetary Science* 49, 1057–1075.
Rubin, A. E. (2015a) Impact features of enstatite-rich meteorites. *Chemie der Erde – Geochemistry* 75, 1–28.
Rubin, A. E. (2015b) Maskelynite in asteroidal, lunar and planetary basaltic meteorites: An indicator of shock pressure during impact ejection from their parent bodies. *Icarus* 257, 221–229.
Rubin, A. E. (2015c) An American on Paris: Extent of aqueous alteration of a CM chondrite and the petrography of its refractory and amoeboid olivine inclusions. *Meteoritics & Planetary Science* 50, 1595–1612.
Rubin, A. E. (2015d) Shock and annealing in aubrites: Implications for parent-body history. *Meteoritics & Planetary Science* 50, 1217–1227.
Rubin, A. E. (2016) Impact melting of the largest known enstatite meteorite: Al Haggounia 001, a fossil EL chondrite. *Meteoritics & Planetary Science* 51, 1576–1587.
Rubin, A. E. (2018) Carbonaceous and non-carbonaceous iron meteorites: Differences in chemical, physical and collective properties. *Meteoritics & Planetary Science* 53, 2357–2371.
Rubin, A. E. and Grossman, J. N. (2010) Meteorite and meteoroid: New comprehensive definitions. *Meteoritics & Planetary Science* 45, 114–122.
Rubin, A. E. and Huber, H. (2005) A weathering index for CK and R chondrites. *Meteoritics & Planetary Science* 40, 1123–1130.
Rubin, A. E. and Jerde, E. A. (1987) Diverse eucritic pebbles in the Vaca Muerta mesosiderite. *Earth and Planetary Science Letters* 84, 1–14.
Rubin, A. E. and Jerde, E. A. (1988) Compositional differences between basaltic and gabbroic clasts in mesosiderites. *Earth and Planetary Science Letters* 87, 485–490.
Rubin, A. E. and Kallemeyn, G. W. (1990) Lewis Cliff 85332: A unique carbonaceous chondrite. *Meteoritics* 25, 215–225.
Rubin, A. E. and Kallemeyn, G. W. (1993) Carlisle Lakes chondrites: Relationship to other chondrite groups (abstract). *Meteoritics* 28, 424–425.
Rubin, A. E. and Keil, K. (1983) Mineralogy and petrology of the Abee enstatite chondrite breccia and its dark inclusions. *Earth & Planetary Science Letters* 62, 118–131.
Rubin, A. E. and Li, Y. (2019) Formation and destruction of magnetite in CO3 chondrites and other chondrite groups. *Chemie der Erde – Geochemistry* 79, 125528.
Rubin, A. E. and Ma, C. (2017) Meteoritic minerals and their origins. *Chemie der Erde – Geochemistry* 77, 325–385.

Rubin, A. E. and Mittlefehldt, D. W. (1992) Classification of mafic clasts from mesosiderites: Implications for endogenous igneous processes. *Geochimica et Cosmochimica Acta* 56, 827–840.

Rubin, A. E. and Mittlefehldt, D. W. (1993) Evolutionary history of the mesosiderite asteroid: A chronologic and petrologic synthesis. *Icarus* 101, 232–252.

Rubin, A. E. and Moore, W. B. (2011) What's up? Preservation of gravitational direction in the LAR 06299 LL impact-melt breccia. *Meteoritics & Planetary Science* 46, 737–747.

Rubin, A. E. and Scott, E. R. D. (1997) Abee and related EH chondrite impact-melt breccias. *Geochimica et Cosmochimica Act* 61, 425–435.

Rubin, A. E. and Wasson, J. T. (2011) Shock effects in "EH6" enstatite chondrites and implications for collisional heating of the EH and EL parent asteroids. *Geochimica et Cosmochimica Acta* 75, 3757–3780.

Rubin, A. E., Scott, E. R. D., and Keil, K. (1982) Microchondrule-bearing clast in the Pinacaldoli LL3 meteorite: A new kind of type 3 chondrites and its relevance to the history of chondrules. *Geochimica et Cosmochimica Acta* 46, 1763–1776.

Rubin, A. E., Scott, E. R. D., Taylor, G. J., Keil, K., Allen, J. S. B., Mayeda, T. K., Clayton, R. N., and Bogard, D. D. (1983) Nature of the H chondrite parent body regolith: Evidence from the Dimmitt breccia. *Proceedings of the Thirteenth Lunar and Planetary Science Conference*, A741–A754.

Rubin, A. E., Jerde, E., Zong, P., Wasson, J. T., Westcott, J. W., Mayeda, T. K., and Clayton, R. N. (1986) Properties of the Guin ungrouped iron meteorite: The origin of Guin and of group-IIE irons. *Earth &Planetary Science Letters* 76, 209–226.

Rubin, A. E., Scott, E. R. D., and Keil, K. (1997) Shock metamorphism of enstatite chondrites. *Geochimica et Cosmochimica Acta* 61, 847–858.

Rubin, A. E., Sailer, A. L., and Wasson, J. T. (1999) Troilite in the chondrules of type-3 ordinary chondrites: Implications for chondrule formation. *Geochimica et Cosmochimica Acta* 63, 2281–2298.

Rubin, A. E., Ulff-Møller, F., Wasson, J. T., and Carlson, W. D. (2001) The Portales Valley meteorite breccia: Evidence for impact-induced melting and metamorphism of an ordinary chondrite. *Geochimica et Cosmochimica Acta* 65, 323–342.

Rubin, A. E., Kallemeyn, G. W., Wasson, J. T., Clayton, R. N., Mayeda, T. K., Grady, M., Verchovsky, A. B., Eugster, O., and Lorenzetti, S. (2003) Formation of metal and silicate nodules in Gujba: A new Bencubbin-like meteorite fall. *Geochimica et Cosmochimica Acta* 67, 3283–3298.

Rubin, A. E., Trigo-Rodríguez, J. M., Kunihiro, T., Kallemeyn, G. W., and Wasson, J. T. (2005) Carbon-rich chondritic clast PV1 from the Plainview H-chondrite regolith breccia: Formation from H3 chondrite material by possible cometary impact. *Geochimica et Cosmochimica Acta* 69, 3419–3430.

Rubin, A. E., Trigo-Rodríguez, J. M., Huber, H., and Wasson, J. T. (2007) Progressive aqueous alteration of CM carbonaceous chondrites. *Geochimica et Cosmochimica Acta* 71, 2361–2382.

Rubin, A. E., Griset, C. D., Choi, B.-G., and Wasson, J. T. (2009) Clastic matrix in EH3 chondrites. *Meteoritics & Planetary Science* 44, 589–601.

Rubin, A. E., Breen, J. P., Wasson, J. T., and Pitt, D. (2015) Shock effects in the Willamette iron meteorite. *Meteoritics & Planetary Science* 50, 1984–1994.

Rubin, A. E., Breen, J. P., Isa, J., and Tutorow, S. (2017) NWA 10214 – An LL3 chondrite breccia with an assortment of metamorphosed, shocked, and unique chondrite clasts. *Meteoritics & Planetary Science* 52, 372–390.

Russell, S. S., Pillinger, C. T., Arden, J. W., Lee, M. R., and Ott, U. (1992) A new type of meteoritic diamond in the enstatite chondrite Abee. *Science* 256, 206–209.

Russell, S. S., Connolly, H. C., and Krot, A. N., eds. (2018) *Chondrules: Records of Protoplanetary Disk Processes*. Cambridge, UK: Cambridge University Press, 450 pp.

Ruzicka, A. (2014) Silicate-bearing iron meteorites and their implications for the evolution of asteroidal parent bodies. *Chemie der Erde – Geochemistry* 74, 3–48.

Ruzicka, A., Killgore, M., Mittlefehldt, D. W., and Fries, M. D. (2005) Portales Valley: Petrology of a metallic-melt meteorite breccia. *Meteoritics & Planetary Science* 40, 261–295.

Ruzicka, A., Hugo, R., and Hutson, M. (2015) Deformation and thermal histories of ordinary chondrites: Evidence for post-deformation annealing and syn-metamorphic shock. *Geochimica et Cosmochimica Acta* 163, 219–233.

Saal, A. E., Hauri, E. H., Cascio, M. L., Van Orman, J. A., Rutherford, M. C., and Cooper, R. F. (2008) Volatile content of lunar volcanic glasses and the presence of water in the Moon's interior. *Nature* 454, 192–195.

Saini-Eidukat, B., Kucha, H., and Keppler, H. (1994) Hibbingite, γ-$Fe_2(OH)_3Cl$, a new mineral from the Duluth Complex, Minnesota, with implications for the oxidation of Fe-bearing compounds and transport of metals. *American Mineralogist* 79, 555–561.

Sakamoto, N., Seto, Y., Itoh, S., Kuramoto, K., Fujino, K., Nagashima, K., Krot, A. N., and Yurimoto, H. (2007) Remnants of the early solar system water enriched in heavy oxygen isotopes. *Science* 317, 231–233.

Sanborn, M. E., Wimpenny, J., Williams, C. D., Yamakawa, A., Amelin, Y., Irving, A. J., and Yin, Q.-Z. (2019) Carbonaceous achondrites Northwest Africa 6704/6693: Milestones for early Solar System chronology and genealogy. *Geochimica et Cosmochimica Acta* 245, 577–596.

Sanders, I. S., Scott, E. R. D., and Delaney, J. S. (2017) Origin of mass-independent oxygen isotope variation among ureilites: Clues from chondrites and primitive achondrites. *Meteoritics & Planetary Science* 52, 690–708.

Sasaki, S., Nakamura, K., Hamabe, Y., Kurahashi, E., and Hiroi, T. (2001) Production of iron nanoparticles by laser irradiation in a simulation of lunar-like space weathering. *Nature* 410, 555–557.

Satterwhite, C., Mason, B., and MacPherson, G. J. (1993) Description of LEW88774. *Antarctic Meteorite Newsletter* 16 (1), 15.

Scambos, T. A., Campbell, G. G., Pope, A., Haran, T., Muto, A., Lazzara, M., Reijmer, C. H., and van den Broeke, M. R. (2018) Ultralow surface temperatures in East Antarctica from satellite thermal infrared mapping: The coldest places on Earth. *Geophysical Research Letters* 45, 6124–6133.

Schmieder, M., Buchner, E., and Kröchert, J. (2009) 'Ballen silica' in impactites and magmatic rocks. *Lunar and Planetary Science Conference* 40, Abstract #1020.

Schmitt, H. H. (2006) *Return to the Moon: Exploration, Enterprise, and Energy in the Human Settlement of Space*. New York: Copernicus Books, 335 pp.

Schmitz, B. (2013) Extraterrestrial spinels and the astronomical perspective on Earth's geological record and evolution of life. *Chemie der Erde – Geochemistry* 73, 117–145.

Schmitz, B., Tassinari, M., and Peucker-Ehrenbrink, B. (2001) A rain of ordinary chondritic meteorites in the early Ordovician. *Earth and Planetary Science Letters* 194, 1–15.

Schmitz, B., Yin, Q.-Z., Sanborn, M.E., Tassinari, M., Caplan, C. E., and Huss, G. R. (2016) A new type of solar-system material recovered from Ordovician marine limestone. *Nature Communications* 7, 1–7.

Schrader, D. L., Lauretta, D. S., Connolly, H. C., Goreva, Y. S., Hill, D. H., Domanik, K. J., Berger, E. L., Yang, H., and Downs, R. T. (2010) Sulfide-rich impact melts from chondritic parent bodies. *Meteoritics & Planetary Science* 45, 743–758.

Schrader, D. L., Franchi, I. A., Connolly, H. C., Greenwood, R. C., Lauretta, D. S., and Gibson, J. M. (2011) The formation and alteration of the Renazzo-like carbonaceous chondrites I: Implications of bulk-oxygen isotopic composition. *Geochimica et Cosmohimica Acta* 75, 308–325.

Schrader, D. L., McCoy, T. J., and Gardner-Vandy, K. (2017) Relict chondrules in primitive achondrites: Remnants from their precursor parent bodies. *Geochimica et Cosmochimica Acta* 205, 295–312.

Schultz, P. H. (1988) Cratering on Mercury: A relook. In *Mercury*, eds. Vilas, F., Chapman, C. R., and Matthews, M. S. Tucson: University of Arizona Press, pp. 274–335.

Schulze, H., Bischoff, A., Palme, H., Spettel, B., Dreibus, G., and Otto, J. (1994) Mineralogy and chemistry of Rumuruti: The first meteorite fall of the new R chondrite group. *Meteoritics* 29, 275–286.

Schwinger, S., Dohmen, R., and Schertle, H.-P. (2016) A combined diffusion and thermal modeling approach to determine peak temperatures of thermal metamorphism experienced by meteorites. *Geochimica et Cosmochimia Acta* 191, 255–276.

Scott, E. R. D. (2007) Chondrites and the protoplanetary disk. *Annual Review of Earth and Planetary Science* 35, 577–620.

Scott, E. R. D. (2020) Iron meteorites: Composition, age, and origin. In *Oxford Research Encyclopedia of Planetary Science*. Oxford: Oxford University Press. DOI: 10.1093/acrefore/9780190647926.013.206

Scott, E. R. D. and Agrell, S. O. (1971) The occurrence of carbides in iron meteorites (abstract). *Meteoritics* 6, 312–313.

Scott, E. R. D. and Jones, R. H. (1990) Disentangling nebular and asteroidal features of CO3 carbonaceous chondrites. *Geochimica et Cosmochimica Acta* 54, 2485–2502.

Scott, E. R. D. and Krot, A. N. (2005) Chondrites and their components. In *Meteorites, Comets, and Planets*, ed. A. M. Davis. Amsterdam: Elsevier, pp. 143–200.

Scott, E. R. D. and Krot, A. N. (2014) Chondrites and their components. In *Meteorites, Comets, and Planets*, 2nd ed., ed. Davis, A. M. Amsterdam: Elsevier, pp. 65–137.

Scott, E. R. D. and Wasson, J. T. (1975) Classification and properties of iron meteorites. *Reviews of Geophysics and Space Physics* 13, 527–546.

Scott, E. R. D., Wasson, J. T., and Buchwald, V. F. (1973) The chemical classification of iron meteorites, 7, A reinvestigation of irons with Ge concentrations between 25 and 80 ppm. *Geochimica et Cosmochimica Acta* 37, 1957–1983.

Scott, E. R. D., Taylor, G. J., Rubin, A. E., Okada, A., and Keil, K. (1981a) Graphite-magnetite aggregates in ordinary chondritic meteorites. *Nature* 291, 544–546.

Scott, E. R. D., Rubin, A. E., Taylor, G. J., and Keil, K. (1981b) New kind of type 3 chondrite with a graphite-magnetite matrix. *Earth & Planetary Science Letters* 56, 19–31.

Scott, E. R. D., Rubin, A. E., Taylor, G. J., and Keil, K. (1984) Matrix material in type 3 chondrites – occurrence, heterogeneity and relationship with chondrules. *Geochimica et Cosmochimica Acta* 48, 1741–1757.

Scott, E. R. D., Brearley, A. J., Keil, K., Grady, M. M., Pillinger, C. T., Calyton, R. N., Mayeda, T. K., Wieler, R., and Signer, P. (1988a) Nature and origin of C-rich ordinary chondrites and chondritic clasts. *Proceedings of the Lunar and Planetary Science Conference* 18, 513–523.

Scott, E. R. D., Barber, D. J., Alexander, C. M. O'D., Hutchison, R., and Peck, J. A. (1988b) Primitive material surviving in chondrites: Matrix. In *Meteorites and the Early Solar System*, eds. Kerridge, J. F. and Matthews, M. S. Tucson: University of Arizona Press, pp. 718–745

Scott, E. R. D., Keil, K., and Stöffler, D. (1992) Shock metamorphism of carbonaceous chondrites. *Geochimica et Cosmochimica Acta* 56, 4281–4293.

Scott, E. R. D., Krot, A. N., and Sanders, I. S. (2018) Isotopic dichotomy among meteorites and its bearing on the protoplanetary disc. *Astrophysical Journal* 854, 164, 12 pp.

Sears, D. W. G. (1980) Formation of E chondrites and aubrites – a thermodynamic model. *Icarus* 43, 184–202.

Sears, D. W. G. (2004) *The Origin of Chondrules and Chondrites*. Cambridge, UK: Cambridge University Press, 209 pp.

Sears, D. W. G. (2018) *Shedding Light: The Luminescent Glow of Meteorites and Moon Rocks*. Published by the author. CreateSpace Independent Publishing Platform, 158 pp.

Sears, D. W. G., Grossman, J. N., Melcher, C. L., Ross, L. M., and Mills, A. A. (1980) Measuring metamorphic history of unequilibrated ordinary chondrites. *Nature* 287, 791–795.

Sears, D. W. G., Grossman, J. N., and Melcher, C. L. (1982) Chemical and physical studies of type 3 chondrites – I: Metamorphism related studies of Antarctic and other type 3 ordinary chondrites. *Geochimica et Cosmochimica Acta*, 46, 2471–2481.

Sears, D. W., Kallemeyn, G. W., and Wasson, J. T. (1983) Composition and origin of clasts and inclusions in the Abee enstatite chondrite breccia. *Earth and Planetary Science Letters* 62, 180–192.

Sears, D. W. G., Batchelor, J. D., Lu, J., and Keck, B. D. (1991a) Metamorphism of CO and CO-like chondrites and comparison with type 3 ordinary chondrites. *Proceedings of the NIPR Symposium on Antarctic Meteorites* 4, 319–343.

Sears, D. W. G., Hasan, E. A., Batchelor, J. D., and Lu, J. (1991b) Chemical and physical studies of type 3 chondrites – XI: Metamorphism, pairing, and brecciation of ordinary chondrites. *Proceedings of the Lunar and Planetary Science Conference* 21, 493–512.

Sears, D. W. G., Ninagawa, K., and Singhvi, A. K. (2018) *Glimmerings from the Past: The Luminescence Properties of Meteorites and Lunar Samples with an Emphasis on Applications.* Published by the author. CreateSpace Independent Publishing Platform, 194 pp.

Seto, Y., Sakamoto, N., Fujino, K., Kaito, T., Oikawa, T., Yurimoto, H. (2008) Mineralogical characterization of a unique material having heavy oxygen isotope anomaly in matrix of the primitive carbonaceous chondrite Acfer 094. *Geochimica et Cosmochimica Acta* 72, 2723–2734.

Sharygin, V. V., Ripp, G. S., Yakovlev, G. A., Seryotkin, Y. V., Karmanov, N. S., Izbrodin, I. A., Grokhovsky, V. I., and Khromova, E. A. (2020) Uakitite, VN, a new mononitride from Uakit iron meteorite (IIAB). *Minerals* 10(2), 150. https://doi.org/10.3390/min10020150.

Shearer, C. K., Burger, P. V., Papike, J. J., Sharp, Z. D., and McKeegan, K. D. (2011) Fluids on differentiated asteroids: Evidence from phosphates in differentiated meteorites GRA 06128 and GRA 06129. *Meteoritics & Planetary Science* 46, 1345–1362.

Shechtman, D., Blech, I., Gratias, D., and Cahn, J. W. (1984) Metallic phase with long-range orientational order and no translational symmetry. *Physical Review Letters* 53, 1951–1953.

Sheng, Y. J., Hutcheon, I. D., and Wasserburg, G. J. (1991) Origin of plagioclase-olivine inclusions in carbonaceous chondrites. *Geochimica et Cosmochimica Acta* 55, 581–599.

Shimizu, M., Yoshida, H., and Mandarino, J. A. (2002) The new mineral species keilite, (Fe, Mg)S, the iron-dominant analogue of niningerite. *The Canadian Mineralogist* 40, 1687–1692.

Shrivastava, J. P. and Rani, N. (2012) *Introduction to Ore Microscopy*. New Delhi: PHI Learning Private Ltd., 112 pp.

Shukolyukov, Y. A., Nazarov, M. A., and Schultz, L. (2002) A new Martian meteorite: The Dhofar 019 shergottite with an exposure age of 20 million years. *Solar System Research* 36, 125–135.

Simon, S. B. and Grossman, L. (1992) Low-temperature exsolution in refractory siderophile element-rich opaque assemblages from the Leoville carbonaceous chondrite. *Earth and Planetary Science Letters* 110, 67–75.

Simon, S. B., Sutton, S. R., Brearley, A. J., Krot, A. N., and Nagashima, K. (2019) The effects of thermal metamorphism as recorded in CO3.0 through CO3.2 chondrites. *Lunar and Planetary Science* 50, Abstract #1444.

Simpson, E. S. (1938) Some new and little-known meteorites found in western Australia. *Mineralogical Magazine* 25, 157–171, 154 plates.

Smith, J. V. (1974) Lunar mineralogy: A heavenly detective story. Presidential Address. Part I. *American Mineralogist* 59, 231–243.

Smith, J. V. and Steele, I. M. (1976) Lunar mineralogy: A heavenly detective story. Part II. *American Mineralogist* 61, 1059–1116.

Snetsinger, K. G. and Keil, K. (1969) Ilmenite in ordinary chondrites. *American Mineralogist* 54, 780–786.

Socolar, J. E. S., Steinhardt, P. J., and Levine, D. (1985) Quasicrystals with arbitary orientational symmetry. *Physical Review B* 32, 5547–5550.

Solomon, S. C., Nittler, L. R., and Anderson, B. J., eds. (2019) *Mercury: The View after MESSENGER*. Cambridge, UK: Cambridge University Press, 596 pp.

Spicuzza, M. J., Day, J. M. D., Taylor, L. A., and Valley, J. W. (2007) Oxygen isotope constraints on the origin and differentiation of the Moon. *Earth and Planetary Science Letters* 253, 254–265.

Spudis, P. D. (2016) *How to Explore, Live, and Prosper in Space Using the Moon's Resources*. Washington, DC: Smithsonian Books, 272 pp.

Squiller, S. F. and Sclar, C. B. (1980) Genesis of the Sterling Hill zinc deposit, Sussex County, New Jersey. In *International Association on the Genesis of Ore Deposits Symposium, 5th*, ed. Ridge, J. D. Vol. 1, 759–766.

Squyres, S. W. (2005) *Roving Mars: Spirit, Opportunity, and the Exploration of the Red Planet*. New York: Hyperion, 422 pp.

Srinivasan, P., Dunlap, D. R., Agee, C. B., Wadhwa, M., Coleff, D., Ziegler, K., Ziegler, R., and McCubbin, F. M. (2018) Silica-rich volcanism in the early solar system dated at 4.565 Ga. *Nature Communications* 9, 3036. https://doi.org/ 10.1038/s41467-018-05501-0.

Stagno, V., Bindi, L., Steinhardt, P. J., and Fei, Y. (2017) Phase equilibria in the nominally $Al_{65}Cu_{23}Fe_{12}$ system at 3, 5 and 21 GPa: Implications for the quasicrystal-bearing Khatyrka meteorite. *Physics of the Earth and Planetary Interiors* 271, 47–56.

Steinhardt, P. J. (2019) *The Second Kind of Impossible: The Extraordinary Quest for a New Form of Matter*. New York: Simon and Schuster, 400 pp.

Steinhardt, P. J. and Bindi, L. (2012) In search of natural quasicrystals. *Reports on Progress in Physics* 75, 092601–092611.

Stevenson, D. J. (1987) Origin of the Moon – The collision hypothesis. *Annual Review of Earth and Planetary Science* 15, 271–315.

Stinchcomb, B. L. (2011) *Meteorites*. Atglen, PA: Schiffer Publishing, 160 pp.

Stöffler, D., Keil, K., and Scott, E. R. D. (1991) Shock metamorphism of ordinary chondrites. *Geochimica et Cosmochimica Acta* 55, 3845–3867.

Stöffler, D., Hamann C., and Metzler K. (2018) Shock metamorphism of planetary silicate rocks and sediments: Proposal for an updated classification system. *Meteoritics & Planetary Science* 53, 5–49.

Suttle, M. D. and Genge, M. J. (2017) Diagenetically altered fossil micrometeorites suggest cosmic dust is common in the geological record. *Earth and Planetary Science Letters* 476, 132–142.

Suttle, M. D., Twegar, K., Nava, J., Spiess, R., Spratt, J., Campanale, F., and Folco, L. (2019) A unique CO-like micrometeorite hosting an exotic Al-Cu-Fe-bearing assemblage – close affinities with the Khatyrka meteorite. *Scientific Reports* 9, 12426.

Suttle, M. D., Greshake, A., King, A. J., Schofield, P. F., Tomkins, A., and Russell, S. S. (2020) The alteration history of the CY chondrites, investigated through analysis of a new member: Dhofar 1988. *Geochimica et Cosmochimica Acta*, doi.org/10.1016/j.gca.2020.11.008.

Takeda, H. and Graham, A. L. (1991) Degree of equilibration of eucritic pyroxenes and thermal metamorphism of the earliest planetary crust. *Meteoritics* 26, 129–134.

Tarbuck, E. J., Lutgens, F. K., Tasa, D. G., and Linneman, S. (2019) *Earth: An Introduction to Physical Geology*, 13th ed. New York: Pearson, 784 pp.

Tarduno, J. A., Cottrell, R. D., Nimmo, F., Hopkins, J., Voronov, J., Erickson, A., Blackman, E., Scott, E. R. D., and McKinley, R. (2012) Evidence for a dynamo in the main group pallasite parent body. *Science* 338, 939–942.

Taylor, G. J. (2009) Ancient lunar crust: Origin, composition, and implications. *Elements* 5, 17–22.

Taylor, G. J., Okada, A., Scott, E. R. D., Rubin, A. E., Huss, G. R., and Keil, K. (1981) The occurrence and implications of carbide-magnetite assemblages in unequilibrated ordinary chondrites (abstract). *Lunar and Planetary Science* 12, 1076–1078.

Taylor, G. J., Warren, P., Ryder, G., Delano, J., Pieters, C., and Lofgren, G. (1991) Lunar rocks. In *Lunar Sourcebook: A User's Guide to the Moon*, eds. Heiken, G. H., Vaniman, D. T., and French, B. M. Cambridge: Cambrdige University Press, pp. 183–284.

Taylor, L. A., Nazarov, M. A., Demidova, S. I., and Patchen, A. D. (2001) Dhofar 287: A new lunar mare basalt from Oman. *Meteoritics & Planetary Science* 36, 204.

Terada, K., Sano, Y., Takahata, N., Ishida, A., Tsuchiyama, A., Nakamura, T., Noguchi, T., Karouji, Y., Uesugi, M., Yada, T., Nakabayashi, M., Fukuda, K., and Nagahara, H. (2018) Thermal and impact histories of 25143 Itokawa recorded in Hayabusa particles. *Scientific Reports* 8, 1–7. Article #11806, https://doi.org/10.1038/s41598-018-30192-4.

Tomeoka, K. (1990) Mineralogy and petrology of Belgica-7904: A new kind of carbonaceous chondrite from Antarctica. *Proceedings of the NIPR Symposium on Antarctic Meteorites* 3, 40–54.

Tomeoka, K. and Buseck, P. R. (1985) Indicators of aqueous alteration in CM carbonaceous chondrites: Microtextures of a layered mieral containing Fe, S, O and Ni. *Geochimica et Cosmochimica Acta* 49, 2149–2163.

Tomeoka, K. and Buseck, P. R. (1988) Matrix mineralogy of the Orgueil CI carbonaceous chondrite. *Geochimica et Cosmochimica Acta* 52, 1627–1640.

Tomeoka, K. and Kojima, T. (1998) Arcuate band texture in a dark inclusion from the Vigarano CV3 chondrite: Possible evidence for early sedimentary processes. *Meteoritics & Planetary Science* 33, 519–525.

Tomioka, N. and Fujino K. (1999) Akimotoite, (Mg, Fe) SiO_3, a new silicate mineral of the ilmenite group in the Tenham chondrite. *American Mineralogist* 84, 267–271.

Tomioka, N. and Miyahara, M. (2017) High-pressure minerals in shocked meteorites. *Meteoritics & Planetary Science* 33, 331–227.

Tomioka, N., Miyahara, M., and Ito M. (2016) Discovery of natural $MgSiO_3$ tetragonal garnet in a shocked chondritic meteorite. *Science Advances* 2, e1501725.

Tomioka, N., Bindi, L., Okuchi, T., Miyahara, M., Iitaka, T., Li, Z., Kawatsu, T., Xie, X., Purevjav, N., Tani, R., and Kodama, Y. (2021) Poirierite, a dense metastable polymorph of magnesium iron silicate in shocked meteorites. *Communications Earth & Environment* 2, Article No. 16.

Tomkins, A. G. (2009) What metal-troilite textures can tell us about post-impact metamorphism in chondrite meteorites. *Meteoritics & Planetary Science* 44, 1133–1149.

Tonui, E., Zolensky, M., Hiroi, T., Nakamura, T., Lipschutz, M. E., Wang, M.-S., and Okudaira, K. (2014) Petrographic, chemical and spectroscopic evidence for thermal metamorphism in carbonaceous chondrites I: CI and CM chondrites. *Geochimica et Cosmochimica Acta* 126, 284–306.

Tornabene, H., Hilton, C. D., Ash, R. D., and Walker, R. J. (2019) New insights to the genetics, age, and crystallization of Group IIC iron meteorites. *Lunar and Planetary Science*, 50, Abstract #1236.

Treiman, A. H. (1985) Amphibole and hercynite spinel in Shergotty and Zagami: Magmatic water, depth of crystallization, and metasomatism. *Meteoritics* 20, 229–243.

Treiman, A. H. (2005) The nakhlite meteorites: Augite-rich igneous rocks from Mars. *Chemie der Erde – Geochemistry* 65, 203–270.

Trigo-Rodriguez, J. M., Llorca, J., Borovička, J., and Fabregat, J. (2003) Chemical abundances determined from meteor spectra: I. Ratios of the main chemical elements. *Meteoritics & Planetary Science* 38, 1283–1294.

Tsai, A.-P., Inoue, A., and Matsumoto, T. (1987) A stable quasicrystal in Al-Cu-Fe system. *Japanese Journal of Applied Physics* 26, L1505–L1507.

Tschauner, O. and Ma, C. (2017) Stöfflerite, IMA 2017-062. CNMNC Newsletter No. 39, October 2017, page 1285; *Mineralogical Magazine* 81, 1279–1286.

Tschauner, O., Ma, C., Beckett, J. R., Prescher, C., Prakapenka, V. B., and Rossman, G. R. (2014) Discovery of bridgmannite, the most abundant mineral in Earth, in a shocked meteorite. *Science* 346, 1100–1102.

Tsuda, Y., Yoshikawa, M., Saiki, T., Nakazawa, S., and Watanabe, S. (2019) Hayabusa 2 – Sample return and kinetic impact mission to near-earth asteroid Ryugu. *Acta Astronautica* 156, 387–393.

Ulff-Møller F., Choi B.-G., Rubin, A. E., Tran J., and Wasson J. T. (1998) Paucity of sulfide in a large slab of Esquel: New perspectives on pallasite formation. *Meteoritics & Planetary Science* 33, 221–227.

Ulyanov, A. A. (1991) The meteorite minerals. In 14th Brown-Vernadsky Microsymposium on Comparative Planetology, Moscow.

Utas, J. A., Rubin, A. E., and Ziegler, K. (2017) Northwest Africa 10085: An equilibrated Kakangari chondrite. *Lunar and Planetary Science*, 48, Abstract #2906.

Vacher, L. G., Truche, L., Faure, F., Tissandier, L., Mosser-Ruck, R., and Marrocchi, Y. (2019) Deciphering the conditions of tochilinite and cronstedtite formation in CM chondrites from low temperature hydrothermal experiments. *Meteoritics & Planetary Science* 54, 1870–1889.

Váci, Z., Yang, S., Humayun, M., and Agee, C. B. (2020a) Petrology and geochemistry of andesitic ungrouped achondrites Northwest Africa 6698 and 11575. *Lunar and Planetary Science* 51, Abstract #1697.

Váci, Z., Agee, C. B., Humayun, M., Ziegler, K., Asmerom, Y., Polyak, V., Busemann, H., Krietsch, D., Heizler, M., Sanborn, M. E., and Yin, Q.-Z. (2020b) Unique achondrite Northwest Africa 11042: Exploring the melting and breakup of the L chondrite parent body. *Meteoritics & Planetary Science* 55, 1–27. https://doi.org/ 1111/maps.13456.

Van Niekerk, D. and Keil, K. (2011) Metal/sulfide-silicate intergrowth textures in EL3 meteorites: Origin by impact melting on the EL parent body. *Meteoritics & Planetary Science* 46, 1484–1497.

Van Roosbroek, N., Debaille, V., Pittarello, L., Goderis, S., Humayun, M., Hecht, L., Jourdan, F., Spicuzza, M. J., Vanhaecke, F., and Claeys, P. (2015) The formation of IIE iron meteorites investigated by the chondrule-bearing Mont Dieu meteorite. *Meteoritics & Planetary Science* 50, 1173–1196.

Van Schmus, W. R. and Wood, J. A. (1967) A chemical-petrologic classification for the chondritic meteorites. *Geochimica et Cosmochimica Acta* 31, 747–765.

Vdovykin, G. P. (1969) New hexagonal modification of carbon in meteorites. *Geochemistry International* 6, 915–918.

Vdovykin, G. P. (1972) Forms of carbon in the new Havero ureilite of Finland. *Meteoritics* 7, 547–552.

Velbel, M. A. (1988) The distribution and significance of evaporitic weathering products on Antarctic meteorites. *Meteoritics* 23, 151–159.

Velbel, M. A. (2012) Aqueous alteration in Martian meteorites: Comparing mineral relations in igneous-rock weathering of Martian meteorites and in the sedimentary cycle of Mars. *Sedimentary Geology of Mars* 102, 97–117.

Vernazza, P., Zanda, B., Nakamura, T. Scott, E., and Russell, S. (2015) The formation and evolution of ordinary chondrite parent bodies. In *Asteroids IV*, eds. Michel, P., Demeo, F. E., and Bottke, W. F. Tucson: University of Arizona Press, pp. 617–634.

Vokrouhlicky, D., Bottke, W. F., and Nesvorny, D. (2016) Capture of trans-Neptunian planetesimals in the main asteroid belt. *Astronomical Journal* 152, 39.

Vondrak, R. R. (1974) Creation of an artificial lunar atmosphere. *Nature* 248, 657–659.

Wampler, J., Thiemens, M., Cheng. S., Zhu, Y., and Schuller, I. K. (2020) Superconductivity found in meteorites. *Proceedings of the National Academy of Sciences* 117, 7645–7649. https://doi.org/10.1073/pnas.1918056117.

Wang, K. (1986) Zhanghengite – a new mineral. *Acta Mineralogica Sinica* 6, 220–223.

Wark, D. A. and Lovering, J. F. (1977) Marker events in the early evolution of the solar system: Evidence from rims on Ca-Al-rich inclusions in carbonaceous chondrites. *Lunar and Planetary Science Conference Proceedings* 8, 95–112.

Wark, D. A. and Lovering, J. F. (1978) Refractory/platinum metals and other opaque phases in Allende Ca-Al-rich inclusions (CAI's) (abstract). *Lunar and Planetary Science* 9, 1214–1216.

Warren, P. H. (2011a) Stable-isotope anomalies and the accretionary assemblage of the Earth and Mars: A subordinate role for carbonaceous chondrites. *Earth and Planetary Science Letters* 331, 93–100.

Warren, P. H. (2011b) Stable isotopes and the noncarbonaceous derivation of ureilites, in common with nearly all differentiated planetary materials. *Geochimica et Cosmochimica Acta* 75, 6912–6926.

Warren, P. H. and Jerde, E. A. (1987) Composition and origin of Nuevo Laredo Trend eucrites. *Geochimica et Cosmochimica Acta* 51, 713–725.

Warren, P. H. and Kallemeyn, G. W. (1989a) Allan Hills 84025: The second Brachinite, far more differentiated than Brachina, and an ultramafic achondritic clast from L chondrite Yamato 75097. *Lunar and Planetary Science Conference Proceedings* 19, 475–486.

Warren, P. H. and Kallemeyn, G. W. (1989b) Geochemistry of polymict ureilite EET83309: And a partially disruptive impact model for ureilite origin. *Meteoritics* 24, 233–246.

Warren, P. H. and Kallemeyn G. W. (1994) Petrology of LEW88774: An extremely Cr-rich ureilite (abstract). *Lunar and Planetery Science Conference* 25, 1465–1466.

Warren, P. H. and Rubin, A. E. (2010) Pigeonite-selective impact smelting in ureilites. *Geochimica et Cosmochimica Acta* 74, 5109–5133.

Warren, P. H. and Rubin, A. E. (2020) Trace element and textural evidence favoring lunar, not terrestrial, origin of the mini-granite in Apollo sample 14321. *Icarus*, 346, Article #113779. https://doi.org/10.1016/j.icarus.2020.113771.

Warren, P. H., Taylor, G. J., Keil, K., Shirley, D. N., and Wasson, J. T. (1983) Petrology and chemistry of two "large" granite clasts from the moon. *Earth & Planetary Science Letters* 64, 175–185.

Warren, P. H., Rubin, A. E., Isa, J., Brittenham, S., Ahn, I., and Choi, B.-G. (2013) Northwest Africa 6693: A new type of FeO-rich, low-$\Delta^{17}O$, poikilitic cumulate achondrite. *Geochimica et Cosmochimica Acta* 107, 135–154.

Warren, P. H., Rubin, A. E., Isa, J., Gessler, N., Ahn, I., and Choi, B.-G. (2014) Northwest Africa 5738: Multistage fluid-driven secondary alteration in an extraordinarily evolved eucrite. *Geochimica et Cosmochimica Acta* 141, 199–227.

Wasserburg, G. J., Sanz, H. G., and Bence, A. E. (1968) Potassium-feldspar phenocrysts in the surface of Colomera, and iron meteorite. *Science* 161, 684–687.

Wasserburg, G. J., Lee, T., and Papanastassiou, D. A. (1977) Correlated O and Mg isotopic anomalies in Allende inclusions: II. Magnesium. *Geophysical Research Letters* 4, 299–302.

Wasson, J. T. (1969) The chemical classification of iron meteorites—III. Hexahedrites and other irons with germanium concentrations between 80 and 200 ppm. *Geochimica et Cosmochimica Acta* 33, 859–876.

Wasson, J. T. (1985) *Meteorites: Classification and Properties*. New York: Springer-Verlag, 316 pp.

Wasson, J. T. (1988) The building stones of the planets. In *Mercury*, eds. Vilas, F., Chapman, C. R., and Matthews, M.S. Tucson: University of Arizona Press, pp. 622–650.

Wasson, J. T. (2017) Formation of non-magmatic iron-meteorite group IIE. *Geochimica et Cosmochimica Acta* 197, 396–416.

Wasson, J. T. and Choe, W.-H. (2009) The IIG iron meteorites: Probable formation in the IIAB core. *Geochimica et Cosmochimica Acta* 73, 4879–4890.

Wasson, J. T. and Choi, B.-G. (2003) Main-group pallasites: Chemical composition, relation to IIIAB irons, and origin. *Geochimica et Cosmochimica Acta* 67, 3079–3096.

Wasson J. T. and Kallemeyn G. W. (2002) The IAB iron-meteorite complex: A group, five subgroups, numerous grouplets, closely related, mainly formed by crystal segregation in rapidly cooling melts. *Geochimica et Cosmochimica Acta* 66, 2445–2473.

Wasson, J. T. and Rubin, A. E. (2009) Composition and matrix in the CR chondrite LAP 02342. *Geochimica et Cosmochimica Acta* 73, 1436–1460.

Wasson, J. T. and Wai, C. M. (1970) Composition of the metal, schreibersite and perryite of enstatite achondrites and the origin of enstatite chondrites and achondrites. *Geochimica et Cosmochimica Acta* 34, 169–184.

Wasson, J. T., Boynton, W. V., Chou, C.-L., and Baedecker, P. A. (1975) Compositional evidence regarding the influx of interplanetary materials onto the lunar surface. *Moon* 13, 121–141.

Wasson J. T., Rubin A. E., and Kallemeyn G. W. (1993) Reduction during metamorphism of four ordinary chondrites. *Geochimica et Cosmochimica Acta* 57, 1867–1878.

Wasson, J. T., Lange, D. E., Francis, C. A., and Ulff-Møller, F. (1999) Massive chromite in the Brenham pallasite and the fractionation of Cr during the crystallization of asteroidal cores. *Geochimica et Cosmochimica Acta* 63, 1219–1239.

Wasson, J. T., Matsunami, Y., and Rubin, A. E. (2006) Silica and pyroxene in IVA irons; possible formation of the IVA magma by impact melting and reduction of L-LL-chondrite materials followed by crystallization and cooling. *Geochimica et Cosmochimica Acta* 70 3149–3172.

Watt, L. E., Bland, P. A., Prior, D. J., and Russell, S. S. (2006) Fabric analysis of Allende matrix using EBSD. *Meteoritics & Planetary Science* 41, 989–1001.

Watters, T. R. and Prinz, M. (1979) Aubrites: Their origin and relationship to chondrites. *Lunar and Planetary Science Conference* 10, 1073–1093.

Weber, D. and Bischoff, A. (1994a) Grossite ($CaAl_4O_7$) – A rare phase in terrestrial rocks and meteorites. *European Journal of Mineralogy* 6, 591–594.

Weber, D. and Bischoff, A. (1994b) The occurrence of grossite ($CaAl_4O_7$) in chonrdites. *Geochimica et Cosmochimica Acta* 58, 3855–3877.

Weber, D., Clayton, R. N., Mayeda, T. K., and Bischoff, A. (1996) Unusual equilibrated carbonaceous chondrites and CO3 meteorites from the Sahara (abstract). *Lunar and Planetary Science* 27, 1395–1396.

Wegener, A. (1924) *The Origin of Continents and Oceans*. London: Methuen Publishing.

Weidenschilling, S. J. (2019) Accretion of the asteroids: Implications for their thermal evolution. *Meteoritics & Planetary Science* 54, 1115–1132.

Weider, S. Z., Nittler, L. R., Starr, R. D., McCoy, T. J., and Solomon, S. C. (2014) Variations in the abundance of iron on Mercury‘s surface from MESSENGER X-ray spectrometer observations. *Icarus* 235, 170–186.

Weisberg, M. K. and Huber, H. (2007) The GRO 95577 CR1chondrite and hydration of the CR parent body. *Meteoritics & Planetary Science* 42, 1495–1503.

Weisberg, M. K. and Kimura, M. (2010) Petrology and Raman spectroscopy of high pressure phases in the Gujba CB chondrite and the shock history of the CB parent body. *Meteoritics & Planetary Science* 45, 873–884.

Weisberg, M. K. and Kimura, M. (2012) The unequilibrated enstatite chondrites. *Chemie der Erde – Geochemistry* 72, 101–115.

Weisberg, M. K., Prinz, M., Clayton, R. N., and Mayeda, T. K. (1993) The CR (Renazzo-type) carbonaceous chondrite group and its implications. *Geochimica et Cosmochimica Acta* 57, 1567–1586.

Weisberg, M. K., Prinz, M., Clayton, R. N., Mayeda, T. K., Grady, M. M., Franchi, I., Pillinger, C. T., and Kallemeyn, G. W. (1996) The K (Kakangari) chondrite grouplet. *Geochimica et Cosmochimica Acta* 60, 4253–4263.

Weisberg, M. K., Connolly, H. C., and Ebel, D. S. (2004) Petrology and origin of amoeboid olivine aggregates in CR chondrites. *Meteoritics & Planetary Science* 39, 1741–1753.

Weisberg, M. K., Connolly, H., Ebel, D. S., and Kimura, M. (2006) Sulfide-metal nodules in EH3 chondrites. *Meteoritics & Planetary Science* 69, 5317.

Weisberg, M. K., Zolensky, M. E., Kimura, M., and Ebel, D. S. (2014) Primitive fine-grained matrix in the unequilibrated enstatite chondrites. *Lunar and Planetary Science* 45, Abstract #1551.

Weisberg, M. K., Ebel, D. S., Nakashima, D., Kita, N. T., and Humayun, M. (2015) Petrology and geochemistry of chondrules and metal in NWA 5492 and GRO 95551: A new type of metal-rich chondrite. *Geochimica et Cosmochimica Acta*, 167, 269–285.

Weisberg, M. K., Zolensky, M. E., Kimura, M., Howard, K. T., Ebel, D. S., and Bolega, Y. (2019) Northwest Africa (NWA) 8785, an EL3 chondrite with FeO-rich matrix. *Meteoritics & Planetary Science*, Abstract #6340.

Weisberg, M. K., Zolensky, M. E., Howard, K. T., Kimura, M., Ebel, D. S., Alexander, C. M. O'D., and Bolega, Y. (2020) Petrologic evidence of hydrothermal activity on the EL3 parent asteroid. *Lunar and Planetary Science* 51, Abstract #1683.

Welten, K. C., Huber, L., Caffee, M. W., Wittmann, A., Kring, D. A., Wieler, R., and Nishiizumi, K. (2014) What heated H/L chondrite LaPaz Icefield 031047 ~0.5 million years ago? 77th Annual Meeting of the Meteoritical Society, held September 7–12, 2014, in Casablanca, Morocco. LPI Contribution No. 1800. Abstract#5422.

Wentworth, S. J. and Gooding, J. L. (1994) Carbonates and sulfates in the Chassigny meteorite: Further evidence for aqueous chemistry on the SNC parent planet. *Meteoritics* 29, 860–863.

Whipple, F. L. (1983) 1983 TB and the Geminid meteors. *International Astronomical Union Circular*, 3881 (Oct.). http://www.cbat.eps.harvard.edu/iauc/03800/03881.html

White, J. S., Henderson, E. P., and Mason, B. (1967) Secondary minerals produced by weathering of the Wolf Creek meteorite. *American Mineralogist* 52, 1190–1197.

Widom, E., Rubin, A. E., and Wasson, J. T. (1986) Composition and formation of metal nodules and veins in ordinary chondrites. *Geochimica et Cosmochimica Acta* 50, 1989–1995.

Wiechert, V., Halliday, A. N., Lee, D.-C., Snyder, G. A., Taylor, L. A., and Rumble, D. (2001) *Science* 294, 345–348.

Wiegert, P. and Galiazzo, M. A. (2017) Meteorites from Phobos and Deimos at Earth? *Planetary and Space Science* 142, 48–52.

Wittmann, A., Friedrich, J. M., Troiano, J., Macke, R. J., Britt, D. T., Swindle, T. D., Weirich, J. R., Rumble, D., Lasue, J., and Kring, D. A. (2011) H/L chondrite LaPaz Icefield 031047 – A feather of Icarus? *Geochimica et Cosmochimica Acta* 75, 6140–6159.

Wittmann, A., Korotev, R. L., Jolliff, B. L., Irving, A. J, Moser, D. E., Barker, I., and Rumble, D. (2015) Petrography and composition of martian regolith breccia meteorite Northwest Africa 7475. *Meteoritics & Planetary Science* 50, 326–352.

Wlotzka, F. (1993) A weathering scale for the ordinary chondrites (abstract). *Meteoritics* 28, 460.

Wlotzka, F. (2005) Cr spinel and chromite as petrogenetic indicators in ordinary chondrites: Equilibration temperatures of petrologic types 3.7 to 6. *Meteoritics & Planetary Science* 40, 1673–1702.

Wojnarowska, A., Dziel, T., and Gałązka-Friedman, J. (2008) New mineralogical phases identified by Mössbauer measurements in Morasko meteorite. *Hyperfine Interactions* 186, 167–171.

Wood, J. A. (1964) The cooling rates and parent planets of several iron meteorites. *Icarus* 3, 429–459.

Wood, J. A., Marvin, U. B., Reed, J. B., Taylor, G. J., Bower, J. F., Powell, B. N., and Dickey, J. S. (1971) Mineralogy and petrology of the Apollo 12 lunar sample. *Smithsonian Astrophysical Observatory Special Report* 333, 177.

Wopenka, B. and Swan, P. D. (1985) Identification of micron-sized phases in meteorites by lasar Raman microprobe spectroscopy (abstract). *Meteoritics*, 20, 788–789.

Wyckoff, S. (1982) Overview of comet observations. In *Comets*, ed. Wilkening, L. L. Tucson: University of Arizona Press, pp. 3–55.

Xie, X., Minitti, M. E., Chen, M., Mao, H. K., Wang, D., Shu, J., and Fei, Y. (2003) Tuite, γ-$Ca_3(PO_4)_2$: A new mineral from the Suizhou L6 chondrite. *European Journal of Mineralogy* 15, 1001–1005.

Xie, X., Gu, X., Yang, H., Chen, M., and Li, K. (2016) Wangdaodeite, IMA 2016-007. CNMNC Newsletter No. 31. June 2016, page 695. *Mineralogical Magazine* 80, 691–697.

Xing, W., Lin, Y., Zhang, C., Zhang, M., Hu, S., Hofmann, B. A., Sekine, T., Xiao, L., and Gu, L. (2020) Discovery of reidite in the lunar meteorite Sayh al Uhaymir 169. *Geophysical Research Letters* 47, e2020GL089583.

Xiong, Y., Zhang, A.-C., Kawasaki, N., Ma, C., Sakamoto, N., Chen, J.-N., Gu, L.-X., and Yurimoto, H. (2020) Mineralogical and oxygen isotopic study of a new ultrarefractory

inclusion in the Northwest Africa 3118 CV3 chondrite. *Meteoritics & Planetary Science* 55, 2164–2205.
Xu, Y., Zinner, E., Gallino, R., Heger, A., Pignatari, M., and Lin, Y. (2015) Sulfur isotopic compositions of submicrometer SiC grains from the Murchison meteorite. *The Astrophysical Journal* 799, 156 (23 pp).
Yagi, K., Lovering, J. F., Shima, M., and Okada, A. (1978) Petrology of the Yamato meteorites (j), (k), (l), and (m) from Antarctica. *Meteoritics* 13, 23–45.
Yamaguchi, A., Taylor, G. J., and Keil, K. (1996) Global crustal metamorphism of the eucrite parent body. *Icarus* 124, 97–112.
Yanai, K. (1981) *Photographic Catalog of the Selected Antarctic Meteorites*. Tokyo: National Institute of Polar Research, 104 pp.
Yanai, K. and Kojima, H. (1987) *Photographic Catalog of the Selected Antarctic Meteorites*. Tokyo: National Institute of Polar Research, 298 pp.
Yang, J. and Goldstein, J. I. (2005) The formation mechanism of the Widmanstätten structure in meteorites. *Meteoritics & Planetary Science* 40, 239–253.
Yang, J., Goldstein, J. I., and Scott, E. R. D. (2007) Iron meteorite evidence for early formation and catastrophic disruption of protoplanets. *Nature* 446, 888–891.
Yang, J., Goldstein, J. I., Scott, E. R. D., Michael, J. R., Kotula, P. G., Grimberg, A., and Leya, I. (2014) Thermal and collisional history of Tishomingo iron meteorite: More evidence for early disruption of differentiated planetesimals. *Geochimica et Cosmohcimica Acta* 124, 34–53.
Yang, T., Zhao, Y. L., Tong, Y., Jiao, Z. B., Wei, J., Cai, J. X., Han, X. D., Chen, D., Hu, A., Kai, J. J., Lu, K., Liu, Y., and Liu, C. T. (2018) Multicomponent intermetallic nanoparticles and superb mechanical behaviors of complex alloys. *Science* 362, 933–937.
Yin, Q.-Z., Sanborn, M. E., and Ziegler, K. (2017) Testing the common source hypothesis for CV and CK chondrites parent body using Δ^{17}O-ε^{54}Cr isotope systematics. *Lunar and Planetary Science* 48, Abstract #1771.
Young, E. D., Kohl, I. E., Warren, P. H., Rubie, D. C., Jacobson, S. A., and Morbidelli, A. (2016) Oxygen isotopic evidence for vigorous mixing during the Moon-forming giant impact. *Science* 351, 493–496.
Yu, Z. (1984) Two new minerals gupeiite and xifengite in cosmic dusts from Yanshan. *Acta Petrologica Mineralogica et Analytica* 3, 230–237.
Yudin, I. A. and Kolomenskiy, V. D. (1987) *Mineralogy of Meteorites* (in Russian). Sverdlovsk, Russia: Academy of Sciences, 200 pp.
Yurimoto, H., Abe, K., Abe, M., Ebihara, M., Fujimura, A., Hashiguchi, M., Hashizume, K., Ireland, T. R., Itoh, S., Katayama, J., Kato, C., Kawaguchi, J., Kawasaki, N., Kitajima, F., Kobayashi, S., Meike, T., Mukai, T., Nagao, K., Nakamura, T., Naraoka, H., Noguchi, T., Okazaki, R., Park, C., Sakamoto, N., Seto, Y., Takei, M., Tsuchiyama, A., Uesugi, M., Wakaki, S., Yada, T., Yamamoto, K., Yoshikawa, M., and Zolensky, M. E. (2011) Oxygen isotopic compositions of asteroidal materials returned from Itokawa by the Hayabusa mission. *Science* 333, 1116–1119.
Zhang, A. C., Ma, C., Sakamoto, N., Wang R. C., Hsu, W.B., and Yurimoto, H. (2015) Mineralogical anatomy and implications of a Ti–Sc-rich ultrarefractory inclusion from Sayh al Uhaymir 290 CH3 chondrite. *Geochimica et Cosmochimica Act* 163, 27–39.
Zhang, A.-C., Pang, R.-L., Sakamoto, N., and Yurimoto, H. (2020) The Cr-Zr-Ca armalcolite in lunar rocks is loveringite: Constraints from electron backscatter diffraction measurements. *American Mineralogist* 105, 1021–1029.
Zinner, E. K. (2005) Presolar grains. In *Meteorites, Comets, and Planets*, ed. Davis, A. M. Oxford: Elsevier, pp. 17–39.
Zinner, E. (2014) Presolar grains. In *Meteorites and Cosmochemical Processes*, eds. Holland, H. D. and Turekian, K. K. *Treatise on Geochemistry*, Vol. 1. Oxford: Elsevier, pp. 181–210.
Zolensky, M. E. (1997) Structural water in the Bench Crater chondrite returned from the Moon. *Meteoritics & Planetary Science* 32, 15–18.

Zolensky, M. E. and Gooding, J. L. (1986) Aqueous alteration on carbonaceous-chondrite parent bodies as inferred from weathering of meteorites in Antarctica (abstract). *Meteoritics* 21, 548–549.

Zolensky, M. and Ivanov, A. V. (2003) The Kaidun microbreccia meteorite: A harvest from the inner and outer asteroid belt. *Chemie der Erde – Geochemistry* 63, 185–246.

Zolensky, M. E. and Krot, A. N. (1996) Mineralogical and compositional study of an Allende dark inclusion (abstract). *Lunar and Planetary Science* 27, 1503–1504.

Zolensky, M. E. and McSween, H. Y. (1988) Aqueous alteration. In *Meteorites and the Early Solar System*, eds. Kerridge, J. F. and Matthews, M. S. Tucson: University of Arizona Press, pp. 114–143.

Zolensky, M. E., Ivanov, A. V., Yang, V., and Oshsumi, K. (1996) The Kaidun meteorite: Mineralogy of an unusual CM1 clast. *Meteoritics & Planetary Science* 31, 484–493.

Zolensky, M. E., Mittlefehldt, D. W., Lipschutz, M. E., Wang, M.-S., Clayton, R. N., Mayeda, T. K., Grady, M. M., Pillinger, C., and Barber, D. J. (1997) CM chondrites exhibit the complete petrologic range from 2 to 1. *Geochimica et Cosmochimica Acta* 61, 5099–5115.

Zolensky, M., Abell, P., and Tonui, E. (2005) Metamorphosed CM and CI cabonaceous chondrites could be from the breakup of the same Earth-crossing asteroid. *Lunar & Planetary Science* 36, Abstract #2084.

Zolensky, M., Gounelle, M., Mikouchi, T., Ohsumi, K., Le, L., Hagiya, K., and Tachikawa, O. (2008) Andreyivanovite: A second new phosphide from the Kaidun meteorite. *American Mineralogist* 93, 1295–1299.

Zolensky, M., Mikouchi, T., Fries, M., Bodnar, R., Jenniskens, P., Yin, Q.-Z., Hagiya, K., Ohsumi, K., Konatsu, M., Colbert, N., Hanna, R., Maisano, J., Ketcham, R., Kebukawa, Y., Nakamura, T., Matsuoka, M., Sasaki, S., Tsuchiyama, A., Gounelle, M., Le, L., Martinez, J., Ross, K., and Rahman, Z. (2014) The Sutter's Mill CM meteorite. *Meteoritics & Planetary Science* 49, 1997–2016.

Zolensky, M. E., Fries, M. D., Chan, Q. H.-S., Kebukawa, Y., Bodnar, R., Burton, A., Callahan, M., Steele, A., and Sandford, S. (2015a) Survival of organic materials in ancient cryovolcanically-produced halite crystals. *Workshop on the Potential for Finding Life in a Europa Plume*, Mofft Field, CA.

Zolensky, M., Mikouchi, T., Hagiya, K., Ohsumi, K., Kumatsu, M., and Le, L. (2015b) Evidence for impact shock melting in CM and CI chondrite regolith samples. *Lunar and Planetary Science* 46, Abstract#2261.

Zolensky, M. E., Bodnar, R. J., Fries, M., Chan, Q. H.-S., Kebukawa, Y., Mikouchi, T., Hagiya, K., Komatsu, M., Ohsumi, K., and Steele, A. (2016a) *Ceres revealed in a grain of salt (abstract).* 26th Goldschmidt Conference 2016; June 26, 2016, to July 1, 2016; Yokohama, Japan.

Zolensky, M., Mikouchi, T., Hagiya, K., Ohsumi, K., Komatsu, M., Chan, Q.H.S., Le, L., Kring, D., Cato, M., Fagan, A. L., Gross, J., Tanaka, A., Takegawa, D., Hoshikawa, T., Yoshida, T., and Sawa, N. (2016b) Unique view of C asteroid regolith from the Jbilet Winselwan CM chondrite. *Lunar and Planetary Science*, 47, Abstract#2148.

Zolotov, M. Y., Mironenko, M. V., and Shock, E. L. (2006) Thermodynamic constraints on fayalite formation on parent bodies of chondrites. *Meteoritics & Planetary Science* 41, 1775–1796.

Zubkova, N. V., Pekov, I.V., Chukanov, N. V., Pushcharovsky, D. Y., and Kazantsev, S. S. (2008) Nickelhexahydrite from the weathered meteorite Dronino: Variations of chemical composition, crystal structure, and genesis. *Doklady Earth Sciences* 422, 1109–1112.

Zubrin, R. (2011) *The Case for Mars: The Plan to Settle the Red Planet and Why We Must.* New York: Free Press, Simon & Schuster, 416 pp.

Zurfluh, F. J., Hofmann, B. A., Gnos, E., Eggenberger, U., and Jull, A. J. (2016) Weathering of ordinary chondrites from Oman: Correlation of weathering parameters with ^{14}C terrestrial ages and a refined weathering scale. *Meteoritics & Planetary Science* 51, 1–16.

Index

2I/Borisov, 104, *See* interstellar interloper

Abbott. *See* meteorite
Abee. *See* meteorite
Acapulco. *See* meteorite
acapulcoites, 107, 173, 179, 291, 303, 309, 314
accretion, 238, 260, 292, 347, 365
acetylene, 230
Acfer 059. *See* meteorite
Acfer 094. *See* meteorite
Acfer 097. *See* meteorite
achondrites, 101, 106–108, 150, 171, 175, 178–179, 182, 226, 253, 283, 291, 294, 303, 309–310, 318, 350, 368, 374
acute bisectrix, 90
adamite, 83
addibischoffite, 116, 167
Adelaide. *See* meteorite
Adhi Kot. *See* meteorite
Admire. *See* meteorite
adrianite, 117, 134, 167, 268
aerogel, 234
Aeschylus, 6
agate, 2
AGB stars. *See* asymptotic giant branch stars
agglutinate, 201, 212, 224, 279, 301–302, 308
Agpalilik. *See* Cape York
Agrell dip, 312
Agrell, Stuart, 312
Agricola, Georgius, 3
Aguas Zarcas. *See* meteorite
ahrensite, 220, 239, 280, 298, 325–326
Ahumada. *See* meteorite
akaganéite, 211, 213, 246, 297, 321, 323
akaogiite, 300
åkermanite, 54, 243
akimotoite, 220, 280, 298, 333
Al Haggounia 001. *See* meteorite
Al Huwaysah 010. *See* meteorite
Al Rais. *See* meteorite
alabandite, 70, 96, 115, 142–143, 151, 170, 174, 177, 181, 187, 189, 306
Alais. *See* meteorite
Albareto. *See* meteorite
Albin. *See* meteorite
Al-Biruni, 3
albite, 68, 70, 72, 76, 78, 87, 92, 98, 136–137, 139, 144, 152, 155, 157–158, 162, 171, 175, 177–178, 189–190, 200, 205–206, 226, 243, 255–257, 261, 272, 279, 295, 306, 309, 347
albite twinning, 68
Aldrin, Buzz, 330
Aletai. *See* meteorite
ALH 77307. *See* meteorite
ALH 78091. *See* meteorite
ALH 78113. *See* meteorite
ALH 81005. *See* meteorite
ALH 82130. *See* meteorite
ALH 83009. *See* meteorite
ALH 83014. *See* meteorite
ALH 83015. *See* meteorite
ALH 83108. *See* meteorite
ALH 84001. *See* meteorite
ALH 84028. *See* meteorite
ALH 85085. *See* meteorite
ALH 85151. *See* meteorite
ALHA76004. *See* meteorite
ALHA77005. *See* meteorite
ALHA77011. *See* meteorite
ALHA77015. *See* meteorite
ALHA77156. *See* meteorite
ALHA77255. *See* meteorite
ALHA77283. *See* meteorite
ALHA77307. *See* meteorite
ALHA78113. *See* meteorite
ALHA78119. *See* meteorite
ALHA81005. *See* meteorite
ALHA81014. *See* meteorite
ALHA81024. *See* meteorite
Alikatnima. *See* meteorite
Aliskerovo. *See* meteorite

Allegan. *See* meteorite
allendeite, 116, 167–168
Almahata Sitta. *See* meteorite
almandine, 213, 355
alpha decay, 82–83
Al-Ti diopside, 114, 125, 134, 166–167, 169
amethyst, 2, 53, 74
amino acids, 259, 276, 316
ammonia, 224, 231, 237, 268
amoeboid olivine aggregates. *See* amoeboid olivine inclusions
amoeboid olivine inclusions, 125–126, 158, 166–167, 169, 171, 248, 258–259, 263, 366
amorphous, xiii, 54, 56, 82, 88, 129, 131, 133–135, 137, 149, 166, 168, 219–220, 233, 239, 253–254, 264–265, 267–268, 274, 301–302, 319, 328
amphibole, 57, 92, 134, 142, 164, 174, 180, 182, 220–221, 261, 269, 272, 275, 333, 341, 358, 366, 372
analbite, 295
anatase, 91, 322
andalusite, 205
andesine, 309
andesite, 174, 179, 218
andradite, 115, 128, 134, 167, 269–270, 342
angle of incidence, 85–86
anglesite, 206
Angra dos Reis. *See* meteorite
angrites, 103, 105, 107, 172, 176, 192, 303, 305
anhydrite, 66–67, 71, 84, 132–133, 166, 204–205, 218, 275
ankerite, 5, 219–220, 275
annealing, 78, 140, 191, 239, 253, 272, 278, 290–292, 328, 365–366
anorthite, 70, 87, 98, 101, 115–117, 124, 126, 134, 137, 139, 144, 157–158, 162, 166–167, 169, 171, 176–178, 182, 206, 234, 243, 248, 268, 270, 308, 351
anorthosite, 56, 207, 210–211, 213
Antarctica, 52, 103, 212, 338, 343, 368, 377–378
antarcticite, 66, 218
antiferromagnetism, 80
antigorite, 225
antiperthite, 174, 189–190, 295
antitaenite, 64
apatite, 66–67, 69, 77, 81, 83, 88, 132, 134, 160, 164, 172, 175, 182, 189–190, 208, 211, 213, 219–220, 225, 271, 275–276, 279, 297, 321, 323, 349
Apollo 11, 208, 214
Apollo 12, 135, 208, 214, 376
Apollo 14, 104, 208, 214, 332
Apollo 15, 208, 214, 345
Apollo 17, xiii, 208, 215, 297
Apollonius of Myndus, 232
Appley Bridge. *See* meteorite
aquamarine, 73
aqueous alteration, 115, 117, 125–126, 128, 130–131, 133, 144, 153, 158–159, 162, 219, 229, 235, 239, 251, 254, 258–260, 263–265, 267–268, 270–273, 275, 328, 331, 335, 343, 345, 352, 366–367, 372–373, 378
aragonite, 53, 81, 83, 132, 205–206, 265, 275
Arietids. *See* meteor shower
Aristarchus, 8
Aristotle, 2, 6, 232
Arltunga. *See* meteorite
armalcolite, 127, 190, 208, 211, 213, 297
Armanty. *See* meteorite
Arrokoth, 236–237
arsenide, 10, 138, 142, 146, 162, 167, 259–261, 322, 328
arsenopyrite, 84
arupite, 321
asbestos, 72, 84
aspidolite, 134
asterism, 76
asteroid, xiii, 7–10, 46–50, 103, 106–108, 117, 150, 153, 155, 195, 222, 224, 229, 235, 239, 253–254, 263, 271, 302, 306, 309, 311, 328, 330, 344, 361
  2008 EV5, 228
  2017 YE5, 226
  Apollo, 222
  Bamberga, 222
  Bennu, 46, 228–229, 345
  Ceres, 7, 46–49, 222, 224–225, 276, 378
  Clarissa, 222
  definition, 46
  Dembowska, 222
  Eros, xiii, 225–226, 341, 352–353, 358
  Gault, 229
  Hidalgo, 222
  Hilda, 222
  Hungaria, 222
  Hygiea, 49
  Icarus, 50, 315, 332, 341, 343, 352–353, 357, 366–367, 369, 375–377
  Ida, 302
  Itokawa, xiii, 9–10, 226–227, 302, 348, 361, 377
  Julia, 326
  Juno, 7, 47–48, 222
  Kassandra, 222
  Lick, 222
  Magellan, 103
  Mathilde, 225
  Nyx, 103
  Pallas, 7, 47, 222
  Phaethon, 50, 229
  Psyche, 222, 229
  Ryugu, xiii, 228
  Themis, 10, 337
  Verenia, 103
  Vesta, 7, 9–10, 47, 103, 108, 175–176, 207, 222–224, 262, 274, 289, 302
  vestoids, 103
Asuka 10164. *See* meteorite
Asuka 12236. *See* meteorite
Asuka 881314. *See* meteorite

Asuka 881371. *See* meteorite
Asuka 881394. *See* meteorite
Asuka 881655. *See* meteorite
asymptotic giant branch stars, 239, 242
atacamite, 206
ataxite, 183
Atlanta. *See* meteorite
atmospheric passage, 46, 50–51, 53, 239, 318, 329
aubrites, 10, 54, 101, 107–108, 177–178, 181, 222, 292, 303, 306, 311, 365, 369, 375, *See* enstatite achondrites
Auburn. *See* meteorite
augite, 101, 112, 149, 162, 169, 172, 175–177, 182, 190, 192, 195, 197–199, 211, 213, 219–221, 224, 252, 262, 290, 309–311, 314–315
Augustus, 238
autunite, 82, 84
awaruite, 94, 115, 127, 133–134, 136, 138–140, 142, 146, 155, 162, 167, 179, 268, 270, 312, 363, 365
Axtell. *See* meteorite
azurite, 1, 71–72, 91

Babb's Mill (Troost's Iron). *See* meteorite
backscattered-electron image, 98
baddeleyite, 125, 166, 175–176, 182, 211, 213, 219–221, 305, 351
baghdadite, 167, 354
Bali. *See* meteorite
ballen silica, 177
Ballinoo. *See* meteorite
Ban Rong Du. *See* meteorite
banded iron formations, 205
bandwidth, kamacite, 183
Barbianello. *See* meteorite
barite, 66, 178, 213, 321–322
Barnard's Star, 49
barroisite, 272
Barwell. *See* meteorite
basalt, xiii, 4, 8, 10, 45, 56, 149, 172, 174–176, 182, 197–200, 203, 205, 207–208, 211, 213–214, 216, 218–219, 223, 262, 273, 289–290, 293–294, 297, 303–305, 307–309, 328, 363, 366, 372
bassanite, 132, 178, 218, 220–221, 275, 321
baumhauerite, 206
beachrock, 56
Becke lines, 88
beckettite, 117, 356
Belgica 7904. *See* meteorite
bementite, 206
Benares. *See* meteorite
Bench Crater. *See* meteorite
Bencubbin. *See* meteorite
bencubbinites, 148
Bendegó. *See* meteorite
Benld. *See* meteorite
BepiColombo spacecraft, 202
Berduc. *See* meteorite
Bergman, Torbern, 5
berlinite, 79
berthierite, 206
beryl, 2, 5, 55, 71, 73
Berzelius, Jöns Jacob, 5, 66
Bessel, Friedrich, 233
beta decay, 82
beusite, 187
Bholghati. *See* meteorite
Bianchini's planisphere, 215
biaxial mineral, 90
Biermann, Ludwig, 232
biopyriboles, 134, 269, 335
biotite, 70, 87, 182, 205, 220–221, 275, 358
bireflectance, 91
birefringence, 87, 89–90
bischofite, 218
Bishop Canyon. *See* meteorite
Bishunpur. *See* meteorite
Bison. *See* meteorite
Black Beauty. *See* NWA 7034
black chondrites, 141, 278
Blake, William, 330
Blithfield. *See* meteorite
blödite, 130, 218, 263, 322
blueberries
  hematite. *See* Mars
Bocaiuva. *See* meteorite
body-centered cubic, 59, 95, 140, 306, 311
böhmite, 323
bolide, 51
Bondoc. *See* meteorite
boninite, 200
borax, 71, 84
bornite, 71, 73, 76, 213, 262, 273, 321
Bournon, Count de, 7
Bouvante. *See* meteorite
Brachina. *See* meteorite
brachinites, 107, 172, 303, 310
Bragg, Lawrence, 5
Bragg, William, 5
Brahin. *See* meteorite
brannerite, 82
Bravais, Auguste, 58
Bravais lattices, 58
breccia
  definition, 57
  dimict, 57
  fragmental, 57, 151, 177, 292, 307
  genomict, 57, 272
  impact-melt, 57
  monomict, 57, 172, 175–177, 219, 289–290, 292, 304
  polymict, 9, 57, 129, 133, 170, 175–177, 180, 182, 208, 213, 224, 244, 274, 283, 289–290, 292, 305, 308, 339
  regolith, 57, 212
Bremervörde. *See* meteorite

Brenham. *See* meteorite
breunnerite, 129–130, 163, 219–220, 263, 265, 267, 352
Brezina, Aristides, 102
Brezina lamellae, 185, 187
brezinaite, 168, 172, 177, 187, 191, 306
brianite, 174, 188–189, 342
bridgmanite, 220, 280, 298
brine, 217, 274–275
bronze, 1–2, 73, 326, 343
brookite, 322
brown dwarf, 48
browneite, 180, 355
brucite, 70, 72, 83, 130
Brunflo. *See* fossil meteorite
BSE image. *See* backscattered electron image
buchwaldite, 187–188, 306
Budulan. *See* meteorite
Bunburra Rockhole. *See* meteorite
bunsenite, 321
Burgers vector, 55
burnettite, 116, 167
buseckite, 180, 355
Bustee. *See* meteorite
butianite, 117, 354
Butler. *See* meteorite

Ca-Al-rich inclusions. See refractory inclusions. *See* CAIs
Ca-armalcolite, 127
Cachari. *See* meteorite
Cachiyuyal. *See* meteorite
Caddo County. *See* meteorite
Caesar, Julius, 238
cahnite, 206
CAIs, 83, 106, 115–117, 123–126, 128–130, 134, 142, 144–145, 147, 155, 158, 162–163, 166–169, 171, 234, 239, 243–248, 251, 256, 259, 263–265, 268, 270, 272, 302, 328, 334, 340, 350–351, 356–357, 366, 373, *See* refractory inclusions
calcite, 4, 54, 57, 66–69, 71, 75, 77, 79, 81, 83, 87, 130, 132–134, 136–137, 163–164, 168, 178, 205, 213, 219–221, 229, 263, 265, 270–271, 275, 321–322, 352
Caloris basin, 200
Campo del Cielo. *See* meteorite
Canon City. *See* meteorite
Canyon Diablo. *See* meteorite
Cape York. *See* meteorite
Ca-pyroxene, 112–115, 117, 136, 140, 155, 158, 163, 167–171, 173, 179–180, 189, 192, 194–196, 198, 211, 248, 257, 280, 293, 297, 314
Carancas. *See* meteorite
Carancas crater, 65
carbide, 10, 66, 99, 115, 137, 139–140, 142–145, 148, 155, 163, 166–167, 171, 174, 184, 188, 191, 235, 241–242, 267, 272, 313, 321, 328, 331–332, 351, 355, 369, 371
carbide-magnetite assemblages, 139, 351
Carbo. *See* meteorite
carbon
  poorly graphitized, 129, 134, 168
carbon stars, 242
carbonaceous chondrites, 10, 54, 67, 78, 97, 101, 106–107, 109, 113, 126–128, 130, 135, 148, 152, 163–164, 169, 192, 214, 222, 224, 227, 243, 252, 254, 257, 259–260, 263–264, 267, 271, 282, 284, 315, 317–318, 331–332, 342, 344–345, 349, 351–352, 356, 360, 367, 369, 372–373
carbonate, 5, 10, 53, 63, 66, 73, 83, 91, 132–134, 136, 163–164, 168–169, 176, 182, 205, 218–220, 225, 233, 257, 263–265, 267, 269–270, 275, 295, 315, 319, 321–322, 328
carbon-rich aggregates, 129–130, 335
Cargo Muchacho Mountains. *See* meteorite
Carlisle Lakes. *See* meteorite
carlsbergite, 98, 184, 186–187
Carlton. *See* meteorite
carnallite, 85
carnelian, 2
carnotite, 72, 82–83
cassidyite, 321
Cassini spacecraft, 231
cassiterite, 1, 91
castorite, 5
caswellsilverite, 115, 142, 170, 177–178, 251, 273, 306, 321
Cat Mountain. *See* meteorite
cathodoluminescence, 78, 155–156, 255, 262
Cavendish, Henry, 8
Cayote Dry Lake 033. *See* meteorite
CB chondrites, 106, 129, 163, 168, 284, 326, *See* bencubbinite
celestine, 66, 72, 213
celsian, 176, 305
center of symmetry, 60, 79
cerargyrite, 72
Ceres. *See* asteroid
Cerro los Calvos. *See* meteorite
cerussite, 71, 206
CH chondrites, 105, 107, 113, 115, 125, 147, 162, 168–169, 232, 284, 350–351, 356, 359, 377
chadacrysts, 151
chalcanthite, 83–84
chalcedony, 1, 72
chalcopyrite, 1, 71–72, 76, 96–97, 136–137, 142, 146, 162, 167, 182, 213, 219–220, 321–322, 333
chalk, 1, 322, 324
Chandrayaan spacecraft, 207, 215
Chang'e spacecraft, 207–208
chaoite, 172, 290, 300
charlesite, 206
Charon, 236–237
chassignites, 101, 107, 182, 221, 308, 314, *See* martian meteorite
Chassigny. *See* meteorite

Château-Renard. *See* meteorite
chatoyancy, 76, 308
chemical bonding, 61, 68, 79
chengbolite, 138, 142, 146, 162
chengdeite, 71
chenmingite, 220, 280, 298
chert, 1, 56, 205
chevkinite-perrierite, 275
Chico. *See* meteorite
Chicxulub crater, 324
Chladni, Ernst, 6
chladniite, 174, 189
chlorapatite, 97, 150, 159, 162, 172, 174, 182, 184, 188–189, 219, 226, 255, 294, 310, 342
chlorite, 57, 72, 128, 133–135, 137, 166, 268–269
chlormayenite, 117
chlorophoenicite, 206
chondrules, 4, 7, 68, 78, 94, 101, 105–106, 109, 112–115, 128–130, 134–136, 138–140, 142, 144, 146–147, 150–151, 155, 157–160, 163–164, 166–167, 169, 171, 173–175, 184, 189, 234, 239, 244, 248–253, 255–256, 258–260, 262–265, 267–271, 273, 283, 286, 289, 293–294, 302, 314, 328, 332, 335, 340, 342, 344, 350–351, 361, 365–369
  aluminum-rich, 113–114, 169
  barred olivine, 109
  bleached, 271
  chromium-rich, 114
  compound, 113, 147, 250, 366
  cryptocrystalline, 112, 250
  droplet, 112–113
  glass-rich, 113–114
  granular olivine-pyroxene, 112, 249
  nonporphyritic, 109, 114
  porphyritic, 109, 112–113, 169, 171, 249, 270
  radial pyroxene, 109, 113
  silica-rich, 113
  Type I, 112, 131, 139, 142, 145–146, 151, 155, 158–159, 166–167, 169, 260, 293
  Type II, 112, 145, 155, 158, 166
Choteau. *See* meteorite
Christy, James, 236
chromian spinel, 114
chromite, 69, 71, 91, 96–98, 112–114, 128–129, 133–134, 137, 139–142, 144, 146–147, 150, 152–153, 157–160, 162, 167, 169, 171–176, 182, 184–189, 191–199, 209, 213, 219–221, 226, 253, 255–260, 263, 278–280, 290, 293–295, 298, 304, 307, 309–310, 314, 317, 322, 351, 365
chromite-plagioclase assemblages, 97, 256, 278, 294
chrysoberyl, 68, 71, 76
chrysotile, 68, 131, 164, 225
CI chondrites, 130, 136, 144, 148, 163–165, 229, 257, 264, 283, 322, 332, 335, 341
cinnabar, 1–2, 68, 71, 75, 91, 99
CK chondrites, 56, 134–135, 146, 167, 259, 283, 316, 320, 340, 342, 377
Claxton. *See* meteorite
clay, 84, 205, 225, 343
claystone, 56
cleavage, 5, 67–68, 76, 88, 91
Clementine spacecraft, 207
cliftonite, 97, 186, 314, *See* graphite
clinobronzite, 187, 192
clinochlore, 258
clinoenstatite, 68, 112, 115, 139, 144, 158, 160–161, 166, 170–171, 177–178, 192, 213, 252, 279–280, 285, 325
clinohedrite, 206
clinopyroxenite, 219, 308
clintonite, 134, 269
closing temperature, 83
cloudy zoned taenite, 95
CM chondrites, 7, 125, 131–132, 144, 163–164, 166, 212, 225, 228–229, 233, 257, 264–265, 283, 316–317, 350, 373
CO chondrites, 81, 109, 126, 133, 139, 144–145, 166, 244, 258–259, 268, 270, 283
Coahuila. *See* meteorite
cobaltite, 322
Coconino sandstone, 67
coercivity, 80
coesite, 67, 168, 170, 176, 180, 202, 280, 284–285, 289, 300, 325–326
coffinite, 82
cohenite, 95, 97, 99, 115, 137, 139–140, 142–145, 148, 155, 166, 171–172, 174, 184, 186, 188, 193, 195, 213, 271, 289, 293, 296, 313, 315
Cold Bay. *See* meteorite
Cold Bokkeveld. *See* meteorite
collinsite, 321
Colomera. *See* meteorite
Colonia Obrera. *See* meteorite
Colony. *See* meteorite
comet, xiii, 9–10, 46–48, 50, 103, 201, 214, 231–233, 235, 237–238, 328
  103P/Hartley, 233, 237
  109P/Swift-Tuttle, 50, 233
  19P/Borrelly, 233, 237
  1P/Halley, 50, 233, 235, 237
  209P/LINEAR, 237
  2P/Encke, 232
  45P/Honda-Mrkos-Pajdušáková, 237
  55P/Tempel-Tuttle, 50, 233
  67P/Churyumov-Gerasimenko, 233
  81P/Wild, xiii
  8P/Tuttle, 237
  9P/Tempel, 233
  definition, 46
  long-period, 47
  short-period, 47
condensation, xiii, 10, 140, 143, 163, 165, 239, 243–246, 280, 301–303, 310, 328, 334
contact twin, 68
Coolidge. *See* meteorite

Coopertown. *See* meteorite
coordination number, 63–65, 92
copper, 1–3, 69, 71–73, 80, 83, 91, 95, 184, 206, 337, 342, 365
coquina, 53, 56
cordierite, 72, 88, 205, 213
corundum, 67, 69, 71, 73, 76, 87, 116, 145–146, 148, 163, 166–167, 172, 206, 242–243, 270, 325, 344
cosmic–ray–exposure age, 52
covalent bonding, 62, 79
covellite, 91, 262, 273
CR chondrites, 107, 109, 115, 125, 129, 133, 146–147, 167–168, 235, 251, 258, 267, 283, 322, 331, 334, 345, 351, 374–375
Crab Orchard. *See* meteorite
Craig, Harmon, 102
Cratheús (1950). *See* meteorite
Cretaceous–Paleogene boundary, 323
cristobalite, 67, 92, 99, 113, 170, 177–178, 180, 185, 208, 211, 213, 220–221, 251, 262
crocoite, 71
Cronstedt, Baron Axel, 7
cronstedtite, 7, 131, 164, 264–265, 373
crossed polarizers, 86–88, 285
crushing strength, 317
cryolite, 72
crystal
  anhedral, 54
  definition, 54
  euhedral, 54, 67
  subhedral, 54, 195
crystal chemistry, 58, 332
crystal classes, 60
crystal face, 55, 61, 67–68, 109
crystal form, 55, 61, 230
crystal symmetry, 66
crystal systems, 54, 59, 68, 86
crystallization, xiii, 52, 103, 112, 143, 155, 184, 191, 198, 212, 239, 245, 248–251, 254–255, 279, 283, 286, 289–290, 293, 296, 304, 306–307, 309, 314, 328, 354–355, 358, 365, 372, 375
crystallography, 4–5, 58, 61
cubanite, 91, 130, 144, 163, 188, 213, 234, 263
cubic closest packing, 64
cubic zirconia, 71, 354
Cumberland Falls. *See* meteorite
cupalite, 325
cuprite, 71, 91
Curie temperature, 80
CV chondrites, 83, 106, 109, 113, 123–125, 128–129, 134, 145, 166–167, 245, 251, 259–260, 268–270, 284, 326
cyanogen, 230
cyclic twins, 68
czochralskiite, 188–189
D'Orbigny. *See* meteorite
dacite, 179, 218
Dana, James, 5
Danebury. *See* meteorite
Dar al Gani 055. *See* meteorite
Dar al Gani 1031. *See* meteorite
Dar al Gani 670. *See* meteorite
Dar al Gani 978. *See* meteorite
dark flight
  definition, 46
dark inclusions, 127–129, 260, 267, 270, 351, 366
daubréelite, 96, 115, 138, 142, 151, 155, 168, 170, 174, 177, 184–188, 191, 214, 252, 262, 273, 289, 293–296, 306, 314, 323
davidite, 82
davisite, 99, 116, 125–126, 166–168
Dawn spacecraft, 9, 223–224, 276
Dayton. *See* meteorite
decagonite, 56, 325
Deep Impact spacecraft, 233
defects
  antisite, 55
  bulk, 55
  line, 55
  planar, 55
  topological, 55
deformation twins, 68
Deimos, 49, 104, 376
Del Rio. *See* meteorite
deliquescence, 85
delocalized electrons, 62, 71, 75, 79
Democritus, 54
dense collection area, 52
descloizite, 206
desert varnish, 53
devitrification, 158, 254, 295
devolatilization, 150, 328
Dhajala. *See* meteorite
Dhofar 1222. *See* meteorite
Dhofar 1514. *See* meteorite
Dhofar 280. *See* meteorite
Dhofar 312. *See* meteorite
Dhofar 778. *See* meteorite
diabase, 219, 309
diamagnetism, 80
diamond, 5, 53–54, 62, 68–69, 71, 77, 79, 85–86, 135, 148, 163, 167–168, 172, 184, 188, 225, 240, 285, 290, 293, 300, 331, 347, 352, 361, 367
diaspore, 72
Diepenveen. *See* meteorite
Dimmitt. *See* meteorite
diogenites, 101, 175–176, 198, 224, 289–290, 304–305, 307, 311
diopside, 54, 76, 99, 114–117, 124–128, 134, 149, 158, 162, 166–169, 171–172, 174–179, 182, 187, 189, 191–192, 200, 205–206, 213, 226, 234, 243, 245, 248, 261, 268, 270, 279, 305–306, 311, 325

dispersion, 76–77, 284–285, 289
Divnoe. *See* meteorite
djerfisherite, 142–143, 170, 177, 187, 262, 273, 306, 342
dmisteinbergite, 167, 355
dmitryivanovite, 284, 359
dolomite, 83, 130, 136, 163, 168, 205, 214, 218, 220, 229, 263, 265, 275, 352
dolostone, 205
DOM 03238. *See* meteorite
DOM 08006. *See* meteorite
DOM 08351. *See* meteorite
DOM 10088. *See* meteorite
DOM 10104. *See* meteorite
Dong Ujimqin Qi. *See* meteorite
Dongtai. *See* meteorite
Doppler radar, 51
Dora. *See* meteorite
dry ice, 216
Duel Hill (1854). *See* meteorite
dumortierite, 74, 76
Dunbogan. *See* meteorite
dunite, 56, 178, 210–211, 221, 306–308, 341
dust
  definition, 46
dust trail
  definition, 46
dustballs, 246
dwarf planet
  definition, 49

Eagle Station. *See* meteorite
Eagle Station pallasites, 107, 192
Earth, xiii–xiv, 2–4, 6–10, 44, 46, 48–50, 52, 54, 68, 81, 83, 104–105, 200–202, 204, 212, 215–216, 221–222, 224, 226, 230, 232–233, 235, 301, 317–318, 321–322, 328, 332, 337, 339–340, 342, 347, 353, 356, 359, 365–366, 368–374, 376, 378
eastonite, 204
EBSD. *See* electron backscatter diffraction
eclogite, 150
edenite, 174, 272
edscottite, 97, 137, 142, 155, 188, 271, 355
EET 83309. *See* meteorite
EET 83311. *See* meteorite
EET 84302. *See* meteorite
EET 90043. *See* meteorite
EET 90102. *See* meteorite
EET 90299. *See* meteorite
EET 90992. *See* meteorite
EET 92105. *See* meteorite
EETA79001. *See* meteorite
efflorescence, 84
Efremovka. *See* meteorite
EH chondrites, 128, 138, 142–143, 148, 170, 262, 273, 340, 367, 375
EL chondrites, 138, 170, 251, 261–262, 273, 340, 353
El Médano 100. *See* meteorite
El Taco. *See* Camp del Cielo
Elbogen. *See* meteorite
electrical conductivity, 62, 79
electron backscatter diffraction, 92, 98
electron capture, 83
electron color center, 73
electron microprobe, 9, 78, 86, 92, 102
electron orbitals, 62, 74, 80
electron shell, 61
electrophonic sounds, 51
electrum, 72, 142, 162
Elga. *See* meteorite
Eltanin. *See* meteorite
emerald, 2, 53, 73
Emery. *See* meteorite
enantiomorphs, 60
enstatite, 54, 67, 92, 96, 98, 101, 107, 116, 127–128, 135, 138, 143–144, 152, 158, 166, 168, 170–172, 174, 177–178, 180–181, 189, 191, 200, 235, 243, 253, 261–262, 270, 280, 285–286, 289–290, 292, 311, 321, 333, 335, 341, 345, 352, 362, 365–367, 370, 375
enstatite achondrites. *See* aubrites
enstatite chondrites, 10, 54, 92, 96–97, 102, 107–108, 115, 125, 128, 143, 148, 151, 163, 170, 178, 181, 192, 246, 250, 252, 261–262, 273, 285–287, 289, 306, 317, 326, 331, 340, 342, 348–350, 352, 361, 364–365, 367, 374
epsomite, 84, 87, 130, 218, 220, 319, 322
Eratosthenes, 8
eringaite, 166–167, 354
Eris, 10, 47
erlichmanite, 138, 142, 162
eskolaite, 130, 144, 163–164, 172, 179, 317
Esquel. *See* meteorite
Estherville. *See* meteorite
ethane, 230
eucrites, 10, 56, 101, 103, 108, 175, 179, 198, 224, 262–263, 273, 289, 304–305, 307, 311, 326, 339
Europa, 10, 378
evaporite, 56, 205
exoplanet, xiii, 47, 49, 330
exsolution, 67, 76, 96, 136, 142–143, 146, 155, 159, 173–178, 185–187, 189–190, 197–199, 239, 251–252, 255, 257, 261–262, 279, 295, 311, 314, 328, 370
exsolution lamellae, 67, 76, 96, 136, 142–143, 146, 155, 159, 173–178, 185–187, 189, 197–199, 251, 262, 279, 295, 311, 314
external iridescence, 76
extinction, 88–90, 114, 158, 160–161, 179, 256, 283, 285, 289–290, 324
extinction angle, 88
extraordinary ray, 90

face-centered cubic, 59, 61, 63–64, 306, 311
facies, 205

fall
definition, 51
falling star, 50
Farmington. *See* meteorite
Farrington, Oliver, 102
farringtonite, 174, 189, 192, 196, 307
fassaite, 114, 116, 124, 127, 134, 166–167, 171, 234, 246, 357, *See* Al-Ti diopside
fast ray, 86–88, 91
fayalite, 70, 72, 86, 98, 104, 113, 115, 134, 180, 182, 213, 220–222, 234, 244, 258–260, 268, 270
Fayetteville. *See* meteorite
Fe-hydroxide, 220
feiite, 221, 300
feitnechtite, 206
feldspar, 57, 76–78, 88–89, 104, 155, 157, 174, 179, 182, 189–190, 205, 208, 211, 213, 219–220, 226, 254–255, 257, 272, 295, 297, 309
feldspathoid, 126, 205
Felix. *See* meteorite
ferdisilicite, 213
ferrihydrite, 130, 134–136, 139, 144, 163, 167, 171–172, 263, 269, 321
ferri-magnesiohornblende, 272
ferrimagnetism, 80–81
ferroan alabandite, 96, 170, 174
ferroan anorthosites, 211, 308
ferroan olivine, 126, 128, 134, 137, 220, 259, 280
ferrochromite, 179
ferrohedenbergite, 104
ferro-kaersutite, 182, 220–221
ferromagnesian alabandite, 177
ferromagnetism, 80, 167, 301
ferropigeonite, 182, 213
ferrosilite, 98
fersilicite, 213
find
definition, 52
Finmarken. *See* meteorite
fireball, 51, 105, 233
fission, 83
fission track dating, 83
fizzed troilite, 278
fluorapatite, 174, 188, 190, 205, 274, 297, 322
fluorescence, 78–79, 203
fluorite, 63, 66–67, 69, 71, 77–79, 206
fluoro-endenite, 174
fluorphlogopite, 204, 287
fluor-richterite, 170, 178, 287, 292
foidolite, 205
foliations, 285
formic acid, 268
forsterite, 70, 72, 78, 86–87, 92, 98, 114, 116, 124, 126, 132–133, 135, 139, 151, 164, 166, 168–171, 174, 177, 189, 191, 200, 213, 234–235, 243, 247–248, 253, 261, 270, 301, 306, 325
fossil meteorite, 321, 324
Fountain Hills. *See* meteorite
fractional condensation, 94, 140, 280, 288
fracture
conchoidal, 68
hackly, 68
splintery, 68
francolite, 322
Franklin, Benjamin, 6
franklinfurnaceite, 206
franklinite, 206
franklinphilite, 206
Fredriksson, Kurt, 102, 362
Fremdlinge, 123, 144, 146, 268, 332, 340
Frenkel defects, 73
Fresnel equations, 86
FRO 90011. *See* meteorite
Fukang. *See* meteorite
fusion crust, 317
definition, 52

gabbro, 108, 174–175, 179, 182, 197–199, 208, 210–211, 213, 219, 292–293, 304, 307, 309, 347
gabbronorite, 211
Gale Crater, 217, 275
galena, 70–71, 91, 322
galileiite, 187, 314
Galileo spacecraft, 203
Galim. *See* meteorite
Ganymede, 49, 230
garnet, 53, 57, 71, 76, 117, 150, 205, 215, 268, 280, 354, 356, 372
gehlenite, 54, 166–167, 169, 234, 243
geikielite, 127, 177, 306
Geminids. *See* meteor shower
GEMS, 235, 253
gersdorffite, 322
Giant Crystal Cave, 55
Giant Impact hypothesis, 208
gibbsite, 72
Gibeon. *See* meteorite
Gilbert, Grove Karl, 8
glauberite, 84, 218
glauconite, 72, 324
glide plane, 60
gliding twins, 68
Glorieta Mountain. *See* meteorite
goethite, 2, 4, 57, 76, 91, 97, 178, 215, 218, 220, 319, 321–323
gold, 1–3, 66, 69, 71–72, 91, 97, 206, 238, 326
Gove. *See* fossil meteorite
Governador Valadares. *See* meteorite
GRA 06100. *See* meteorite
GRA 06129. *See* meteorite
GRA 95205. *See* meteorite
GRA 95209. *See* meteorite
GRA 98025. *See* meteorite
graftonite, 187, 314

Grand Rapids. *See* meteorite
granite, 4, 45, 57, 104, 205, 211, 374
Grant. *See* meteorite
granulite, 274
graphite, 10, 57, 62, 67, 70, 79, 84, 91, 97, 99, 128–129, 142–144, 148, 150–151, 156, 163, 168, 170–174, 177, 180, 184–188, 191, 193, 200–201, 206, 229, 235, 242, 245, 252, 285–286, 290, 293–295, 306, 309, 314–315, 321, 331, 333, 353, 359, 369
greenalite, 177, 205
greigite, 177
Grein 002. *See* meteorite
GRO 03116. *See* meteorite
GRO 95551. *See* meteorite
GRO 95574. *See* meteorite
GRO 95577. *See* meteorite
Grosnaja. *See* meteorite
grossite, 116, 125–126, 166–169, 243, 270, 375
grossmanite, 116, 167
grossular, 115, 117, 134, 167–168, 268–269
Guangmingshan. *See* meteorite
Guidder. *See* meteorite
Guin. *See* meteorite
Gujba. *See* meteorite
gwihabaite, 85
gypsum, 69, 76, 84, 87, 90–91, 130, 132, 205, 213, 218–221, 263, 275, 319, 321–322

H chondrites, 52, 105, 129–130, 140–141, 150–151, 155, 173–174, 198, 225, 255, 276, 280, 365, 367
habit, 54, 66–67
Hadley Rille. *See* meteorite
Haig. *See* meteorite
half-life, 81
halide, 5, 10, 66, 246, 328
halite, 54–55, 61, 63, 67, 69, 83–85, 132, 164, 172, 205, 218–220, 225, 274–276, 321, 378
halogens, 159, 246, 261, 274
halotrichite, 83
Hammadah al Hamra 073. *See* meteorite
Hammadah al Hamra 193. *See* meteorite
Hammadah al Hamra 237. *See* meteorite
Hammond. *See* meteorite
hanksite, 84
hapkeite, 213, 302, 331
hardness, 2–4, 62, 66, 68–69, 75, 85, 98
harzburgite, 149, 180, 224
hastingsite, 275
Haumea, 10
Haüy, Abbé, 4–5
haxonite, 97, 99, 115, 137, 139–140, 142, 144–145, 155, 166, 174, 184–188, 271, 295, 313
Hayabusa 2 spacecraft, 227
Hayabusa spacecraft, 226
heazlewoodite, 147, 168, 179, 321
Hebrew Scriptures, 2
hedenbergite, 106, 115, 126–127, 129, 134, 167, 175, 182, 213, 220, 224, 268–270
HEDs, 9, 103, 107–108, 129, 222–224, 303, 305
heideite, 142, 170, 177, 306
heliodor, 73
hematite, 1–2, 10, 45, 57, 71, 74–76, 79–80, 91, 132, 144, 147, 164, 204–205, 215, 218, 220, 258, 275, 321, 323–324, 358
hemimorphite, 206
hemleyite, 280
hercynite, 126, 133, 142, 144, 166, 176–177, 182, 213, 219–220, 305, 309, 372
Herschel, Alexander, 77
Herschel, William, 7, 77, 216
hetaerolite, 206
Hex River Mountains. *See* meteorite
hexaferrum, 167
hexagonal, 58–59, 69, 76, 86, 95, 97, 113, 148, 178, 184, 230, 293, 300, 355, 373
hexagonal closest packing, 64
hexahedrite, 56, 96, 102, 183, 315, *See* iron meteorites
hexahydrite, 218
hexamethylenetetramine, 267
hexamolybdenum, 116, 167, 355
hibbingite, 321
hibonite, 116–117, 124–125, 145, 148, 162–163, 166–169, 171, 242, 244, 247–248, 270
highlands, lunar, 200, 207, 213, 308
hiroseite, 280
hisingerite, 322
Hoba. *See* meteorite
Holbrook. *See* meteorite
hole color center, 74
hollandite, 280, 298, 318, 343
hollisterite, 325
Holmes, Sherlock, 327
honessite, 321
Hooke, Robert, 7
Hoover, Herbert, 3
Hoover, Lou, 3
hornblende, 57, 66, 71, 92, 162, 213, 261, 272
Horse Creek. *See* meteorite
Howard, Edward, 7, 101
howardites, 101, 103, 175–176, 224, 289–290, 302, 305, 311
howiite, 71
H-R diagram, 242
Huckitta. *See* meteorite
Hughes 026. *See* meteorite
hutcheonite, 117, 167, 268, 354
Huygens, Christian, 216
Hvittis. *See* meteorite
hydrogen bonding, 62
hydrogen cyanide, 230
hydromagnesite, 221, 319
hydrophilite, 85
hydrothermal alteration, 130, 133, 139, 205, 239, 254, 261, 263, 267, 272, 328, 339

hydroxide, 10, 71, 83, 320–321, 328
hydroxylapatite, 162, 265, 272
hydroxysulfide, 10, 328
hypersthene, 66, 87–88, 210
hypochlorite, 217

IAB silicates, 107, 174, 189, 293–294
Ibitira. *See* meteorite
icosahedrite, 56, 325
idaite, 262, 273
idiochromatic, 72
igneous rocks, 10, 56, 149, 176, 205, 218, 289, 305–306, 308, 372
Ilafegh (009). *See* meteorite
ILD 83500. *See* meteorite
illite, 57, 205, 221, 322
ilmenite, 57, 71, 96–97, 104, 114, 127, 129, 134, 136, 138–139, 142, 146, 153, 158–159, 162, 166–167, 175–176, 180, 182, 190, 197–199, 208–209, 211, 213, 219–221, 257, 259, 261, 267, 279–280, 293–295, 297–298, 304, 308–309, 314, 317, 342, 370, 372
Imilac. *See* meteorite
impact craters, 57, 83, 200, 203–204, 214–215, 221, 224, 230, 237, 290, 301
impact plumes, 239, 302, 329
impact-melt breccias, 97, 107, 141, 177–178, 211, 226, 261, 279–280, 286–287, 292–293, 349, 367
impact-melt pockets, 152
incandescence, 77
inclined extinction, 88
Indarch. *See* meteorite
inderite, 83
index of refraction, 75, 77, 85–86, 88–90, 92
Indian Valley. *See* meteorite
interference colors, 76, 86–87
internal iridescence, 76
internal reflections, 77, 91
International Astronomical Union, 44
interplanetary dust particles, 9, 46, 103, 233, 235, 318
interstellar interloper, 104, 237
inversion, 60, 68, 158, 239, 311, 314, 329
inverted pigeonite, 173, 175
ionic bonding, 61, 63, 65, 69
irarsite, 138, 142, 162
iron, 2
iron meteorites, 9–10, 44, 56, 68–70, 80, 92, 94, 97, 101, 103, 108, 183–184, 190, 196, 222, 246, 293, 295, 303, 306, 312, 321, 323, 333, 347–349, 355, 357–358, 363
  ungrouped, 46, 190–191, 196
iron-wüstite buffer, 305
Isheyevo. *See* meteorite
Isna. *See* meteorite
isocubanite, 188, 321
isogyre, 90–91
isometric, 59, 68, 86, 89, 91
isotope, 81–83, 150, 155, 235, 240, 259, 270, 374
Itqiy. *See* meteorite
Itzawisis. *See* meteorite
Ivuna. *See* meteorite

jade, 1, 72
jadeite, 1, 72, 280, 300
Jajh deh Kot Lalu. *See* meteorite
jarosewichite, 206
jarosite, 218, 220, 275, 321, 358
jasper, 2
Jbilet Winselwan. *See* meteorite
Jelica. *See* meteorite
jimthompsonite, 134
Jodzie. *See* meteorite
joegoldsteinite, 96, 187, 296, 347
johnsomervilleite, 187, 314
Johnson, Samuel, 205, 238, 348
Juancheng. *See* meteorite
Juno. *See* asteroid
Jupiter, 9, 47–48, 108, 163, 230, 235
Juvinas. *See* meteorite

K chondrites, 138
Kaba. *See* meteorite
kaersutite, 66, 182, 220–221, 294
Kaidun. *See* meteorite
Kainsaz. *See* meteorite
kaitianite, 116, 243, 354
Kakangari. *See* meteorite
kamacite, 56, 68–69, 80, 91, 94–95, 97, 99, 106, 109, 112, 115, 133–136, 138–140, 142–148, 151, 153, 155, 157–160, 162, 166–167, 169–171, 173, 175–177, 179–189, 191–196, 198, 213, 222, 226, 234, 246, 252–253, 255, 257–259, 262, 271, 283, 286–287, 289–290, 293, 295–296, 303–304, 306, 311, 315, 321, 323, 352
  cobalt-rich, 115, 128, 139
  polycrystalline, 95
  swathing, 94, 185–186, 189, 198
kamiokite, 123
Kamiomi. *See* meteorite
Kanada, 54
kangite, 54, 116, 166–167
kaolinite, 57, 72, 84, 205, 219, 229
Kapoeta. *See* meteorite
Karavannoe. *See* meteorite
Karoonda. *See* meteorite
K-Ba feldspar, 211, 213
Keil, Klaus, 49, 102, 143, 154, 331–333, 336, 341, 343, 346, 349, 351, 353, 358–359, 362, 366–367, 369–371, 373–374, 377
keilite, 66, 96, 143, 170, 180, 200, 286, 349, 370
Kernouvé. *See* meteorite
K-feldspar, 57, 174, 179, 190, 211, 220
khamrabaevite, 148, 243
Khatyrka. *See* meteorite
khatyrkite, 325–326, 346
Khor Temiki. *See* meteorite
kieserite, 218

King Tut, 2, 70
kirschsteinite, 115, 128, 134, 167, 176–177, 269, 305, 311, 314
Kirwan, Richard, 5
Klaproth, Martin, 101
Kodaikanal. *See* meteorite
Kokstad. *See* meteorite
komatiite, 200
kosmochlor, 186, 188, 234
Kota-Kota. *See* meteorite
kraisslite, 206
Krasnojarsk. *See* meteorite
KREEP, 210–211, 308, *See* Moon
krinovite, 188
krotite, 116, 167, 243, 270, 284, 355
kryachkoite, 325, 356
Kuiper Belt, 46, 235, 237, 348
kushiroite, 116, 167
kyanite, 68

L chondrites, 141, 150, 152, 282, 322, 359
L'Aigle. *See* meteorite
La Garita. *See* meteorite
labradorite, 76
laihunite, 318
Lakangaon. *See* meteorite
Lampadius, Wilhelm, 5
lamproite, 205
Lancé. *See* meteorite
Landes. *See* meteorite
LAP 02342. *See* meteorite
LAP 031047. *See* meteorite
LAP 03587. *See* meteorite
LAP 03719. *See* meteorite
LAP 04840. *See* meteorite
LAR 04315. *See* meteorite
LAR 06299. *See* meteorite
LAR 06317. *See* meteorite
Larned. *See* meteorite
larnite, 117, 176, 268
lattice, 54–55, 57–59, 61, 63, 68, 71, 74, 79, 160, 214, 292, 322
laumontite, 205
launch pairs, 52
laurite, 138, 142, 162
Lavoisier, Antoine-Laurent, 5–6
lawrencite, 170, 246
Lazarev. *See* meteorite
Lea County 002. *See* meteorite
lead, 2–3, 206
Leonids. *See* meteor shower
Leoville. *See* meteorite
lepidocrocite, 97, 225, 321–322
Leucippus, 54
leucite, 87, 205
LEW 85332. *See* meteorite
LEW 85440. *See* meteorite
LEW 86010. *See* meteorite
LEW 86216. *See* meteorite
LEW 86220. *See* meteorite
LEW 87051. *See* meteorite
LEW 87119. *See* meteorite
LEW 87223. *See* meteorite
LEW 87232. *See* meteorite
LEW 88774. *See* meteorite
lherzolite, 149, 219, 309
liebermannite, 221, 300
LIME olivine, 133, 137, 234
limestone, 205
limonite, 2, 76, 97, 130, 144, 171, 175
lingunite, 202, 220, 280, 298, 300
Linnaeus, Carl, 3–4
Lipovsky. *See* meteorite
lipscombite, 321
liuite, 221, 300
lizardite, 225
LL chondrites, 9, 105, 139, 141, 148, 155, 226–227, 282, 302, 361, 375
Lockne-Målingen crater, 322
lodestone, 81
Lodran. *See* meteorite
lodranites, 107, 173, 179, 291, 303, 314
lonsdaleite, 148, 172, 290
Loongana 001. *See* meteorite
Los Vientos 263. *See* meteorite
loseyite, 206
Lost City. *See* meteorite
Losttown. *See* meteorite
low-Ca pyroxene, 98, 112–114, 126, 130, 134, 155, 158–159, 162, 168–169, 171, 179, 187, 190, 196, 211, 225, 249–250, 257, 260, 262, 285, 297, 314
Lueders. *See* meteorite
luminescence, 78, 370
Luna 16, 208, 214
Luna 24, 208
lunar atmosphere, 214
lunar chondrules, 212
lunar meteorites, 8, 10, 44, 49, 67, 102–103, 107, 182, 208, 212–214, 275, 297, 301, 303, 308, 314
Lunar Prospector spacecraft, 207, 215
Lunar Reconnaissance Orbiter, 207, 215
lunar sample
  14321, 104
  69715, 274
  79215, 274
Luotolax. *See* meteorite
luster, 2, 4, 71, 86

MAC 02635. *See* meteorite
MAC 02675. *See* meteorite
MAC 88107. *See* meteorite
MAC 88136. *See* meteorite
macedonite, 79
machiite, 116, 166

mackinawite, 97, 139, 188, 193, 213, 257
Magellan spacecraft, 203
Magellan-Elcano expedition, 8
maghemite, 81, 137, 204, 318, 321
Magnéli phases, 244
magnesio-arfvedsonite, 213
magnesiochromite, 130, 148, 163, 174, 344
magnesiowüstite, 280
magnesite, 83, 219–221, 229, 275
magnetic moment, 80–81
magnetic susceptibility, 167, 338
magnetism, 66
magnetite, 1–2, 10, 45, 57, 67–68, 71, 75, 81, 96, 106, 115, 123, 126, 129–130, 132–137, 139–140, 142, 144–147, 155, 157, 162–164, 166–169, 172, 182, 184, 204–206, 213, 215, 218–220, 222, 225, 229, 234, 257–260, 263–265, 267–272, 275, 284, 309, 314, 317–318, 321, 323–326, 340, 348, 350–351, 359, 365, 369, 371
Magnus, Albertus, 3
main-group pallasites, 107, 307
majindeite, 123
majorite, 161, 168, 176, 202, 220, 280, 284, 300, 318, 338
malachite, 1, 71–72, 91
Malakialina pegmatite, 55
malleability, 3, 70
manganite, 71, 206
Manych. *See* meteorite
maohokite, 300
Maraldi, Giacomo, 216
Maralinga. *See* meteorite
marcasite, 71, 85, 91, 182, 219–221
Mare Serenitatis, xiii
margarite, 134, 269
maria, lunar, 8, 200, 207, 212–213
Maribo. *See* meteorite
maricite, 187, 189, 306
Marius Hills, 8
Marjalahti. *See* meteorite
Mars, xiii, 9–10, 44, 47, 49, 51–52, 102–104, 182, 214–217, 219, 221, 233, 275, 297, 309, 328, 330, 343, 356, 359, 363, 372–373, 378
martensite, 95, 128, 138, 140, 142, 146, 151, 162, 167, 180, 184–185, 190–191, 194, 256, 278, 283, 293, 312
martian meteorites, 52, 103, 107, 182, 192, 218–220, 275, 297, 303, 326, 332, 356, 359
maskelynite, 9, 54, 150, 160–161, 182, 199, 214, 219–220, 253, 280, 285, 289, 295, 297, 300, 309, 333
Mason, Brian, 86, 102
matrix material, 7, 57, 78, 97, 113, 128–140, 142, 144, 146–148, 150, 155–156, 158–159, 162–164, 166–169, 171–172, 176, 190, 194, 197, 228, 244, 246, 253, 255–259, 262–265, 267–273, 276, 283–284, 289–290, 292–293, 317, 322, 331, 335, 337, 339–340, 342, 344, 347, 351, 356, 367, 369, 374
maucherite, 322
Mayo Belwa. *See* meteorite
Mbale. *See* meteorite
mechanical twins, 68, 301
melanterite, 83, 85, 321
melatope, 90
melilite, 54, 116, 123–126, 128, 166–169, 246–248, 259, 264–265, 268, 270
Mercury, 10, 49, 104, 179, 200–202, 230, 315, 348, 363–364, 369, 374–375
Mercury Magnetospheric Orbiter, 202
Mercury Planet Orbiter, 202
merrihueite, 113, 339, 351
Merrill, George, 102
merrillite, 112–114, 130, 137, 146, 150, 158–159, 162–163, 166–168, 172, 174–177, 179, 182, 184, 188–189, 192, 194–198, 208, 211, 213, 219–220, 226, 234, 255, 270, 279, 293–294, 297–298, 305, 307, 335
mesosiderites, 80, 94, 101, 107–108, 192, 197–198, 222–229, 254, 257, 260–264, 267–268, 270–276, 282, 285, 289–293, 297, 300, 304–310, 314, 367
mesostasis, 109, 112, 114–115, 139, 151, 157–159, 166, 169, 171, 182, 190, 208, 211, 219, 226, 234, 249–250, 256, 263, 268, 271, 275
Messenger spacecraft, 200–201
MET 00426. *See* meteorite
MET 01070. *See* meteorite
MET 01075. *See* meteorite
MET 01149. *See* meteorite
MET 78008. *See* meteorite
meta-autunite, 84
metallic bonding, 62, 69, 71, 79, 84
metallic copper, 73, 95, 99, 140, 142, 145–146, 162, 173–174, 177–178, 187, 189, 193, 213, 278, 291, 294, 306, 321
metallic Fe-Ni, 10, 46, 70, 80, 93–95, 97–99, 101–102, 112–115, 123–124, 126, 128–129, 131, 133–134, 136–148, 150–155, 159–160, 162–164, 166–169, 171–178, 180, 182, 184, 187, 189, 192–196, 198–199, 211, 213, 224–226, 229, 235, 243, 250, 252, 255–259, 263–265, 267–268, 272, 278–280, 283, 285, 290–291, 293–297, 303, 305–307, 309, 311, 314, 317, 319, 321, 324, 326
metamiction, 72, 82
metasomatism, 115, 129, 134, 239, 263, 268, 272, 274, 329, 340, 353, 372
meta-torbernite, 84
meteor, 6–8, 46, 50, 229, 233
  definition, 49
Meteor Crater, 67, 337
meteor shower
  Arietids, 50
  definition, 50
  Leonids, 50
  Orionids, 50
  Perseids, 50
meteor train
  definition, 50

meteoric smoke
definition, 46
meteorite
Abbott, 57, 129
Abee, 127, 143, 285–287, 289, 365–367, 370
Acapulco, 173
Acfer 059, 267
Acfer 094, 106, 135, 169, 344, 361
Acfer 097, 129
Adelaide, 106, 127, 135, 169, 243, 335
Adhi Kot, 152, 285–287, 365
Admire, 193
Aguas Zarcas, 51, 163
Ahumada, 193
Al Haggounia 001, 321
Al Huwaysah 010, 106
Al Rais, 147, 258, 267
Alais, 101, 264, 332
Albareto, 141
Albin, 193
Aletai, 295
ALH 77307, 144
ALH 78091, 172
ALH 78113, 177
ALH 81005, 102
ALH 81101, 290
ALH 81189, 285
ALH 82130, 315
ALH 83009, 177
ALH 83014, 172
ALH 83015, 292
ALH 83108, 145
ALH 84001, 52, 182, 219, 275, 308, 314
ALH 84028, 106
ALH 85085, 113, 125, 129, 147, 168, 334, 350
ALH 85151, 142, 162, 260
ALHA76004, 137
ALHA77005, 298
ALHA77011, 129, 359
ALHA77015, 137
ALHA77156, 261–262
ALHA77255, 190
ALHA77283, 293
ALHA77307, 133, 335
ALHA78113, 177
ALHA78119, 129
ALH A81005, 301
ALHA81014, 190
ALHA81024, 129
Alikatnima, 106
Aliskerovo, 296
Allegan, 159
Allende, 44, 97, 99, 106, 115, 117, 127–129, 134, 145–146, 167, 243, 260, 268, 270, 334–335, 340, 342, 350–351, 353–356, 365–366, 373, 378
Almahata Sitta, 9, 104, 180, 245, 291, 340, 353
Angra dos Reis, 103, 176, 305
Appley Bridge, 140
Arch, 106
Arltunga, 185
Armanty, 295
Asuka 10164, 170, 285, 350
Asuka 12236, 131
Asuka 881314, 143
Asuka 881371, 305
Asuka 881394, 108
Asuka 881655, 258
Atlanta, 143, 261, 285
Auburn, 186
Axtell, 106
Babb's Mill (Troost's Iron), 196
Bali, 106, 128, 134, 145–146, 167, 260, 268, 351
Ballinoo, 185
Ban Rong Du, 190
Barbianello, 190
Barwell, 150
Belgica 7904, 169, 258, 283
Benares, 7
Bench Crater, 135, 214, 377
Bencubbin, 106, 148, 168, 284, 302
Bendegó, 186
Benld, 51
Berduc, 51
Bholghati, 176
Bishop Canyon, 187, 310
Bishopville, 101, 292
Bishunpur, 137, 160, 331
Bison, 282
Blithfield, 143, 285, 287, 289
Bocaiuva, 190
Bondoc, 293
Bouvante, 304
Brachina, 172
Brahin, 194
Bremervörde, 113
Brenham, 193–194, 307
Brunflo, 322, 362
Budulan, 274, 353
Bunburra Rockhole, 51, 103, 224
Bustee, 178, 349
Butler, 106
Cachari, 290
Cachiyuyal, 296
Caddo County, 174, 294
Campo del Cielo, 70, 175, 189, 293
Canon City, 51
Canyon Diablo, 188, 293–294, 352
Cape York, 56, 70, 187, 306
Carancas, 64
Carbo, 185, 357
Cargo Muchacho Mountains, 144–145
Carlisle Lakes, 162, 366
Carlton, 174, 189, 358
Cat Mountain, 282

meteorite (cont.)
Cayote Dry Lake 033, 52
Cerro los Calvos, 257
Chassigny, 103, 314
Château-Renard, 280, 332
Chico, 282
Choteau, 193, 195
Claxton, 51
Coahuila, 56, 186
Cold Bay, 194
Cold Bokkeveld, 101, 264
Colomera, 190, 295
Colonia Obrera, 295
Colony, 133, 145
Coolidge, 106, 169, 257, 260
Coopertown, 295
Crab Orchard, 293
Cratheús (1950), 185
Cumberland Falls, 177, 292
D'Orbigny, 176, 306
Danebury, 322
Dar al Gani 055, 170
Dar al Gani 1031, 143
Dar al Gani 670, 300
Dar al Gani 978, 170
Dayton, 174, 189, 342
definition, 44
Del Rio, 185
Dhajala, 156
Dhofar 1222, 173
Dhofar 1514, 142
Dhofar 280, 301
Dhofar 312, 173
Dhofar 778, 180
Diepenveen, 283
Dimmitt, 57, 129, 140, 151, 367
Divnoe, 106–107
DOM 03238, 318
DOM 08006, 81, 133, 139, 267–268
DOM 08351, 145
DOM 10088, 318
DOM 10104, 144
Dong Ujimqin Qi, 51
Dongtai, 141
Dora, 193
Duel Hill (1854), 187
Dunbogan, 51
Eagle Station, 192, 194
EET 83309, 274
EET 83311, 284
EET 84302, 173
EET 90043, 258
EET 90102, 178, 261
EET 90299, 143
EET 90992, 285
EET 92105, 248
EETA79001, 103, 219, 298, 300
Efremovka, 83, 106, 128, 152, 260, 270, 284, 351
El Médano 100, 106
Elbogen, 185
Elga, 190, 294
Eltanin, 199, 352
Emery, 10, 198, 293, 345, 365
Esquel, 193
Estherville, 293
fall, 51
Farmington, 141, 159
Fayetteville, 279, 302
Felix, 145
Finmarken, 193
Fountain Hills, 106, 148
FRO 90011, 173
Fukang, 192, 194
Galim, 170
Gibeon, 187, 310
Glorieta Mountain, 193
Gove, 323, 333
Governador Valadares, 314
GRA 06100, 258
GRA 06129, 180
GRA 95205, 172
GRA 95209, 173
GRA 98025, 170, 270
Grand Rapids, 190
Grant, 187
Grein 002, 286
GRO 03116, 258
GRO 95551, 106
GRO 95574, 290
GRO 95577, 133, 147, 168, 375
Grosnaja, 106
Guangmingshan, 51
Guidder, 141
Guin, 106, 190, 293
Gujba, 106, 148, 168, 284, 302–303, 362, 375
Hadley Rille, 143, 214, 345, 365
Haig, 293, 296, 349
Hammadah al Hamra 073, 169
Hammadah al Hamra 193, 174
Hammadah al Hamra 237, 106, 129, 148, 168, 302
Hammond, 190
Hex River Mountains, 186
Hoba, 46, 186
Holbrook, 52
Horse Creek, 190
Huckitta, 193
Hughes 026, 172
Hvittis, 151, 285, 365
Ibitira, 108
Ilafegh (009), 289
ILD 83500, 196
Imilac, 193
Indarch, 96, 171, 246, 262, 342
Indian Valley, 315

Isheyevo, 125, 169
Isna, 133, 145
Itqiy, 181
Itzawisis, 193–194
Ivuna, 144, 163
Jajh deh Kot Lalu, 128, 143, 285
Jbilet Winselwan, 283, 378
Jelica, 94, 279
Jodzie, 176
Juancheng, 51
Juvinas, 175
Kaba, 106
Kaidun, 133, 270, 283, 347, 378
Kainsaz, 145
Kakangari, 106–107, 138, 144, 171, 335, 358
Kamiomi, 51
Kapoeta, 57, 129, 176, 224, 290, 302
Karavannoe, 194
Karoonda, 134, 167, 349, 357
Kernouvé, 277
Khatyrka, 56, 325–326, 332–333, 346, 348, 353, 356–357, 371
Khor Temiki, 292
Kodaikanal, 295
Kokstad, 296
Kota-Kota, 262
Krasnojarsk, 193
La Garita, 190
L'Aigle, 7
Lakangaon, 304
Lancé, 145, 283
Landes, 175
LAP 02342, 133, 235, 374
LAP 031047, 316
LAP 03587, 291
LAP 03719, 292
LAP 04840, 142, 162, 272
LAR 04315, 291
LAR 06299, 141, 160, 280, 367
LAR 06317, 152, 284
Larned, 177
Lazarev, 106
Lea County 002, 144, 171
Leoville, 106, 127, 152, 251, 370
LEW 85332, 136, 270, 335
LEW 85440, 172, 315
LEW 86010, 176, 305, 311
LEW 86216, 290
LEW 86220, 173
LEW 87051, 305
LEW 87119, 285
LEW 87223, 171
LEW 87232, 171
LEW 88774, 172, 315
Lipovsky, 193
Lodran, 173
Loongana 001, 106, 169, 257, 260
Los Vientos 263, 195–196
Lost City, 51
Losttown, 185
Lueders, 175
Luotolax, 101
MAC 02635, 143, 285
MAC 02675, 106, 148
MAC 88107, 270
MAC 88136, 138, 361
Manych, 150
Maralinga, 146
Maribo, 265
Marjalahti, 193
Mayo Belwa, 178, 292, 333
Mbale, 51
MET 00426, 133, 331
MET 01070, 144, 265
MET 01075, 132
MET 01149, 142
MET 78008, 290
Middlesbrough, 77
Mifflin, 51
Mighei, 163
MIL 03346, 275
MIL 07440, 162, 272
MIL 090206, 107
MIL 090543, 144
MIL 090807, 289
MIL 11207, 162, 272
MIL 11213, 144
MIL 99301, 277
Miles, 294
Milton, 195–196, 337, 358
Moama, 175
Mokoia, 106, 115
Molong, 193
Monahans (1938), 185
Monahans (1998), 225, 276
Mont Dieu, 107, 155, 295
Monument Draw, 173
Moonbi, 185
Moorabie, 257
Moore County, 175
Morasko, 188
Morokweng, 160
Morristown, 293
Moss, 145
Mount Egerton, 178, 306
Mount Morris (Wisconsin), 294
Mount Padbury, 307
Mundrabilla, 188
Muonionalusta, 293
Murchison, 132, 333, 345, 355
Murray, 163, 264
Nakhla, 275, 314, 343
Neagari, 51
Nedagolla, 190

meteorite (cont.)
Needles, 185
Nelson County, 185
Netschaëvo, 107, 155, 189, 294
Newport, 193, 337
Ngawi, 271
Nilpena, 129, 172, 244
Ningqiang, 127, 134, 170, 346, 353
Nocoleche, 186
Nogata, 51
Nogoya, 132, 144, 283
Norton County, 177–178, 292, 306, 335, 362
Nothing, 185
Nova 001, 172
Novato, 51
Nuevo Laredo, 304, 307
Nulles, 141
NWA 011, 108
NWA 10019, 193, 195
NWA 10085, 144, 171
NWA 10214, 107, 155
NWA 10588, 259
NWA 10986, 308
NWA 11042, 322
NWA 1109, 175, 224
NWA 11119, 179–180
NWA 1152, 170
NWA 11561, 258
NWA 11575, 179
NWA 12774, 176
NWA 1463, 174, 294
NWA 1559, 135, 366
NWA 176, 190
NWA 1814, 106, 148
NWA 1878, 197
NWA 1911, 195–196
NWA 2085, 282
NWA 2131, 133
NWA 2196, 147, 258
NWA 2646, 182, 219
NWA 2650, 290
NWA 2653, 260
NWA 2743, 186
NWA 2824, 108
NWA 2994, 283
NWA 3250, 283
NWA 4284, 322
NWA 4301, 289
NWA 4590, 177
NWA 468, 175
NWA 470, 284
NWA 4704, 295
NWA 4705, 190
NWA 5000, 52
NWA 5028, 147, 258
NWA 5202, 133
NWA 5343, 135
NWA 5400, 106
NWA 5492, 106, 250
NWA 5738, 175, 273–274
NWA 5790, 275
NWA 6112, 107
NWA 6163, 190
NWA 6258, 289
NWA 6259, 80
NWA 6693, 179
NWA 6698, 179
NWA 6901, 147, 283
NWA 6921, 147, 258, 283
NWA 7034, 219, 275, 297, 353
NWA 7317, 147, 258, 283
NWA 7325, 106, 179–180, 348
NWA 7531, 258, 283
NWA 7533, 275
NWA 7821, 106
NWA 8003, 289
NWA 801, 150
NWA 8535, 176
NWA 856, 300
NWA 859, 106
NWA 869, 57
Ocotillo, 174, 294
Odessa, 174
Old Woman, 186
Orgueil, 263–264, 283, 332, 344, 372
Ornans, 145, 166–167, 352, 359
Orvinio, 282
Österplana, 323
Oued Bourdim 001, 194
Padvarninkai, 289
Paloduro, 296
Parambu, 140
Paris, 131–132, 144, 164, 265, 366
Parsa, 151, 285, 333
PAT 91501, 141, 160, 258, 280, 283, 293
Pavlodar, 193
PCA 91002, 162, 260
PCA 91008, 258
PCA 91020, 285
PCA 91241, 142, 272
Peekskill, 51
Peña Blanca Spring, 177
Perryville, 185
Phillips County, 194, 196
Piancaldoli, 256
Pine River, 294
Plainview, 129, 341, 367
Pontlyfni, 294
Portales Valley, 141, 277, 367–368
Porto Alegre, 295
Portugal, 7
Prambanan, 190
Puente del Zacate, 187
Pultusk, 52

Qingzhen, 115, 262, 273, 331, 340
QUE 93005, 144
QUE 93429, 106
QUE 94201, 103, 314
QUE 94204, 289
QUE 94368, 286
QUE 94411, 106, 148, 168, 302
QUE 97289, 177
QUE 99038, 170, 270
QUE 99177, 133, 331
Quebrada Chimborazo 001, 106, 333
Raguli, 97, 280
Ramsdorf, 160, 280
Rawlinna 001, 193–194
RBT 04143, 152, 284
Renazzo, 147, 167, 258, 331
Revelstoke, 51
Rhine Villa, 295
RKPA80259, 152, 285
Rose City, 141, 280
Rumuruti, 107, 162, 334, 369
Sahara 00182, 169
Sahara 987645, 129
Sahara 99555, 306
Saint-Sauveur, 286–287
Salles, 7
San Cristobal, 174, 189
Santa Clara, 186
Santa Rosa, 186
Santa Rosalia, 193
São João Nepomuceno, 187
Sariçiçek, 176, 224
Sayama, 51
Sayh al Uhaymir 169, 308
Semarkona, 56, 78, 81, 114, 126, 137, 139, 142, 155, 158–160, 245, 252, 255, 271, 331–332, 347, 365–366
Sericho, 194
Seymchan, 193
Shallowater, 178, 292, 306
Sharps, 129
Shaw, 282
Shergotty, 300, 340, 355, 372
Siena, 7
Sierra Gorda 010, 142
Sierra Gorda 013, 168, 348
Sikhote-Alin, 186
Sixiangkou, 95, 343
Smyer, 141, 365
Social Circle, 96, 187, 293, 296, 315, 347
Sombrerete, 294
South Bend, 193
South Byron, 196, 358
Springwater, 193–194, 196
St. Genevieve County, 186
St. Mary's County, 256
St. Mesmin, 57
Stannern, 101, 304
Staunton, 295
Steinbach, 187
STG 07009, 190
Superior Valley 014, 173
Sutter's Mill, 315
Suwahib (Buwah), 257
Sylacauga, 51
Tagish Lake, 108, 136, 170, 270
Tahara, 51
Tanokami Mountain, 295
Tarahumara, 295
Tawallah Valley, 186
Techado, 189
Thiel Mountains, 193
Tibooburra, 106
Tieschitz, 99, 137, 150, 256, 272, 339
Tishomingo, 190, 293
Tissint, 298, 314, 332, 337, 356
Tlacotepec, 186
Tobychan, 295
Toluca, 174, 189
Trysil, 95
Tucson, 106, 191, 293, 335, 350, 369, 374, 376
Ulasitai, 295
Unter-Mässing, 185
Vaca Muerta, 198, 293
Valera, 51
Verkhne Dnieprovsk, 295
Vermillion, 192–195, 334
Vigarano, 106, 127, 134, 146, 167, 247, 269, 351, 355
Vilna, 51
Wallapai, 185
Warrenton, 133, 145
Washington County, 191
Watson 001, 189–190
Weatherford, 106, 148, 302
Wedderburn, 97, 188, 355
Weekeroo Station, 294–295
Wethersfield (1982), 51
Wiley, 185, 342, 346, 350
Willamette, 46, 191, 293
Willaroy, 257
Willow Creek, 296
Winona, 174, 294, 358
Wold Cottage, 7
Woodbine, 175
Woolgorong, 51
Xinjiang, 295
Y-691, 115
Y-693, 146
Y-74063, 294
Y-74160, 280, 283
Y-75032, 290
Y-75097, 150
Y-75267, 318
Y-790143, 282

meteorite (cont.)
Y-790964, 282
Y-791199, 290
Y-791200, 290
Y-791510, 171
Y-791790, 285
Y-793225, 171
Y-793241, 150
Y-793321, 258
Y-81020, 133
Y-82104, 146
Y-82189, 287, 289
Y-8414, 289
Y-8451, 192, 194–195
Y-86029, 257
Y-86720, 258
Y-86789, 258
Y-980433, 175
Yamato 86720, 258
Yamato 86789, 258
Yanzhuang, 282
Zacatecas (1792), 190
Zag, 225, 276, 349
Zagami, 300, 340, 356, 372
Zaisho, 194, 196
Zakłodzie, 99, 180, 289, 355
Zinder, 195–196, 334
meteorite clan, 106, 302
meteorite class, 107, 163
meteorite group, 52, 104–106, 184, 188
meteorite shower, 52
definition, 52
Meteoritical Bulletin Database, 103, 192, 212–213, 258, 262, 283
meteoroid, 7–8, 46, 48–52, 104, 201, 315–317
definition, 46
meteoroid stream
definition, 51
meteorwrong, 45, 101
definition, 45
methanol, 237
mica, 57, 83, 88, 134, 142, 205, 261, 270, 272, 319, 366
microchondrules, 113, 136–137, 250, 267
microgabbro, 219, 309
micrometeorite, 9, 44, 46, 103, 113, 148, 215, 226, 235, 290, 301, 318, 324, 326, 328, 371
definition, 44
micrometeoroid
definition, 46
Middlesbrough. *See* meteorite
Mifflin. *See* meteorite
Mighei. *See* meteorite
MIL 03346. *See* meteorite
MIL 07440. *See* meteorite
MIL 090206. *See* meteorite
MIL 090543. *See* meteorite
MIL 090807. *See* meteorite
MIL 11207. *See* meteorite
MIL 11213. *See* meteorite
MIL 99301. *See* meteorite
Miles. *See* meteorite
Miller, William, 60
Miller Indices, 60–61
millerite, 67, 146, 167, 235
Milton. *See* meteorite
mineral
definition, 53
mineraloid, xiii, 54, 56
definition, 54
mini-Neptune, 49
mirror planes, 60
mirror symmetry, 55
Moama. *See* meteorite
Mohs, Fredrich, 68
Mohs scale, 68
moissanite, 148, 241
Mokoia. *See* meteorite
molecular clouds, 240
Molong. *See* meteorite
molybdate, 10, 328
molybdenite, 69–70, 91, 123
Monahans (1938). *See* meteorite
Monahans (1998). *See* meteorite
monazite, 72, 82, 213, 220, 275
moncheite, 138, 142, 162
monipite, 123
monoclinic, 59, 68, 84, 86, 136–137, 157–158, 162, 168, 170, 175, 220–221, 231, 280, 289, 300
monosulfide solid solution, 142, 146, 167, 259
Mont Dieu. *See* meteorite
monticellite, 55, 117, 134, 166–167, 268, 270
montmorillonite, 57, 69, 134, 205, 269
Monument Draw. *See* meteorite
monzodiorite, 211
moon
definition, 49
Moon, xiii, 7–10, 44–45, 49, 57, 72, 103–104, 136, 143, 182, 201–202, 206–208, 210, 213–214, 217, 231, 290, 297, 301, 308, 328, 330, 339, 346, 354, 363, 365, 369, 371–372, 374, 377
Moonbi. *See* meteorite
Moorabie. *See* meteorite
Moore County. *See* meteorite
Moore, Patrick, 8
Morasko. *See* meteorite
moraskoite, 188
morganite, 73
Morokweng. *See* meteorite
Morokweng impact crater, 160
Morristown. *See* meteorite
mosaic extinction, 285
Moss. *See* meteorite
Mount Egerton. *See* meteorite

Mount Morris (Wisconsin). *See* meteorite
Mount Padbury. *See* meteorite
MSS. *See* monosulfide solid solution
mudstone, 205, 218
mullite, 205, 243
Mundrabilla. *See* meteorite
Muonionalusta. *See* meteorite
murchisite, 164
Murchison. *See* meteorite
Murray. *See* meteorite
muscovite, 70, 87–88

nahcolite, 83
Nakhla. *See* meteorite
nakhlites, 102, 107, 182, 219, 275, 308, 314, *See* martian meteorite
nanomineralogy, 100
nanophase iron particles, 301–302
native elements, 5, 10, 66, 69, 328
Neagari. *See* meteorite
NEAR-Shoemaker spacecraft, 225, 352
Nedagolla. *See* meteorite
Needles. *See* meteorite
Nelson County. *See* meteorite
NEOs, 315
nepheline, 87, 106, 114–115, 117, 125–127, 129, 134, 137, 162, 166–167, 176, 268, 270, 272, 325, 350, *See* feldspathoid
nephrite, 1, 72
neptunism, 4
nesquehonite, 319
Netschaëvo. *See* meteorite
Neumann bands. *See* Neumann lines
Neumann lines, 68, 95, 296, 315
New Horizons spacecraft, 236
newberyite, 206
Newport. *See* meteorite
Newton, Isaac, 6, 8, 233
Ngawi. *See* meteorite
niahite, 206
Ni-rich blödite, 130
nickeline, 322
Nicol, William, 4
nierite, 148, 171, 242
Nilpena. *See* meteorite
Ningqiang. *See* meteorite
niningerite, 66, 96, 115, 142–143, 170, 177, 181, 200, 214, 251, 262, 273, 286, 306, 370
Nital, 95
niter, 71, 83–84
nitride, 10, 66, 171, 184, 191, 242, 328, 331, 336, 361, 370
noble-metal nuggets, 123–124, 146, 234
Nocoleche. *See* meteorite
Nogata. *See* meteorite
Nogoya. *See* meteorite
norite, 149, 210–211, 308
Norton County. *See* meteorite
Nothing. *See* meteorite
Nova 001. *See* meteorite
Novato. *See* meteorite
Nuevo Laredo. *See* meteorite
Nulles. *See* meteorite
nuwaite, 66, 117
NWA 011. *See* meteorite
NWA 10019. *See* meteorite
NWA 10085. *See* meteorite
NWA 10214. *See* meteorite
NWA 10588. *See* meteorite
NWA 10986. *See* meteorite
NWA 11042. *See* meteorite
NWA 1109. *See* meteorite
NWA 11119. *See* meteorite
NWA 1152. *See* meteorite
NWA 11561. *See* meteorite
NWA 11575. *See* meteorite
NWA 12774. *See* meteorite
NWA 1463. *See* meteorite
NWA 1559. *See* meteorite
NWA 176. *See* meteorite
NWA 1814. *See* meteorite
NWA 1878. *See* meteorite
NWA 1911. *See* meteorite
NWA 2085. *See* meteorite
NWA 2131. *See* meteorite
NWA 2196. *See* meteorite
NWA 2646. *See* meteorite
NWA 2650. *See* meteorite
NWA 2653. *See* meteorite
NWA 2743. *See* meteorite
NWA 2824. *See* meteorite
NWA 2994. *See* meteorite
NWA 3250. *See* meteorite
NWA 4284. *See* meteorite
NWA 4301. *See* meteorite
NWA 4590. *See* meteorite
NWA 468. *See* meteorite
NWA 470. *See* meteorite
NWA 4704. *See* meteorite
NWA 4705. *See* meteorite
NWA 5000. *See* meteorite
NWA 5028. *See* meteorite
NWA 5202. *See* meteorite
NWA 5343. *See* meteorite
NWA 5400. *See* meteorite
NWA 5492. *See* meteorite
NWA 5738. *See* meteorite
NWA 5790. *See* meteorite
NWA 6112. *See* meteorite
NWA 6163. *See* meteorite
NWA 6258. *See* meteorite
NWA 6259. *See* meteorite
NWA 6693. *See* meteorite
NWA 6698. *See* meteorite

NWA 6901. *See* meteorite
NWA 6921. *See* meteorite
NWA 7034. *See* meteorite
NWA 7317. *See* meteorite
NWA 7325. *See* meteorite
NWA 7531. *See* meteorite
NWA 7533. *See* meteorite
NWA 7821. *See* meteorite
NWA 8003. *See* meteorite
NWA 8535. *See* meteorite
NWA 856. *See* meteorite
NWA 859. *See* meteorite
NWA 869. *See* meteorite

obsidian, 54
obtuse bisectrix, 91
Ocotillo. *See* meteorite
octahedral sites, 55, 63, 114, 194
octahedrite, 96, 102, 183, 189, 295, 312, *See* iron meteorites
Odessa. *See* meteorite
odor, 4, 84
ogdensburgite, 206
oikocrysts, 151, 179
Olbers, Wilhelm, 7
Old Woman. *See* meteorite
oldhamite, 66, 96, 115, 127–128, 142–143, 151, 170, 177–178, 180, 200, 245, 251, 273, 285–286, 306, 321, 340, 353, 362
oligoclase, 158, 162, 190, 256
olivine, 57, 63, 70, 72, 78, 80, 86–89, 92, 94, 97–98, 102, 105–106, 109, 112–115, 125–131, 133–137, 139, 144, 146, 148–153, 155–159, 161–164, 166–169, 171–177, 179–180, 182, 186–187, 189, 192–199, 201–202, 204–205, 208–209, 211–213, 216, 218–222, 224–226, 233–235, 239, 244, 248, 250, 253, 255–260, 264, 269, 271–272, 274–275, 279–280, 283–285, 290, 292–293, 295, 297–298, 301–305, 307–309, 311, 314–316, 318–320, 326, 347, 351, 353, 357, 365–366
omphacite, 150, 205, 280
onyx, 2
Oort Cloud, 47, 236–237
opal, 1, 54, 68, 76, 79, 218–220, 274–275, 321
opaque phases, 76, 91–93, 109, 112, 140, 142, 144–146, 155, 166, 168, 180, 212, 250, 271, 373
opaque veins, 278, 283, 285, 291
optic sign, 89–91
optical indicatrix, 89
orcelite, 322
ordinary chondrites, 78, 95–97, 99, 101–102, 107, 112–113, 129, 136, 139–142, 145, 149, 152–153, 155–156, 158, 160, 177, 192, 225–226, 249, 278, 295, 302, 319–320, 326, 331, 333, 335, 338–339, 343–344, 347, 349, 351, 361, 365, 367, 369–371, 375–376, 378
ordinary ray, 90
Orgueil. *See* meteorite
Orionids. *See* meteor shower
Ornans. *See* meteorite
orpiment, 91
orthobronzite, 187
orthoclase, 67–69, 76, 82, 87, 205–206, 213
orthoenstatite, 114, 132, 135, 163–164, 170, 262
orthopyroxene, 54, 71, 88, 102, 112, 115, 149–151, 158, 160, 172–178, 180, 182, 187, 190, 192, 194–198, 208, 213, 218–222, 224, 226, 229, 254, 256–257, 263, 280, 289–290, 292–295, 297, 300, 304, 309–311, 314
orthopyroxenite, 56, 107, 149, 176, 178, 219, 224, 275, 306–308, 314
orthorhombic, 59, 86, 97, 114, 136, 157–158, 168, 170, 175, 221, 280, 300, 340
Orvinio. *See* meteorite
osbornite, 97, 125, 143, 171, 177, 234, 306
oscillatory zoning, 249, 318
OSIRIS-REx spacecraft, 228
Österplana. *See* fossil meteorite
osumilite, 113
Oued Bourdim 001. *See* meteorite
Oumuamua, 104, 237, *See* interstellar interloper
oxalate, 10, 328
oxidation, 10, 73, 76, 88, 97, 102, 107, 128–129, 139–140, 145, 153, 159, 170, 192, 216, 244, 251, 257, 260–261, 264, 267–268, 280, 284, 310, 314, 317, 319–322, 337, 354, 358, 368
oxide, 5, 10, 53, 66, 71, 75–76, 83, 114, 126, 131, 137, 139–140, 142, 146–148, 162, 170, 182, 191, 205, 211–212, 219–220, 244, 257, 265, 270, 280, 289, 318, 320–321, 328, 335, 347
oxynitride, 10, 171, 328

Padvarninkai. *See* meteorite
pairing
  definition, 52
Pallas. *See* asteroid
pallasites, 80, 94, 101, 107–108, 192–196, 307–308, 314, 334, 374
Paloduro. *See* meteorite
panethite, 174, 189, 342
panguite, 54, 66, 116, 126, 166–167
paqueite, 116, 167, 354
parallel extinction, 88
paramagnetism, 80
Parambu. *See* meteorite
pararealgar, 84
Paris. *See* meteorite
Parker Solar Probe, 50
Parsa. *See* meteorite
parting, 67–68
Partsch, Paul, 101
PAT 91501. *See* meteorite
Pauli exclusion principle, 74, 79
Pauling, Linus, 65
Pauling's Rules, 65
Pavlodar. *See* meteorite
PCA 91002. *See* meteorite

PCA 91008. *See* meteorite
PCA 91020. *See* meteorite
PCA 91241. *See* meteorite
Peekskill. *See* meteorite
Peña Blanca Spring. *See* meteorite
penetration twins, 68
pentlandite, 71, 97, 115, 126, 130–131, 133–137, 139–140, 142, 144–147, 150, 155, 160, 162–164, 166–170, 172, 179–180, 182, 187, 190, 193, 213, 219–220, 234–235, 252, 257, 261, 263, 265, 267–268, 270–271, 283–284, 320–321, 325, 332, 337
perchlorate, 217
periclase, 130, 163, 275, 344
perihelion, 50, 201, 216, 239, 315–316, 328–329
perovskite, 116, 124–128, 142, 144–145, 162, 166–169, 171, 177, 243, 259, 270, 280, 298, 306
perryite, 115, 143, 171, 177, 262, 306, 314, 352, 374
Perryville. *See* meteorite
Perseids. *See* meteor shower
petedunnite, 206
petrofabric, 265, 270, 285, 316
Phase T, 127
Phillips County. *See* meteorite
phlogopite, 70, 115, 134, 150, 162, 170, 182, 204–205, 220–221, 261, 269–270, 272
Phobos, 49, 104, 376
phosphate, 5, 10, 66, 76, 83, 123, 128–129, 134, 140, 145, 157–159, 162, 164, 166, 169, 171–172, 174–177, 182, 184, 187, 189, 192–194, 196–197, 206, 208, 211, 213, 219–221, 226, 255, 267, 270, 272, 294–295, 297, 304–307, 310, 314, 321, 324, 328, 342, 344, 347, 349, 362
phosphide, 10, 66, 95, 99, 140, 142, 144, 171, 173, 184, 255, 295, 328, 347, 355, 364, 378
phosphorescence, 66, 78
phosphorite, 205
photoluminescence, 78
phyllosilicate, 10, 84, 106, 115, 126, 128–131, 133–138, 160, 164, 166–169, 220, 222, 224–225, 228–229, 233, 235, 257, 260, 263–265, 267–270, 272, 275, 283, 315, 321–322, 332, 335
Piancaldoli. *See* meteorite
piezoelectricity, 79
pigeonite, 112, 127, 149, 158, 162, 166, 171–172, 175–177, 179, 182, 190, 205, 210, 213, 219–221, 224, 234, 252, 262, 290, 304, 311, 314–315
Pine River. *See* meteorite
plagioclase, 54, 57, 68, 70–71, 89, 97, 112, 114–115, 127–128, 134, 136, 141, 144, 149–152, 154, 158–162, 169–182, 187, 189–191, 193, 195, 197–199, 202–203, 205, 208, 210–211, 213–214, 218–221, 223, 226, 248, 256, 268, 270, 272, 274, 278–280, 285, 288–289, 291–295, 297, 301, 304–309, 318, 334, 344, 353, 356, 365
Plainview. *See* meteorite
planar fractures, 160–161, 285
planet
  definition, 47
planet, dwarf, 10, 46
planetary nebula, 61
planetary-type rare gases, 294
pleochroism, 82, 88, 91–92, 97
pleonaste, 166–167, 176, 182, 213, 357
plessite, 95, 142, 147, 168, 175, 185–189, 192, 194–195, 255–256, 258–259, 278, 283, 293, 295–296
plessitic octahedrite, 183, 185
Pliny the Elder, 2, 6, 68
Pluto, 10, 47, 236–237
point groups, 60
polychromatic, 76, 87
polycyclic aromatic hydrocarbons, 230
polymorphs, 67, 138, 148, 188, 202, 221, 231, 280, 290, 298, 300, 350
polysynthetic twinning, 68, 257
Pontlyfni. *See* meteorite
Portales Valley. *See* meteorite
portlandite, 66, 178, 321
Porto Alegre. *See* meteorite
Portugal. *See* meteorite
Posidonius of Apameia, 232
positron emission, 82
powellite, 66, 206
Prambanan. *See* meteorite
prehnite, 205
presolar grains, 9, 104, 130, 135, 138, 144, 148, 155, 158, 163, 166–167, 234, 240, 254, 260, 267–268, 377
Prior, George, 102, 153, 333
propane, 230
protoenstatite, 114, 158, 279
protoplanetary disk, 164, 263, 369
protopyroxene, 68
pseudobrookite, 317
pseudochromatic, 75
pseudotachylite, 290
Psyche spacecraft, 229
Puente del Zacate. *See* meteorite
Pultusk. *See* meteorite
purpurite, 83
pyrargyrite, 71
pyrite, 7, 66–68, 71, 76, 84–85, 91, 96–97, 137, 142, 146, 162, 167, 182, 204, 219–220, 263, 322
pyrochlore, 211
pyroclastic rocks, 208
pyroelectricity, 79
pyromorphite, 72
pyrope, 150, 168, 205, 215, 338
pyrophanite, 97, 206, 280, 351
pyrophyllite, 84

pyroxene, 10, 57, 68, 89, 92–93, 98, 106, 109, 112–115, 117, 124, 126, 130–137, 139, 144, 148, 150–154, 157–164, 166–180, 182, 187, 189, 192–199, 201–203, 208–213, 216, 218–223, 225–227, 233–235, 243–244, 248–250, 252–254, 256–258, 260, 262, 264, 267, 269–272, 274, 279–280, 285, 289–291, 293, 295, 297–298, 301–304, 307–309, 311, 314, 316, 318, 326, 332, 334, 350, 354–355, 358, 365, 375
pyroxene pallasites, 107, 192
pyroxenite, 178, 182, 306
pyroxferroite, 175, 182, 213, 220–221, 224
pyrrhotite, 71, 91, 96, 126, 130–131, 133–137, 142, 144, 146–147, 160, 162–164, 166–169, 172, 175, 182, 214, 219–220, 234–235, 252, 260–261, 263, 265, 267, 270, 283, 320, 322, 332

Qingzhen. *See* meteorite
quartz, 3, 5, 55, 57, 67–69, 71, 74, 76, 79–80, 87–89, 99, 104, 170, 175, 180, 200, 205, 211, 213, 220–221, 262, 267, 301, 322, 344, 355
  rose, 74
  rutilated, 74
quartzolite, 205
quasicrystal, 56, 325, 332–333, 346, 348
QUE 93005. *See* meteorite
QUE 93429. *See* meteorite
QUE 94201. *See* meteorite
QUE 94204. *See* meteorite
QUE 94368. *See* meteorite
QUE 94411. *See* meteorite
QUE 97289. *See* meteorite
QUE 99038. *See* meteorite
QUE 99177. *See* meteorite
Quebrada Chimborazo 001. *See* meteorite

R chondrites, 92, 94, 107, 130, 137, 142, 162, 192, 222, 260–261, 272, 319, 321, 366
radioactivity, 66, 81
radiometric dating, 83, 227
radius ratio, 63, 65
Raguli. *See* meteorite
Raman spectroscopy, 260, 355, 375–376
Ramdohr, Paul, 92, 273, 340, 364
rammelsbergite, 322
Ramsdorf. *See* meteorite
Rawlinna 001. *See* meteorite
RBT 04143. *See* meteorite
realgar, 2, 84, 91
redox effects, 239, 311, 318, 329
reevesite, 321
reflectance spectrum, 108, 227
reflection pleochroism, 91
refractive index, 72, 85–86, 90
refractory inclusions. *See* CAIs
refractory metal nuggets, 123–124, 146, 234, 243
regolith breccias, 9–10, 57, 113, 129, 141, 177, 191, 212, 276, 279, 289, 302, 308
Reichenbach lamellae, 187, 314
reidite, 290, 300
relief, 88, 98
remanent magnetization, 81, 167
Renazzo. *See* meteorite
retardation, 86–88
Revelstoke. *See* meteorite
rhabdite, 96, 185–188, 313, 315, *See* schreibersite
Rheasilvia crater, 103
Rhine Villa. *See* meteorite
rhodonite, 72
rhombohedral, 59, 67
rhönite, 124, 176, 243, 246
richterite, 205
riebeckite, 66, 84
riesite, 300
ringwoodite, 161, 180, 202, 220–221, 239, 280, 298, 301, 318
RKPA80259. *See* meteorite
rock
  definition, 56
  igneous, 56
  metamorphic, 56, 81, 205
  sedimentary, 56, 205, 217–218
roedderite, 113, 170, 174, 177, 189, 306, 351
Rose, Gustave, 101–102
Rose City. *See* meteorite
Rosetta spacecraft, 233
rotation, 59, 68, 200, 203, 214
rotoinversion, 60
roweite, 206
rubinite, 116, 167, 356
ruby, 2, 53, 73, 76, 79, 215
Rumuruti. *See* meteorite
rustenburgite, 142, 162
ruthenosmiridim, 142, 162
rutile, 54, 68, 74, 76, 88, 91, 97, 127, 139, 148, 158–159, 175, 182, 188, 190, 193, 206, 211, 213, 219–221, 243, 257, 294, 314, 344

safflorite, 322
Sahara 00182. *See* meteorite
Sahara 987645. *See* meteorite
Sahara 99555. *See* meteorite
Saint-Sauveur. *See* meteorite
Salles. *See* meteorite
samfowlerite, 206
San Cristobal. *See* meteorite
sandstone, 1, 56, 205, 218
sanidine, 182, 190, 205, 220–221, 295
Santa Clara. *See* meteorite
Santa Rosa. *See* meteorite
Santa Rosalia. *See* meteorite
São João Nepomuceno. *See* meteorite
saponite, 115, 130, 133–134, 136, 163, 167–169, 172, 214, 263, 267, 269–270, 275, 322, 324, 335
sapphire, 5, 53, 73, 76, 85

sapphirine, 127
sarcopside, 185, 187, 306, 314, 344
Sariçiçek. *See* meteorite
Sayama. *See* meteorite
Sayh al Uhaymir 169. *See* meteorite
scanning electron microscope, 78, 98
scapolite, 71, 162, 331
scheelite, 66, 91, 206
Schiaparelli, Giovanni, 233
schist, 57
Schmitt, Harrison, xiii, 215
schöllhornite, 178, 321
Schottky defects, 55
schreibersite, 94–95, 99, 115, 128–129, 136, 140, 142–143, 152, 164, 170–175, 177, 180, 184–189, 191–198, 211, 213, 234, 252, 262, 267, 274, 293, 297, 306–307, 311, 313–314, 321, 335, 352, 374
screw axis, 60
screw dislocations, 55
Sears, Derek, 77
seifertite, 67, 220–221, 300, 340
SELENE spacecraft, 207
SEM. *See* scanning electron microscope
Semarkona. *See* meteorite
Sericho. *See* meteorite
serpentine, 115, 130–131, 133–136, 163–164, 166–169, 172, 219–220, 225, 229, 258, 263, 265, 267–268, 270, 275, 283
Seymchan. *See* meteorite
Shakespeare, William, 238
shale, 1, 56
Shallowater. *See* meteorite
Sharps. *See* meteorite
Shaw. *See* meteorite
shergottites, 88, 102, 107, 182, 214, 219, 297, 300, 309, 314, 335, *See* martian meteorite
Shergotty. *See* meteorite
shock darkening, 93, 141, 278
shock metamorphism, 67–68, 153, 160, 239, 274, 276, 289, 292–293, 297, 328–329
shock pressure, 276, 293, 297, 366
shock stages, 95, 283–285, 292
shooting star, 50
siderite, 2, 71, 130, 136, 163, 168, 205–206, 219–220, 263, 267, 275, 321, 323
Siena. *See* meteorite
Sierra Gorda 010. *See* meteorite
Sierra Gorda 013. *See* meteorite
sign of elongation, 89
Sikhote-Alin. *See* meteorite
silica, 2, 54, 76, 92, 98–99, 112–113, 115, 148, 167, 169–172, 174–180, 182, 186–187, 192, 197–199, 202–203, 208, 211–212, 219–221, 224, 246, 255, 261, 267, 274, 280, 285–286, 291, 293, 295, 297, 300, 304, 306, 326, 340, 350, 365
silicate, 5, 10, 63, 66, 71, 91–92, 108, 112, 130–131, 135–136, 138, 141, 145, 148, 157, 166, 170–171, 173, 175, 179, 184, 186, 188, 191, 195, 197–198, 205, 219, 221–222, 232–233, 235, 245, 258–259, 265, 268, 270, 278, 289–290, 293–295, 301, 303, 319–321, 325–326, 328, 355
silicate darkening, 93, 141, 278, 284, 290
silicide, 10, 171, 200, 255, 328
silico phosphate, 10, 328
sillimanite, 205
silver, 2–3, 70–72, 74, 91, 206, 238, 344
sinoite, 152, 171, 180, 245, 286, 353
Sixiangkou. *See* meteorite
slow ray, 86, 88–89, 91
smectite, 137, 214, 218–220, 267, 271, 275, 319, 324
smelting, 2–3, 274, 291
Smithsonian Institution, 76, 86, 102
smithsonite, 71, 206
Smyer. *See* meteorite
SNCs. *See* martian meteorites
Snell's Law, 85
snow line, 263
Social Circle. *See* meteorite
sodalite, 106, 115, 117, 125–127, 129, 134, 162, 166–167, 171, 190, 204, 268, 270, 272, 325, 350, *See* feldspathoid
sodic anorthosites, 211
sodium sulfate, 135
solar heating, 239, 315, 329
solar nebula, xiii, 10, 44, 56, 106, 108, 145, 150, 153, 180, 239, 243–244, 246, 253, 284, 303, 328, 332, 334, 350, 354–356
solar wind, 201, 214–215, 232, 301
solid solution, 70
solubility, 66, 83, 252
solvus, 296, 311
Sombrerete. *See* meteorite
Sorby, Henry Clifton, 4
South Bend. *See* meteorite
South Byron. *See* meteorite
South Pole-Aitken, 208, *See* Moon
space group, 60, 66, 70, 216
space weathering, 224, 226, 301–302
specific gravity, 3–4, 8, 66, 70, 85–86, 92, 102
sperrylite, 138, 142, 162
sphalerite, 67, 72, 76, 91, 142, 170, 174, 177, 181, 187–188, 213, 234–235, 245, 262, 322, 335, 355
spinel, 55, 68, 71, 96, 114, 116, 124, 126–129, 134, 137, 139, 144–148, 162–163, 166–169, 171, 175–176, 179–180, 191, 208, 211, 213, 222, 234, 242–243, 248, 259, 267, 270, 274, 280, 297, 301, 305, 314, 342, 372
Springwater. *See* meteorite
Sputnik, 44
sputtering, 301
St. Genevieve County. *See* meteorite

St. Mary's County. *See* meteorite
St. Mesmin. *See* meteorite
stacking faults, 55
stanfieldite, 192, 194, 196, 307
Stannern. *See* meteorite
Stardust spacecraft, 125, 234
starkeyite, 319
Staunton. *See* meteorite
staurolite, 68
Steinbach. *See* meteorite
steinhardite, 325
sterlinghillite, 206
STG 07009. *See* meteorite
stibnite, 91
stilbite, 72
stilpnomelane, 219–220
stishovite, 67, 70, 176, 191, 202, 220–221, 280, 290, 300, 325–326
stöfflerite, 221, 300, 372
stolperite, 325–326, 346, 356
Story-Maskelyne, Nevil, 9, 102
streak, 75, 255
strewn field
  definition, 52
strong nuclear force, 82
strontianite, 83, 206
subcalcic augite, 182, 213, 219–220
suessite, 172, 274, 290, 324
sulfarsenide, 10, 138, 142, 162, 322, 328
sulfate, 5, 10, 66, 83–84, 168, 205, 217–221, 235, 263, 267, 319, 321–322, 328, 376
sulfidation, 251, 253, 261, 268, 286
sulfide, 5, 10, 45, 66, 71, 76, 79, 93, 96–97, 99, 106, 115, 123, 128–131, 133–134, 136–140, 142–144, 146–147, 150–151, 153, 155, 162–164, 166–170, 172, 177–182, 184, 187, 191–192, 196, 200–201, 205, 218–220, 224–226, 232–233, 235, 246, 250–253, 257, 259–261, 263–265, 267, 270, 272–274, 278, 280, 283–286, 289–290, 292–296, 306, 309, 318–319, 321–323, 325–326, 328, 332, 337, 342, 347–350, 352
sulfur, 2, 10, 72, 130, 144, 163, 206, 278, 321, 353, 365
superbolide, 52
superclass, 107–108
superconductivity, 172, 188
super-Earth, 49
Superior Valley 014. *See* meteorite
supernovae, 10, 148, 232, 240–242
super-silicic garnet, 290
Surveyor 5, xiii
Sutter's Mill. *See* meteorite
Suwahib (Buwah). *See* meteorite
syenite, 205
Sylacauga. *See* meteorite
sylvite, 63, 71, 82–85, 132, 164, 172, 274–275, 321
symmetry operations, 60–61, 68
symplectite, 180
tachyhydrite, 85
taenite, 56, 63–64, 80, 94–95, 99, 115, 128, 133, 136, 139–140, 144–147, 154–155, 157–159, 166–169, 171–173, 175, 177–179, 183–192, 194–196, 198, 213, 222, 226, 251–253, 256, 258–259, 279, 293, 296, 306, 311–314, 323, 325
Tagish Lake. *See* meteorite
Tahara. *See* meteorite
talc, 57, 68–70, 72, 84, 134, 269
Tanokami Mountain. *See* meteorite
Tarahumara. *See* meteorite
tarnish, 73
taste, 3, 84
Taurus-Littrow Valley, xiii
Tawallah Valley. *See* meteorite
tazheranite, 125, 167, 354
Techado. *See* meteorite
tektite, 54, 83
telluride, 10, 138, 142, 146, 162, 167, 259–261, 328
tenacity, 69
tennantite, 206
tenorite, 83
terminal velocity, 317
terrestrial rocks, 6, 45, 54, 56, 101, 104, 192, 203, 375
Tertullian, 327
tetragonal, 59–61, 86, 221, 231, 280, 300, 355–356, 372
tetrahedra, 55, 63, 113
tetrahedral sites, 114, 194, 265
tetrataenite, 94–95, 115, 134, 139–140, 142, 145–147, 166–168, 172, 184, 196, 198, 256, 259, 268, 279, 312, 364
Thales of Miletus, 81
Themis. *See* asteroid
thenardite, 132
Theophrastus, 2
thermal metamorphism, 94, 143, 153, 155, 158, 239, 253–255, 257–263, 267, 271, 273, 315, 328, 340, 360, 371
thermoluminescence, 77–78, 155–156, 254–255, 259
Thiel Mountains. *See* meteorite
Tholen, David, 222
Thomson, William, 102
thorianite, 82
thorite, 213–214
thortveitite, 166–167
Tibooburra. *See* meteorite
Ti-chromite, 182, 213
Tieschitz. *See* meteorite
tilleyite, 269
tin, 2
Tishomingo. *See* meteorite
Tissint. *See* meteorite
tissintite, 176, 220, 290, 298, 300, 356
tistarite, 116, 243
Titan, xiii, 10, 49–50, 230
titanite, 213
titano-chromite, 314

titanomagnetite, 81, 176, 182, 219–220, 305, 309, 314
titan-phlogopite, 162, 261, 272
TL sensitivity, 77, 259, *See* thermoluminescence
Tlacotepec. *See* meteorite
TNOs, 46–47
tobernite, 82
Tobychan. *See* meteorite
tochilinite, 131, 258
tochilinite-cronstedtite intergrowths, 131, 164, 264
Toluca. *See* meteorite
Tombaugh, Clyde, 236
topaz, 2, 69, 71, 79
torbernite, 84
torreyite, 206
tourmaline, 71, 79, 87–88
trachyandesite, 180
tranquillityite, 180, 211, 213, 297
translucent minerals, 75
transparency, 2, 75
trevorite, 321, 325
triboluminescence, 79
triclinic, 59–60, 86
tridymite, 67, 70, 99, 170, 175, 177, 180, 182, 187, 190, 192, 197–198, 205, 208, 213, 234, 251, 262, 294–295, 310
trigonal, 59, 63–64, 86, 178, 280
troctolite, 149–150, 174, 210–211, 213, 294, 308
troilite, 7, 54, 94–99, 112–115, 128, 130, 133–134, 136–137, 139–147, 150–152, 155, 157–159, 162, 166–182, 184–197, 199–200, 208, 211, 213, 220–221, 224, 226, 229, 235, 245, 251–252, 255–257, 260, 262, 268, 270–271, 273–274, 278–280, 283, 285–286, 289, 291, 293–297, 302–307, 314–315, 317, 320–321, 325, 372
troilite nodules, 139–140, 185–186, 188, 191, 314
trona, 83
Trysil. *See* meteorite
tsangpoite, 180
tschaunerite, 221, 300
Tschermak, Gustav, 9, 102, 355
Tucson. *See* meteorite
tuite, 220, 280, 294, 298, 335
tungstate, 10, 66, 123, 328
turneaureite, 206
turquoise, 2, 72
twin boundaries, 55
twin plane, 68
twinning, 67–68, 112, 162, 175, 187, 190, 198, 256, 279, 293
  polysynthetic, 158, 166, 169, 195, 297
Tycho crater, 301

Ulasitai. *See* meteorite
ulexite, 72, 84
Ultima Thule, 237
ulvöspinel, 57, 96, 142, 176–177, 180, 182, 211, 213, 219–220, 297, 305, 309, 314
undulose extinction, 89, 285
uniaxial mineral, 89
unit cell, 55, 58–61, 64, 295
Unter-Mässing. *See* meteorite
uraninite, 72, 82, 206
uranophane, 82
ureilites, 97, 107, 129, 171–172, 179–180, 192, 274, 290, 303, 310, 315, 332, 361, 365, 374
Urey, Harold, 102

Vaca Muerta. *See* meteorite
vacancy-rich clinopyroxene, 290
valence electron, 61–62, 75, 80
Valera. *See* meteorite
van der Waals bonding, 62, 69, 84
van der Waals, Johannes, 62
vanthoffite, 218
vaterite, 178, 321
Vega 2, 203
Venera 10, 204
Venera 13, 203
Venera 14, 203
Venera 9, 204
Venus, xiii, 49, 51, 104, 202–204, 341, 343
Venus Express spacecraft, 203
Verkhne Dnieprovsk. *See* meteorite
Vermillion. *See* meteorite
Vesta, 103, *See* asteroid
vestaite, 176, 289, 362
vestoids. *See* asteroid
Vickers hardness, 69
Vigarano. *See* meteorite
Vilna. *See* meteorite
vivianite, 321
von Goethe, Johann Wolfgang, 4
von Humboldt, Alexander, 4
von Lang, Viktor, 9
von Reichenbach, Karl, 9
von Widmanstätten, Alois, 102
Vredefort crater, 300
V-rich magnetite, 146

Wabar craters, 300
wadalite, 117, 134, 268, 347
wadsleyite, 168, 180, 280, 284, 333
wairakite, 205
wairauite, 128, 346
Wallapai. *See* meteorite
warkite, 116, 125, 166–168, 355
Wark-Lovering rims, 124, 246–247, 269
Warren, Paul, 163
Warrenton. *See* meteorite
Washington County. *See* meteorite
Wasson, John, 103, 154, 331–335, 345, 347, 349, 351, 367, 370, 374–376
wassonite, 56, 66, 115, 142, 170, 200, 251, 321
Watson 001. *See* meteorite

Weatherford. *See* meteorite
weathering, xiii, 10, 52, 101, 135, 153, 171, 175, 178, 196, 205, 213, 219, 224, 226–227, 239, 260, 295, 318–319, 321–324, 328–329, 331, 343, 362, 366, 373, 376, 378
weathering scale, 97, 319, 321, 378
websterite, 177
Wedderburn. *See* meteorite
Weekeroo Station. *See* meteorite
Wegener, Alfred, 327
wehrlite, 149
Werner, Abraham, 4
Wethersfield (1982). *See* meteorite
Whipple, Fred, 232
white matrix, 137, 272
whitlockite, 234
Widmanstätten pattern, 56, 102, 183, 193–194, 198, 295, 311
Wiley. *See* meteorite
Willamette. *See* meteorite
Willaroy. *See* meteorite
willemite, 79, 206
Willow Creek. *See* meteorite
winchite, 272
Winona. *See* meteorite
winonaites, 107, 174, 293–294
Wold Cottage. *See* meteorite
wolframite, 91
wollastonite, 66, 79, 115, 129, 134, 167, 205, 269, 342
Woodbine. *See* meteorite
woodruffite, 206
Woolgorong. *See* meteorite
wüstite, 10, 137, 155, 220, 298, 317–318

xenotime, 220, 275
xieite, 220, 280, 298
Xinjiang. *See* meteorite
X-ray maps, 98

Y-691. *See* meteorite
Y-693. *See* meteorite
Y-74063. *See* meteorite
Y-74160. *See* meteorite
Y-75032. *See* meteorite
Y-75097. *See* meteorite
Y-75267. *See* meteorite
Y-790143. *See* meteorite
Y-790964. *See* meteorite
Y-791199. *See* meteorite
Y-791200. *See* meteorite
Y-791510. *See* meteorite
Y-791790. *See* meteorite
Y-793225. *See* meteorite
Y-793321. *See* meteorite
Y-793241. *See* meteorite
Y-81020. *See* meteorite
Y-82104. *See* meteorite
Y-82189. *See* meteorite
Y-8414. *See* meteorite
Y-8451. *See* meteorite
Y-86029. *See* meteorite
Y-86720. *See* meteorite
Y-86789. *See* meteorite
Y-980433. *See* meteorite
yagiite, 190, 295
Yanzhuang. *See* meteorite
Yarkovsky effect, 224
yeatmanite, 206
yoshiokaite, 213

Zacatecas (1792). *See* meteorite
Zag. *See* meteorite
Zagami. *See* meteorite
zagamiite, 220, 300, 356
Zaisho. *See* meteorite
Zakłodzie. *See* meteorite
zaratite, 321
zeolite, 166, 205, 225
zeunerite, 82
zincite, 71, 206
Zinder. *See* meteorite
zinkenite, 206
zircon, 67, 69, 72, 82–83, 88, 98, 104, 175, 180, 208, 211, 213, 220, 332, 343, 358
zirconolite, 125, 127, 167, 211, 213
zoneless plessite, 256
zoning
  oscillatory, 318
zorite, 72

www.ingramcontent.com/pod-product-compliance
Lightning Source LLC
LaVergne TN
LVHW022027051126
851858LV00004B/62
* 9 7 9 8 9 9 7 2 0 7 6 0 1 *